Hua-Jie Zhu

Organic Stereochemistry

Related Titles

Joswig, J.

Theoretical Chemistry
Electronic Structure Methods and the Chemical Bond

2014
Print ISBN: 978-3-527-33191-8; also available in electronic formats

Ananikov, V.P. (ed.)

Understanding Organometallic Reaction Mechanisms and Catalysis
Computational and Experimental Tools

2014
Print ISBN: 978-3-527-33562-6; also available in electronic formats

Glaser, R.

Symmetry, Spectroscopy, and Crystallography
The Structural Nexus

2015
Print ISBN: 978-3-527-33749-1; also available in electronic formats

Günther, H.

NMR Spectroscopy
Basic Principles, Concepts, and Applications in Chemistry

Third Edition
2013
Print ISBN: 978-3-527-33004-1; also available in electronic formats

Hua-Jie Zhu

Organic Stereochemistry

Experimental and Computational Methods

WILEY-VCH
Verlag GmbH & Co. KGaA

The Author

Prof. Dr. Hua-Jie Zhu
Hebei University
Chinese Center for Chirality
180# Wusi Rd.
071002 Baoding, Hebei Province
China

Library of Congress Card No.: applied for

British Library Cataloguing-in-Publication Data
A catalogue record for this book is available from the British Library.

Bibliographic information published by the Deutsche Nationalbibliothek
The Deutsche Nationalbibliothek lists this publication in the Deutsche Nationalbibliografie; detailed bibliographic data are available on the Internet at <http://dnb.d-nb.de>.

Print ISBN: 978-3-527-33822-1
ePDF ISBN: 978-3-527-68815-9
ePub ISBN: 978-3-527-68817-3
Mobi ISBN: 978-3-527-68818-0
oBook ISBN: 978-3-527-68816-6

Typesetting Laserwords Private Limited, Chennai, India

MIX
Paper from responsible sources
FSC® C013604

Contents

Preface *IX*
Acknowledgments *XI*
List of Abbreviations *XIII*

Part I Fundamentals *1*

1 Chirality *3*
1.1 Introduction *3*
1.2 Tetrahedron of Carbon *9*
1.2.1 Terpenoids *9*
1.2.2 Flavonoids *13*
1.2.3 Alkaloids *14*
1.2.4 Steroids *16*
1.2.5 Glycosides *17*
1.2.6 Others *18*
1.3 Other Stereogenic Centers *19*
1.4 Optical Characteristics *23*
1.4.1 Measurement of OR *24*
1.4.2 ECD and Its Definition *25*
1.4.3 Outline of VCD *26*
1.4.4 Outline of ROA *27*
References *28*

2 Non-optical Method in Configuration Study *31*
2.1 ^{13}C NMR Spectra *31*
2.1.1 NMR and Atomic Structure *31*
2.1.2 ^{13}C NMR Calculation *32*
2.1.3 ^{1}H NMR *34*
2.1.4 ^{13}C NMR Prediction and Conformational Search *34*
2.2 X-Ray Diffraction and Mosher Method *41*
2.2.1 X-Ray Diffraction *41*
2.2.2 Mosher Method *44*
2.3 Transition State Energy and Chirality Selectivity *51*

2.4 Separation of Chiral Compounds *53*
2.4.1 Chiral Organic Bases *53*
2.4.2 Chiral Organic Acids *53*
2.4.3 Chiral Organic Alcohols *53*
2.4.4 Others *54*
References *55*

Part II Techniques *57*

3 Optical Rotation (Rotatory Dispersion, ORD) *59*
3.1 Introduction *59*
3.2 Quantum Theory *59*
3.3 Matrix Model *63*
3.3.1 Matrix Basis *65*
3.3.2 Explanation of General OR Characteristics *68*
3.3.2.1 Sample Calculations *68*
3.3.2.2 Calculated Values in Same Series of Compounds *73*
3.4 ORD *77*
3.5 Application *77*
3.5.1 AC Assignment for Mono-Stereogenic Center Compounds *79*
3.5.2 Matrix Model Application *80*
3.5.3 AC Assignment for Poly-Stereogenic Center Compounds *82*
3.5.4 Using ORD Method *83*
References *84*

4 Electronic Circular Dichroism *87*
4.1 Exciton Chirality CD *87*
4.2 ECD Characteristics for Chiral Metallic Compounds *92*
4.3 Quantum Theory Basis *94*
4.4 Principle Using ECD *95*
4.5 Application *97*
4.5.1 Procedure to Do ECD *97*
4.5.2 ECD Application *97*
4.5.3 UV Correction *102*
References *105*

5 Vibrational Circular Dichroism and Raman Optical Activity *107*
5.1 Exciton Chirality *108*
5.2 Quantum Theory Basis *109*
5.2.1 VCD and IR *109*
5.2.2 ROA and Raman Scattering *110*
5.3 Principles Using VCD and ROA *113*
5.4 Application *115*
5.4.1 VCD Application *115*

5.4.2 ROA Application 124
References 126

6 Combinational Use of Different Methods 129
6.1 Tactics to Select Methods 129
6.1.1 ^{13}C NMR Methods 130
6.1.2 OR and ORD 131
6.1.3 Matrix 132
6.1.4 ECD 132
6.1.5 VCD Method 133
6.2 Examples and Discussion 133
6.3 Revised Structures 138
6.3.1 ORD Method 139
6.3.2 Combinational Use of OR and ECD 144
6.3.3 VCD and ECD 146
6.3.4 Comprehensive Use of OR, ECD, and VCD 147
References 160

Part III Reactions 163

7 Enantioselective Reaction 165
7.1 Enantioselective Addition 165
7.1.1 Organic Zn- or Zn-Ti Reagent 165
7.1.2 Organic Cu–Zn, Cu–Li Reagent 169
7.1.3 Organo-Fe Complexes 173
7.1.4 Other Organo-Metallic Complexes 175
7.1.5 Organo-Si Reagents 178
7.2 Enantioselective Reduction 178
7.2.1 Green Chemistry 181
7.3 Enantioselective Oxidation 184
7.4 Prediction of ee Using Calculations 185
7.5 Catalyst Types 189
7.5.1 Amino Alcohols 189
7.5.2 Chiral Ligands Containing N–O Group 190
7.5.3 Chiral Axial Catalysts 191
7.5.4 Solid-Supported Chiral Compounds 192
7.5.5 Spiral Chiral Compounds 193
7.5.6 Asymmetric-Axle-Supported Chiral Catalyst 194
7.5.7 Chiral Shiff-Base Ligands 195
7.5.8 Some Asymmetric Lewis Acids 196
7.5.9 Organic P-Containing Ligands 197
7.6 Three Phenomena 198
7.6.1 Chirality Amplification (Nonlinear Effect) 198
7.6.2 Auto-Self Catalysis 199

7.6.3 Odd–Even Carbon Effect *199*
References *201*

8 Chemoselective Reaction *205*
8.1 Chemoselective Additions *205*
8.1.1 Addition to C=O or C=C Groups *205*
8.2 Chemoselective Reduction *210*
8.3 Chemoselective Oxidation *225*
8.4 Other Chemoselective Reactions *231*
References *237*

9 Stereoselective Reaction *241*
9.1 Conformational Study *241*
9.2 Effect of Conformation on Reactions *247*
9.3 Regioselective Reactions *251*
9.4 Diastereoselective Reactions *260*
9.4.1 Diastereoselective Additions *260*
9.4.2 Other Diastereoselective Reactions *270*
9.5 Calculation Using Theoretical Protocol *272*
References *277*

10 Total Organic Synthesis *279*
10.1 Introduction *279*
10.2 Retrosynthesis Strategies *279*
10.3 Examples in Synthesis *285*
10.3.1 (+)-Hirsutene *285*
10.3.2 (2R,3S)-Rubiginone A_2 and Its Analog *287*
10.3.3 (+)-Brefeldin A *290*
10.3.4 Malyngamide U and Its AC Reassignment *292*
10.3.5 Taxol Derivatives *296*
10.3.6 Amphidinolide T2 and Its Derivatives *298*
10.3.7 (+)-Vindoline *303*
10.4 Calculation in Total Synthesis *306*
References *309*

Index ***313***

Preface

Chirality exists throughout the micro world and the macro world. To know the handedness of matter is the key aim in many research areas. Stereochemistry widely exists in organic, inorganic, polymer, and computational chemistry. It can be seen that the construction of stereogenic centers involves (chiral) organic compounds in most cases. Therefore, organic stereochemistry undoubtedly is the key science in this area.

There are two main issues involved in organic stereochemistry. One is to compute molecular characteristics such as ^{13}C NMR, optical rotation, and electronic and vibrational circular dichroism including Raman optical activity. Another is to investigate the reaction pathway, such as the transition state energy. Computational methods are widely applied in these two topics. Now they have become a very powerful tool to explain the experimental results and direct the reactions. It shows that suitable computational methods can provide experimental chemists a very valuable tool in their studies.

With the development of supercomputer technology, computational accuracy is increasing, and reliability is also increasing at the same time. For example, the basis set 6-311+G(d) was an expensive and luxurious tool for most computational chemists almost a decade ago (nearly about 2000), but now it has become a very general requirement. The use of a higher basis set such as 6-311+G(2d,p) and others has also become popular. Therefore, a tight combination of experimental and theoretical methods has become more and more important in modern stereochemistry.

This is the beginning of the wide use of valuable and reliable computational methods to assist experiments. What we can compute now includes magnetic shielding constants, optical rotation and its dispersion, electronic and vibrational circular dichroism, and Raman optical activity, all of them giving us the evidence to assign the absolute configuration of a chiral molecule. Investigation of the correct transition state barrier not only shows us the reaction pathway but also allows us to realize the absolute configuration changes in the reactions.

This book records my experience in the combined use of experimental and theoretical methods. I hope this can be helpful for modern readers who want to master the skills in the combinational use in their experimental study. Because of the limitation of my own knowledge, the book may have some defects or errors inside, but I sincerely hope that readers will point them out to me for improvement in future editions.

Baoding 2015

Hua-Jie Zhu
Hebei University

Acknowledgments

I greatly appreciate the valid comments and suggestions on VCD and ROA theory from Prof. Laurence Nafie (Syracuse University, USA) and Prof. Laurence Barron (Glasgow University, UK). I also sincerely appreciate the useful comments on the coordination chemistry material from Prof. Hui Zhang (Xiamen University, China). Dr. James Cheeseman (Gaussian, Inc. USA) is thanked for his useful suggestions in writing programs. I am grateful to Prof. Shi-Gang Shen, Prof. Jiang-Zhong Xu, Prof. Jun Fen, and Prof. Gui-Fen Pei for their valid help at Hebei University and to my students He Yu, Wen-Xin Li, Qin Yang, Yan Xu, Sha-Sha Tian, Hui-Jun Wang, Shuang-Shuang Ding, Can-Can Zhang, Jing-Chen Wang, and Wen-Si Shi at Hebei University and Dong-Bao Hu at Kunming Institute of Botany (CAS) for their assistance. I would like to thank the staff of Gaussian, Inc Dr. Fernando R. Clemente and Dr. Douglas J. Fox for their computation assistance.

The Wiley-VCH editors (Dr. Anne Brennführer and Lesley Fenske) and staff (Srinivasan Swapna) are greatly thanked for their patience and assistance during the production of this book. Especially, Dr. Anne Brennführer gave me many valid suggestions, and I am indebted to her for her help.

List of Abbreviations

(DHQ)2PHAL	hydroquinine 1,4-phthalazinediyl diether
(DHQD)2AQN	hydroquinidine (anthraquinone-1,4-diyl) diether
1 a.u.	627.5 kcal mol^{-1}
AC	absolute configuration
acac	acetylacetonate
BAIB	formula is PhI(OAc)2
BINAP	2,2′-bis(diphenylphosphino)-1,1′-binaphthyl
BINOL	1,1′-bi-2,2′-naphthol
Bn	benzyl
BOC	*t*-butoxycarbonyl
Bz	benzoyl
CAN	ceric ammonium nitrate
CDI	carbonyldiimidazole
chiroptical	chiral optical
CID	circularly intensity difference
cod	1,5-cyclooctadiene
CPL	circularly polarized light
CSGT	continuous set of gauge transformations
Cy(cy)	cyclohexyl
DBU	1,8-diazobicyclo[5,4,0]undec-7-ene
DCC	1,3-dicyclohexylcarbodiimide
DCM	dichloromethane
DCP	dual circular polarization
DIBAL-H	diisobutylaluminum hydride
DMAP	4-dimethylaminopyridine
DMF	*N,N*-dimethyl formamide
DMP	Dess–Martin periodinane
DMSO	dimethyl sulfoxide
DPPA	diphenylphosphoryl azide
DRCD	diffuse reflectance circular dichroism
ECD	electronic circular dichroism
EDCI	1-(3-dimethylaminopropyl)-3-ethylcarbodiimide hydrochloride

FFR	far from resonance
GFE	Gibbs free energy
Hartree	$4.35974381 \times 10^{-18}$ (=J Ha^{-1})
HBSS	Hanks balanced salt solution
HMPA	hexamethyl phosphoric triamide
HPLC	high performance liquid chromatography
IBX	2-iodobenzoic acid
ICP	incident circular polarization
IDCS	iodotrichlorosilane
IEFPCM	integral equation formalism of polarizable continuum model
LDA	lithium diisopropylamide
Light speed	29 979 245 800.0 cm s^{-1}
LORG	localized orbitals/localized origins
MBH	Morita–Baylis–Hillman (MBH) products, the densely functionalized β-hydroxyl α-methylene carbonyl compounds
m-CPBA	*m*-chloroperbenzoic acid
MNBA	2-methyl-6-nitrobenzoic anhydride, mixed with DMAP
MNCB	2-(2′-methoxy-1,1′-naphthyl)-3,5-dichlorobenzoic acid
MOM	methoxymethyl
MPA	methoxyphenylacetic acid
MTBE	methyl *t*-butyl ether
MTPA	α-methoxyl-trifluoromethylphenyl acetic acid
NBS	*N*-bromosuccinimide (*N*-bromobutanimide)
NMO	4-methyl morpholine *N*-oxide
NMR	nuclear magnetic resonance
NOBIN	2-amino-2′-hydroxy-1,1′-binaphthyl
OR	optical rotation
ORD	optical rotation dispersion
PCC	pyridinium chlorochromate
PCM	polarizable continuum model
PDC	pyridinium dichromate
PEA	phenylethyl amine
PES	potential energy scan
Planck constant	$6.62606876 \times 10^{-34}$ (Js)
PMB	phthaloyl
PMP	pentamethylpiperidine or *p*-methoxyphenyl
QSSR	quantitative structure selectivity relationship
RC	relative configuration
RCM	ring-closing metathesis
ROA	Raman optical activity
SAR	structure–activity relationship
SCI-PCM	self-consistent isodensity polarized model
SCP	scattered circular polarization
SFC	supercritical fluid chromatography
TBDPS	*tert*-butyldiphenylsilyl

TBS	*tert*-butyldimethylsilyl
TBSOTf	*tert*-butyl dimethylsilyl trifluoromethane sulfonate
TCTATO	1,3,5-trichloro-1,3,5-triazinane-2,4,6-trione
TD	time-dependent
TEE	total electronic energy
TEMPO	2,2,6,6-tetramethyl-piperidine 1-oxyl
TESOTf	triethylsilyl trifluoromethane sulfonate
TFA	trifluoroacetic acid
THF	tetrahydrofuran
ToE	$4.8032041969 \times 10^{-10}$ (=C per electron – multiplied by speed of light/10 to produce electrostatic units (ESUs) per electron)
TS	transition state
TsCl	4-toluenesulfonyl chloride
VCD	vibrational circular dichroism
VCI	vibrational configuration interaction
VOA	vibrational optical activity
ZPE	zero point energy correction

Part I
Fundamentals

Organic Stereochemistry: Experimental and Computational Methods, First Edition. Hua-Jie Zhu.

1
Chirality

1.1
Introduction

A simple chiral molecule just contains one sp^3-hybridized carbon that connects four different groups, such as chiral molecule **1** ($R^1 \neq R^2 \neq R^3 \neq R^4$, Figure 1.1). Its mirror image structure, namely the enantiomer, is **2**, which has the same relative configuration (RC) but different absolute configuration (AC); for example, when $R^1 < R^2 < R^3 < R^4$, chiral molecule **1** has (*R*)-AC and **2** has (*S*)-AC. Both have the same chemical characteristics, such as the transition state (TS) barrier (reaction activation energy) in reactions in the absence of a chiral catalyst or a chiral auxiliary compound, and physical characteristics such as melting and boiling point and magnetic shielding constant for each corresponding atom. However, in a beam of circularly polarized light (CPL), the enantiomers exhibit opposite characteristics. For example, both have the same absolute optical rotation (OR) values, but their OR signs are reversed. The different optical characteristics are the basis allowing us to identify their AC.

As a widely known phenomenon, chirality, also named as handedness, brings out many mysteries that we have not known until now. For example, the L-amino acids were selected in peptide formation for the origin of life instead of D-amino acids. D-sugars were used instead of L-sugars. That is the amazing natural choice. Although we do not know why Nature selected different chiral molecules in life, it does not affect us in the study of the handedness of molecules and other materials.

Absolutely, because of the natural selection of L-amino acids and D-sugars in life formation, different chiral compounds must, logically, have different effects on the life process. Historically, a mixture of the (*R*) and (*S*) enantiomers (**3** and **4**) was used as a medicine for preventing vomiting in pregnant women in European countries – but not in America – in the 1960s. Many infants with deformed limbs were born. This was caused by a chiral chemical compound, (*S*)-thalidomide (**4**), which can be isomerized from (*R*)-**3** (Scheme 1.1) [1a–c]. In America, this tragic issue was avoided because Frances O. Kelsey in the Food and Drug Administration (FDA) could not find evidences from the documents handed by the pharmaceutical company to confirm that it was not harmless to the central nervous system.

Organic Stereochemistry: Experimental and Computational Methods, First Edition. Hua-Jie Zhu.

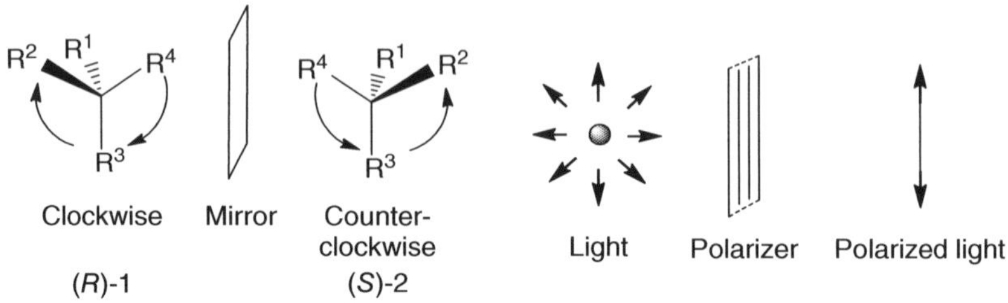

Figure 1.1 Enantiomers and polarized light.

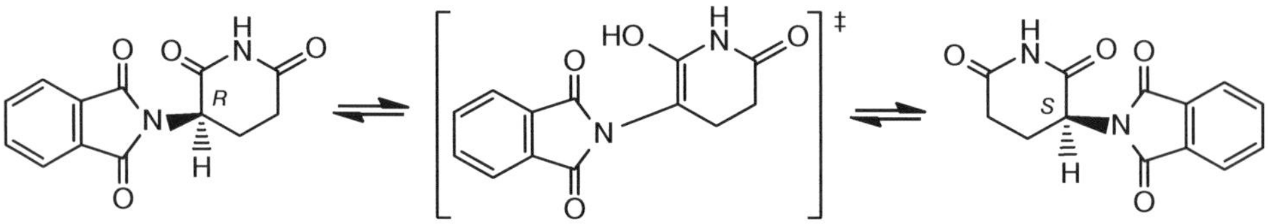

3 4

Scheme 1.1 Isomerization of (*R*)-thalidomide to its (*S*)-enantiomer [1].

This doubt delayed the approval of this medicine for use in the United States, and just because of this action it saved many infants and families in the United States.

Tertodotoxin (**5**) is a strong toxic chiral compound that is isolated from globefish [1d]. The strong toxicity comes from the (*S*) AC of C9. Once C9 becomes (*R*) AC, its severe toxicity almost disappears.

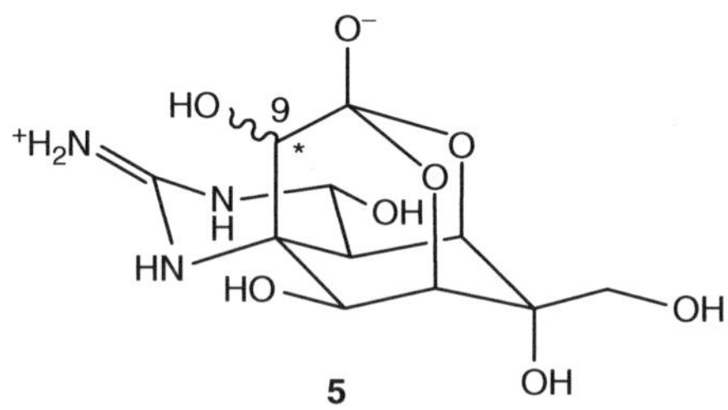

5

More examples can be found with different bioactivities. Some pairs of enantiomers and their bioactivities are listed after their structures (Figure 1.2) [1e].

The different ACs of **6–19** result in different bioactivities. Obviously, it should be a big challenge to obtain various chiral compounds in organic stereochemistry. Generally, we can design and synthesize different chiral catalysts or auxiliary reagents to control the formation of stereogenic centers of a chiral compound. A well-known example is the use a chiral catalyst to control the asymmetric epoxidation of the C=C bond. This is called Sharpless epoxidation. For example, when L-diisopropyltartrate (**20**) was used as a catalyst, the product **22** could be achieved in 98% yield and 68% ee (Eq. (1.1)) [2].

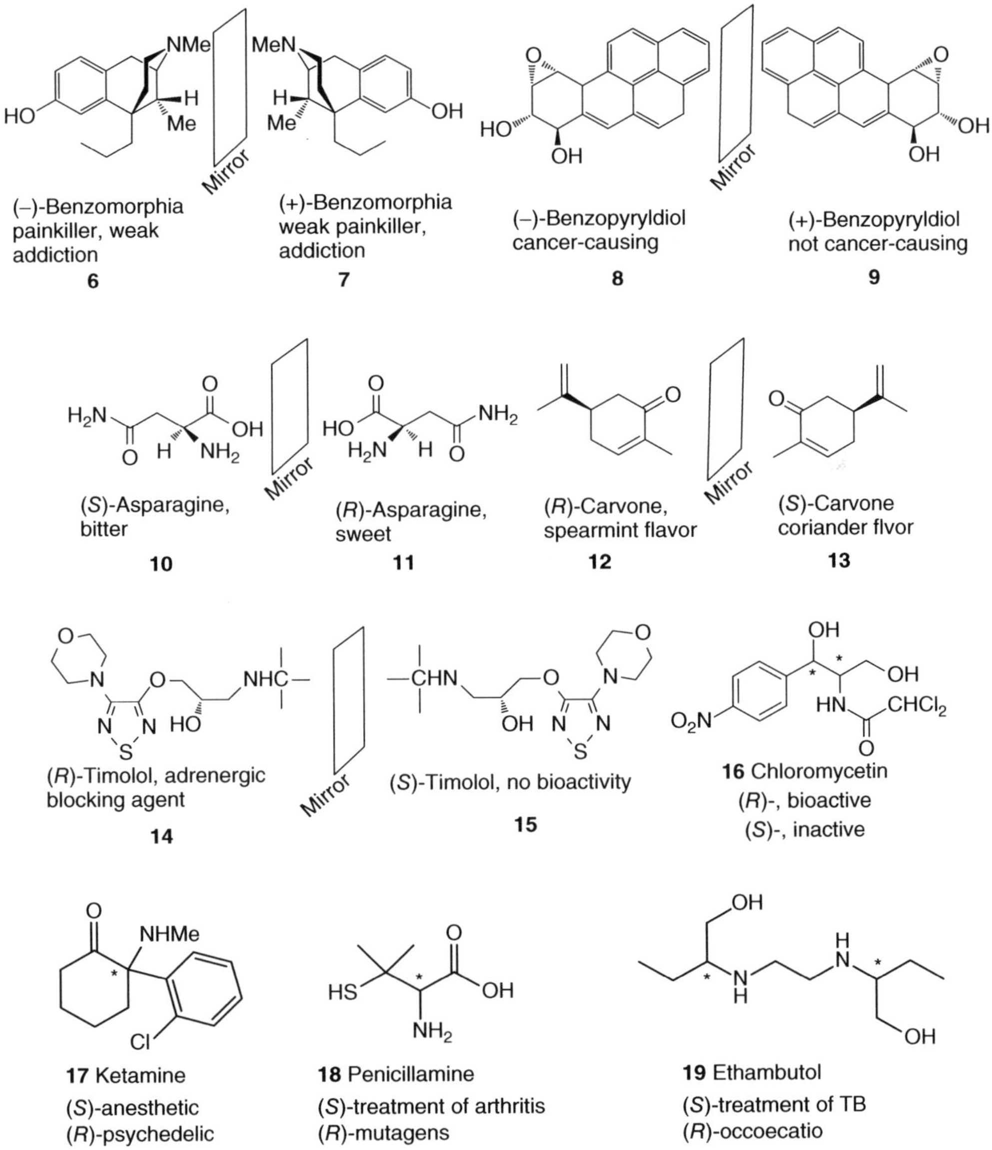

Figure 1.2 Examples of enantiomers and their different bioactivities.

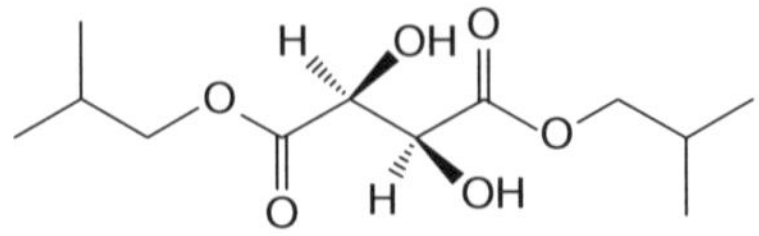

20

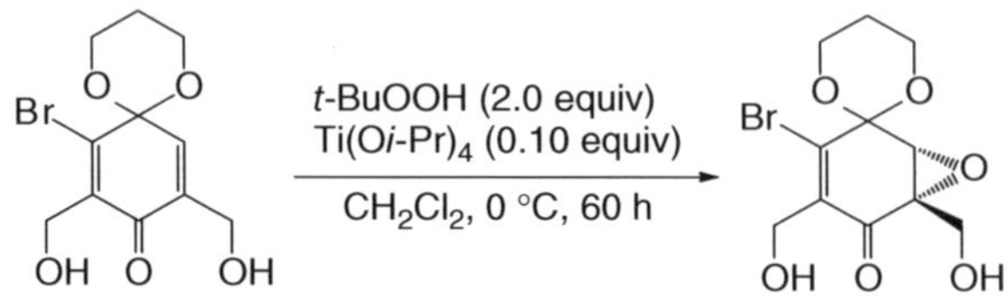

21 **22** (1.1)

Another big challenge in organic stereochemistry is the identification of the AC of chiral compounds. The most convenient method is comparing the theoretical chiroptical spectrum to the experimental one. For example, the RC of compound **23** has been estimated by X-ray study as illustrated below; the experimental OR was +57.4 in methanol [3]. The computed OR for **23** with the same AC as the illustrated one was +74.3 in the gas phase, and this value decreased to 66.2 in methanol. This prediction is close to the recorded value of +57.4. Therefore, its AC was assigned as that illustrated below. Another example is that the predicted OR for (1*R*,5*S*,8*S*,9*S*,10*S*)-oruwacin (**24**) was −193. The experimental value was 193. Therefore, it was assigned as (1*S*,5*R*,8*R*,9*R*,10*R*) based on its OR value [4].

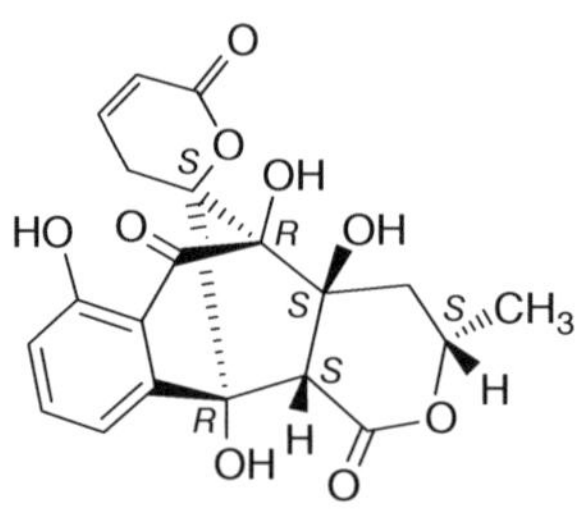

23

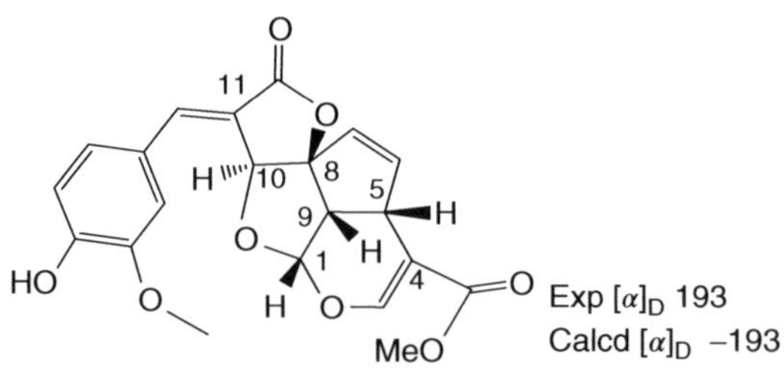

24 Oruwacin

The study of chiral materials is an important branch in stereochemistry. By the reaction of a chiral molecule with a monomer, a chiral polymer can be formed. For example, the widely used (−)-menthyl methacrylate (MnMA, **25**, Eq. (1.2)) can undergo polymerization and afford the chiral polymer *P*-(−)-MnMA (**26**) [5].

25 (−)-MnMA **26** *P*- (−)-MnMA (1.2)

The last but important issue is to understand the chirality selection for the formation of life in prebiotic Earth. That is about the origin of life. The process leading to its origin is extremely long. Where is the first chiral compound coming from in the long time? Evidence comes from the presence of chiral α-methyl amino acids **27–29** in meteorites [6]. A reasonable hypothesis is that the chiral compounds led to the formation of other chiral molecules in the prebiotic Earth, and due to the effect of amplification reactions and other factors on the formation of chiral compounds, finally it led to the formation of L-amino acids that form the basis of the origin of life.

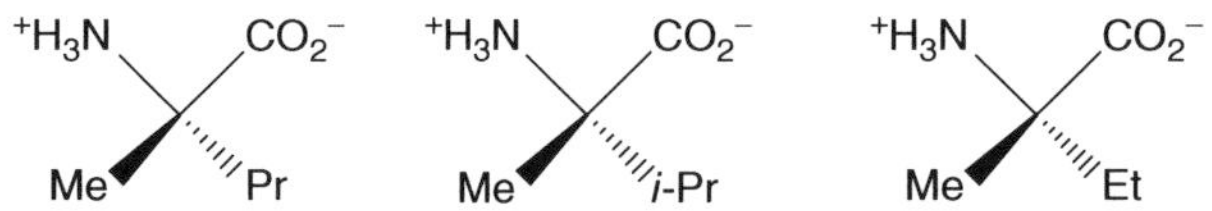

27 2.8% ee, (*S*) **28** 2.8% ee, (*S*) **29** 15.2% ee, (*S*)

As a general rule in organic chemistry, the formation of D- and L-amino acids without any catalyst would be in the ratio 1 : 1. This ratio, however, might not have been strictly 1 : 1 in the prebiotic Earth for some unknown reasons. A mixture of D- and L-amino acids with a tiny excess of the L-form could dissolve in water.

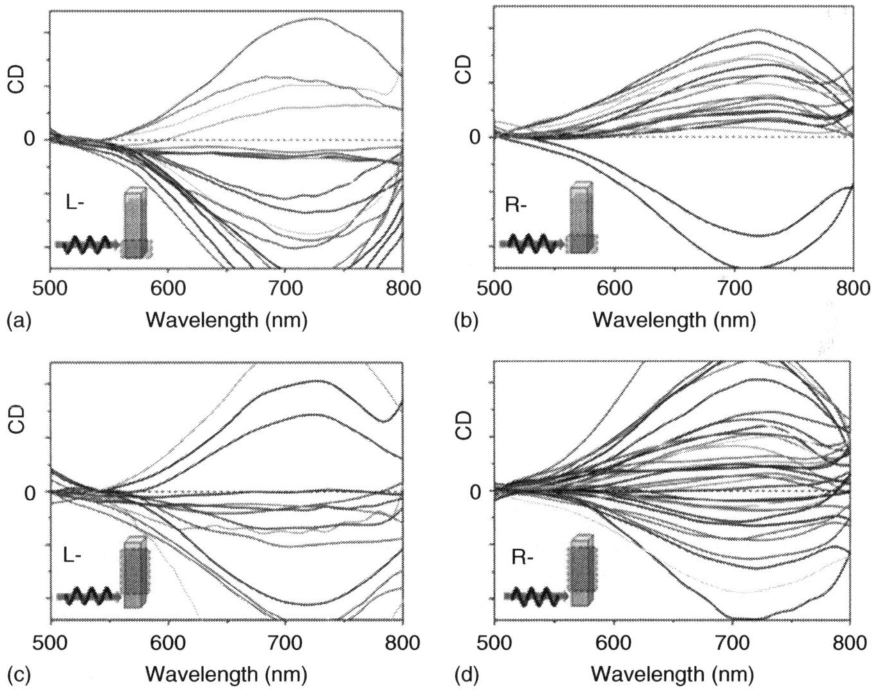

Figure 1.3 Times of ECD experiments in solid state for (a) light-left circularly polarized light, (b) light-right CPL, (c) dark-left CPL, and (d) dark-right CPL. (Solution contains $[Cu(NH_3)_4]^{2+}$ (0.5 mmol), succinate (1.5 mmol), 4,4′-bipyridine (0.5 mmol), water (10 ml), and ethanol (10 ml).) [8]

Because of the different solubility of L- and D- salts from the respective amino acids, the mixing of D- and L- salts resulted in precipitation. This resulted in a solution with concentrated L-amino acid [7]. Thus, starting with a small excess of the L-amino acid, the final L-/D- ratio would increase.

Another example is that irradiation with right-CPL results in the possibility growing crystals with right-handed helical structure more than with left-handed helical structure from a $[Cu(NH_3)_4]^{2+}$-containing solution found through a statistical analysis of the Cotton effect of crystals (Figure 1.3) [8]. This might hint that the crystal could abstract one enantiomer and enrich its enantiomer in the solution.

In view point of chemists, a chemical procedure for choosing L-amino acids should be much more interesting in the study of the origin of life because this procedure is irreversible. A reversible procedure such as the deposition of salts of racemic amino acids is too short for the formation of chiral peptides.

Recently, it was reported that there is a chirality pairing recognition reaction, or molecular sex recognition reaction, which may enrich L-tryptophan if it has a small excessive quantity in a mixture of D and L-tryptophan. When (*R*)- and (*S*)-tryptophan methyl esters were used as the starting materials, L-tryptophan methyl ester tended to react with D-tryptophan methyl ester to form L-/D- products instead of the L-tryptophan methyl ester (Eq. (1.3)) [7].

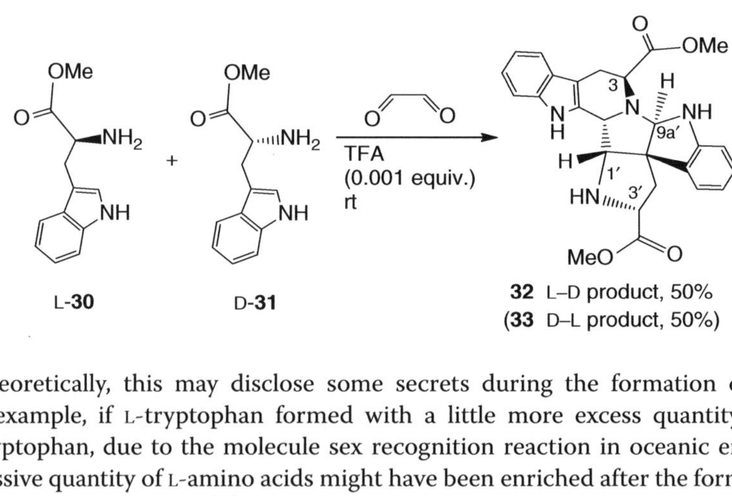

L-**30** D-**31** **32** L–D product, 50% (**33** D–L product, 50%) (1.3)

Theoretically, this may disclose some secrets during the formation of life. For example, if L-tryptophan formed with a little more excess quantity than D-tryptophan, due to the molecule sex recognition reaction in oceanic era, the excessive quantity of L-amino acids might have been enriched after the formation of **32** and **33**. The enriched L-tryptophan may act as a first chiral source to promote the formation of other chiral compounds that were involved in life formation. This is the chemical procedure. It may be a starting point of a new chemistry: "Evolution Chemistry".

In this book, the key points include the (i) methods used for RC and AC assignment, (ii) design and synthesis of chiral catalysts, (iii) chemoselective and (iv) regioselective reactions, (v) diastereoselective reactions, and (vi) the total synthesis of bioactive natural products. Theoretical methods used in the experimental study are introduced in each later chapter.

1.2
Tetrahedron of Carbon

The simple chiral molecule just contains one sp^3-hybridized carbon that connects four different groups, such as the chiral molecule **1** ($R^1 \neq R^2 \neq R^3 \neq R^4$). Because of different connections of the asymmetric centers to various atoms or groups, it forms various chiral compounds, such as terpenoids, flavonoids, alkaloids, glycosides, alkaloids, and so on. This is the basis of life evolution and survival.

1.2.1
Terpenoids

This is a large family of compounds showing various bioactivities. The basic unit is isopentadiene (**34**, also called isoprene) [9]. However, isopentadiene itself does not take part in the reaction directly in the body of plants. When it is converted into isopentenyl pyrophosphate (IPP) and dimethylallyl pyrophosphate (DMAPP), they can form various natural terpenoids. Structurally, the different connections, such as "head-to-head," "tail-to-tail," or "head-to-tail," are called the "isoprene rule." Two isopentadienes can form a linear (acyclic) monoterpenoid without a stereogenic center (**35**, "head-to-tail" connection) or a cyclic monoterpene, which generally has one or more stereogenic centers, such as L-menthol (**36**). Most monoterpenes with different functional groups are also the major components of essential oils (volatile oils) and exhibit different bioactivities.

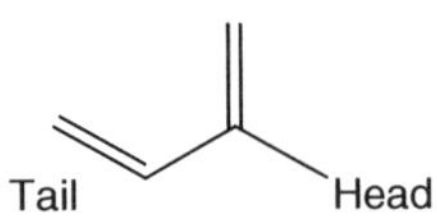

34

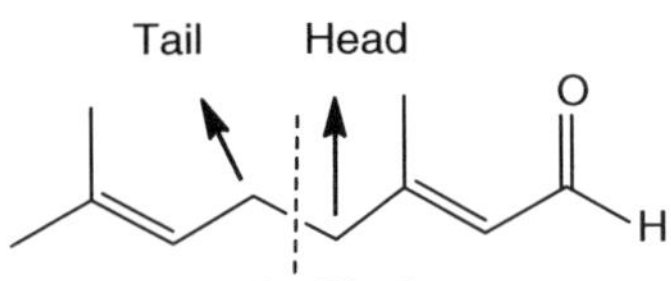

35 Citral

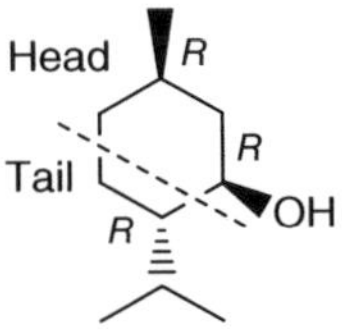

36 L-Menthol

37 *α*-Phellandrene

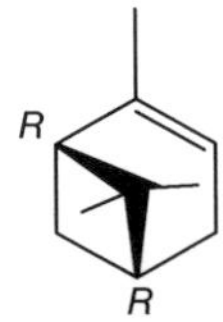

38 *α*-Pinene

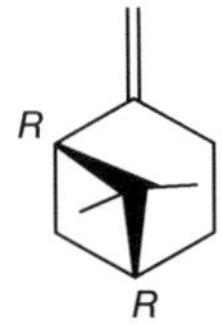

39 *β*-Pinene

The prefix *α* or *β* in the above does not refer to their stereochemistry; they are used to show the positions of double bond: for example, *α*-pinene **38** and *β*-pinene **39**. However, in some textbooks – or some journals until now – both are used to express the configuration of a stereogenic center. One should be careful in reading reports from different journals. In this book, all stereogenic centers will be labeled as (*R*) or (*S*) except for the introduction of some structures from reports. In addition, L- and D- are used for the AC expression of amino acids or sugars.

When dealing with helical compounds, capital letters of *M* and *P* are used to show their stereochemistry. This will be introduced in the "Other Stereogenic Centers" section.

Some monoterpenes may contain two or more –CHO groups together so that are close enough in space like the paederoside **40** [9] and harpagoside **41**. Both aldehyde groups may undergo reaction with each other and form an iridoid, which can further react with a glucose to afford the iridoid glycoside (**41**). The AC of C1 may be controlled by the C9's AC in both structures (obeying the Cram rule).

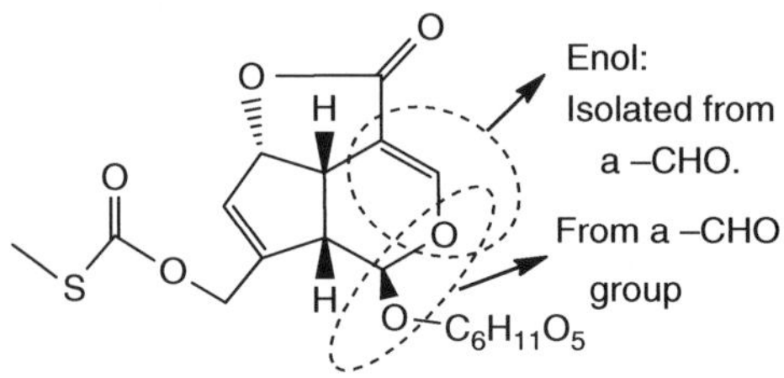

40 paederoside

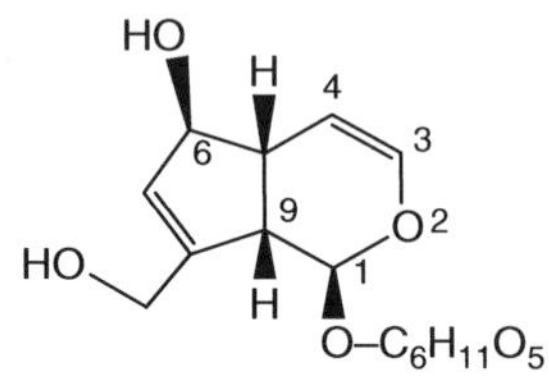

41 Aucubin

Three isopentadiene units can form the corresponding sesquiterpenes. They are widely distributed in nature with thousands of types. As with terpenoids, linear and cyclic sesquiterpenes with one or more rings are found in different species. Some representatives are the compounds **42–48** [9].

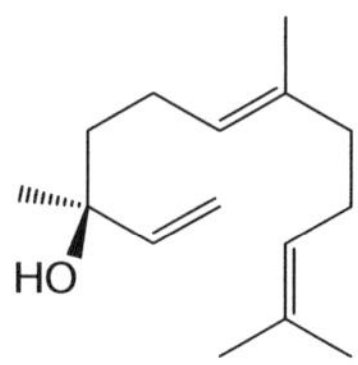

42 Nerolidol

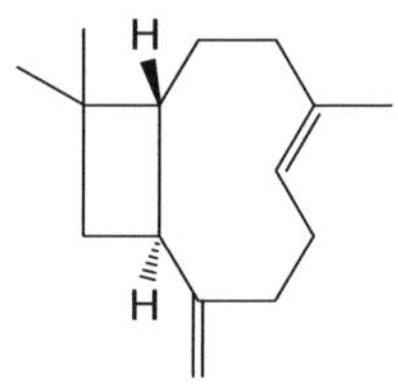

43 *β*-Caryophyllene

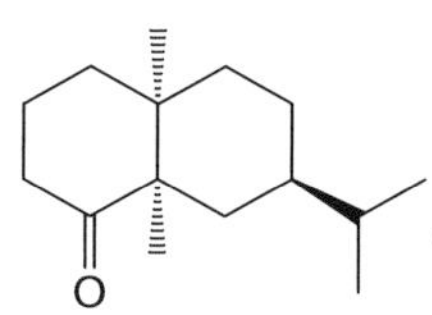

44 Valeranone

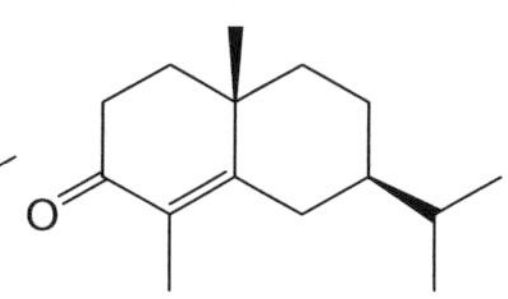

45 Cyperone

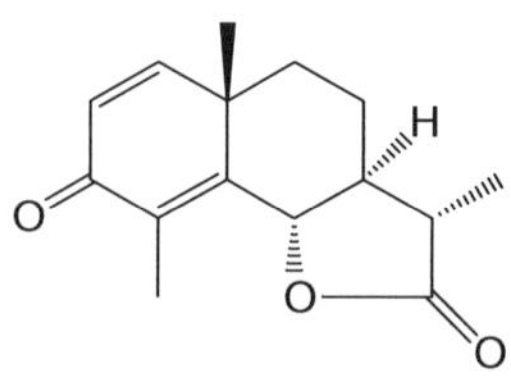

46 1-*α*-Santonin
Strong ascarifuge

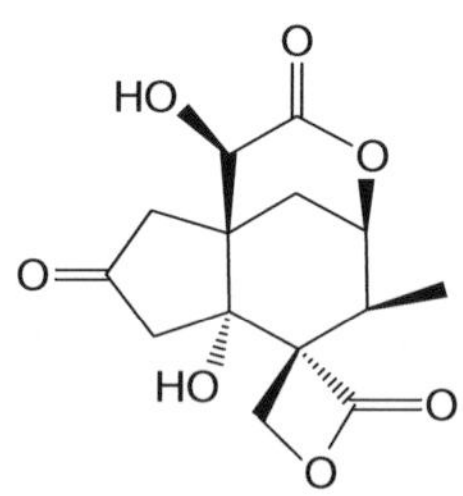

47 *α*-Oxo-6-deoxyneo-anisatin
Strong toxic, LD_{50} near 0.94 mg kg^{-1}

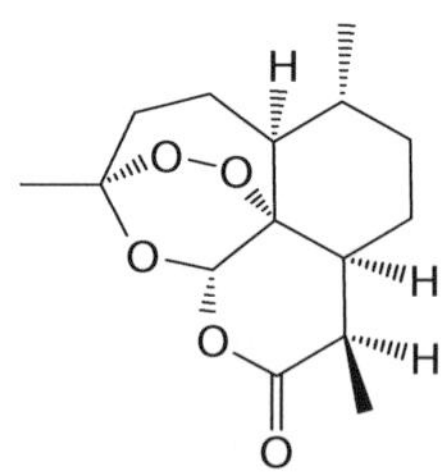

48 Qinhaosu
Anti malaria

With the development of separation materials and technology, more and more skeletons were found from different materials. For example, the sterostreins A (**49**) and D (**50**) are furan-containing illudalanes, which were isolated from fungi [10]. Its dimeric compound stereostreins A exhibited antimalarial activity with an inhibitory concentration (IC_{50}) of 2.3 μg ml^{-1}.

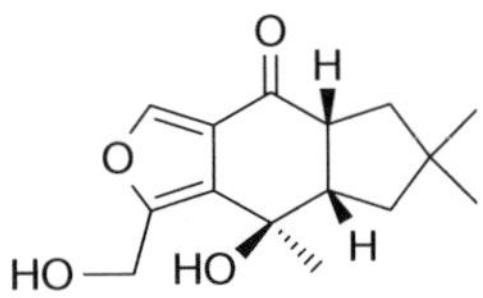

49 Sterostreins D

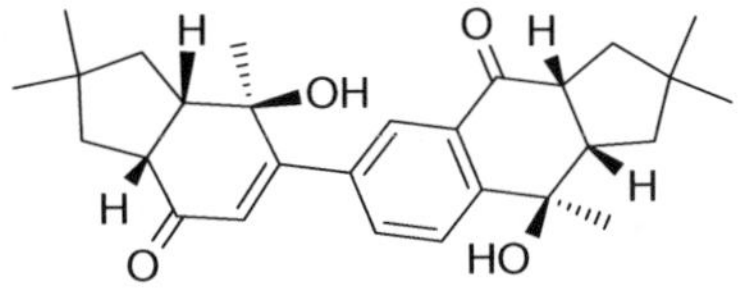

50 Sterostreins A

Diterpenoids contain four isopentadiene units, which are classified as linear and cyclic structures. For example, an acyclic phytol (**51**), widely existing in chlorophyll, could be used as a starting material for the synthesis of vitamins E and K_1.

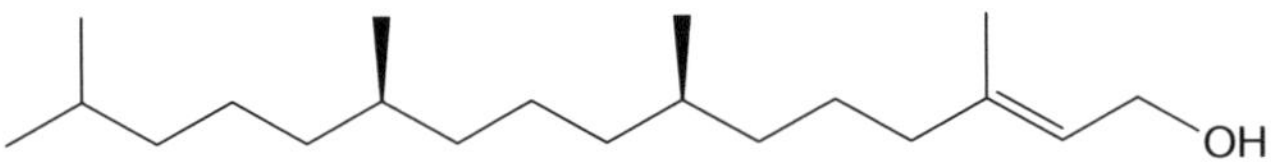

51 Phytol

Because of the four isopentadiene units in diterpenes, their structural complexity is more than that of sesquiterpenes. Many bioactive compounds have been found, such as taxol (**54**), with strong antitumor activity. Some representatives are illustrated below (**52–58**).

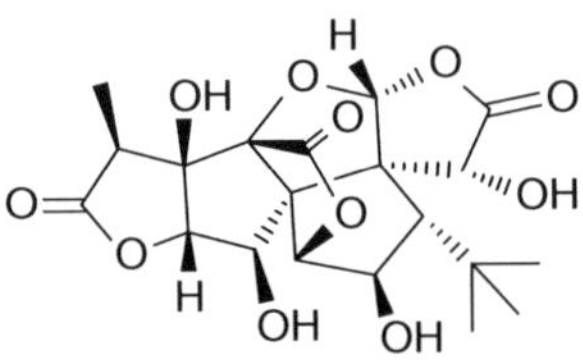

52 Ginkgolide C
Cerebrovascular drug

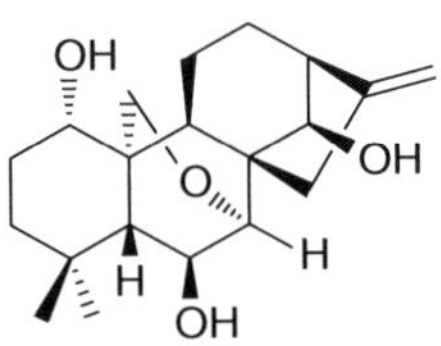

53 Oridnin
Anti-cancer

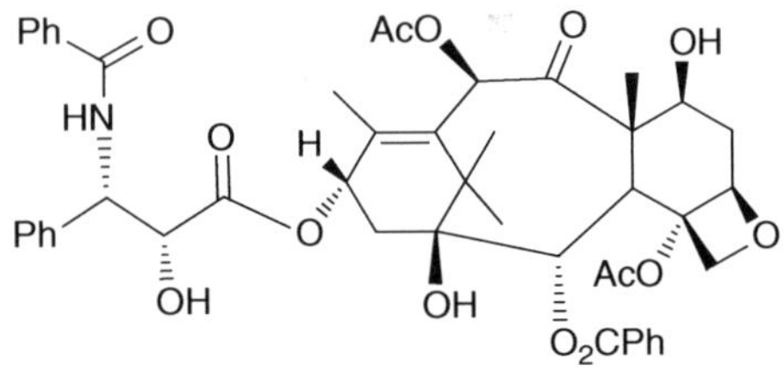

54 Taxol, anti-cancer

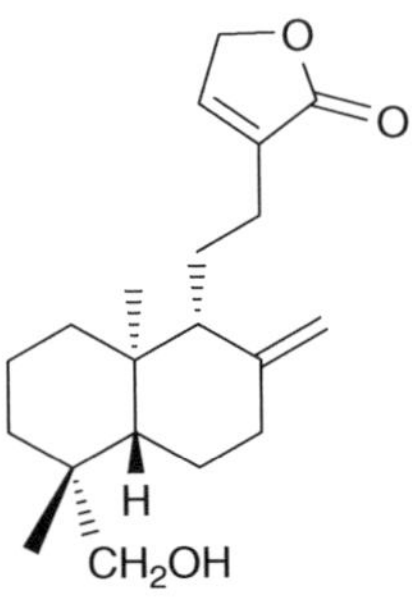

55
Neo-andrographolide

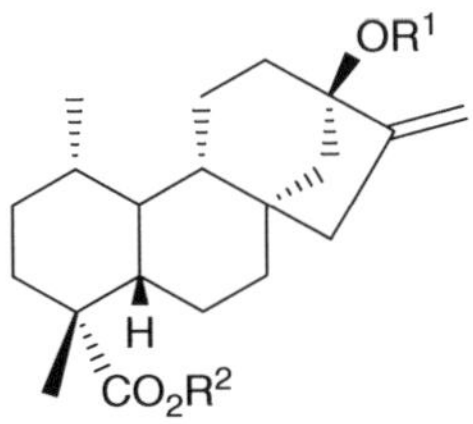

Intense sweetener

	R^1	R^2
56 Stevioside	glc $^{2-1*}$ glc	glc
57 Stevioside A	glc 2-1 glc; glc 3-1 glc	glc
58 Stevioside D	glc 2-1 glc; glc 3-1 glc	glc$^{2-1}$ glc

*Note: glc$^{2-1}$ glc means that the –OH on C2 of the first glucose connects to the –OH on C1 of next glucose to form an ether (–O–) structure.

New skeletons of diterpenes have been found recently, such as a macrocyclic diterpene with a C_2 symmetric axis, and schaffnerine (**59**) [11], which has slight bioactivity. A smaller structure, actinoranone (**60**), was obtained that exhibited significant cytotoxicity to the HCT-116 human colon cancer cells, with an LD_{50} of 2.0 μg ml^{-1} [12]. The structures here belong to diterpenoids. However, it is not easy to find the connection model, either "head to head" or "head to tail" or "tail to tail" at first sight.

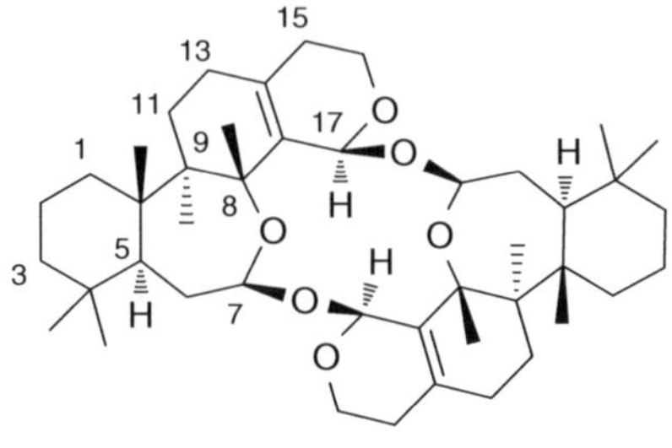

59 Schaffnerine

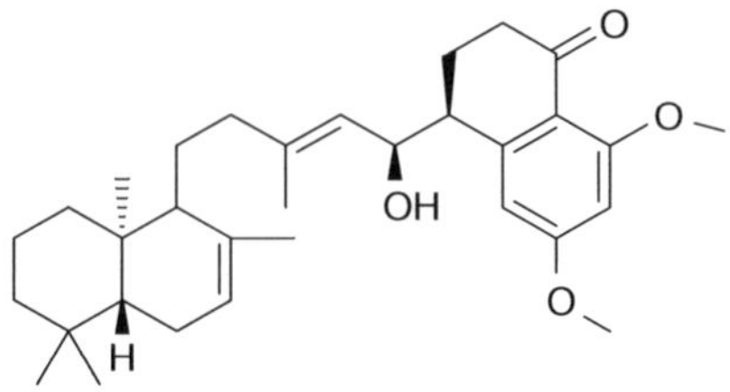

60 Actinoranone

Sesterterpenoids consist of five isopentadiene units. They are not frequently reported. A traditional sesterterpenoid **61** and a recently reported compound **62** [13] are shown below.

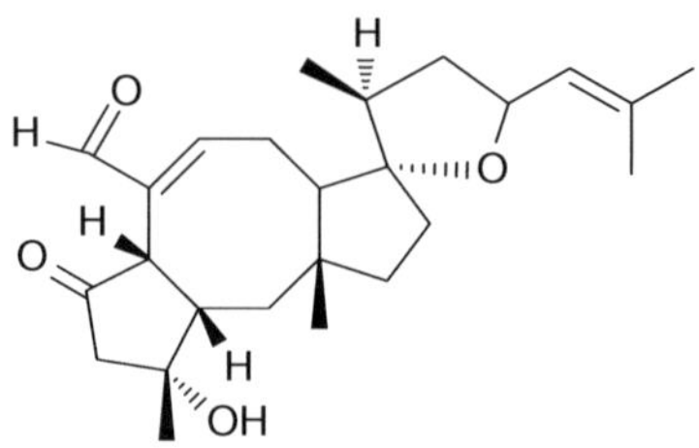

61 Ophiobolin A

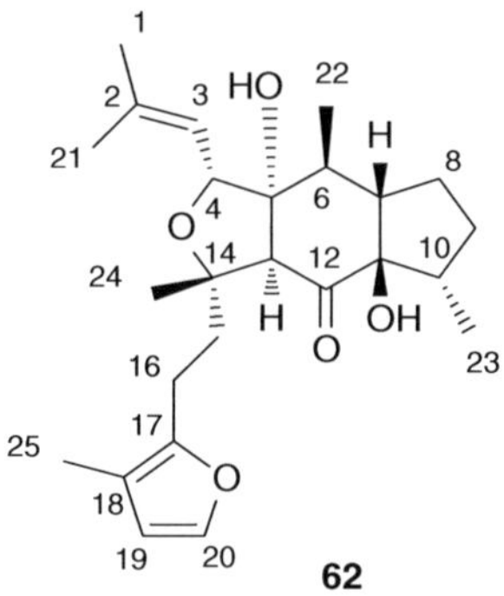

62

Triterpenoids have six isopentadiene units. They include various backbones with different bioactivities. Some of them have been developed as medicines. Some typical structures are illustrated below (**63–67**).

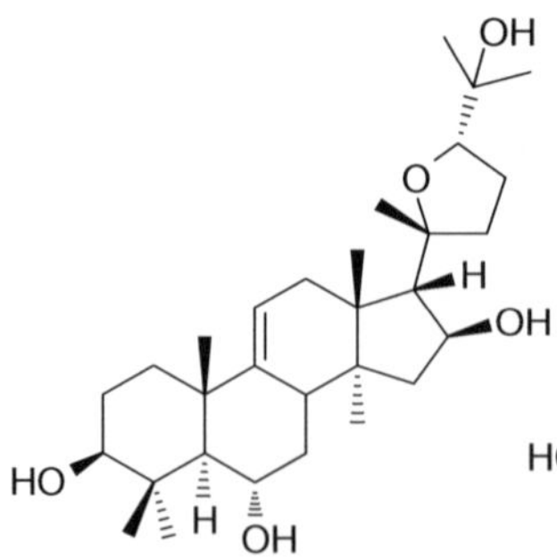

63 Astragenol

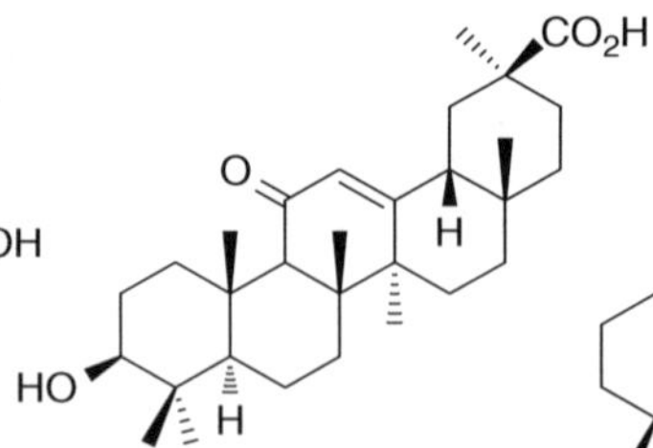

64 Glycyrrhetinic acid
Anti-inflammation

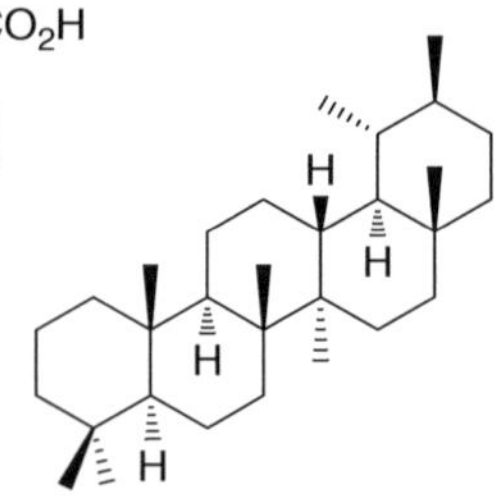

65 Tetraxastane

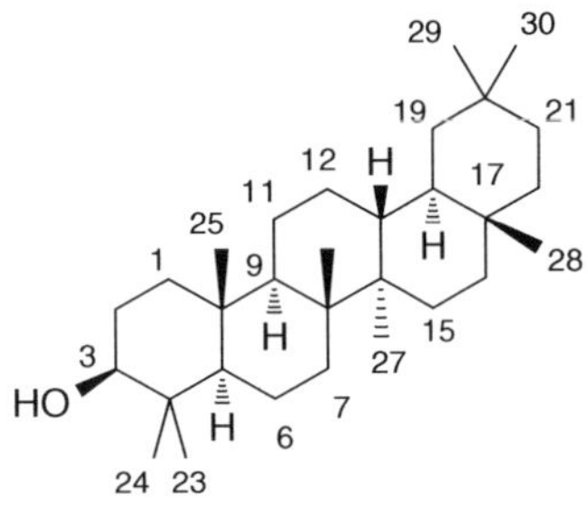

66 oleana-3β-ol

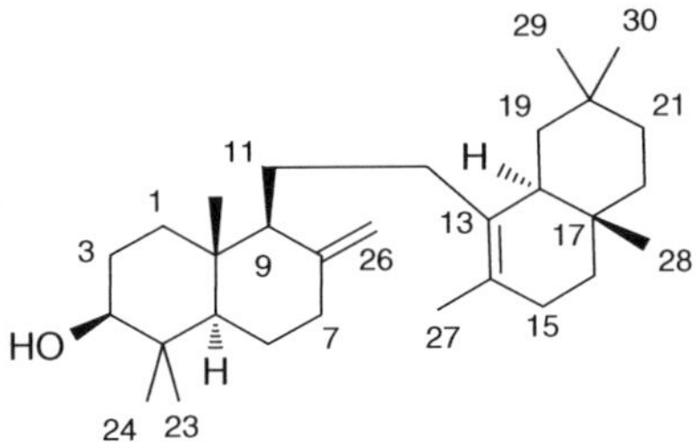

67 8,14-*seco*-oleana-8(26),13-dien-3β-ol

In order to name a triterpene, we need to look at its precursor. For example, as a typical triterpene, oleane-3*β*-ol (**66**) must be the precursor of **67** [14], which has the skeleton name "oleana" inside. By analyzing the disconnected positions (C8 and C14), it can be named as 8,14-*seco*-oleana-8(26),13-dien-3*β*-ol.

Be careful that "*β*" here is used to name the configuration of C3. To avoid mistaking the same symbol with different concepts in structural study, it is suggested that you use (*R*) or (*S*) to assign the C3's configuration. For the example of **67**, C3 has the (*S*) configuration.

1.2.2
Flavonoids

Flavonoids, which generally have a characteristic yellow color, are widely distributed in nature. The general skeleton contains two phenyl rings, as in **68**. Most flavonoids have no chirality. The double bond at C2 and C3 may be hydrogenated to afford flavano or flavanone. The stereogenic centers may form in unequal quantities if the hydrogenation happens in a chiral environment. Its basic characteristic is six carbons on ring A, three carbons on ring B, and six carbons on ring C, and it forms C6–C3–C6 unit connections. When there is a –OH on C3 in **68**, it belongs to flavanol. Once it is on C3 in **69**, it is named as flavanonol.

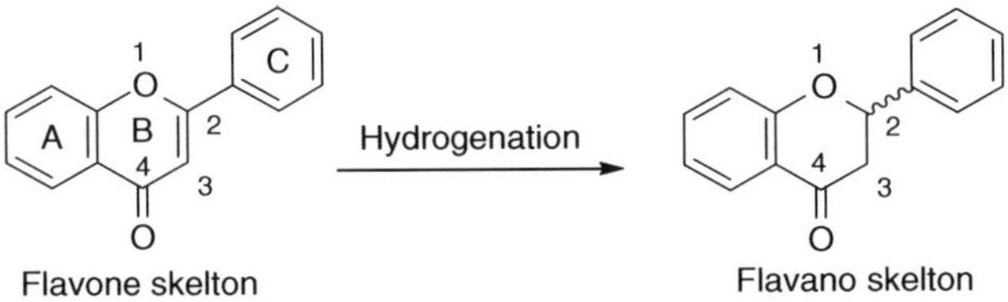

Flavone skelton
68

Flavano skelton
69

(1.4)

Many flavanonols show different but good activity in humans. For example, (+)-catechin (**70**) is used as a liver protection medicine (its commercial name in European market is Categen) its isomer **71** was also reported with similar bioactivity. The study of their AC is difficult because the isomers easily convert to each other. To report its structure with RC or racemic structures is also acceptable in studies.

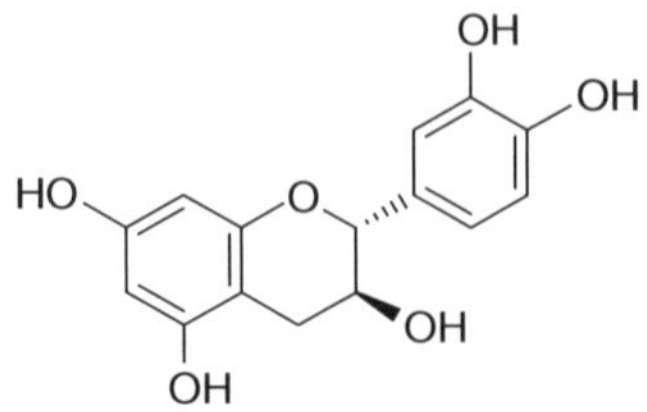

70 (+)-Catechin

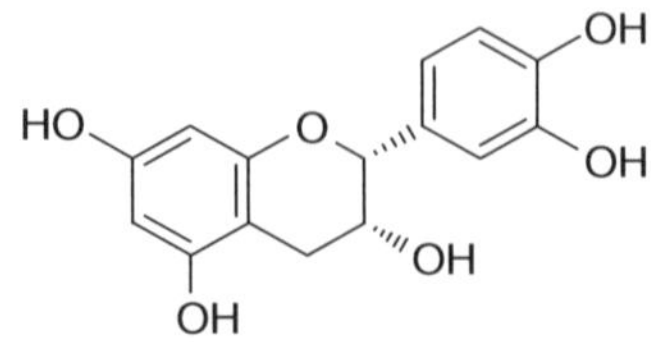

71 (−)-Epicatechin

There are many flavanols derived from the "standard" flavano skeleton. For example, rotenoids have a basic skeleton (**72**) [9] with a fixed AC, whereas the others, such as rotenone (**73**), have almost the same backbone of **72**. It has strong toxicity against fish (when the concentration of rutenone in water reaches ~0.08 ppm, it will lead to fish stupor or death) and flies. But it is harmless to humans and animals.

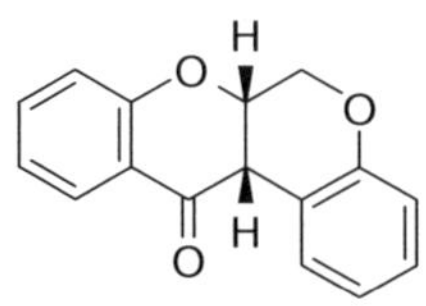

Skeleton of rotenoids
72

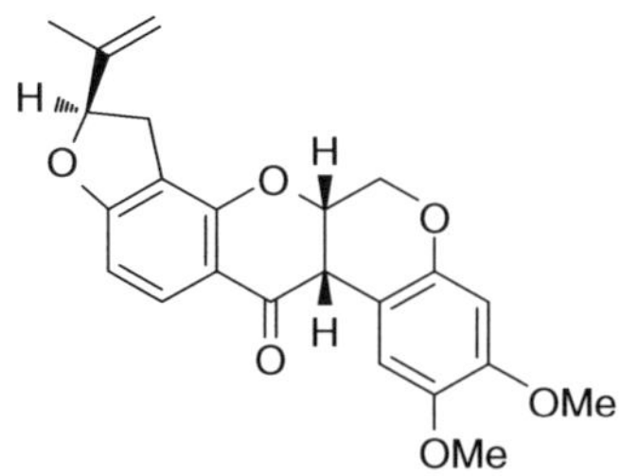

73 Rotenone

The hydroxyl group of a flavone or flavanone undergoes condensation reaction with a sugar to form glycoside, which derivatizes with chirality. For example, the flavone rutin is a derivative of rutinose. It can be used to assist in antihypertension treatment.

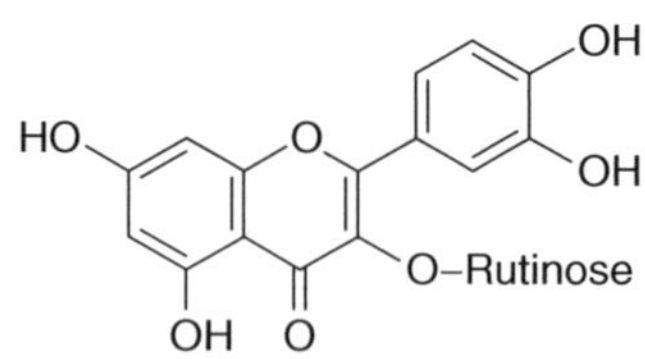

74 Rutin

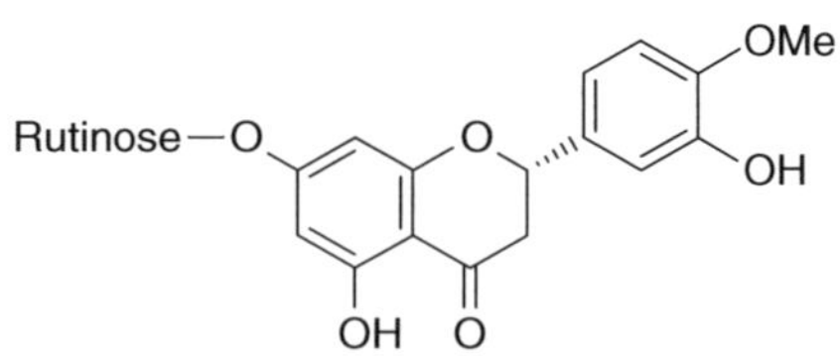

75 Hesperidin

1.2.3 Alkaloids

Alkaloids are very important compounds in modern organic chemistry. Many alkaloids have strong bioactivity; for example, ephedrine (**76**) is well known for its effect in relieving cough and asthma. Others, such as morphine (**78**), are also famous for their bioactivities. Alkaloids bring many benefits for modern pharmaceutical industry.

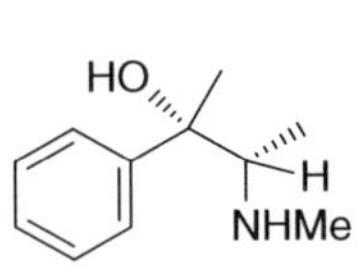

76 L-ephedrine

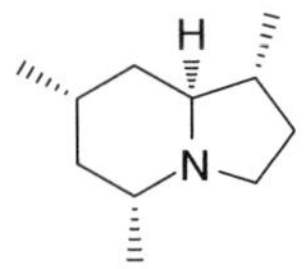

77 Dendroprimine

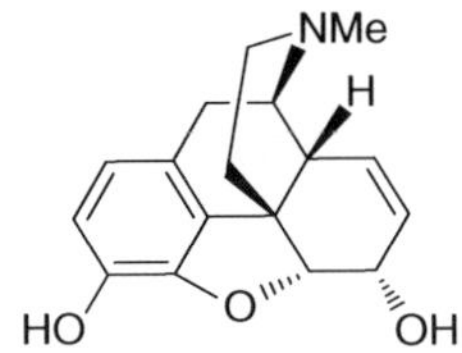

78 Morphine

In many cases, alkaloids have very complex structures. For example, acutumine (**79**) has five stereogenic centers; it is not usual for a Cl atom in the molecule. Other alkaloids, such as conessine (**80**), which are derived from steroids, belong to steroidal alkaloids.

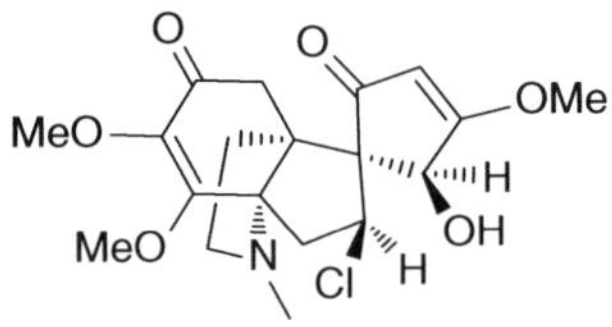

79 Acutumine

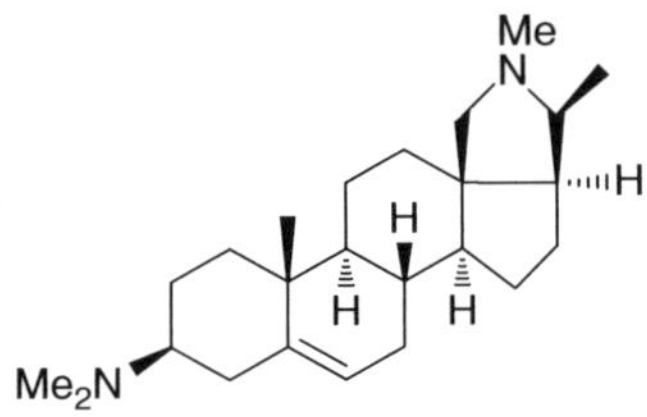

80 Conessine

Alkaloids with a terpenoid skeleton are widely reported. Their structural complexity is a big stumbling block for organic chemists when assigning their AC. Two traditional structures are illustrated below [9].

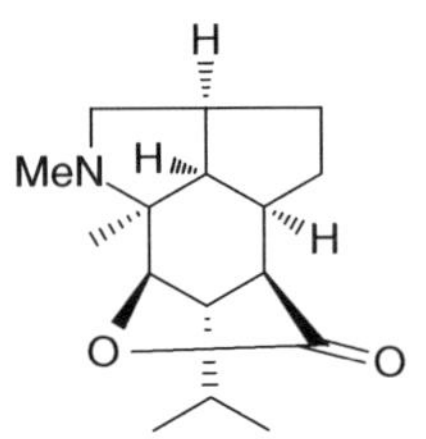

81 Dendrobine
Sesquiterpene alkaloid

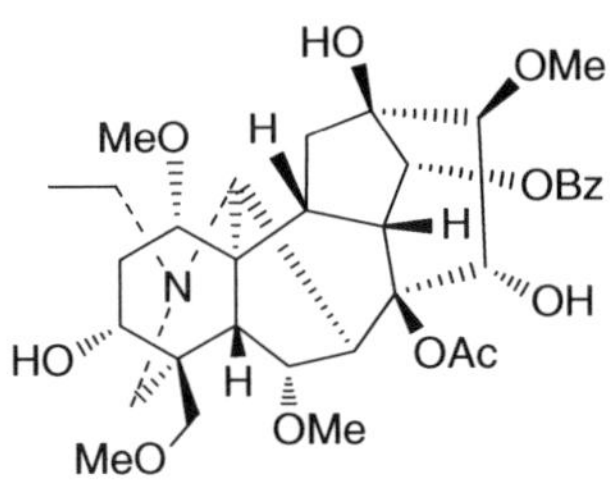

82 Aconitine
Diterpene alkaloid

There are hundreds of alkaloids reported until now. Some of them are valuable lead compounds for the pharmaceutical study. Some alkaloid structures are illustrated below.

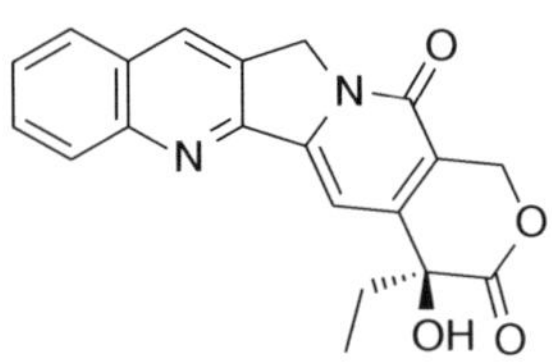

83 Camptothecine
Antitumor

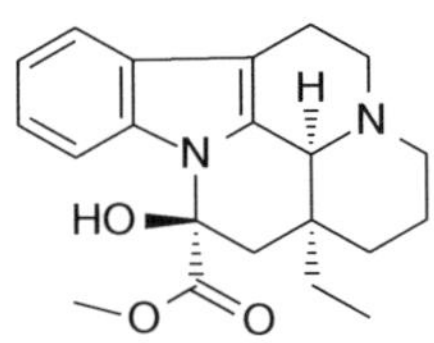

84 Vincamine
Peripheral asodilator

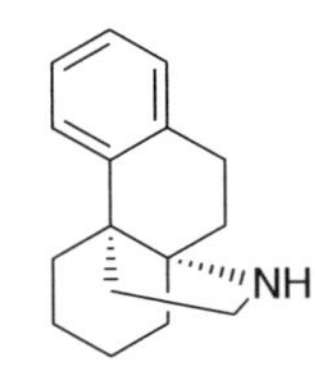

85 Hasubanane
Anti-plasmodial

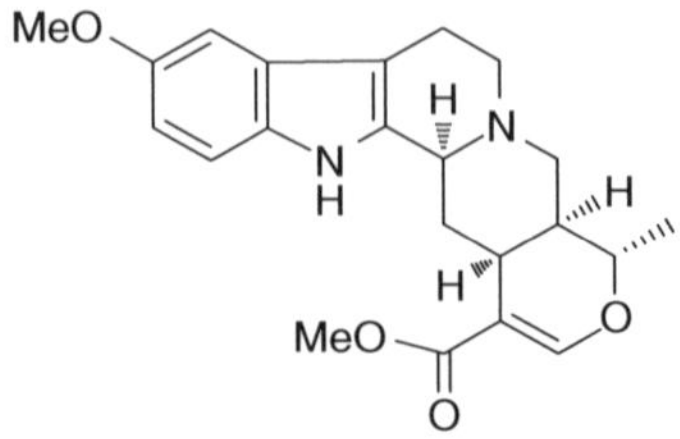

86 Arisine

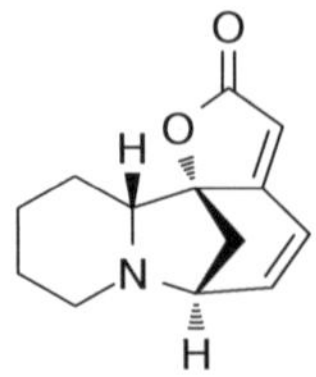

87 Securinine
Central stimulant

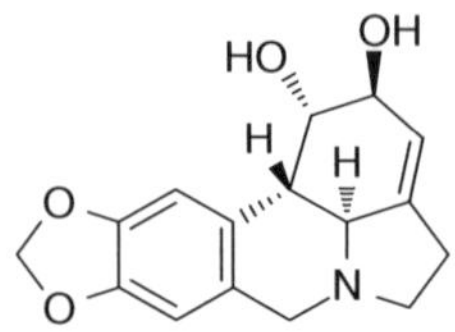

88 Lycorine
Antitumor

1.2.4
Steroids

Steroids form an important kind of natural compounds with various bioactivities. They can be mainly classified as C_{21}-steroids, steroidal saponin, and so on.

C_{21}-steroids, as the name implies, have 21 carbons in a basic skeleton of pregnane or its isomer. As pointed out earlier, *β* was used to label the configuration of **91**. If it is labeled as (*R*) or (*S*), it should be (3*S*,8*R*,9*S*,10*R*,13*R*,14*S*,15*R*,17*S*).

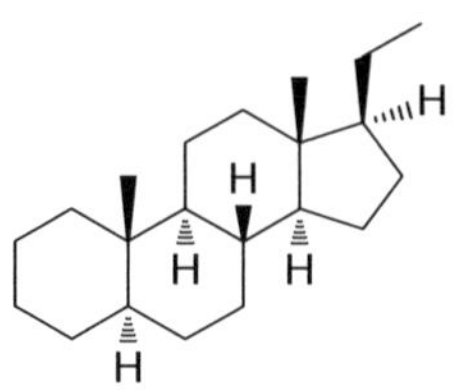

89 Pregnane

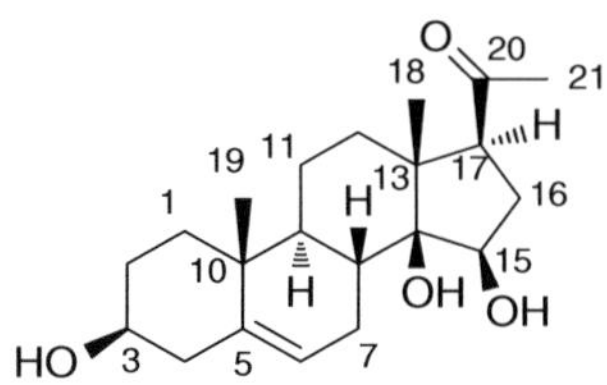

90 3*β*,14*β*,15*β*-Trihydroxypregn-5-en-20-one

Other steroids, such as androstan-3-one (**91**), cholesterol (**92**), ursodeoxycholic acid (**93**), mestanolone (**94**), formestane (**95**), and enicillitone (**96**), are listed below [15]. These are some representatives among the huge number of steroids.

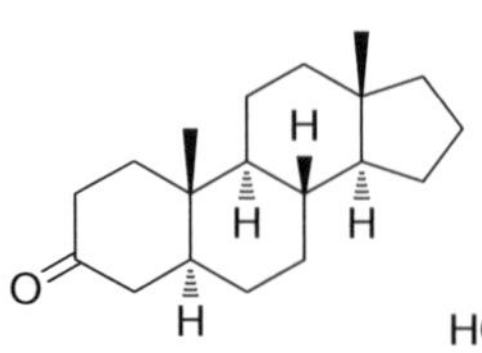

91 Androstan-3-one

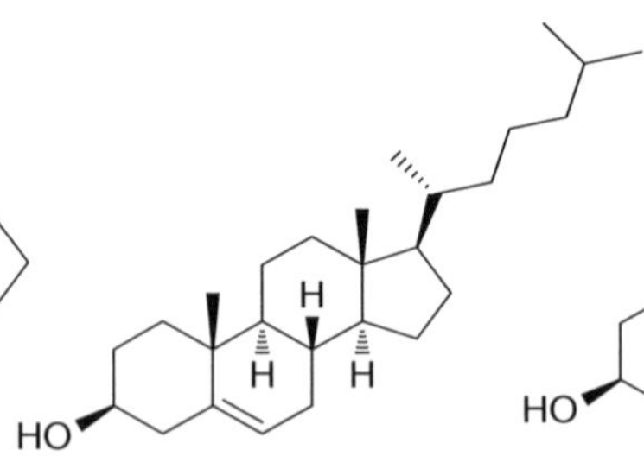

92 Cholesterol

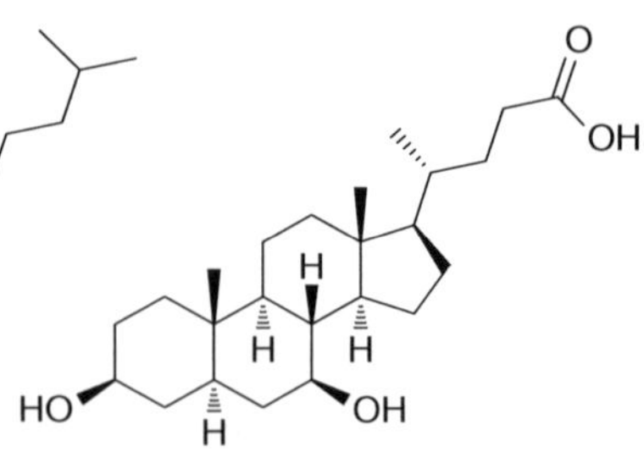

93 Ursodeoxycholic acid

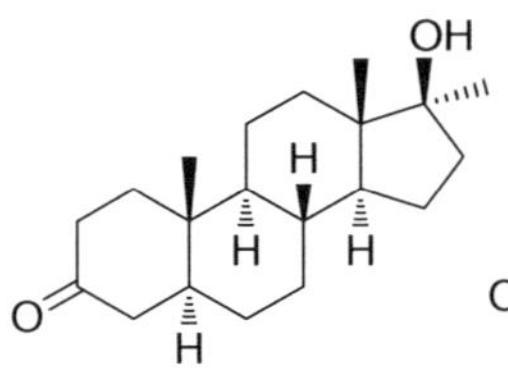

94 Mestanolone

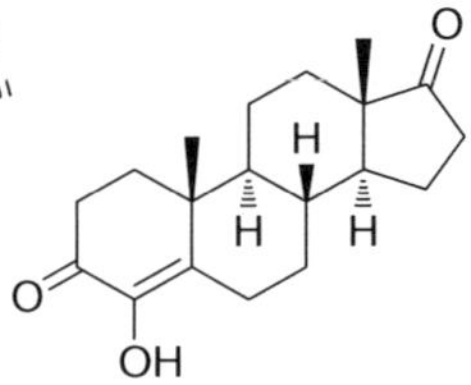

95 Formestane

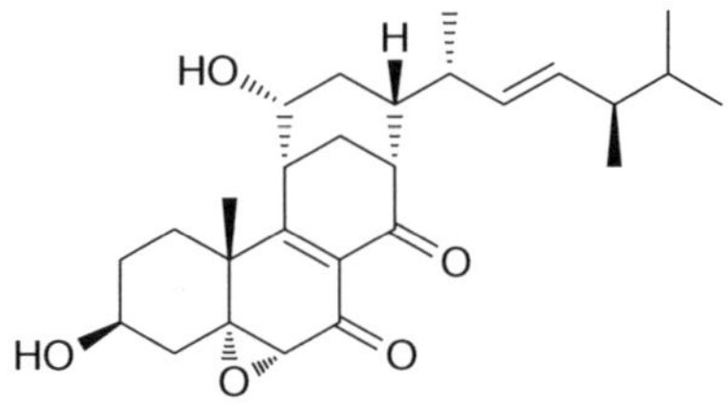

96 Penicillitone

1.2.5
Glycosides

When a sugar reacts with an –OH-containing compound (such as a terpenoid, a teroid, and so on), the product is called a glycoside. Therefore, glycosides contain two moieties; one is an aglycone (like terpene, alkaloid), and the other is a sugar moiety. Until now, hundreds and thousands of glycosides have been found and reported all over the world. Most of them have definite bioactivities. Because of their interesting activities, researchers have paid much attention to this kind of products.

The group –OH can be replaced by one –NH, –CH, or –SH. After the replacement, it forms the corresponding *N*-glycosides and *C*-glycosides, respectively, such as crotonide (**98**) and isovitexin (**99**). It clearly shows that there are two moieties: a sugar and an aglycone.

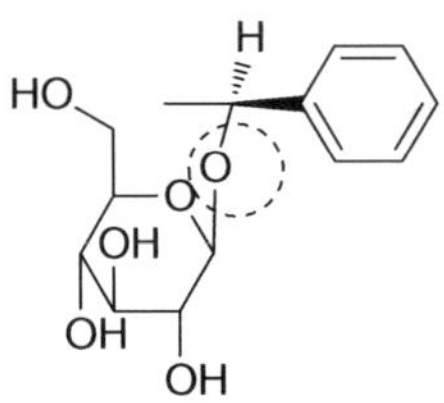

97 Prunasin
O-glycoside

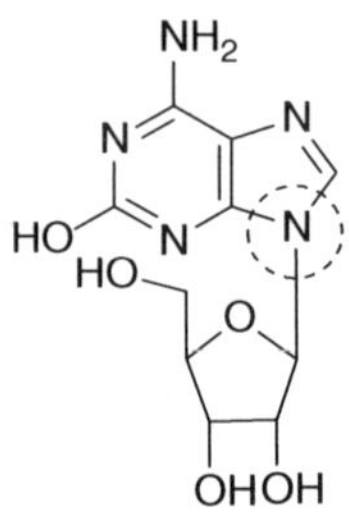

98 Crotonide
N-glycoside

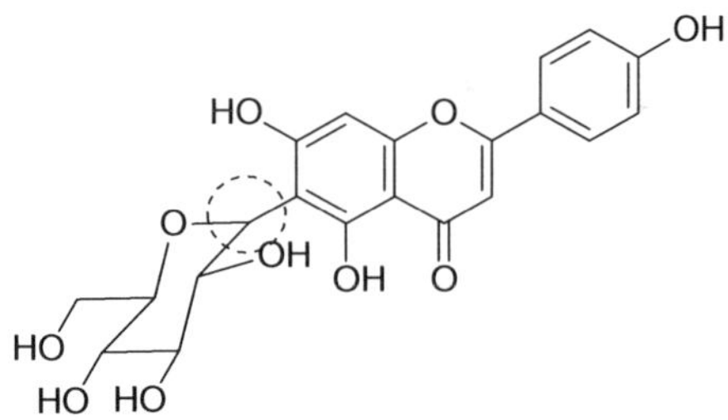

99 iso-Vitexin
C-glycoside

The –OH is hidden for clarity in the structures in some reports (101). As one of the most important steroidal glycosides, cardiac glycosides have been attracting the attention of many researchers. Traditionally, the configuration of glycosides in some positions is named as α or β. For example, naturally, most cardiac glycosides have β configuration at C3 and C17. If C17 has β orientation, H17 will have α orientation (**100**). However, a glycoside such as **101** has β orientation on H17, which is named as 17β-*H*-neriifolin.

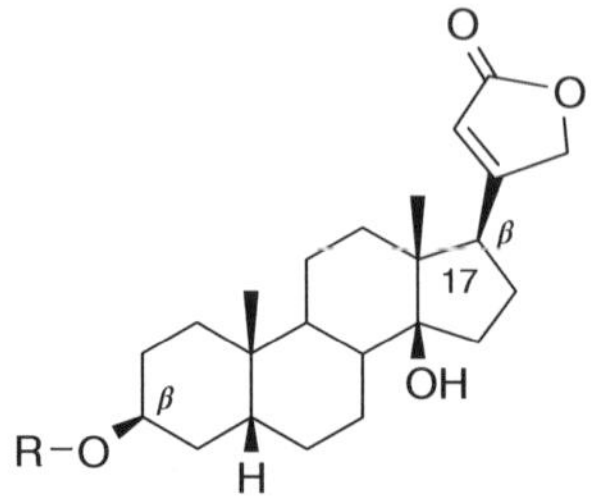

100 R = differnet glucose moiety

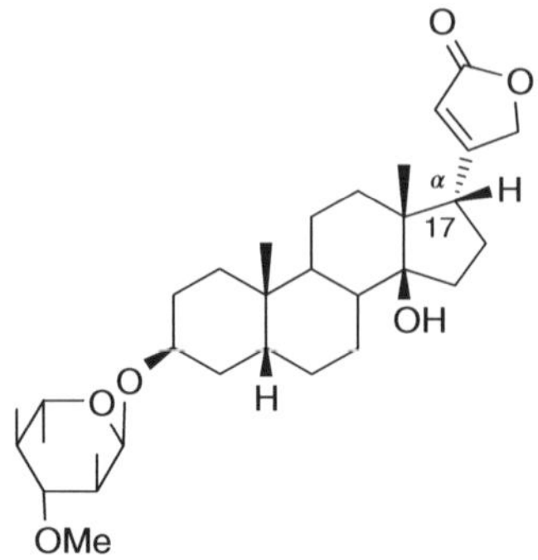

101 17βH-neriifolin

The difficulty is to convert the α or β configuration to the corresponding (*R*) or (*S*) at the beginning. Researchers must understand that the use of α or β is out of tradition. With the development of organic stereochemistry, it is possible to change traditional habits to use (*R*) or (*S*) in AC study, although it may take a lot of time.

1.2.6 Others

There are many different chiral compounds found in nature. However, amino acids and sugars are not included in the text. They are important organic matter. There are some special books dealing with them.

For a quick look at the chiral compounds, their structures are illustrated below.

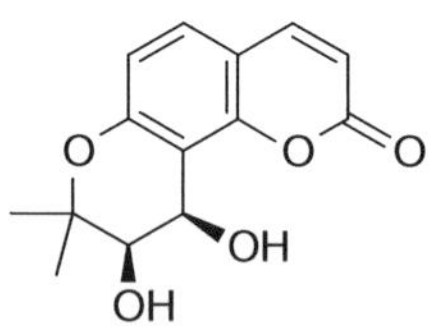

102 Khellactone
Coumarinoids

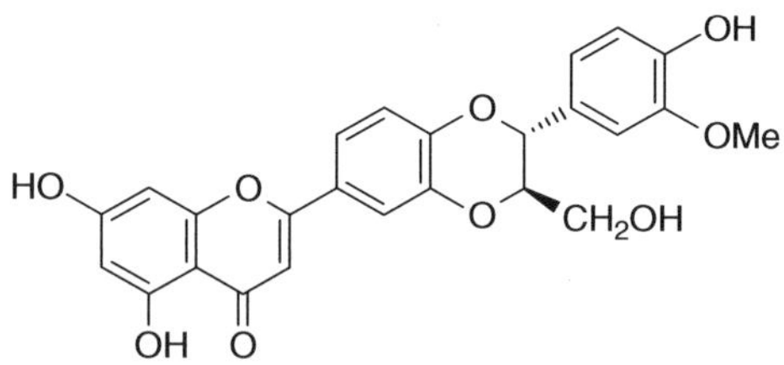

103 Silymarin
New lignanoids

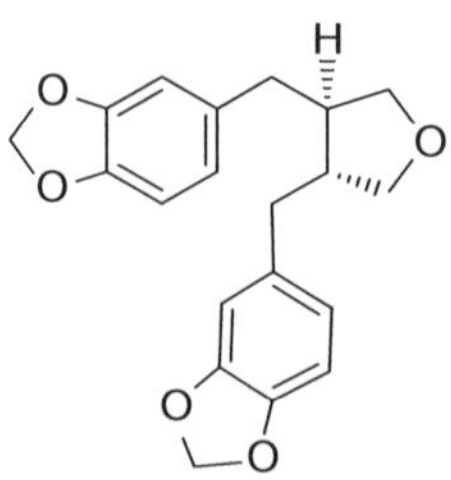

104 Cobebin
Lignanoids

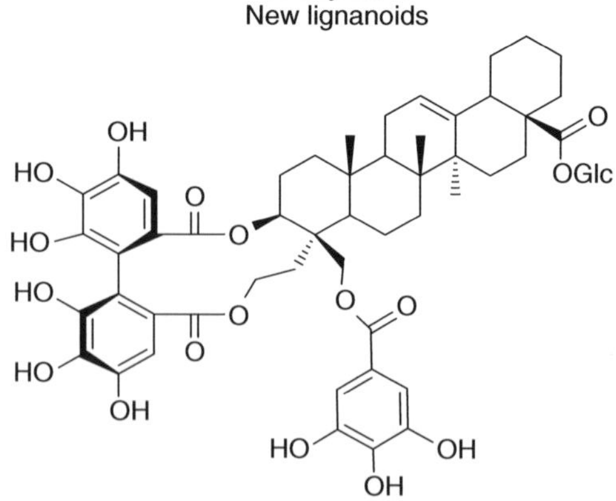

105 Castanopsinins B

The pretty chiral materials exhibit a very colorful picture in front of chemists. Finding newer chiral compounds and their uses is an interesting task. The assignment of their AC then becomes a big challenge in modern science.

1.3
Other Stereogenic Centers

The earliest discovery of chirality is the chiral sp^3 C atom. It is also regarded as C-chirality. There is another kind of chiral compounds in which the stereogenic centers are located on the N, P, or S atom. For example, when the lone electrons on the N atom are fixed, like quaternary ammonium salts (the substituents on N are not equivalent) or N–O compounds, chirality of N atom is formed.

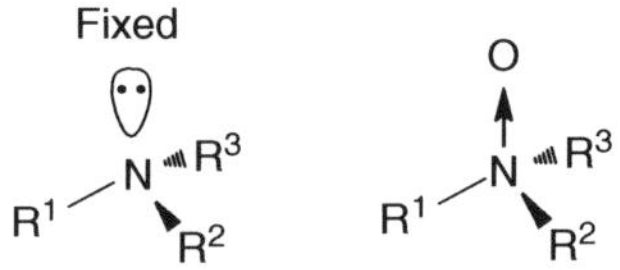

106 **107**

$R^1 \neq R^2 \neq R^3$ in the N-chirality compounds shown above. Compared to the instability of the N-chirality in absence of some acidic reagents, sulfoxide's chirality (S-chirality) can exist stably at room temperature and can be used widely in asymmetric synthesis. Other atoms, such as P, can also form the corresponding P-chirality. Their AC nomenclature is the same as that in C-chirality compounds.

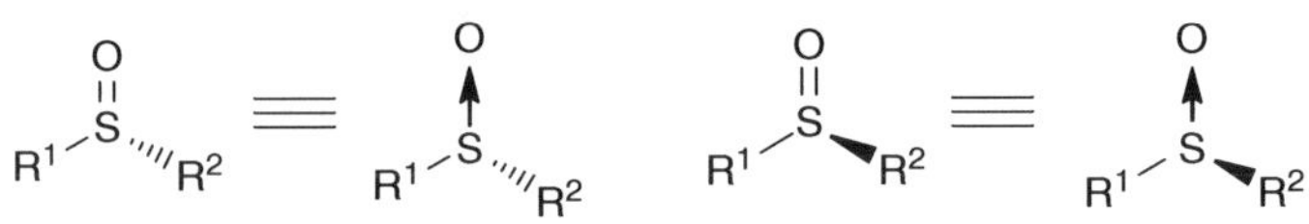

108 **109**

Among the different chiral compounds, one kind of compounds has chirality due to the rotation restriction (limitation) of the C–C single bond [16]; the early reported racemic 6,6′-dinitro-[1,1′-bipheyl]-2,2′-dicarboxylic acid has two ACs in solution, which can be separated from each other. This is axis chirality or axial chirality, and the compounds are atropisomers.

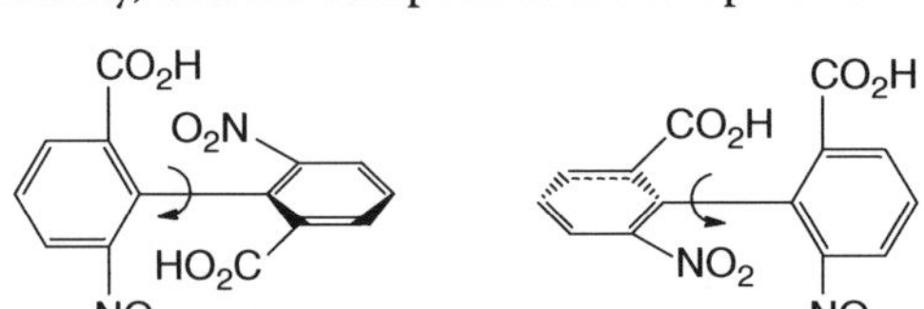

110 **111**

The widely used single-bond rotation-limited chiral compounds are derived from [1,1′-binaphthalene]-2,2′-diol (**112**). The number of chiral catalysts derived from it may be over a hundred, and some of them, such as (*R*)-2′-((dimethylamino)methyl)-[1,1′-binaphthalen]-2-ol (**113**), exhibit very high enantioselectivity in different reactions, giving up to 90% ee in the enantioselective addition of diethylzinc to aldehydes [17].

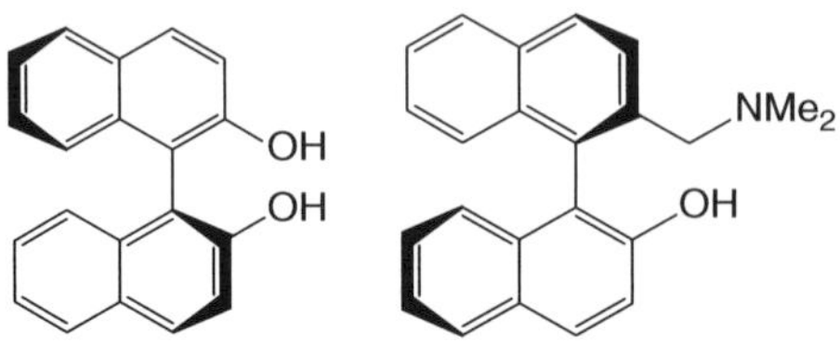

(S)-**112** (R)-**113**

The rule to name the AC for binaphthyl-containing chiral axial compounds is almost the same as used for chiral carbon tetrahedron structures. Chiral compound **112** was used as an example [18]. First, we can "concentrate" on the corresponding part as a point. For **112**, it can be labeled as the four points as in structure **A**. Then to convert it to a tetrahedron (**B**), obviously, the group point a should be smaller than the group point b (a < b) and c < d (Figure 1.4).

Then, we can put point a in a "rotation" circle center (**D**). In this case, point a is regarded as the smallest, and the point b, which is located on the same molecular moiety, will be regarded as the largest. Therefore, b > d > c. It forms a counterclockwise configuration and named as (*S*) for **112** (Figure 1.5).

Of course, we can regard point c as the smallest one; then point d is the largest. We can draw another "rotation circle" (**F**). The order is d > b > a. It is also the counterclockwise configuration, and assigned as (*S*) for structure **112**.

There is another method to name its AC. Compound **112** could be "seen" along the axial direction to make sure its AC. Interested readers can find more details in

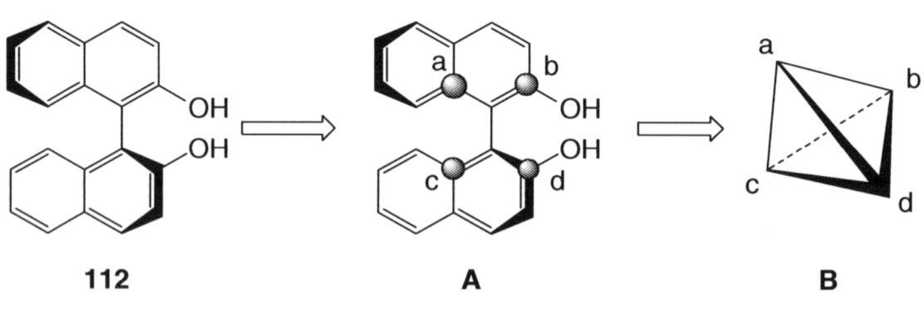

Figure 1.4 Conversion of axial binapthanol to a tetrahedron structure [18].

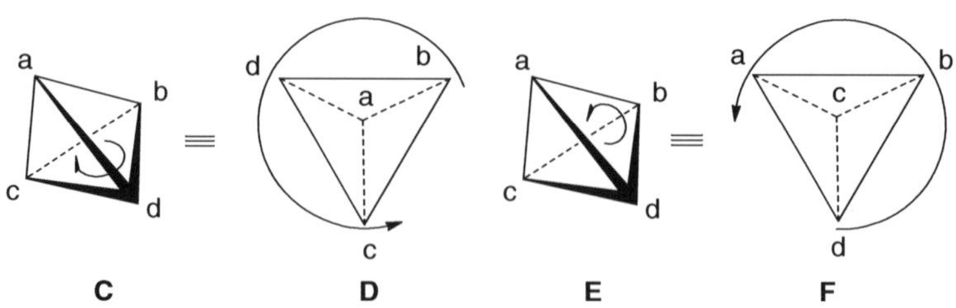

Figure 1.5 Two equal tetrahedron structure conversion.

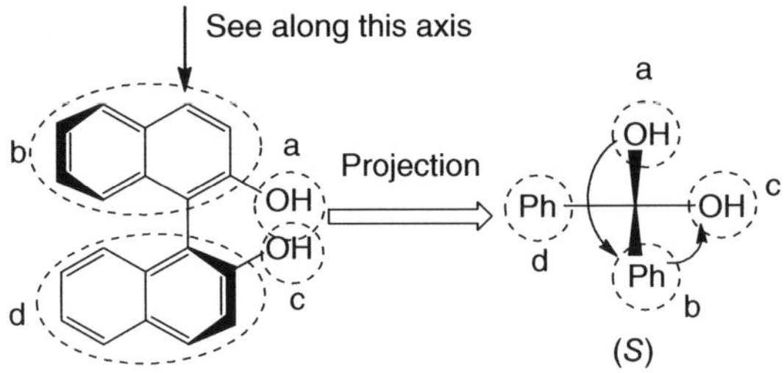

Figure 1.6 Diagrammatic sketch of the projection structure from **112** [18].

Ref. [19]. Although the nomenclature is different, the final result is the same. For example, **112** can be "seen" as in Figure 1.6.

In this rule, for example, if the starting point is a hydroxyl group a, then it needs to go to phenyl group at point b, and then to another hydroxyl group at point c. The final result is the same: (*S*)-AC.

However, practically, it may be easy to go from point a, then to c, and finally to point b since the two hydroxyl groups are equal, and lead to a wrong (*R*)-AC. To avoid this error, we can apply the rule mentioned above to this case. We use **112** again as an example. If point a is regarded as the biggest, then point b must be regarded as the smallest, the second hydroxyl group c is the second largest, and point d is the smaller one. It forms a circle with the direction from point a to point c and finally to point d (Figure 1.7).

Substituted propadiene derivatives also have chirality. However, because of their instability at room temperature, they are rarely studied alone. However, as a kind of chiral compound, its academic significance cannot be ignored. Its nomenclature can be illustrated as in Figure 1.8. Here, point a is regarded as the largest, then, automatically, point b on the same atom C is the smallest, and the rotation direction is from a to c to d. It is (*R*)-AC in this case.

Spiro compounds have similar nomenclature. We need to "see" the structure from the top through spiro point. It can be found that the chiral compound **115** has (*R*)-AC. The largest point and the smallest point should be on the same side (Figure 1.9) [18].

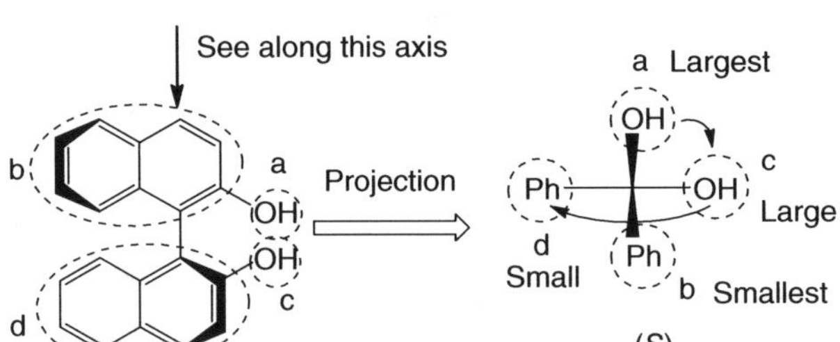

Figure 1.7 Diagrammatic sketch of the projection structure from **112** [18].

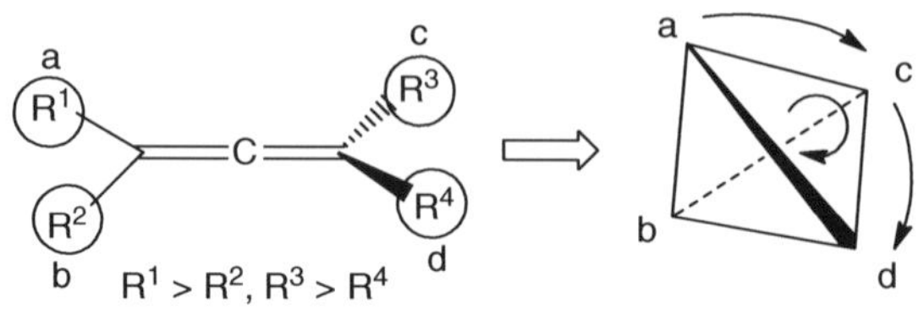

114

Figure 1.8 Diagrammatic sketch of projection structure from **114** [18].

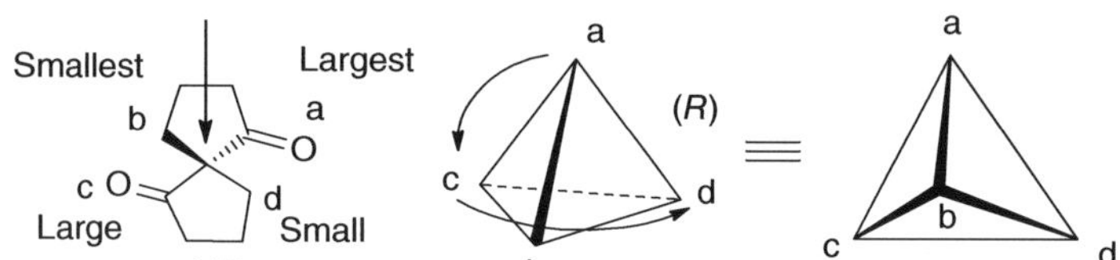

115

Figure 1.9 Diagrammatic sketch of **115** to the tetrahedron structure [18].

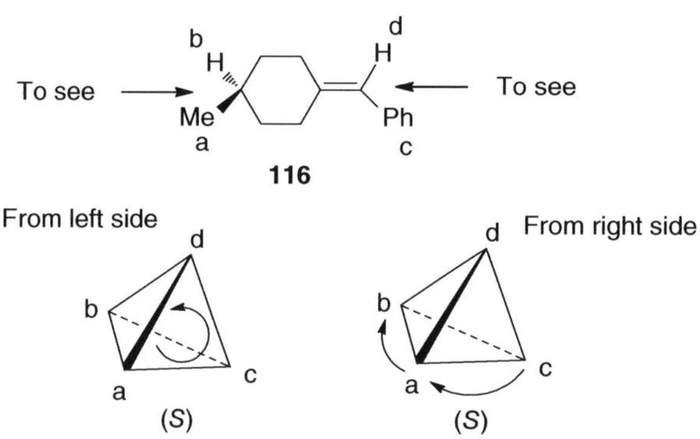

Figure 1.10 Diagrammatic sketch of **116** to the tetrahedron structure [18].

Other similar chiral structures may derive from it, for example, (*S*)-((4-methyl cyclohexylidene)methyl)benzene **116**. In this case, Ph is larger than Me, However, if we see it from the left, point a can be regarded as the largest and point b as the smallest; then the rotation direction is from a to c to d, and it has (*S*) AC (Figure 1.10). We could see it from the right side too: then the rotation direction is from c to a to b, and it has (*S*) AC. The conclusion is the same.

Other chiral molecules include planar chirality compounds (**117**), chiral ferrocenes (**118**), and other chiral metallic complexes. This is a very interesting area that involves the chelation of different metallic ions with various organic ligands. This is not a major topic in organic stereochemistry; the corresponding research belongs to coordination chemistry.

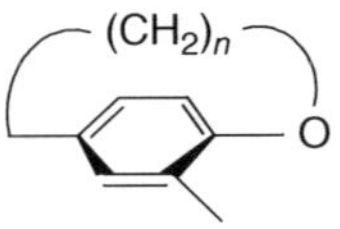

117

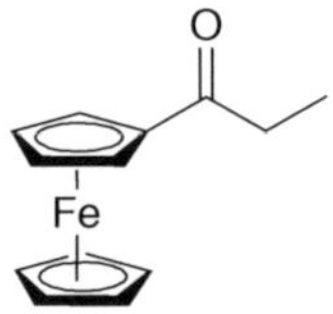

118

The final nonstandard chiral molecule is the helix structure. The study was well reported in the 1960s. This type of chiral compound has different nomenclatures: *M* or *P*. One needs to see it from the top. If the rotation direction is clockwise, it is labeled as *P*, and if it is counterclockwise the direction is *M*. This kind of chiral compound generally has large OR values. For example, the following two helix hexahelicenes have ORs of ~3700°. The difference may be due to measurement errors.

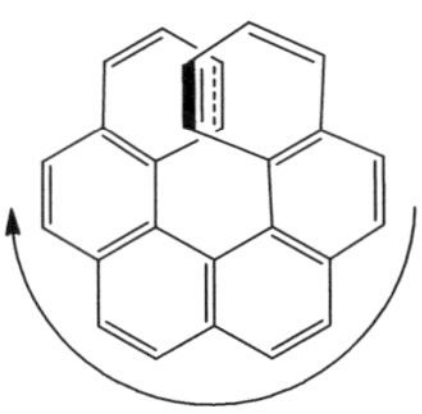

119 *P*-(−)-Hexahelicene
$[\alpha]_D$ − 3640° [18b]

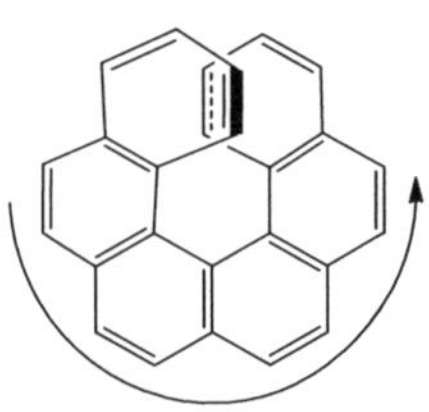

120 *M*-(+)-Hexahelicene
$[\alpha]_D$ + 3707° [18c]

The DNA chain is a well-known helix biopolymer; it is the most important representative among all helical chiral compounds although it is a biopolymer. The synthesis of the similar bioactive compounds may be an interesting but challenging task.

1.4
Optical Characteristics

In polarized light, enantiomers exhibit reversed optical characteristics. Just like their OR, others like electronic circular dichroism (ECD), vibrational circular dichroism (VCD), or Raman optical activity (ROA) also exhibits reversed characteristics.

OR is one of the earliest discoveries in organic chemistry. Since its discovery, many excellent chemists have developed this knowledge. Now, OR is one of the very general parameters for characterizing chiral compounds in various reports not only because of its easy measurement but also its wide catholicity. Most compounds have definite OR values that can be easily recorded with reliable resolution. It is absolutely necessary to know more about these characteristics.

1.4.1
Measurement of OR

Since the first separation of the left-handed enantiomer of sodium ammonium tartrate from the right-handed one by Louis Pasteur, many enantiomers have been prepared. The optically pure chiral compounds provided researchers great opportunities to study the relationship between OR and the molecular configuration. Many theories have been proposed to explain the OR phenomenon. These will be introduced in the following chapters. The general setup for determining OR is shown in Figure 1.11.

The change in the angle α of a beam of polarized light depends on the wavelength of light. At the same time, its size also depends on the length of the cell and the solution concentration when the temperature and solvent are fixed. The specific rotatory power $[\alpha]^t$ is proportional the values of the measured rotation α (Eq. (1.5)) as

$$[\alpha]^t = \frac{\alpha}{l \times c} \tag{1.5}$$

where c is the concentration with unit in g per 100 ml in most cases, α is the measured rotation value, l is the length of cell (10.0 cm in most OR setups), and t represents the temperature. When the D line of sodium light (589.6 nm) is used for OR determination, the specific rotatory power is written as $[\alpha]_D^t$. In general, the specific rotatory power is simply called OR.

The molar rotation $[M]_D^t$ has the following relationship with $[\alpha]_D^t$:

$$[M]_D^t = \frac{[\alpha]_D^t \times M}{100} \tag{1.6}$$

where M is the molecular weight.

In many asymmetric synthesis studies, "optical purity" is used to characterize the chiral compound's content in percentage. It needs a 100% pure chirality compound as the OR standard value. For example, if one enantiomer has the OR value of 100, and if the synthesized product has the OR value of 80, its optical purity is 80% o.p. or 80% op. This is popular in many studies. It is equal to the enantiomer excess (ee%) theoretically. However, because of measurement errors, both values will not be equal but very close.

Determination of OR is inexpensive and convenient. Most chiral compounds have large enough OR values, and therefore it is more convenient to use OR for AC

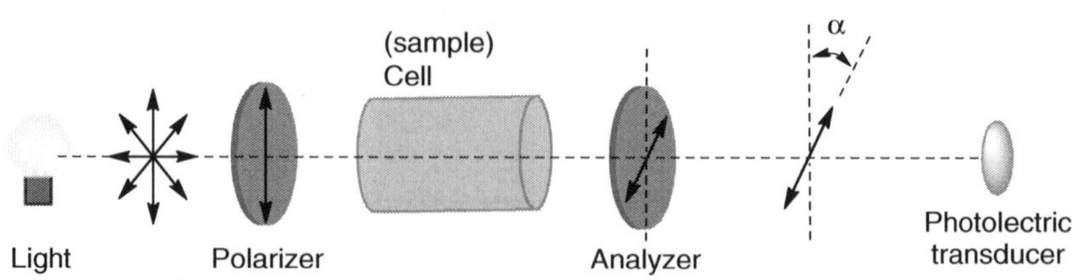

Figure 1.11 General optical setup of OR.

determination. However, some chiral molecules may have a very small OR value. In some cases, it is too small to be examined correctly. For example, 5-ethyl 5-propyl-undecane (**121**) has an extremely small OR; its absolute OR value is <0.001 between 280 and 580 nm [20].

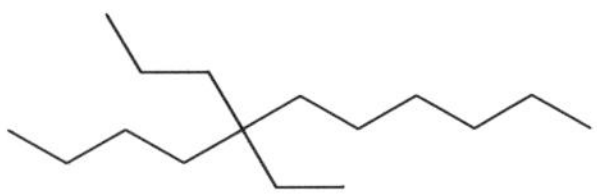

121

When the wavelength of light is changed from that of the D line of sodium to others, the recorded OR values under different wavelengths provide (ORD). The used wavelengths include 578, 546, 436, and 365 nm. Different setups may use different wavelengths for ORD determination.

The OR magnitudes may undergo a small change when one or more stereogenic centers change in some chiral molecules with two more such centers. This leads to difficulty in identifying the AC by comparing its experimental OR with the predicted OR. In this case, if a compound has double bonds that have UV absorption near the stereogenic centers, it is a good way to measure its ECD and compare its experimental ECD with the calculated ECD.

1.4.2 ECD and Its Definition

Left- and right-CPL may cause the phase of polarized light to change, and this produces the OR. On the other hand, the absorption of the left-handed and right-handed enantiomers of polarized light may be different, and this absorption difference ($\Delta\varepsilon$) can be measured as ECD.

$$\Delta\varepsilon = \varepsilon_L - \varepsilon_R \tag{1.7}$$

In the UV–vis region, where the wavelength of polarized light extends from 180 to 800 nm, mostly from 200 to 500 nm is used for determining the ECD of organic chiral compounds. It simply shows that the contribution to ECD is from the electrons (excited state).

ECD is used for AC study of chiral compounds with stereogenic centers close to functional groups, such as C=O and C=C, that have UV absorptions. Therefore, if a chiral molecule has no such UV-absorbing functional groups, it is extremely difficult to use ECD to determine its AC correctly. Even if there is a double bond in a molecule, if the stereogenic center is too far away from this double bond, for example, **122**, it is still difficult to use ECD to assign its AC for the carbon connected to the –OH.

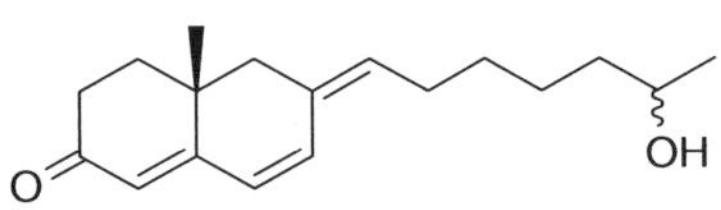

122

This looks like a shortcoming in AC determination. However, it may bring us some benefits in the AC study in some cases. For example, it is possible to use ECD to assign the AC for **123.** In the normal case, the number of theoretical conformations may be up to (3^6) 486 when the rings are assumed not to change. We can use model **124** to study the AC of **123** in practice (Scheme 1.2). In this case, the theoretical number may decrease to 9. This could greatly reduce the computation time.

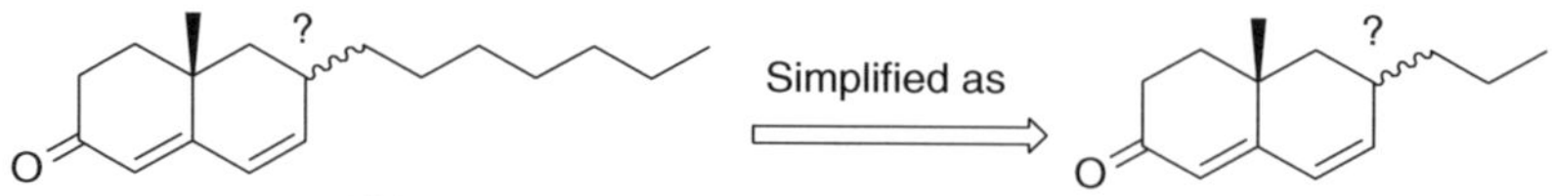

123 **124**

Scheme 1.2 Simplified model for chiral compound **123** with long side chain.

1.4.3
Outline of VCD

When a molecule has no UV–vis absorption and its OR value is not large enough, it is difficult to assign its AC if there is no assistance from VCD . As a recently developed technique, its application is not broad enough as ECD's. However, it has big application potential in this area.

As mentioned above, absorption of the left- and right-handed enantiomers of polarized light is different in a chiral environment. This absorption difference happens mostly from 500 to 8000 cm^{-1} producing VCD. Theoretically, it can be used for all chiral compounds' AC assignment since all molecules have various vibrations. The difference of the vibration of bonds near the stereogenic centers gives rise to different VCD spectra due to different vibration vectors. The illustrative diagram of a VCD setup is shown in Figure 1.12.

The vibrational modes of bonds in chiral molecules are different based on their conformational structures in solution, including the interactions among several

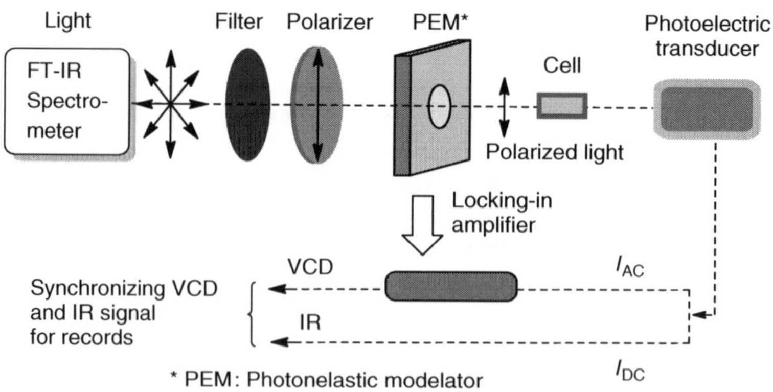

Figure 1.12 Diagrammatic sketch of VCD equipment [21].

molecules. For example, if two or two more chiral compounds interact in a solution, it is a very good way to study their interaction conformations.

On the other hand, just because of this characteristic, it is sometimes found that the predicted VCD does not agree well with the experimental VCD when a chiral molecule has two or two more hydroxyl groups, especially when the two –OH groups are close enough. The –OH may form strong H-bonds between two molecules or among several molecules. These "dimers" or "trimers" might have a different VCD structure from that of their monomer. This causes a disagreement between the calculated and experimental VCD spectra. Therefore, the enriched VCD signal is an advantage for intermolecular interaction studies, but sometimes it is disadvantageous to understand the differences between the computed and experimental VCD results.

1.4.4
Outline of ROA

ROA clearly states that the optical characteristics recorded in experiments involve Raman scattering. The physical variable is the *scattering cross section* when polarized light passes through a chiral sample in a solution or solid [21]. The formula for computing the difference of scattering cross section is very similar to those in ECD or VCD computations:

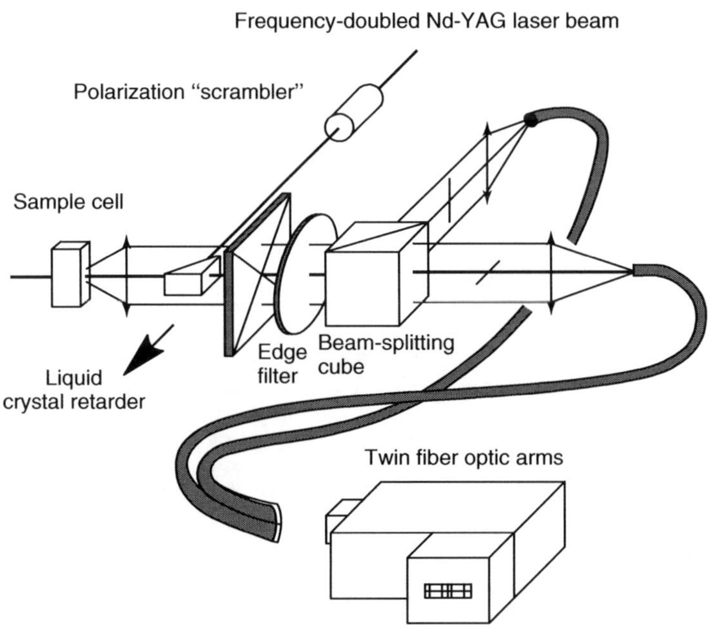

Figure 1.13 Diagrammatic sketch for Hug's SCP (scattered circular polarization) backscattering ROA instrument upon which the BioTools *ChiralRAMAN* instrument is based. The double-headed arrows represent lenses. Copied from Ref. [22].

$$\Delta A = A_L - A_R \tag{1.8}$$

where A is the scattering cross section, A_L and A_R are the scattering cross sections to the corresponding left-handed (circularly) and right-handed (circularly) polarized light, respectively.

The wavelength range in ROA also is mostly 200–4000 cm^{-1} as used in organic chemistry. This is also depends on the requirement of measuring a sample. The traditional instrument for ROA measurements is illustrated in Figure 1.13.

Both VCD and ROA involve vibrational transitions in molecules involving internal normal modes of vibrational motion, such as the stretching and angle-bending of chemical bonds, and as forms of optical activity. Indeed, both VCD and ROA are also intensity differences between left- and right- circularly polarized IR. The quantum theory of VCD and ROA can be found in many specific references, and comprehensive descriptions of all aspects of VOA can be found in the books by Nafie, Barron, and Stephens, Devlin and Cheeseman. See more details in the corresponding chapters.

References

1. (a) McBride, W.G. (1961) *Lancet*, **2**, 1358; (b) Lenz, W. (1962) *Lancet*, **1**, 45; (c) D'amato, R.J., Loughnan, M.S., Flynn, E., and Folkman, J. (1994) *Proc. Nati. Acad. Sci. USA*, **91**, 4082–4085; (d) Structure information for **6** to **19** can be found in Lin, G.Q., Chen, Y.Q. Chen, X.Z. Li, Y.M. (eds.) (2000) *Asymmetric Synthesis*, Science Press of China, the citations are found there.
2. Stephen, S.S. and Porco, J.A. Jr., (2011) *Angew. Chem. Int. Ed.*, **50**, 9722–9726.
3. Ding, Z.-G., Ren, J., Li, M.-G., Zhao, J.-Y., Huang, R., Cui, X.-L., Zhu, H.-J., and Wen, M.-L. (2010) *Chem. Eur. J.*, **16**, 3902–3905.
4. Stephens, P.J., Pan, J.J., Devlin, F.J., Krohn, K., and Kurtan, T. (2007) *J. Org. Chem.*, **72**, 3521–3536.
5. Min, E.H., Wong, K.H., Setijadi, E., Ladouceur, F., Straton, M., and Argyros, A. (2011) *Polym. Chem.*, **2**, 2045–2051.
6. Breslow, R. (2011) *Tetrahedron Lett.*, **52**, 2028–2032.
7. Bai, B., Li, D.S., Huang, S.Z., Ren, J., and Zhu, H.J. (2012) *Nat. Prod. Bioprospect.*, **2**, 53–58.
8. Figure 1.2 drawn after Wu, S.T., Cai, Z.W., Ye, Q.Y., Weng, C.H., Huang, X.H., Hu, X.L., Huang, C.C., and Zhuang, N.F. (2014) *Angew. Chem. Int. Ed.*, **53**, 12860–12864.
9. Structure information for **34** to **105** except for the noticed ones were cited inYao, X.S. (ed.) (1988) *Natural Medicinal Chemistry*, People's Medical Publishing House.
10. Isaka, M., Srisanoh, U., Choowong, W., and Boonpratuang, T. (2011) *Org. Lett.*, **13**, 4886–4889.
11. Manrıquez-Torres, J.J., Torres-Valencia, J.M., Velazquez-Jimenez, R., Valdez-Calderon, A., Alvarado-Rodrıguez, J.G., Cerda-Garcıa-Rojas, C.M., and Joseph-Nathan, P. (2013) *Org. Lett.*, **15**, 4658–4661.
12. Nam, S., Kauffman, C.A., Paul, L.A., Jensen, P.R., and Fenical, W. (2013) *Org. Lett.*, **15**, 5400–5403.
13. Luo, S.-H., Luo, Q., Niu, X.-M., Xie, M.-J., Zhao, X., Schneider, B., Gershenzon, J., and Li, S.-H. (2010) *Angew. Chem. Int. Ed.*, **49**, 4471–4475.
14. Roman, L.U., Guerra-Ramirez, D., Moran, G., Martinez, I., Hernandez, J.D., Cerda-Garcıa-Rojas, C.M., Torres-Valencia, J.M., and Joseph-Nathan, P. (2004) *Org. Lett.*, **6**, 173–176.

15. Xue, J., Wu, P., Xu, L., and Wei, X. (2014) *Org. Lett.*, **16**, 1518–1561.
16. Kagan, H.B. (1977) *Stereochemistry*, **1**, 65–66.
17. Ko, D.H., Kim, K.H., and Ha, D.C. (2002) *Org. Lett.*, **4**, 3759–3762.
18. (a) Zhu, H.J. (2009) *Modern Organic Stereochemistry*, Science Press of China; (b) Newman, M.S., Lutz, W.B., and Lednicer, D. (1955) *J. Am. Chem. Soc.*, **77**, 3420–3421; (c) Newman, M.S., Darlak, R.S., and Tsai, L. (1967) *J. Am Chem. Soc*, **89**, 6191–6193.
19. Buxton, S.R. and Roberts, S.M. (1997) *Guide to Organic Stereochemistry*, Prentice Hall PTR.
20. (a) Brewster, J.H. (1959) *J. Am. Chem. Soc.*, **81**, 5475; (b) Kawasaki, T., Tanaka, H., Tsutsumi, T., Kasahara, T., Sato, I., and Soai, K. (2006) *J. Am. Chem. Soc.*, **128**, 6032–6033. The data is cited there.
21. (a) Berova, N., Polavarapu, P., Nakanishi, K., and Woody, R.W. (2012) *Comprehensive Chiroptical Spectroscopy*, John Wiley & Sons, Inc., Hoboken, NJ, p. 147; (b) Figure 1.12 drawn after Gan, L.S. and Zhou, C.X. (2009) *Chin. J. Org. Chem.*, **29**, 848–857; (c) Wang, B. (1996) *Am. Lab. (Shelton, Conn.)*, **28**, 36C cited in 21b.
22. Figure drawn after Hug, W. and Hangartner, G.A. (1999) *J. Raman Spectrosc.*, **30**, 841.

2
Non-optical Method in Configuration Study

As one of the widely existing natural phenomena, chirality has attracted the attention of many researchers, not only because of its importance in pharmaceutical industry and chemistry but also because of the evolution study of life's origin, even including astronomy. In organic chemistry, researchers now focus on the two related fields: the identification of the absolute configuration (AC) of a specific chiral compound (including relative configuration (RC)), and the control of the formation for a stereogenic center during the synthesis of a molecule.

The first one involves various experimental and theoretical methods used for AC determination; the second one involves the design and synthesis of effective chiral catalysts to control AC formation, which is a major task for organic chemists. In this chapter, some major methods for AC and RC assignment will be discussed.

2.1
^{13}C NMR Spectra

NMR spectra has no relationship with circularly polarized light (CPL). However, the relative position among various atoms, like C, H, allows the determination of RC. In some cases, this is helpful for AC assignment.

2.1.1
NMR and Atomic Structure

A nucleus has a positive charge and spin, so, in the same way as an electron, it can have (i) a nuclear magnetic moment, (ii) spin angular momentum, and (iii) a spin quantum number. Nuclear magnetic moment is proportional to the angular momentum, and their vectors (→) are the same.

$$\vec{\mu} = \gamma \vec{P} \tag{2.1}$$

where γ is the ratio of the nuclear magnetic moment to the spin angular momentum. Different nuclei have different γ. μ is the magnetic moment, and P is the angular momentum.

Organic Stereochemistry: Experimental and Computational Methods, First Edition. Hua-Jie Zhu.

The spin angular momentum is related to the spin quantum number as

$$\rho = \frac{-h\sqrt{I(I+1)}}{2\pi} \tag{2.2}$$

where ρ is the spin angular momentum, h is the Planck constant (6.63×10^{-34} J s), and I is the spin quantum number, which is decided by the atomic mass number (Z) and the atomic number (A).

Any atom that has an even atomic number (A) with both even number of protons and even number of neutrons has its spin quantum number zero, so no nuclear magnetic resonance (NMR) signal can be recorded, for example, ^{12}C and ^{16}O. When the mass number (Z) of a nucleus is odd, its spin quantum number is a half-integer: for example, ^{1}H, ^{13}C, and ^{15}N have 1/2; ^{17}O and ^{33}S have 3/2 and 5/2, respectively. If a nucleus has an even mass number (Z) with both odd numbers of protons and neutrons, its spin quantum number is an integer, such as ^{2}H and ^{14}N.

It is found that the electronic cloud looks like a sphere in all nuclei with spin quantum number 1/2. Thus, its resonance frequency has a very narrow distribution and can be easily observed. This makes the NMR equipment not too complex and easy to use in practice. Until now, most NMR studies have focused on nuclei with spin quantum number 1/2, such as ^{1}H, ^{13}C, and ^{15}N.

The spin leads to the precession of the nucleus under an external magnetic field. The frequency of precession is proportional to the external magnetic field strength. When there is a transfer of the radio frequency signal to the molecule in solution (or in solid state), and frequency of precession is the same as the frequency of the radio signal, the nucleus will resonate with the absorption of the radio signal energy. This signal absorbed by nucleus can be recorded. Different atoms (nuclei) have different frequencies, and the same atom may have a different frequency when it is located in a different position in a molecule. This is the general basis that allows us to use NMR to identify molecular structure. There are many books introducing NMR spectroscopy that can be easily found [1a].

2.1.2 ^{13}C NMR Calculation

When a nucleus is located in a magnetic field, its resonance frequency caused by precession can be estimated as

$$\nu = \frac{\gamma \mathbf{B}_0}{2\pi} \tag{2.3}$$

where ν is the resonance frequency and $\mathbf{B}_0$ is the magnetic field strength [T] [1a].

In a molecule, every atom has a different position in space. Therefore, the magnetic fields from electrons around the center of the nucleus will produce a new magnetic field that has a direction opposite to that of the external magnetic field. Therefore, each atom in a molecule will have a different resonance frequency:

$$\nu_i = \frac{\nu_i \mathbf{B}_0(1-\sigma_i)}{2\pi} \tag{2.4}$$

where ν_i is the frequency for atom i, and σ_i is its magnetic shielding constant.

To easily use NMR spectra, the chemical shift is introduced, given by

$$\delta_i = \frac{(\mathbf{B}_{0i} - \mathbf{B}_{\text{oRef}})}{\mathbf{B}_{\text{oRef}}} \tag{2.5}$$

where δ_i is chemical shift for atom i, $\mathbf{B}_{0i}$ is the magnetic field strength, and $\boldsymbol{B}_{\text{oRef}}$ is the magnetic field strength for a reference atom.

Therefore, after replacing $\mathbf{B}_0$ using the formula $\nu = \gamma\mathbf{B}_0/2\pi$, the chemical shift can be written as

$$\delta_i = \frac{(\nu_{0i} - \nu_{\text{oRef}})}{\nu_{\text{oRef}}} \tag{2.6}$$

And then the chemical shift δ_i in ppm is

$$\delta_i(\text{ppm}) = \frac{(\nu_{0i} - \nu_{\text{oRef}})}{\nu_{\text{oRef}}} \times 10^6 \tag{2.7}$$

For example, if chiral (*R*)-butan 2-ol (**1**, Figure 2.1) is located in a magnetic field $\mathbf{B}_0$ (for clarity all H atoms are hidden (right)), all ^{13}C atoms and one O atom will produce different shift values with a new magnetic field $\mathbf{B}_i$. Finally, the sum of different magnetic fields will bring every atom different shift values. Therefore, the computation of the chemical shifts can be used to compute the corresponding frequency, namely the magnetic shielding constant.

In NMR study, usually tetramethylsilane (TMS) is used as an internal reference, and the signals recorded are used as the *Y*-axis. To conveniently discuss the NMR spectra, signals to the left of the TMS signal is regarded as downfield and those to the right side as upfield. The farther away the signal is from the TMS signal on the left side, the smaller is the field. For example, a proton of $-CH_3$ of toluene may have shifts of 2.1–2.2 ppm. Once this CH_3 connects with an O, like CH_3–O–, its shift may move downfield, from 3.6 to 3.9 ppm. Before we introduce the application of ^{13}C NMR in configuration assignment, it is important to briefly to look back at ^{1}H NMR.

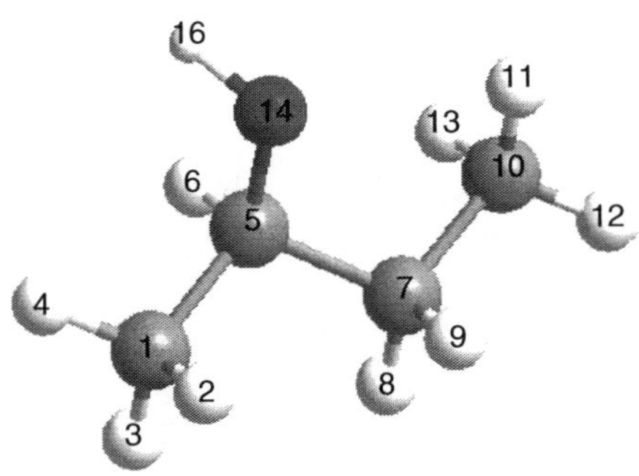

1

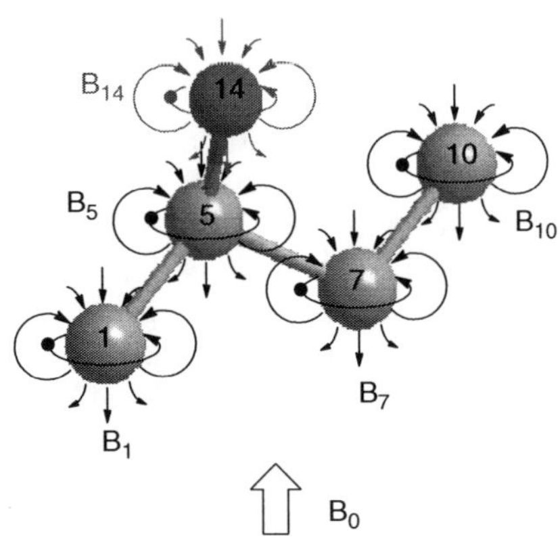

Figure 2.1 The 3D structure **1** and effect of magnetic field on its nucleus (dark sphere is O atom, light grey is H, and grey is C).

2.1.3 ^{1}H NMR

^{1}H NMR can be well recorded and used for configuration assignment. This technique is called "rotating frame Overhauser enhancement spectroscopy" (ROESY) [1a,b]. It is based on the technique of "nuclear Overhauser enhancement spectroscopy" (NOESY). The major aim of ROESY or NOESY is to study the relationship between protons that are close together in space; the distance is generally less 5.0 Å and mostly below 4.0 Å. This could be used to determine the RC, including the *cis* or *trans* configuration of >C=C< bond structures. After the RC assignment, ^{13}C NMR can be used for further configuration study.

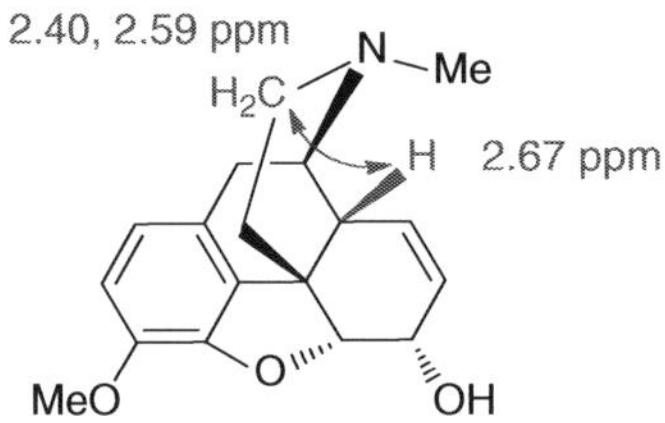

2 Codeine

For example, the ^{1}H signals of codeine (**2**) [1c] are well isolated, especially in the range 2.0–5.0 ppm. It was found that the signal at 2.67 ppm had crosslink with those at 2.40 and 2.59 ppm in NOESY. The signals belong to, correspondingly, the protons labeled using a double arrow. It shows that the protons are close in space. Therefore, this method is used to assign the RC. However, when a chiral molecule is very complex, especially when the ^{1}H NMR signals overlap seriously, it is difficult to assign its configuration. Even in this case, ^{13}C NMR can be used as a tool to assist in the assignment.

2.1.4 ^{13}C NMR Prediction and Conformational Search

Currently, gauge-independent atomic orbital (GIAO) is used for NMR computations [2]. Others such as continuous set of gauge transformations (CSGT), localized orbitals/localized origins (LORG) [3] using Hatree–Fock (HF), density functional theory (DFT), or MP2 methods have also been reported. Up to now, researchers such as Bifulco [4], Forsyth [5], Autschbach [6], Besley *et al.* [7], Reddy and Prakash [8], Ruud *et al.* [9], Oldfield [10], Frenking [11], and others [12] have reported significant progress in this field. The basis set for NMR prediction may be at the DFT/B3LYP/6-311++G(2d,p) level in the gas phase or in solution.

In order to correctly compute the ^{13}C NMR for a chiral molecule, one needs to do a conformational search for this molecule first. A single bond has three degrees of freedom in space. If a chiral compound has a long acyclic chain, it may have a large number of stable conformations. Carrying out the ^{13}C NMR shift computations is not easy then. Therefore, only chiral compounds with relatively rigid structures are considered in practice. For example, there is almost no way to predict the ^{13}C NMR shifts for **3** [13]. In contrast, calculating ^{13}C NMR shifts for **4** is not difficult [14a].

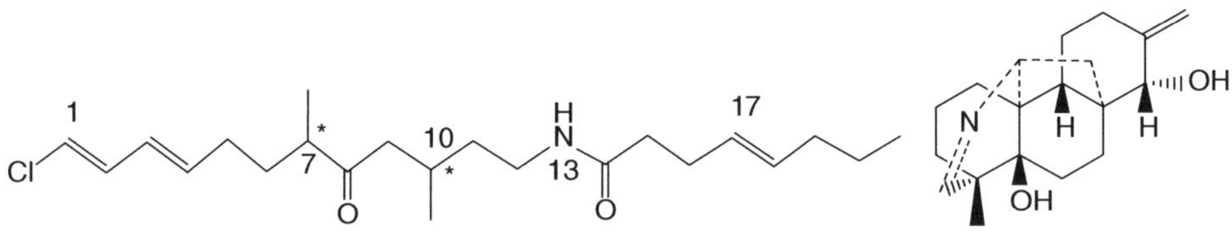

3 Pitiamide A **4**

Conformational searches for a specific chiral compound are required before any computations. There are many software packages available for this.[1)] The different packages use different force fields. One developed molecular mechanical force field is MMFF94 or its improved version MMFF94S [14b,c]. Other force fields can be used in this procedure. Indeed, even if the same force field is applied for conformational search, the search results may have differences if a different software package is used. For relatively rigid chiral compounds, the conformations that can be found with relative energy from 0 to 10 kcal mol^{-1} may be limited, for example, 10–20 conformers, thus, further optimizations are easily performed at the B3LYP/6-311+G(d) level. However, if it is a linear chiral compound, the energy window from 0 to 10 kcal mol^{-1} may be too wide to be easily used for the conformational search; so it must be narrowed, for example, from 0 to 5 kcal mol^{-1}, to reduce the number of conformations. This energy window cannot be narrowed from 0 to 1.0 kcal mol^{-1} in conformational search, since it might lose many stable conformers with low energy in this case, thus resulting in a wrong result in the AC assignment. The use of energy window of 0–5 kcal mol^{-1} for a linear chiral compound is therefore recommended. Even if the number of the conformations is hundreds or up to thousands for a chiral compound, it may still be necessary to use this energy window of 0–5 kcal mol^{-1}.

After the conformational search, the selected conformations should be further optimized using quantum theory. If the number is too large, for example, it reaches 1000, one should use a low basis set for initial optimizations, such as the HF/6-G(d) method, for all the selected conformations. This method is very economical because of its short computational time and relatively accurate energy sequence. The HF/6-31G(d)-optimized conformations with relative energy 0–2.5 kcal mol^{-1} should be used for optimization again at the B3LYP/6-31G(d) level at least. Of course, if the number of conformations is limited, computation at the HF/6-31G* level is not necessary. One can perform the computation directly at the B3LYP/6-311+G(d) level in optimizations.

The second step is to choose the conformations with relative energy 0–2.5 or up to 3.0 kcal mol^{-1} (at the HF/or B3LYP/6-31G(d) level) for further optimization. The calculation should be performed at a higher basis sets, for example, 6-311++G(2d,p) using DFT in the gas phase or in solution. Finally, these

1) The commercially available softwares for conformational searches contain Hyper-Chem, Barista, ComputeVOA, Sparton and so on.

B3LYP/6-311++G(2d,p)-optimized geometries with relative energy 0–2.0 (or 2.5) kcal mol^{-1} are used for ^{13}C NMR shift calculations at, for example, the B3LYP/6-311++G(2d,p) level. If the geometries in range of relative energy 0–2.5 kcal mol^{-1} are too many to be considered for computations, the energy window can be narrowed to at least 0–1.5 kcal mol^{-1}. But be careful! Although it is mentioned as ^{13}C NMR computations, here the computed data are the magnetic shielding constants and not the chemical shifts if no calculated shift reference such as Me_4Si is used.

Boltzmann statistics should be then performed after all magnetic shielding constants are recorded, in an Excel file by hand at the present time. Three energy magnitudes can be used for the statistics if frequency calculations are to be computed. They are total electronic energy (TEE), zero-point energy correction (ZPE), or Gibbs free energy (GFE). If the frequency is not to be analyzed, only TEE can be used for the Boltzmann sum.

Notice

Be careful, the novice researcher may make a mistake. Optimization for a molecule may be performed, for example, at the B3LYP/6-31G(d) level, but the frequency calculation may be carried out at another basis set, such as B3LYP/6-31+G(d). Only the computations using the same basis sets in both optimization and frequency analysis are reliable and acceptable.

At the present time, it is not easy for experimental chemists to treat the computational results since the data processing is a little difficult for experimentalists to handle. However, a new package will be available from our research group soon. It can do the data processing to produce the ^{13}C NMR shifts, optical rotation (OR), electronic circular dichroism (ECD), vibrational circular dichroism (VCD), and Raman optical activity (ROA) using any selected conformations using, for example, Gaussian output files. This package just needs 5–8 min to provide high-quality computational results even if the number of conformations is 500. It can save precious time for all chemists.

There are three possible methods to estimate the correctness of the structure. The first one is to convert each conformational structure's magnetic shielding constants to the corresponding chemical shifts using Me_4Si as reference. The shielding shift of $SiMe_4$ must be computed at the same level as that of the chiral molecule. Then, the experimental ^{13}C NMR and the computed ^{13}C NMR can be organized in a line graph. The slope and intercept of the least-squares correlation line are applied for the correction of the computed ^{13}C NMR [15] to afford a new set of ^{13}C NMR values for this conformation. We continue this procedure until all conformational magnetic shielding constants are converted. Boltzmann statistics is now used to compute the final chemical shifts for the whole molecule. This corrected ^{13}C NMR shifts can be used for testing whether the chiral molecule is correct or not.

In the second method, Boltzmann statistics is used to produce the chemical shifts for all conformations. After every conformational structure's magnetic shielding constants are converted to the corresponding chemical shifts, the slope and intercept of least-squares correlation line are used to correct the ^{13}C NMR shifts to afford a new set of ^{13}C NMR shifts for the whole molecule. The new ^{13}C NMR shifts can be used to test whether the structure is correct or not.

The third one is to use Boltzmann statistics to compute the whole molecular magnetic shielding constants, which can be used to form a line with the experimental shifts. The line's slope and intercept are used to produce a new set of ^{13}C NMR shift values. The data can be used to test whether the structure is correct or not. This is relatively simple because it does not require the computation of the magnetic shielding constant of Me_4Si in same cases.

After all ^{13}C NMR data are obtained, there are two methods to judge whether the structure is correct or not. The first one is to use to maximum of the $\Delta\delta$ values. If the maximum of $\Delta\delta$ is over 8.0 ppm, the structure is not reliable. If it is less 8.0 ppm, this structure is located in the reliable structure range [5]. However, more evidences are required for further confirmation. Another one is to compare the coefficients of two structures. The larger the coefficient, the more reliable will be the structure [16a].

For example, chiral compound **4** was obtained with the original structure of either **5** or **6** since no interaction signal was obtained for the first time between the protons H5 and H8 [16b].

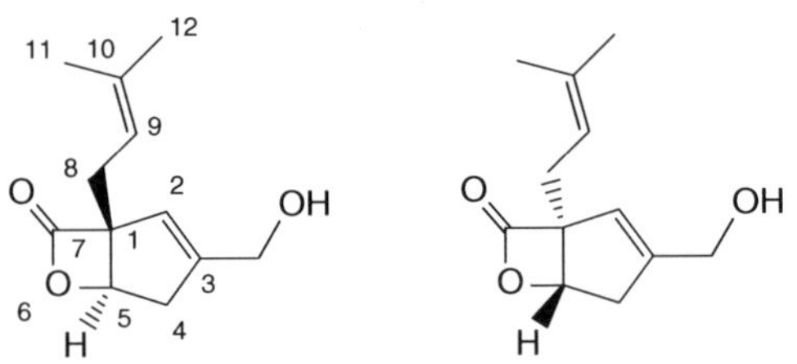

5 **6**

Totally 40 conformations were found using the MMFF94S force field, after optimization at the B3LYP/6-31G(d) level, but only 7 or 8 conformations were recorded for **5** and **6**, respectively, with a relative energy of 0–2.5 kcal mol^{-1}. The geometries were then used for magnetic shielding calculations at the B3LYP/6-311+G(2d,p) level in the gas phase. Finally, after removing the same conformations, all six conformations were used for the Boltzmann sum.

We select the first method to discuss structure of (1*R*,5*R*)-**5**. After using $SiMe_4$ as the reference, all magnetic shielding constants were recorded as the corresponding chemical shifts (Table 2.1).[2)]

For example, the experimental ^{13}C NMR shifts in conformation **1** and the recorded original ^{13}C NMR shifts are used to produce a line, whose slope can be computed as 0.940 and intercept as −1.98. The correction function to compute

2) It is summarized in supporting information of Ref. 16b.

Table 2.1 Experimental and computed ^{13}C NMR and relative energy for **5**.[2)]

Conformer	**1**		**2**		**3**	
ΔE (kcal mol^{-1})	**0.000**		**0.254**		**2.368**	
C-Number (^{13}C)[a)]	**Original δ[b)]**	**Corrected δ[c)]**	**Original δ**	**Corrected δ**	**Original δ**	**Corrected δ**
7 (173)	178	165.3	177.7	166.2	177.4	164.9
1 (75.1)	76.3	69.7	76.5	70.1	77.3	70.6
5 (78.5)	86.3	79.1	85.3	78.4	87.6	80.3
4 (37.3)	35.9	31.8	36.2	31.8	36.6	32.3
3 (146.6)	159.3	147.7	157.3	146.8	160.8	149.3
2 (122.4)	143.8	133.1	138.6	129.1	142.7	132.2
13 (61.3)	67.2	61.2	67.9	61.9	67.2	61.1
8 (27.6)	43.6	39.0	43.4	38.6	42.6	37.9
9 (117.2)	128	118.3	124.8	116.0	129.1	119.4
10 (136)	144.3	133.6	149.6	139.5	142.3	131.8
11 (25.8)	28.1	24.4	28.3	24.3	27.9	24.1
12 (18)	18.6	15.5	19.4	15.8	18.5	15.2
Slope (k)	0.940	—	0.950	—	0.942	—
Intercept (b)	−1.98	—	−2.59	—	−2.21	—

4		**5**		**6**		**Final ^{13}C NMR**
1.863		**3.368**		**2.767**		
Original δ	**Corrected δ**	**Original δ**	**Corrected δ**	**Original δ**	**Corrected δ**	
177.9	164.7	179.1	165.7	178.6	165.8	165.7
79.4	72.6	80.1	72.9	77.8	71.2	69.9
87.1	79.8	88.8	81.1	88.9	81.6	78.8
35.6	31.7	38.1	33.6	36.4	32.4	31.8
160.4	148.4	160.6	148.4	161.5	149.7	147.3
141.8	131.0	139	128.1	137.8	127.5	131.5
67.2	61.2	67	60.7	67	61.1	61.5
39.7	35.5	40.4	35.7	39.6	35.4	38.8
126.7	116.9	127.1	117.0	129.9	120.1	117.4
147.5	136.3	148.2	136.7	144.3	133.6	135.9
28.3	24.9	27.7	23.8	28.2	24.7	24.4
18.5	15.7	17.8	14.6	18.2	15.3	15.6
0.935	—	0.937	—	0.938	—	—
−1.59	—	−2.12	—	−1.74	—	—

a) The figures in parentheses are the experimental ^{13}C NMR values.
b) The original ^{13}C NMR from the computed data.
c) Corrected ^{13}C NMR for each conformer from the formula Corrected $\delta_i = k \times$ Original $\delta_i + b$.

Table 2.2 Corrected ^{13}C NMR and their relative errors and coefficients for structures **5** and **6**.[2)]

C-Number (experimental ^{13}C)	δ for 5	δ for 6	$\Delta\delta$ for 5	$\Delta\delta$ for 6
7 (173.0)	165.7	165.1	7.3	7.9
1 (75.1)	69.9	69.6	5.2	5.5
5 (78.5)	78.8	79.1	−0.3	−0.6
4 (37.3)	31.8	31.6	5.5	5.7
3 (146.6)	147.3	147.8	−0.7	−1.2
2 (122.4)	131.5	132.6	−9.1	−10.2
13 (61.3)	61.5	61.2	−0.2	0.1
8 (27.6)	38.8	39.0	−11.2	−11.4
9 (117.2)	117.4	116.8	−0.2	0.4
10 (136)	135.9	135.7	0.1	0.3
11 (25.8)	24.4	24.5	1.4	1.3
12 (18)	15.6	15.8	2.4	2.2
Coefficient	0.99468	0.99402	—	—

the new set of corrected ^{13}C NMR shifts then was organized as

$$\text{Corr. } \delta_i = 0.940 \times \text{orig. } \delta_i - 1.98 \tag{2.8}$$

For example, if the predicted shift is 178 ppm for C7, the corrected ^{13}C NMR shift should be 165.3 ppm using Eq. (2.8). Similarly, a set of new ^{13}C NMR values for all carbons can be computed for all six conformations. Then, using Boltzmann statistics, the final ^{13}C NMR values for the whole chiral compound **5** can be obtained (Table 2.1).

Similarly, the ^{13}C NMR values for (1*S*,5*S*)-**6** could be computed. The maximum errors and coefficients are computed at the same time for clear comparison. The results are listed in Table 2.2.

If the coefficient is used as a standard – namely the larger the coefficient, the more reliable the structure – structure **5** would be the reliable one. However, the coefficients are so close that it is not reliable to use them to judge the structure. It is recommended to consider other factors, the major one being the maximum error control principle. It is found that the maxima are −11.2 for **5** and −11.4 for **6**. Thus, both structures **5** and **6** are not the best candidates. It must be one of two *cis* structures instead of the *trans* structures of **5** and **6**.

The other two *cis* structures (**7** and **8**) were used for ^{13}C NMR calculations. The same methods were used in the procedure, and the predicted ^{13}C NMR shifts and relative shift errors are listed in Table 2.3. It is found that both structures have small shift errors (much <8.0 ppm). At the same time, both *cis* structures have larger coefficients (over 0.999) than the *trans* structures (<0.995). It reflects that the *cis* structures are more reliable than the trans structures. However, both *cis* structures have close coefficients. In this case, it could not be used for RC assignment via comparing the magnitudes of the coefficients.

Table 2.3 Predicted ^{13}C NMR shifts for **7** and **8**.[2)]

C-Number (experimental ^{13}C)	δ for 7	δ for 8	$\Delta\delta$ for 7	$\Delta\delta$ for 8
7 (173.0)	170.7	170.6	2.3	2.4
1 (75.1)	77.9	77.3	−2.8	−2.2
5 (78.5)	76.4	76.4	2.1	2.1
4 (37.3)	37.5	37.5	−0.2	−0.2
3 (146.6)	144.4	144.3	2.2	2.3
2 (122.4)	123.1	123.1	−0.7	−0.7
13 (61.3)	62.0	61.9	−0.7	−0.6
8 (27.6)	29.3	29.3	−1.7	−1.7
9 (117.2)	119.4	119.4	−2.2	−2.2
10 (136)	138.3	138.2	−2.3	−2.2
11 (25.8)	24.8	24.8	1.0	1.0
12 (18)	15.8	15.8	2.2	2.2
Coefficient	0.99932	0.999358	—	—

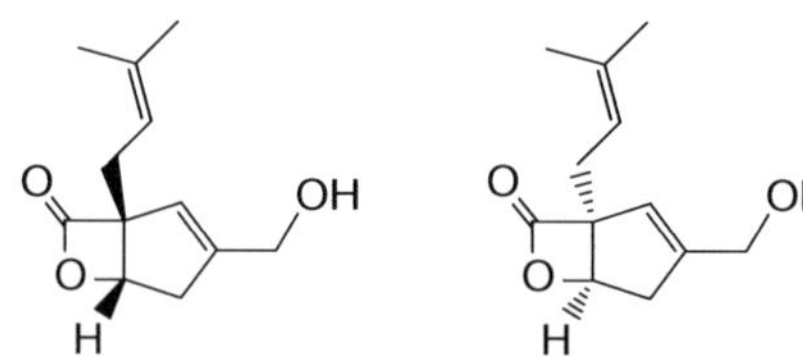

7 8

When ORs were computed for the both cis structures, it was found that the correct structure was *cis*-7. The experimental OR was −135, and the predicted OR for 7 was −127 at the B3LYP/6-311++G(2d,p) level and its enantiomer **8** had +127. Therefore, the correct structure is 7. Clearly, the best way is to use both the maximum error control principle and the coefficient when judging the correctness of a molecular structure.

Notice

Clearly, the maximum error principle must be applied for the examination of structures. If the difference in coefficients of two structures is very small, they cannot be used as evidence for structural evaluation. In some cases, more evidence will be needed for further confirmation of the structures.

In practice, if the stereogenic centers are located on a chain, it is very difficult to compute its ^{13}C NMR value for assignment of its configuration. The reasons are the difficulty in carrying out the calculation as well as that of judging which configuration is more reliable because the shift difference ($\Delta\delta$) between the calculated and experimental ^{13}C NMR values is not large enough to make the selection.

It must be clearly stated here that ^{13}C NMR for configuration assignment can be used for AC study only if a stereogenic center is known. In this case, the other stereogenic centers can be assigned by comparing the computed and experimental ^{13}C NMR. For example, when structures **9** and **10** were obtained

Table 2.4 Comparison of ^{13}C NMR shifts of compound **9** and **10** [17].

C-Number 9 and 10	Original $\Delta\delta_{cal}$/corrected $\Delta\delta_{cal}$ (ppm)[a)]				Experimental $\Delta\delta$[a)]
	Method A[b)]	Method B[b)]	Method C[b)]	Method D[b)]	
C-12	−0.1/−0.1	+0.1/0	−0.3/−0.3	−0.2/−0.2	−0.4
C-13	−0.1/−0.1	−0.2/−0.2	−0.4/−0.4	−0.3/−0.3	−0.2
C-14	+0.1/+0.1	0/0	+0.2/+0.2	+0.1/+0.1	+0.1
C-16	+1.7/+1.7	+1.3/+1.3	+1.5/+1.5	+1.2/+1.3	+1.0
C-17	−0.7/−0.7	−0.1/+0.6	+0.2/+0.2	+0.3/+0.4	0.0
C-20	+0.2/+1.2	+0.6/+1.1	+1.1/+1.1	+0.6/+0.7	+0.4
C-21	−0.9/−0.8	−0.6/−0.6	−1.0/−1.0	−0.7/−0.8	−0.4
C-22	+0.4/+0.5	+0.3/+0.3	+0.7/+0.7	+0.6/0	0.0
C-23	−0.6/−0.6	−0.9/−0.9	−0.7/−0.7	−0.7/−0.6	−0.6

a) $\Delta\delta = \delta(\mathbf{9}) - \delta(\mathbf{10})$.
b) Method A: B3LYP/6-31G(d)-optimized structure was used for ^{13}C NMR computations at the 3LYP/6-31+G(d,p) level. Method B: B3LYP/6-31+G(d,p)-optimized structure was used for ^{13}C NMR computations at the B3LYP/6-311+G(2d,p) level. Method C: HF/6-31G(d)-optimized structure was used for ^{13}C NMR computations at the B3LYP/6-311+G(2d,p) level. Method D: HF/6-31G(d)-optimized structure was used for ^{13}C NMR computations at the HF/6-31G(d) level.
It is recommended that the first method is used to do the estimation of 13C NMR. SiMe4 must be used in 13C NMR calculation first, and then to do the least-squares correlation.

from a mixture, two sets of ^{13}C NMR results showed that they were epimers [17a]. By computing the shift difference of ^{13}C NMR using the methods above, the AC for C-20 was assigned [17].

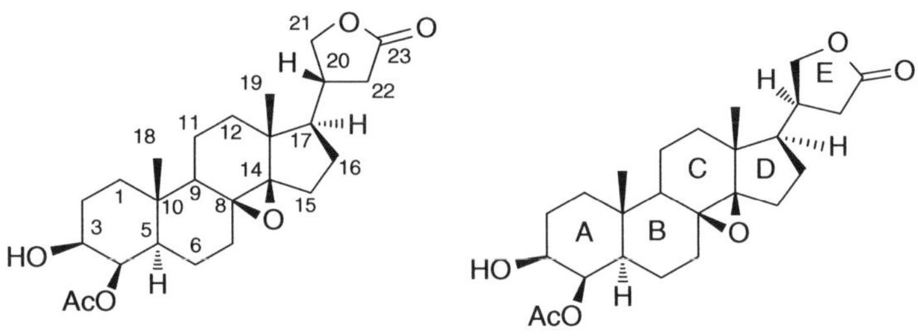

9 **10**

Four methods were used for ^{13}C NMR computations for **9** and **10** [17a]. Only the differences of ^{13}C NMR with 0.1 ppm or more are summarized. The differences between **9** and **10** are also computed and listed in Table 2.4.

2.2 X-Ray Diffraction and Mosher Method

2.2.1 X-Ray Diffraction

X-ray experiments have provided many benefits in AC assignment for structural identification including those of chiral compounds. Its theoretical basis is the

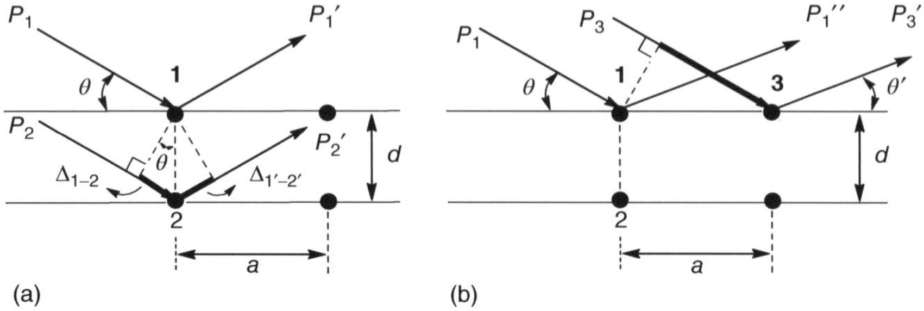

Figure 2.2 (a, b) Geometric figure for construction of Bragg and Laue diffraction equations [17b,c].

quantum formula of the Bragg equation

$$2d\sin\theta = n\lambda,\ n = 1, 2\ \ldots \tag{2.9}$$

where d is the neighboring interplanar spacing, θ is the angle between the crystal plane and the incident direction of X-rays, and λ is wavelength of X-rays (Figure 2.2a) [17b,c].

It can be easily concluded from the geometric figure that the optical path difference between two photons P_1 and P_2 arriving atoms 1 and 2 (Δ_{1-2}) can be computed as

$$\Delta_{1-2} = d \times \sin\theta \tag{2.10}$$

To ensure the largest scattering intensity, the angle of reflection (θ) must be the same as the angle of incidence. Thus, $\Delta_{1'-2'}$ can be the same as $d\sin\theta$. Thus, the total optical path difference is $2d\sin\theta$. For two beams of light to have strong diffraction, the optical path difference must be an integer multiple of the wavelength (λ). This is the Bragg equation. It considers the two neighboring planar spacings [17b,c].

If we consider the diffraction of two beams of photons from the same plane, such as P_1 and P_3 (Figure 2.2b), the optical path difference (Δ_{1-3}) can be derived as

$$\Delta_{1-3} = a \times \cos\theta \tag{2.11}$$

Similarly, two beams of photons P_1'' and P_3' may have a different optical path difference Δ_{1-3} because the angle of incidence is not equal to the angle of diffraction. In this case, the total difference becomes

$$\Delta = \Delta_{1-3} + \Delta_{1''-3'} = a \times (\cos + \cos\theta')$$

To observe a large scattering intensity, the difference also needs to be an integer multiple of the wavelength (λ). Thus, for this one-dimensional case, we get

$$a \times (\cos + \cos\theta') = h\lambda,\quad h = 1, 2, 3\ \ldots \tag{2.12}$$

Similarly, for a three-dimensional crystal, two other formulas can be written as

$$b \times (\cos_2 + \cos_2{}') = k\lambda,\ k = 1, 2, 3\ \ldots \tag{2.13}$$

$$c \times (\cos_3 + \cos_{3'}) = l\lambda,\ l = 1, 2, 3\ \ldots \tag{2.14}$$

When all the conditions are satisfied in the three equations, the diffraction spots can form well and be recorded in experiments.

It is noted that virtually all equations are the same. For example, if we regard atoms 1 and 2 in one plane and atoms 1 and 3 in the neighboring plane, then both the equations will change their roles.

To treat the data recorded in the experiments is a big challenge. There are several different methods to handle the data [17b,c]. It should be noted that the determination of the position of protons using an X-ray study is not accurate because it is difficult to record useful information. So, the positions of H atoms are mostly estimated by general bond lengths.

Cu and Mo radiations are used in an X-ray diffraction equipment. In stereochemistry study, an equipment with Mo radiation can give the RC structure, while Cu radiation can predict the AC structure [17b,c]. For example, a bioactive alkaloid, brevianamide M (**9**), was assigned as (2*R*,13*R*) or (2*S*,13*S*) using Mo radiation [18a]. Although its AC was assigned as (2*S*,13*S*) by its hydrolysis to afford (*S*)-phenylalanine, its real AC is (2*R*,13*R*) [18b]. This will be discussed in Chapter 6.

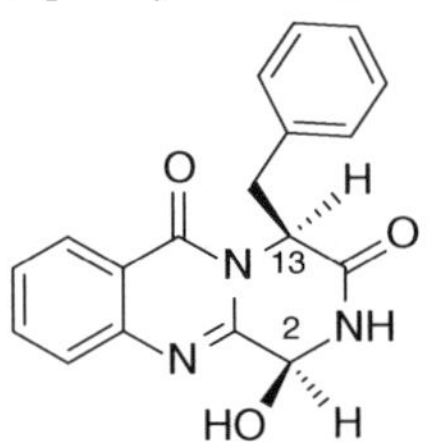

9 Brevianamide M

A heavy atom could be led into a chiral molecule for assignment of AC, [19a,b] heavy atoms, such as S, Br, Cu atoms, could are relatively to the light atoms C, O.N *et al.* Hydrogen atom is directly called as hydrogen atom. The difficulty is using X-ray for AC study is due to the difficulty in the preparation of suitable crystals for experiments. There are many methods that can be cited. Here we list three methods that are frequently used in organic chemistries [19c].

In most cases, we dissolve the sample in alcohol (or in a mixture of solvents) in a small container like a clean tube and cover it with a plastic paper with a tiny hole for slow evaporation of the solvent. This is inexpensive and convenient. However, it may not work.

The first one is the temperature gradient method [19]. Selecting a solvent that can dissolve the sample at room temperature is not good enough. Then smallest quantity of solvent is used to dissolve the sample under a high temperature, for example, 40–50 °C, until the solid sample dissolved completely by adding a small quantity of the solvent. Then, the solution is stored under N_2 at room temperature (Figure 2.3a) [19c]. If this procedure cannot produce crystals within 1 week, place it in a cold environment, such as in a refrigerator at 3–6 °C. If this temperature cannot produce crystals in 1–2 weeks, still lower temperature (like −15 to −20 °C) could be selected for some time, or another solvent should be selected for this procedure.

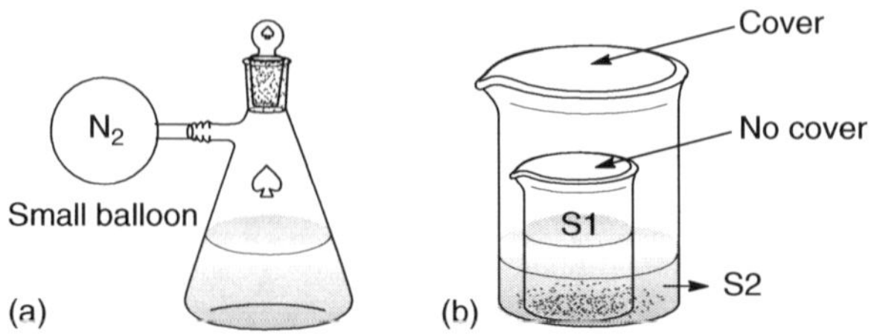

Figure 2.3 Schematic diagrams (a) for temperature gradient method and (b) for vapor diffusion method [19].

The second one is the vapor diffusion method (Figure 2.3b). If S1 is a good solvent and S2 is not one for an organic compound, the sample is dissolved in S1 in a small vessel and then it is put into a larger vessel that contains solvent S2, as illustrated above, after covering with a plastic paper (or other paper). The vapor of S2 would diffuse into S1, and vice versa. The ratio of S2 in S1 solution will increase very slowly until the crystals are formed.

The final one is a chemical method. For some compounds with free hydroxyl groups, such as **10**, they can be converted into the corresponding acetates. Many methods were tried for the preparation of the crystals of **10**. However, all failed. After conversion into acetate (–OH on C-20) using pyridine and acetic anhydride at room temperature overnight, the acetate dissolved in methanol and this was stored under room temperature until the crystals were obtained [20].

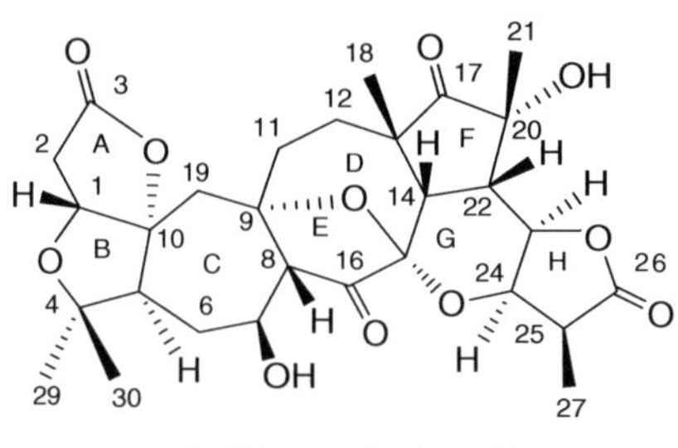

10 Micrandilactone D

Preparation of a high-quality crystal is a big challenge in organic chemistry. The methods introduced above are the usually the used ones. One should look for other used methods in this field.

2.2.2
Mosher Method

The Mosher ester has become a widely used method in organic chemistry. Its construction shows the great talent of organic chemists in this area. It is a type of NMR method for the determination of the AC of secondary alcohols or amines. It was introduced by Mislow [21], and further developed by Mosher and other chemists [22, 23]. Even now, it still plays an important role in AC assignment.

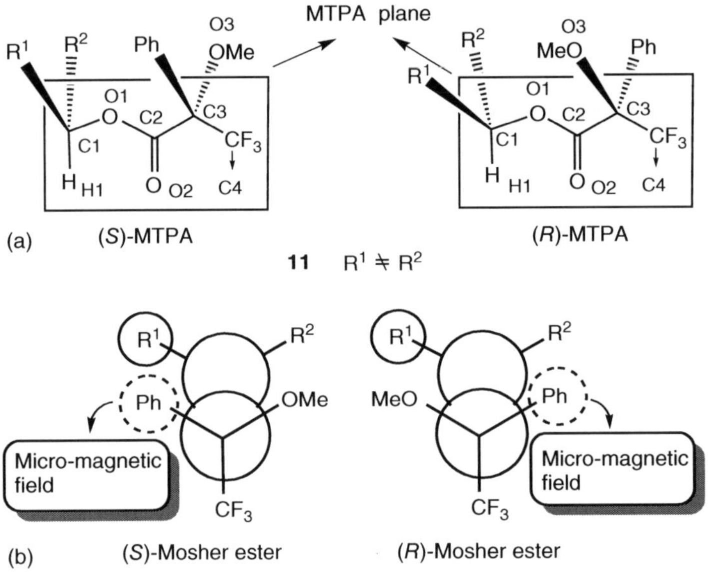

Figure 2.4 (a) Mosher esters and their assumed MTPA plane. (b) Newman projection structures for the observation of the effect of micro-magnetic field produced by the phenyl ring on R^1 (b) [21–23].

A typical Mosher ester preparation needs the sample to react with the (*R*)- and (*S*)-Mosher acid, α-methoxyl-trifluoromethylphenyl acetic acid (MTPA). Then ^{1}H NMR spectra are measured for the two Mosher esters. The difference ($\Delta\delta$) in ^{1}H NMR shifts of the (*R*)- and (*S*)-Mosher esters would show regular changes, which can be used for AC assignment [21–23].

In the existing protocol, predicting AC requires the following two conditions: (i) atoms H1, C1, O1, C2, O2, C3, and C4 should be in the same plane (MTPA plane) as illustrated below, for example, **11** in Figure 2.4, and (ii) H1 and O2 must be at the same side (direction) of the major conformer so that the phenyl ring has a proper effect on the electronic-current-induced proton resonance. These are the general condition for the correct use of the Mosher method in AC determinations.

For the (*S*)-Mosher ester in **11,** the ^{1}H NMR differences are defined as the shift of the (*S*)-Mosher ester subtracted from that of the (*R*)-Mosher ester, namely $\delta = \delta_S - \delta_R$. If a group has protons with $\Delta\delta < 0$, it should be located on the left side of the MPTA plane, and the group with $\Delta\delta > 0$ should be on the right side. After both groups are located, its AC can be identified based on the substituent size.

As an example, we can use modern computational methods to find out the evidence that an MTPA plane exists in a linear chiral alcohol and the reason why the Mosher method works. Compound **11** (R^1 = Me, R^2 = Et) is used as an example. To correctly predict the AC for **11** to be either (*R*) or (*S*), it first needs to react with (*R*)- and (*S*)-Mosher acids, respectively. For example, if it has current AC (*R*), the major conformer should be (*R*)-**11a1** and the minor one is (*R*)-**11a2**. Once it reacts with (*S*)-Mosher acid to afford (*S*)-Mosher **11b**, the major conformer predicted by the Mosher method should be (*S*)-Mosher **11b1**. By comparing the difference in ^{1}H NMR shifts between (*R*)-Mosher **11a** and (*S*)-Mosher **11b**, its AC can be identified [24].

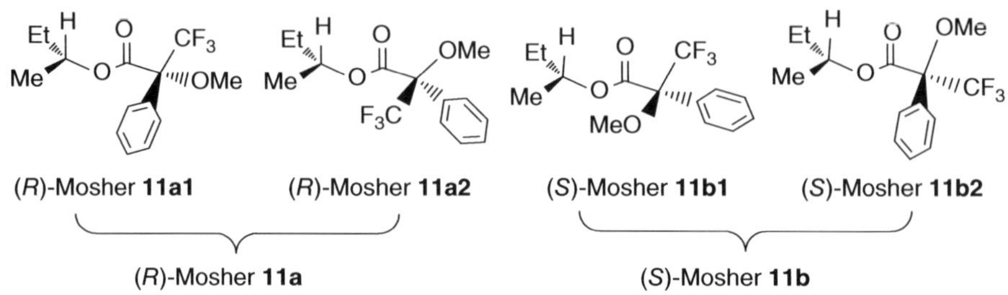

The 6-31G(d) basis set is a very general set used in geometry optimizations when B3LYP theory is used. This theory cannot correctly predict the lowest energy conformation structure in some cases. For this example, the major conformation for (*R*)-Mosher ester (**11a**) was not correctly predicted [24]. But it was correct in predicting the (*S*)-Mosher ester (**11b**) irrespective of whether TEE or GFE was used (Table 2.5, italic data, ΔE_1 and ΔG_1). Both TEE and GFE (ΔE_2 and ΔG_2) could predict the major and minor conformations well. When B3LYP/6-311+G(d) was used, a higher computation level, such as B3LYP/6-311++G(2d,p), also predicted it well. However, this is also expensive. Therefore, the B3LYP/6-311+G(d) method is suggested for the lowest energy conformation calculations.

Its applications have been widely reported. Some assignments may involve organic conversion, as in the case of, for example, Saccopetrin A (**12**) [25]. Its RC was well established using ROESY experiment. However, its AC needed to

Table 2.5 The predicted percentages for the four conformers using DFT methods [25].

	(*R*)-Mosher 11a1	(*R*)-Mosher 11a2	(*S*)-Mosher 11b1	(*S*)-Mosher 11b2
ΔE_1 (%)	0.279 (38%)	0.000 (62%)	0.000 (68%)	0.443 (32%)
ΔE_2 (%)	0.000 (59%)	0.211 (41%)	0.000 (71%)	0.518 (29%)
ΔE_3 (%)	0.000 (70%)	0.488 (30%)	0.000 (75%)	0.650 (25%)
ΔG_1 (%)	*0.193 (42%)*	*0.000 (58%)*	*0.236 (40%)*	*0.000 (60%)*
ΔG_2 (%)	0.000 (68%)	0.442 (32%)	0.000 (84%)	0.975 (16%)
ΔG_3 (%)	0.000 (69%)	0.468 (31%)	0.711 (23%)	0.000 (77%)

ΔE_1 and ΔG_1 were obtained at the B3LYP/6-31G(d) level. ΔE_2 and ΔG_2 were obtained at the B3LYP/6-311+G(d) level. ΔE_3 and ΔG_3 were obtained at the B3LYP/6-311++G(2d,p) level.

be assigned. Clearly, once the AC of C-24 was determined, the whole molecular stereochemistry could be identified [25].

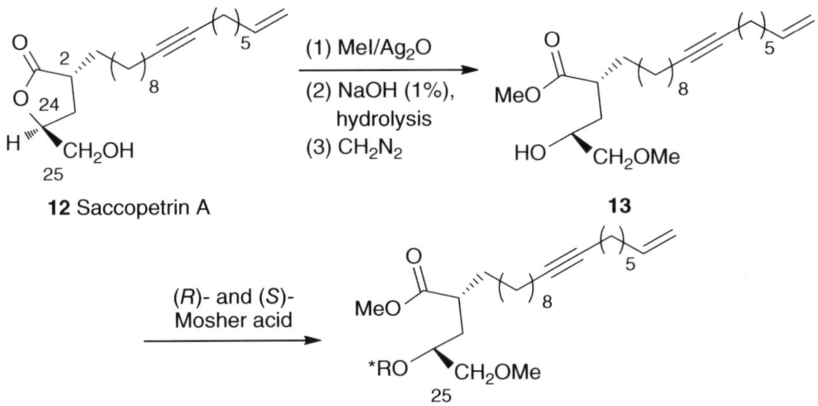

12 Saccopetrin A

13

14 R* = (*R*)-MTPA
15 R* = (*S*)-MTPA

Compound **13** was converted to **14** and **15**, respectively; both were tested using ^{1}H NMR. The ^{1}H NMR differences ($\Delta\delta$) between the (*S*)- and (*R*)-Mosher esters were computed (Table 2.6). Based on the principles that the substituent (R^1) with $\Delta\delta < 0$ should be located on the left side of MPTA plane, and the group (R^2) with $\Delta\delta > 0$ should be the right side, it could be concluded that the group with C-2 and C-23 would be on the left side and the group that contains C-25 and C-25-OMe will be on the right side. This orientation is illustrated as the Newman structure **16**. Obviously, the AC for C-24 was assigned as (*S*). Therefore, AC for C-2 was assigned as (*R*).

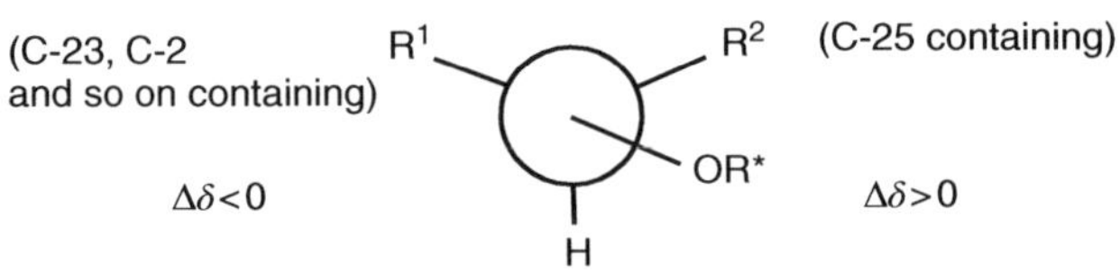

16

Table 2.6 The ^{1}H NMR shift differences between the (*S*)- and (*R*)-Mosher esters [25].

Proton	δ_S (ppm)	δ_R (ppm)	$\Delta\delta$ (ppm)
H-2	2.25	2.44	−0.19
H-23a	1.84	1.88	−0.04
H-23b	1.68	1.77	−0.09
H-25	3.48	3.41	+0.07
H-25-OMe	3.35	3.24	+0.11

There have been many successful predictions using the Mosher method [26]. Examples such as the di-Mosher ester structure **17** [27], the macro-lactone structure **18** [28], the poly-ring structure **19** [29], the and side-chain alcohol **20** [30] are selected from recent reports and listed below (R* = (*S*)- and (*R*)-MTPA, respectively).

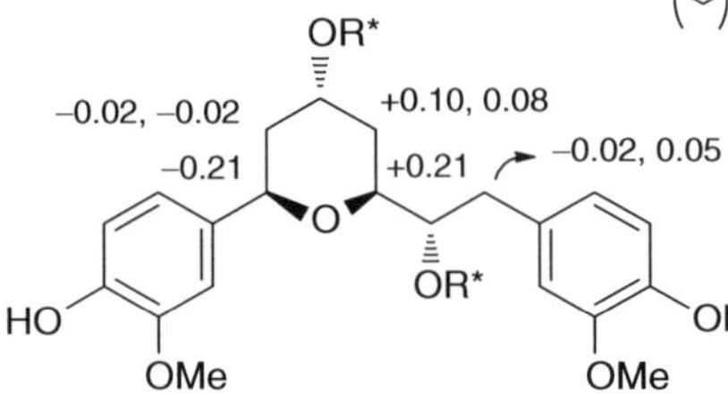

17

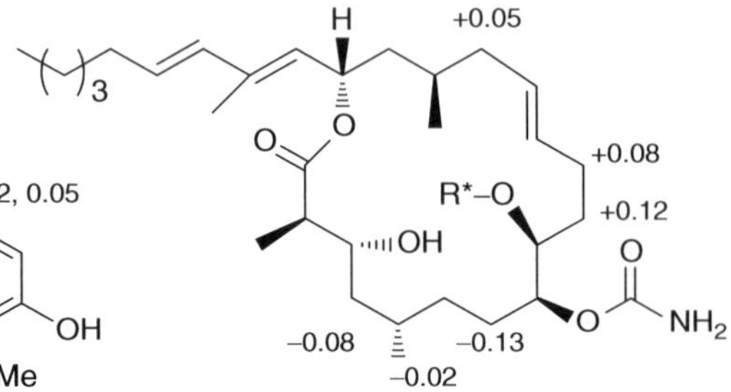

18

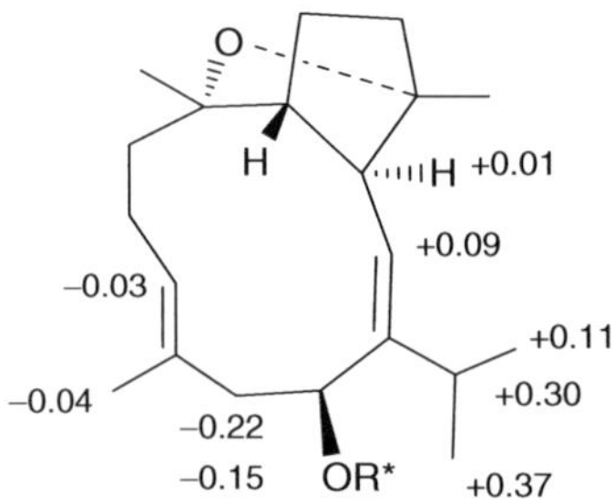

19

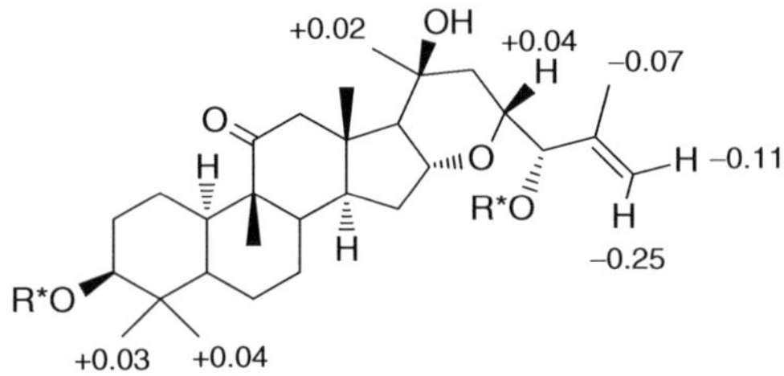

20

In ^{31}P NMR (*R*)-Mosher measurement, the reaction of (*R*)-**21** and (*S*)-**21** with one (*R*)-Mosher acid was used to form two isomers and the $\Delta\delta$ values were determined [31]. This is different from the general method used. In the traditional method, one substrate reacts with (*R*)- and (*S*)-Mosher acids, respectively.

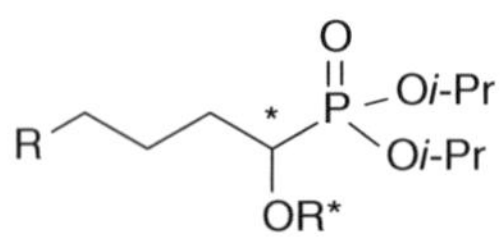

R* = (*R*)-MTPA
21

		(*S*)-**21**	(*R*)-**21**	$\Delta\delta = \delta_S - \delta_R$
R = Br–	$\Delta\delta$	13.59	12.95	0.64
R = $BrCH_2^-$	$\Delta\delta$	13.01	12.39	0.62
R = $ClCH_2^-$	$\Delta\delta$	13.35	12.74	0.61

The Mosher method can be used for secondary amines [32a]. Tertiary alcohols and amines were also used for extending the use of Mosher methods (^{13}C NMR) [32b]. Generally, a known chiral analog should be used as a standard model first when it is used for the same kind of chiral molecular AC assignment.

When chiral compounds are available only in very limited quantity, the use of Mosher method is not easy. In this case, a single Mosher ester with either (*R*)- or (*S*)-methoxyphenyl acetic ester derivative (MPA) is recommended [33]. ^{1}H NMR spectra were determined at two different temperatures. At low temperature, the relative population of the most stable conformer (major) increased, as well as the effect from the phenyl ring, so the increased micro-magnetic field from the phenyl ring led the protons on the left side of MPA plane upfield. At the same time, those protons under the shielding cone in the less populated conformer (minor) shifted downfield. It could be established by comparing the difference of the ^{1}H NMR ($\delta^{T1,T2} = \delta^{T1} - \delta^{T2}$) spectra both at room temperature and low temperatures [33].

For example, (*R*)-MPA of diacetone D-glucose (**22**) was determined at three different temperatures 303, 223, and 193 K in a mixture of CS_2/CH_2Cl_2 (4 : 1) [33]. The recorded ^{1}H NMR spectrum is illustrated in Figure 2.5. The shifts of protons (such as H-1′, H-2′) on the left of (*R*)-MPA plane moved downfield ($\delta^{T1,T2} < 0$), and the shifts of H-4′ and H-5′ moved upfield ($\delta^{T1,T2} > 0$). Obviously, once (*S*)-MPA is used, all the results should be reversed.

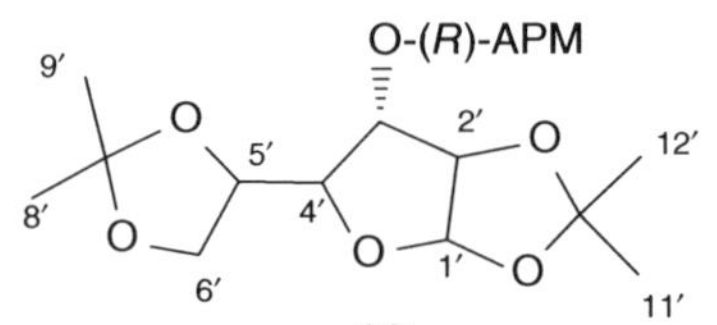

22

Indeed, if the size of a phenyl group in the Mosher acid is not large, its electronic effect on ^{1}H NMR shift changes is not large enough to cause reasonable ^{1}H NMR differences in experiments in some cases. Improved Mosher acids such as 9-anthranylmethoxyacetic acid (9-ATMA) **23** [34], axial acid 2-(2′-methyoxy-1,1′-naphthyl)-3,5-di-chlorobenzoic acid (MNCB) **24** [35], and 1,5-di-F,2,6-di-NO_2-benzene **25** [36] are now used. Mosher acids with a large

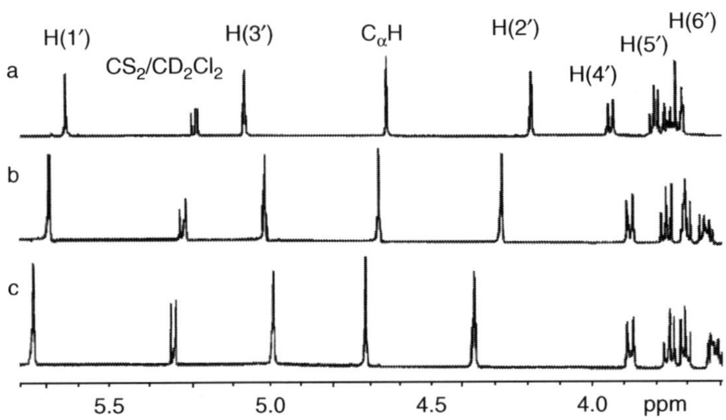

Figure 2.5 Partial NMR spectra of the (*R*)-MPA ester of diacetone D-glucose (**22**) at (a) 303 K, (b) 223 K, and (c) 193 K [33].

aromatic ring system, such as **23**, have big shielding effects on ^{1}H NMR, and those effects can be recorded well [34–36].

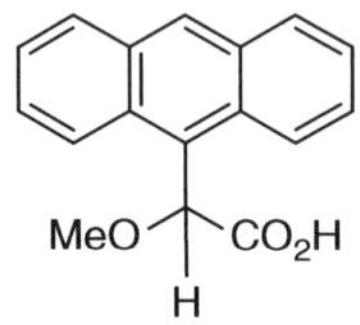

23

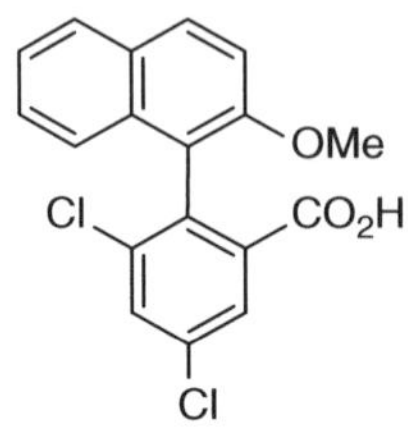

24

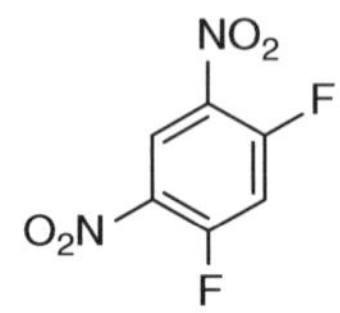

25

For example, (−)-methanol can react with **23** to afford (*R*)- and (*S*)-Mosher esters **26**. The recorded $\Delta\delta$ magnitudes are illustrated [34–36]. In contrast, the $\Delta\delta$ values recorded using MTPA are recorded and summarized below. Clearly, use of 9-ATMA brought about large shift differences [34].

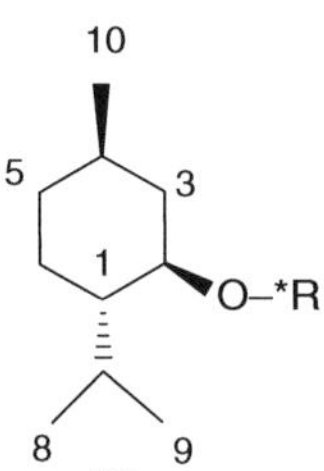

26

	R* = (*R*)- and (*S*)-MTPA	R* = (*R*)- and (*S*)-9-ATMA
C-1:	−0.25	−0.42
C-6:	−0.02, −0.03	−0.26, −0.23
C-7:	−0.24	−1.81
C-3:	+0.13, +0.06	+0.51, +0.50
C-4:	+0.04	+0.04
C-10:	+0.03	+0.19

If one uses the axial compound **24** as a Mosher acid, it will afford an achiral ester like **27**; it needs to make a model such as **28** below to orient the groups at both sides of the MNCB plane [35].

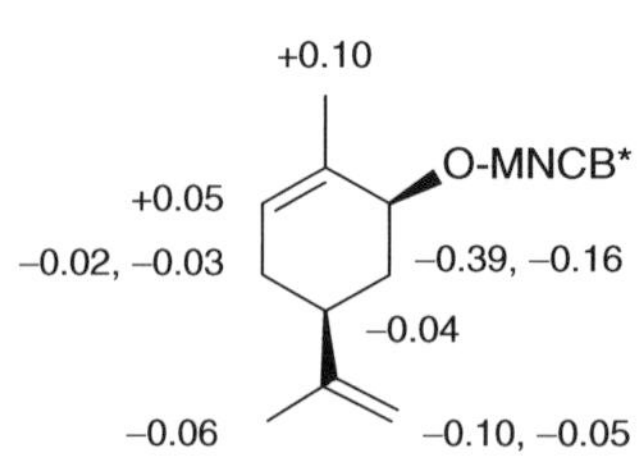

27

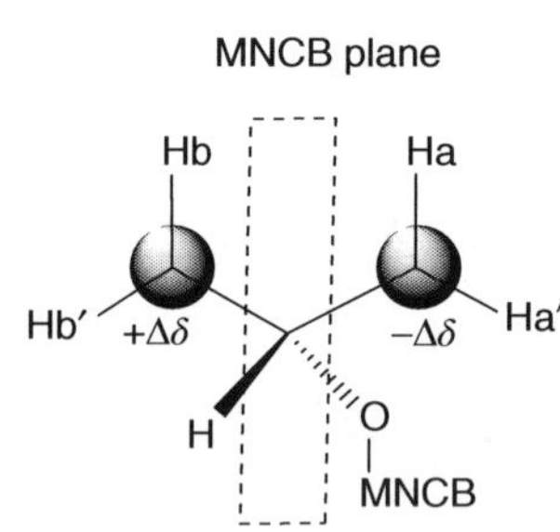

28

It needs to react with a “standard” chiral amine or other chiral compounds that play the role of Mosher acids when **25** is used as an auxiliary. For example, when it was used to assign the AC of the chiral alcohol **29**, (*R*)- or (*S*)-phenylethyl amine (PEA, **30**) was used as the role Mosher acid [36].

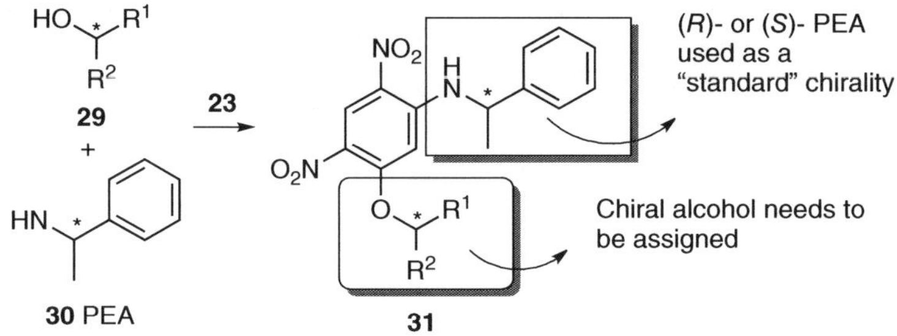

However, the Mosher method is an empirical method and has its own range of use. For example, when there is a big group close to a stereogenic center, the distribution of positive $\Delta\delta$ or negative $\Delta\delta$ of ^{1}H NMR may have no regular changes. For example, in compounds such as **32** and **33**, the appearance of $+\Delta\delta$ and $-\Delta\delta$ has random distribution around the chiral centers [37].

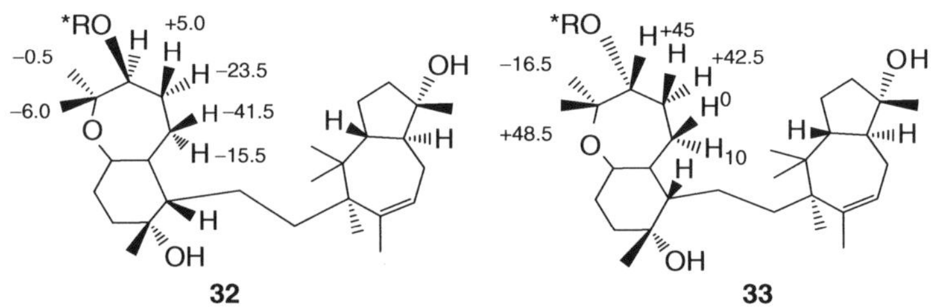

R* = (*R*)-or(*S*)-Mosher ester moiety, the number labelled near atom is the $\Delta\delta$ value.

Most of –OH, –OS, and $-NH_2$-containing chiral compounds can form the corresponding Mosher esters or amides. If the structures of the major and minor conformation are not the ones that the Mosher method requires, the prediction may be wrong. The most reliable way is to use chiral compounds whose structures are close to the model compounds tested in Mosher methods. Or, one can compute the distribution of (*R*)- or (*S*)-Mosher esters (amide) using the B3LYP/6-311+G(d) method after conformational searches.

2.3
Transition State Energy and Chirality Selectivity

In a specific reaction, such as enantioselective additions or reductions, the different transition state (TS) energy barriers decide the formation of the (*R*) or (*S*)-products. This procedure generally is more complex than experimentalists expected.

In this procedure, theoretical predictions of the energy barrier are necessary. First of all, the construction of a reasonable TS geometry is the key step. For example, in a normal reduction of the hydride of sodium borohydride to a carbon, the selections of all atoms of $NaBH_4$ and ketone substrate are necessary (**TS-1**). If this procedure is simplified as the hydride attacking the carbon (**TS-2**), the prediction may not give a good result.

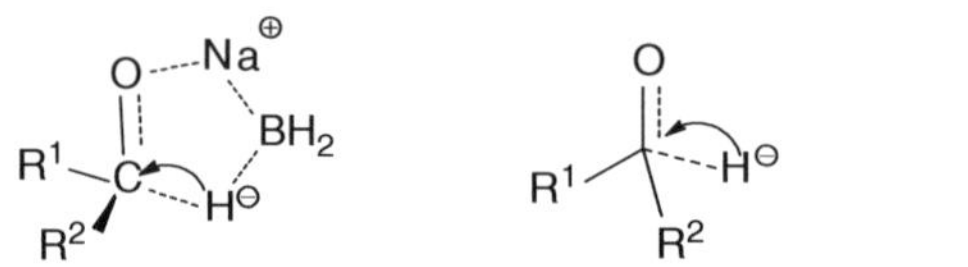

TS-1 **TS-2** (simplified TS model)

In an enantioselective reduction, the catalyst should be added to the TS structure (**TS-3**). Different (*R*) or (*S*) AC TS structures should be analyzed.

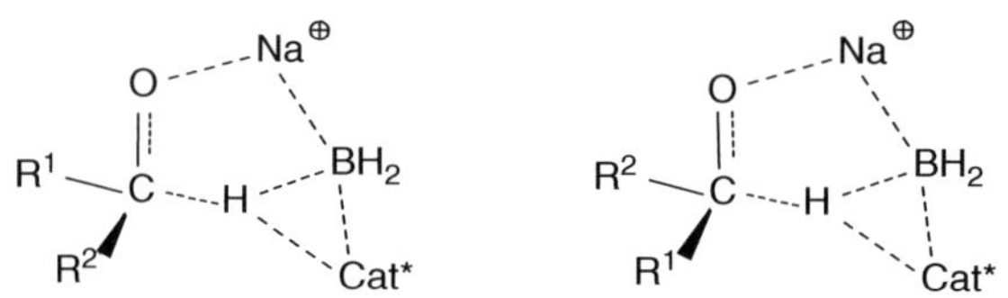

TS-3 **TS-4**

Second, the effect of the conformation of R^1 and R^2 on the TS barrier magnitudes must be investigated. After all the geometries are studied, the lowest TS barriers could be concluded.

The $\Delta(\Delta E)$ obtained from $\Delta E_{(S)}$ and $\Delta E_{(R)}$, which are the barriers to the (*S*)-isomer and (*R*)-isomer, respectively, could be used for the prediction of the magnitude of the enantiomeric excess (ee) using the following formula. The percentage of *R* isomer (*R*%) can be computed using formula [15].

$$R\% = \frac{e^{\frac{-\Delta[(\Delta E)_{(R-S)}]}{RT}}}{1+e^{\frac{-\Delta[(\Delta E)_{(R-S)}]}{RT}}} \times 100\% = \frac{e^{-1.69\times(\Delta E)_{(R-S)}}}{1+e^{-1.69\times(\Delta E)_{(R-S)}}} \times 100\% \quad (2.15)$$

where *R* is the gas constant and *T* is the temperature in kelvin.

Thus, the ee for the (*S*) isomer is

$$S\,\text{ee}\% = (1 - 2\times R)\times 100\% = \left[1 - \frac{2e^{-1.69\times(\Delta E)_{(R-S)}}}{1+e^{-1.69\times(\Delta E)_{(R-S)}}}\right]\times 100\%$$
$$= \frac{1-e^{-1.69[\Delta(\Delta E)_{(R-S)}]}}{1+e^{-1.69[\Delta(\Delta E)_{(R-S)}]}} \times 100\% \quad (2.16)$$

Theoretically, for example, when $\Delta(\Delta E)$ changes from 0.2 to 2.0 kcal mol^{-1}, the predicted ee may increase from 28 to 94%. However, the predicted energy difference may be overestimated in calculations, and the real ee may be lower than the prediction. Anyway, the more the difference, the larger the ee in experiments.

2.4 Separation of Chiral Compounds

Separation of the (*R*)- from the (*S*)-enantiomer is an important task. In many cases, researchers need to use a racemic mixture as the starting material. If it can form a crystal, then a small quantity of (*R*)- or (*S*)-crystals can be added into the solution. First, the (*R*)- or (*S*)-enantiomer may crystallize from the solution. After filtration or translation, the overconcentrated (*S*)- or (*R*)-enantiomer solution may crystallize the (*S*)- or (*R*)-enantiomer later. With this procedure, the (*R*)- and (*S*)-enantiomers could be separated from each other. However, this is an ideal case. In many cases, one has to use different methods to separate the (*R*)-enantiomer from the (*S*)-enantiomer. Some cases are described below [38].

2.4.1 Chiral Organic Bases

It is usual to use chiral organic bases to react with the organic acids to afford the corresponding salts. For example, the usually used organic bases include (−)-strychnine, (−)-quinine, and (−)-brucine. The salt could form in crystalline form from the solution.

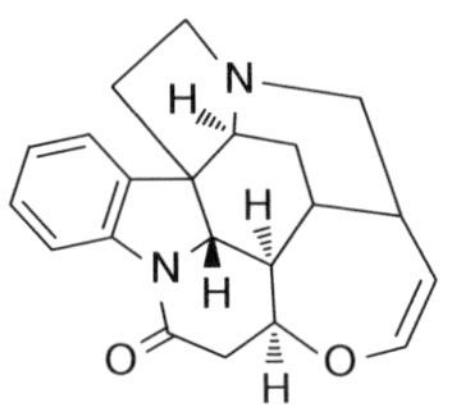

34 (–)-Strychnine

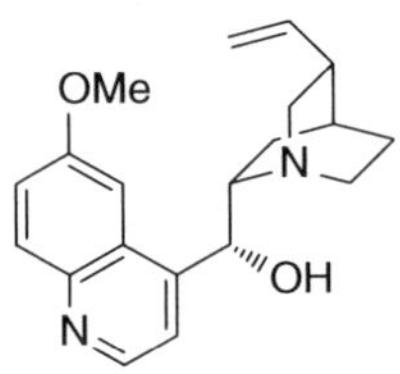

35 (–)-Quinine

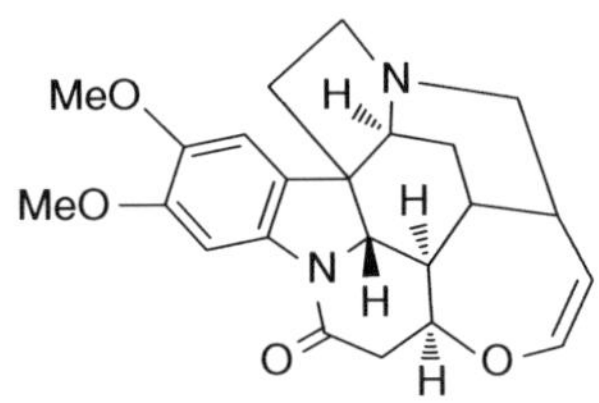

36 (–)-Brucine

2.4.2 Chiral Organic Acids

The often used organic acids are D-(+)-tartaric acid, (*S*)-(−) malic acid, and others. Similarly, the salt would form as crystals and deposit.

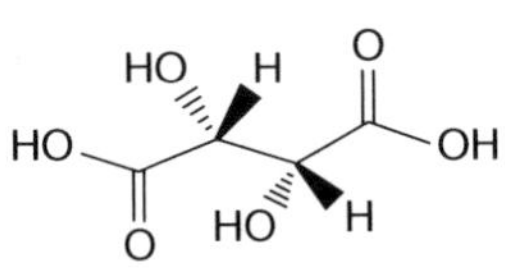

37 (D)-(+)-Tataric acid

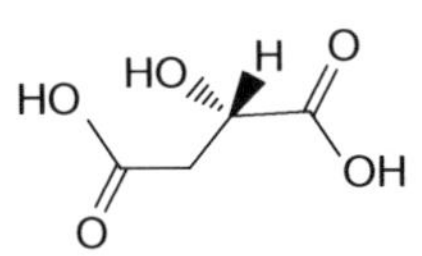

38 (S)-(–)-Malic acid

2.4.3 Chiral Organic Alcohols

The use of menthyl isocyanate, tartaric acid mono-anilide, or other chiral compounds to react with the corresponding chiral alcohols could afford different esters, and they can be isolated and finally hydrolyzed to the corresponding alcohols.

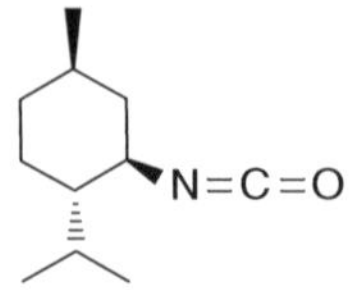

39 Menthyl isocyanate

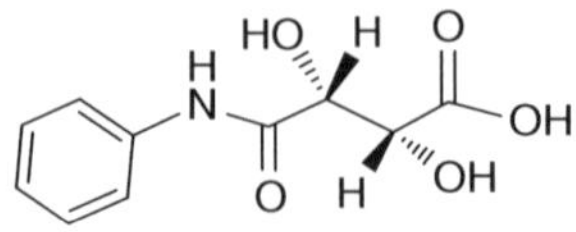

40 Tartaric acid mono-anilide

2.4.4 Others

Hydrazine derivatives, such as that of tartaric amide or menthyl hydrazine, can be used to separate compounds containing aldehyde or ketone groups. The C=O group can easily react with the $-NH_2$ group of hydrazine and form different stereoisomers which can be easily isolated from each other. After isolation, their hydrolysis can afford the corresponding ketone or aldehyde.

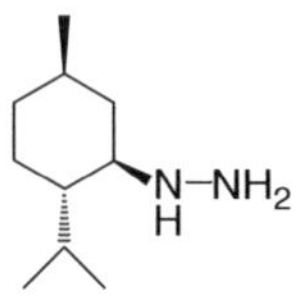

41 Menthyl hydrazine

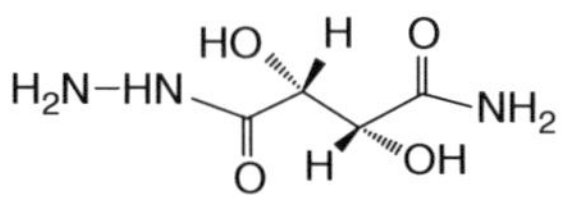

42 (2*R*,3*R*)-4-hydrazinyl-2,3-dihydroxy-4-oxobutanamide

As a useful example, the separation of Binol catalysts is elaborated here. 2-Amino-2′-hydroxy-1,1′-binaphthyl (NOBIN) **43** is one of the useful mother structures for the synthesis of other axial catalysts. The separation of the (*S*)- from the (*R*)-enantiomer can be performed efficiently using (1*S*)-(+)-10-camphorsulfonic acid (**44**) [39]. The procedure is illustrated below.

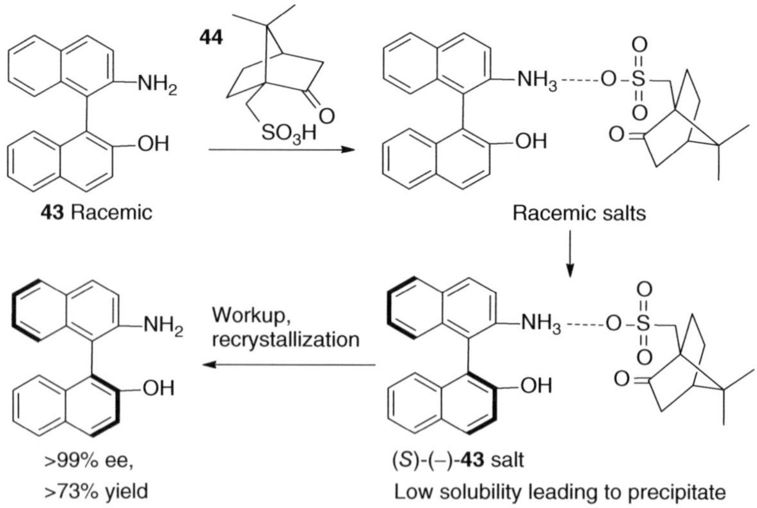

The separation of different chiral materials has become very popular. This is a very active area. Generally, high-performance liquid chromatography (HPLC) is used for this. The widely used commercial materials include modified silica gel, whose surface is covered with a layer of chiral material, like a sugar molecule. Or, directly, an organic reaction is used to form a layer of chiral material on the silica surface.

This is a very interesting area. There are many commercial columns available for the separation of different chiral compounds. To know more about this, the reader may consult the corresponding specialized references.

References

1. (a) Bax, A. and Davis, D.G. (1985) *J. Magn. Reson.*, **63**, 207; (b) Williams, D.H. and Fleming, I. (2001) *Spectroscopic Methods in Organic Chemistry*, translated by Wnag, J.B. and Shi, W.F., Published by Peking University Press, McGraw-Hill; (c) Rice, K.C. (1977) *J. Med. Chem.*, **20**, 164–165.
2. (a) Wolinski, K., Hilton, J.F., and Pulay, P. (1990) *J. Am. Chem. Soc.*, **112**, 8251; (b) Keith, T.A. and Bader, R.F.W. (1992) *Chem. Phys. Lett.*, **194**, 1; (c) Keith, T.A. and Bader, R.F.W. (1993) *Chem. Phys. Lett.*, **210**, 233; (d) Ruud, K., Helgaker, T., Bak, K.L., Jørgensen, P., and Jensen, H.J.A. (1993) *J. Chem. Phys.*, **99**, 3847; (e) Gauss, J. (1993) *J. Chem. Phys.*, **99**, 3629; (f) Gauss, J. (1995) *Phys. Chem. Chem. Phys.*, **99**, 1001.
3. (a) Helgaker, T., Jørgensen, P., and Olsen, J. (2000) *Molecular Electronic Structure Theory*, John Wiley & Sons, Inc., New York; For reviews, see: (b) Helgaker, T., Jaszunski, M., and Ruud, K. (1999) *Chem. Rev.*, **99**, 293; (c) Blanton, W.B.J. (2003) *Magn. Reson.*, **162**, 269.
4. Rodriquez, M., Terracciano, S., Cini, E., Settembrini, G., Bruno, I., Bifulco, G., Taddei, M., and Gomez-Paloma, L. (2006) *Angew. Chem. Int. Ed.*, **45**, 423.
5. Kirss, R.U., Forsyth, D.A., and Plante, M.A. (2003) *J. Org. Chem.*, **68**, 206.
6. Zurek, E. and Autschbach, J. (2004) *J. Am. Chem. Soc.*, **126**, 13079.
7. Besley, N.A., Titman, J.J., and Wright, M.D. (2005) *J. Am. Chem. Soc.*, **127**, 17948.
8. Reddy, V.P., Rasul, G., Prakash, G.K.S., and Olah, G.A. (2003) *J. Org. Chem.*, **68**, 3507.
9. Cheng, M., Li, Q., Lin, B., Sha, Y., Ren, J., He, Y., Wang, Q., Hua, H., and Ruud, K. (2006) *Tetrahedron: Asymmetry*, **17**, 179.
10. Havlin, R.H., Laws, D.D., Bitter, H.M.L., Sanders, L.K., Sun, H., Grimley, J.S., Wemmer, D.E., Pines, A., and Oldfield, E. (2001) *J. Am. Chem. Soc.*, **123**, 10362.
11. Stahl, M., Schopfer, U., Frenking, G., and Hoffmann, R.W. (1996) *J. Org. Chem.*, **61**, 8083.
12. (a) Estrada, E., Perdomo-López, I., and Torres-Labandeira, J.J. (2000) *J. Org. Chem.*, **65**, 8510; (b) Belostotskii, A.M., Goren, Z., and Gottlieb, H.E. (2004) *J. Nat. Prod.*, **67**, 1842; (c) Brown, R.E. and Mendenhall, G.D. (1998) *J. Phys. Chem. A*, **102**, 8537; (d) Rickard, G.A., Karadakov, P.B., Webb, G.A., and Morokuma, K. (2003) *J. Phys. Chem. A*, **107**, 292; (e) Stærk, D., Norrby, P.O., and Jaroszewski, J.W. (2001) *J. Org. Chem.*, **66**, 2217; (f) Price, D.R. and Stanton, J.F. (2002) *Org. Lett.*, **4**, 2809; (g) Seco, J.M., Quinňoa, E., and Riguera, R. (2004) *Chem. Rev.*, **104**, 17–117.
13. (a) Nagle, D.G., Park, P.U., and Paul, V.J. (1997) *Tetrahedron Lett.*, **38**, 6969; (b) Ribe, S., Kondru, R.K., Beratan, D.N., and Wipf, P. (2000) *J. Am. Chem. Soc.*, **122**, 4608.
14. (a) Cao, J.X., Li, L.-B., Ren, J., Jiang, S.-P., Tian, R.-R., Chen, X., Peng, S.-L., Zhang, J., and Zhu, H.-J. (2008) *Helv. Chem. Acta*, **91**, 1954–1960; (b) Thomas, A.H. (1996) *J. Comput. chem.*, **17**, 490–519; (c) Thomas, A.H. (1999) *Comput. Chem.*, **20**, 720–729.
15. (a) Forsyth, D.A. and Sebag, A.B. (1997) *J. Am. Chem. Soc.*, **119**, 9483; (b) Barone, G., Gomez-Paloma, L., Duca, D., Silvestri, A., Riccio, R., and Bifulco, G. (2002) *Chem. Eur. J.*, **8**, 3233.
16. (a) Yang, C.-S., Wang, X.-B., Wang, J.-S., Luo, J.-G., Luo, J., and Kong, L.-Y. (2011) *Org. Lett.*, **13**, 3380–3383; (b) Liu, D.-Z., Wang, F., Liao, T.-G., Tang, J.-G., Steglich, W., Zhu, H.J., and Liu, J.-K. (2006) *Org. Lett.*, **8**, 5749–5752.
17. (a) Hua, Y., Ren, J., Chen, C.X., and Zhu, H.J. (2007) *Chem. Res. Chin. Univ.*, **23**, 592–596; (b) Zhou, G.D. (1981)

Crystal Structure Determination, Science Presses of China; (c) Clegg, W., Blaka, A.J., Gould, R.O., and Main, P. (2001) *Crystal Structure Analysis: Principle and Practices*, Oxford University Press.

18. (a) Li, G.Y., Yang, T., Luo, Y.G., Chen, X.Z., Fang, D.M., and Zhang, G.L. (2009) *Org. Lett.*, **11**, 3714–3717; (b) Ren, J., Li, G.-L., Shen, L., Zhang, G.-L., Nafie, L., and Zhu, H.J. (2013) *Tetrahedron*, **69**, 10351–10356.

19. (a) Bijvoet, J.M., Peerdeman, A.F., and Bommel, A.J. (1951) *Nature*, **168**, 271–272; (b) Zhang, H. *Coordination Chemistry—Principles and Application*, Chemical Industrial Press of China, 2009; (c) Chen, X.M. and Cai, J.W. (2007) *Principle and Practice in Crystal Structure Analysis*, 2nd edn, Science Press of China.

20. Xiao, W.L., Lei, C., Ren, J., Liao, T.G., Pu, J.X., Pittman, C.U., Lu, Y., Zheng, Y.T., Zhu, H.J., and Sun, H.D. (2008) *Chem. Eur. J.*, **14**, 11584–11592.

21. Raban, M. and Mislow, K. (1965) *Tetrahedron Lett.*, **48**, 4249–4253.

22. (a) Sullivan, G.R., Dale, J.A., and Mosher, H.S. (1973) *J. Org. Chem.*, **38**, 2143–2147; (b) Lewis, R.A., Korpium, O., and Milosn, K. (1968) *J. Am. Chem. Soc.*, **90**, 4847–4853; (c) Trost, B.M., Belletire, J.L., Godleski, S., McDougal, P.G., and Balkovec, J.M. (1986) *J. Org. Chem.*, **51**, 2370–2374; (d) Latypov, S.K., Seco, J.M., Quinoa, E., and Riguera, R.J. (1996) *J. Org. Chem.*, **61**, 8569–8577.

23. Ohtani, I., Kusumi, T., Kashman, Y., and Kakisawa, H. (1991) *J. Am. Chem. Soc.*, **113**, 4029–4096.

24. Li, Q.M., Ren, J., Shen, L., Bai, B., Liu, X.C., Wen, M.L., and Zhu, H.J. (2013) *Tetrahedron*, **69**, 3067–3074.

25. Wang, M.L., Du, J., Wang, T., Chen, R.Y., and Yu, D.Q. (2001) *Org. Chem. (Chin)*, **21**, 341–345.

26. Reviews: (a) Błaźewska, K.M. and Gajda, T. (2009) *Tetrahedron: Asymmetry*, **20**, 1337–1361; (b) Wenzel, T.J. and Chisholm, C.D. (2011) *Chirality*, **23**, 190–214.

27. Yin, H., Luo, J.G., and Kong, L.Y. (2013) *Phytochem. Lett.*, **6**, 403–406.

28. Bishara, A., Rudi, A., Goldberg, I., Aknin, M., and Kashman, Y. (2009) *Tetrahedron Lett.*, **50**, 3820–3822.

29. Xi, Z.F., Bie, W., Chen, W., Liu, D., Ofwegen, L.V., Proksch, P., and Lin, W.H. (2013) *Mar. Drugs*, **11**, 3186–3196.

30. Pan, L., Yong, Y., Deng, Y., Lantvit, D.D., Ninh, T.N., Chai, H., Esperanza, J., Blanco, C., Soejarto, D.D., Swanson, S.M., and Kinghorn, A.D. (2012) *J. Nat. Prod.*, **75**, 444–452.

31. Wuggenig, F., Schweifer, A., Mereiter, K., and Hammerschmidt, F. (2011) *Eur. J. Org. Chem.*, 1870–1879.

32. (a) Allen, D.A., Tomaso, A.E. J.r., Priest, O.P., Hindson, D.F., and Hurlburt, J.L. (2008) *J. Chem. Educ.*, **85**, 698; (b) Kobayashi, M. (1998) *Tetrahedron*, **54**, 10987–10998.

33. Latypov, S.K., Seco, J.M., Quińoá, E., and Riguera, R. (1998) *J. Am. Chem. Soc.*, **120**, 877–882.

34. Kusumi, T., Takahashi, H., Xu, P., Fukushimab, T., Asakawac, Y., Hashimotoc, T., Kanc, Y., and Inouye, Y. (1994) *Tetrahedron Lett.*, **35**, 4397.

35. Fukushi, Y., Yajima, C., and Mizutani, J. (1994) *Tetrahedron Lett.*, **35**, 599–601.

36. (a) Kingston, D., Zhou, B.N., Baj, N.J., Glass, T.E., Malone, S., Werkhoven, M.C.M., Troon, F., and Wisse, J.H. (1997) *J. Nat. Prod.*, **60**, 1287; (b) Harada, K., Shimizu, Y., Kawakami, A., and Fujii, K. (1999) *Tetrahedron Lett.*, **40**, 9081.

37. (a) Doesburg, H.M., Petit, G.H., and Merckx, E.M. (1982) *Acta Crystallogr.*, **B38**, 1181; (b) Merckx, E.M., Vanhoeck, L., Lepoivre, J.A., Alderweireldt, F.C., van Der Weken, B.J., Tollenaere, J.P., and Raymaekers, L.A. (1983) *Spectrosc. Int. J.*, **2**, 30.

38. Zhu, H.J. (2009) *Modern Organic Stereochemistry*, Chapter 1, Science Presses of China.

39. Smrcin, M., Vyskocil, S., Polivkova, J., Polakova, J., and Kocovsky, P. (1996) *Collect. Czech. Chem. Commun.*, **61**, 1520.

Part II
Techniques

Organic Stereochemistry: Experimental and Computational Methods, First Edition. Hua-Jie Zhu.

3
Optical Rotation (Rotatory Dispersion, ORD)

3.1
Introduction

Optical rotation (OR) and optical rotatory dispersion (ORD) are widely used in chirality study. As an early representative empirical method, the Brewster model attracted wide attention in the 1960s and many studies were carried out [1]. The Brewster model predicted the OR using a screw pattern of electron polarizability. It hypothesized that the magnitude of specific rotation is related to the refractions of the atoms and depends on different conformations of the atoms/substituents. Before the 1990s, many other methods were also explored, including some quantum methods [2].

Since the1990s, quantum theory developed well. For example, Hatree–Fock (HF) calculations were reported by Polavarapu [3]. The first application of density functional theory (DFT) in OR calculations was reported in Ref. [4]. Then, static-limit OR calculations [5] as well as coupled-cluster (CC) [6] and coupled-cluster single doubles (CCSD) methods [7] were reported from different research groups. Many researchers in this area explored several computational approaches in OR calculations [8]. At the same time, some OR computation programs, such as Gaussian, were developed [9]. Recent developments in OR calculations have made the AC assignment of acyclic chiral compounds relatively easy [10]. All these trials bring a colorful world in OR study. By the way, OR computation results accompany NMR calculation in the Gaussian program.

3.2
Quantum Theory

Quantum mechanical calculations relate OR to a molecular parameter $\beta(\nu)$, which is concerned with the frequency-dependent electronic dipole-magnetic polarizability tensor $\beta_{\alpha,\beta(\nu)}$ [11]. Computations of $\beta_{\alpha,\beta}$ at the HF level of theory were introduced by Amos for the static limit ($\nu = 0$) [12] and by Helgaker for any frequency ν [13]. This method can simply be described as follows:

Organic Stereochemistry: Experimental and Computational Methods, First Edition. Hua-Jie Zhu.

$$[\alpha]_\nu = \frac{28\,800\,\pi^2 N_A{}^2 \nu^2}{c^2 M} \gamma_{s,\nu} [\beta(\nu)]_0 \tag{3.1}$$

where N_A is the Avogadro number, M is the molecular weight, c is the speed of light in vacuum, and γ is the correction for the solvent, which is either neglected ($\gamma = 1$) or approximated by the equation $\gamma = (n^2 + 2)/3$ [14]. Thus, the value of $[\alpha]_\nu$ is proportional to magnitude of the tensor $\beta(\nu)$.

OR calculation is very useful for AC assignment. This procedure uses the corresponding program. There are many different computational methods in the same program, for example, in the Gaussian software, different methods lead to different OR values for the same chiral molecule. Therefore, for understandable reasons, one needs to use several different methods for OR calculations when dealing with the same chiral compound. For example, if the optimization for all low-energy conformations are carried out at the B3LYP/6-311+G(d) level in the gas phase, the computations of OR may be performed at the B3LYP/6-311++G(2d,p) level [14b,c]. The predicted OR values must be close to the experimental results.

When a chiral molecule has OR values between −57.8 and +57.8, the reliability may decrease using current DFT methods [15]. However, this conclusion was reached on the basis of limited samples. More samples for the statistics are required for reaching more reliable conclusions. Anyway, this result hints at the fact that it is not easy to predict a chiral compound's AC correctly when it has a small OR value.

One method to increase reliability is to look for a known chiral molecule that has a structure similar to that of the target molecule as a model molecule. Then, one can compute the OR values for both the chiral target molecule and the model compound. If the predicted AC for the known chiral compound is correct, the same method should be relatively reliable to predict the OR magnitudes for AC prediction for the target molecule.

For example, benzopyrenomycin (**1**) has an OR of +38 ($CHCl_3$), which is close to the OR value of (2*R*,3*S*)-**2** (+50 in $CHCl_3$). Empirically, (+)-**1** may also have the (2*R*,3*S*) configuration [16] because the two compounds are similar. However, their predicted OR sign predicted reversed when using DFT methods [17]. The predicted OR values were from +22 to +88 for **2**, so the predicted OR and its sign matched the experiments well. It could be the model compound. At the same time, the same methods were used to predict the ORs for (2*R*,3*S*)-**1**; they were from −28 to −50. The OR sign is opposite to that in the experiments. This gave the hint that (+)-**1** should be (2*S*,3*R*) [17].

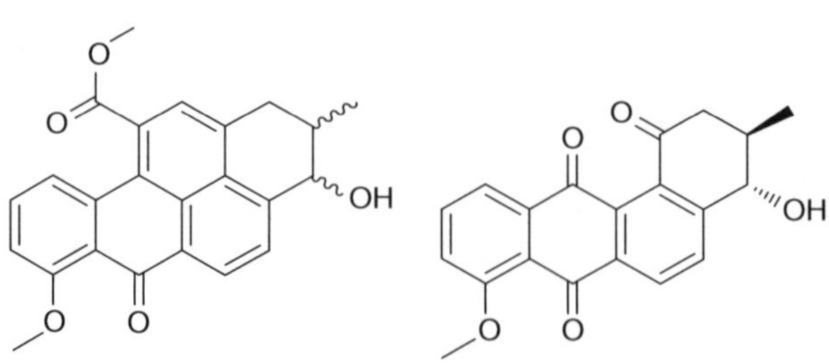

1 Benzopyrenomycin (2*R*,3*S*)-**2** Rubiginone A_2

As is well known, the OR values are very sensitive to the molecular geometry. For a non-chiral molecule such as **3**, there are four conformations **a–d** theoretically. Conformer **b** may not exist because of the large repulsive force between C1 and C4. Conformations **a** and **b** have a symmetric plane, therefore their OR equals zero. Conformation **c** has no symmetric element, so its OR will be nonzero, and so for **d** also. Since **c** and **d** are mirror-symmetric, their relative energy will be the same, their OR magnitudes will be same, but the OR signs will be reversed. Thus the net OR value from **c** and **d** is zero. This analysis could be confirmed. The calculated OR for **c** was −2.61 at the B3LYP/6-31G(d)//B3LYP/6-31G(d) level, and the OR for **d** was +2.58 recorded at the same level and their relative energetics were almost the same. Considering computation errors, the calculated OR values of −2.61 and +2.58 could be regarded as the same in absolute terms. The net OR from **c** and **d** conformations is zero. Therefore, the total OR for **3** will be zero.

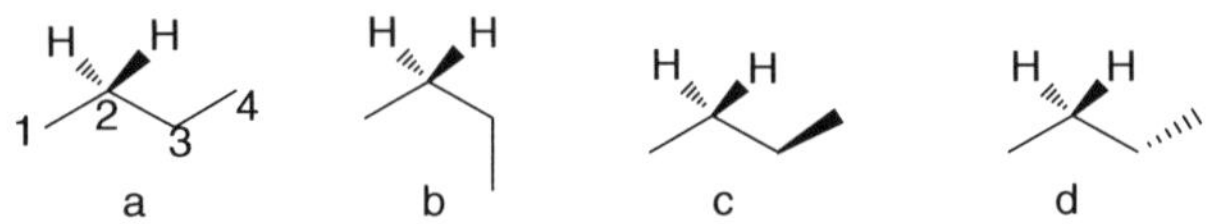

3

Considering the a chiral compound **4**, for example, it could be predicted that the OR sign for **c** and **d** would be reversed and their relative energy would be close. Thus, the major contribution to OR will be from the conformation **a**. This conclusion can be confirmed by OR computation at the B3LYP/6-31G(d)//B3LYP/6-31G(d) level in the gas phase. The results are listed in Table 3.1.

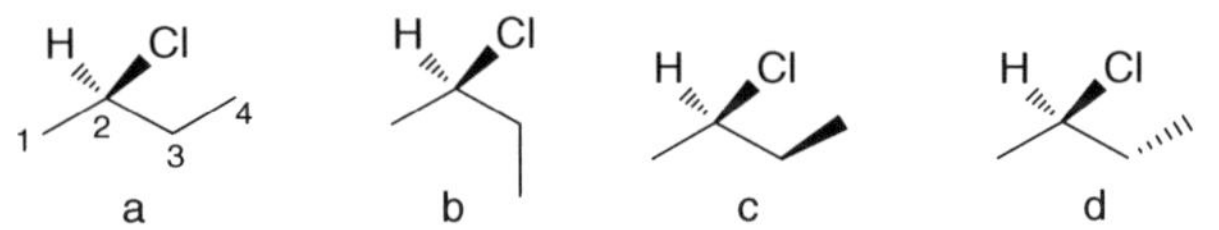

4

As predicted above, conformer **c** has a relative energy of 0.838 kcal mol^{-1} and OR of +24.36; for conformation **d** the corresponding values are 0.506 kcal mol^{-1} and −17.9. The net OR is only −1.7 after the Boltzmann sum. The total OR calculated for (*R*)-**4** is −30. The major contribution is absolutely from geometry **a** (48.26/(48.26 + 1.7) = 96.6%). The net OR contribution from **c** to **d** is only 3.4%.

Table 3.1 The predicted relative energy, $[\alpha]_D$, and net OR from pairs of conformers.

	4a	**4c**		**4d**
E (a.u.)	−618.057082	−618.055746		−618.056275
ΔE (kcal mol^{-1})	0.000	0.838		0.506
Fraction	1.000	0.242		0.425
$[\alpha]_D$	−48.26	24.36		−17.90
Fraction of $[\alpha]_D$	−48.26	5.9		−7.6
Net OR	−48.26		−1.7	

A chiral compound with a longer side chain, such as (*R*)-2-chloropentane (**5**), was tested. Because of the rotation of the single bonds C2–C3 and C3–C4, totally nine conformations were found.

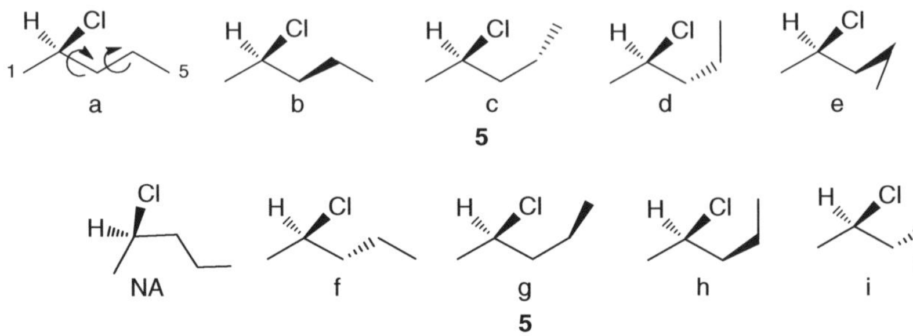

It can be seen that side chains of **b** and **f**, **c** and **g**, **d** and **h**, and **e** and **i** are mirror images. The 3D structures for each pair of conformations are superimposed when the C1–C2 bond is fixed (Figure 3.1). The net OR for each pair of conformers could be calculated, and they are summarized behind the 3D structures.

The net ORs from the pairs **b–d** and **c–f** are very large. However, their OR signs are reversed. For example, the net OR from the **b–d** pair of conformers is +15.6, it is −28.8 from the **c–f** pair. Once all pairs of conformers were used for Boltzmann sum, the total net OR decreased to −14.5. The whole molecule has an OR value of −47.7. Therefore, the contribution from **5a** is 85.8% and from the pair of conformers is only 14.2%. Indeed, when the length increases, it may not be easy to obtain the most reliable conformation for the side chain using the B3LYP/6-31G(d) method, which causes computational errors. If higher basis sets are used, such as B3LYP/6-311+G(d), the averaged computational time for a single optimization job increases almost 10 times that using the B3LYP/6-31G(d) method. Once a higher basis set is used, such as B3LYP/6-311++G(2d,p), which is more popular in modern computations, the computational time for the OR calculation job becomes almost 20 times that of the B3LYP/6-31G(d) method. Therefore, the computation of an acyclic chiral compound's OR is a big challenge when it has a long chain. The ORs of **5** using the B3LYP/6-311++G(2d,p)//B3LYP/6-311+G(d)

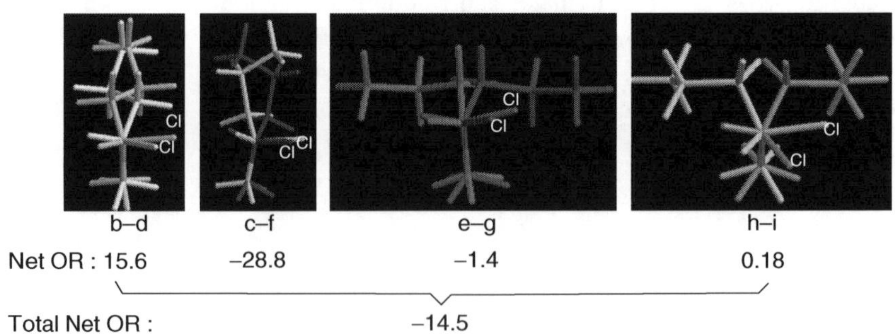

Figure 3.1 The 3D structures for each pair of conformers (similar with mirror image) and their net OR values.

Table 3.2 The net OR values for the four pair of conformers at the B3LYP/6-311++G(2d,p)//B3LYP/6-311+G(d) level.

Pairs	b–d	c–f	e–g	h–i
$\Delta E/\Delta E$	0.933/0.591	0.667/2.710	1.549/3.385	3.250/3.00
$[\alpha]_D/[\alpha]_D$	44.8/20.7	−131.6/−27.5	42.3/14.6	114.0/−44.9
Net OR[a)]	8.5	−15.6	1.6	0.09
Total net OR	−5.4			

a) Using Boltzmann statistics.

method are summarized in Table 3.2. The total net OR is −5.4. The final OR for (*R*)-**5** is −51.0. It is seen that the total net OR decreases from −14.5 to −5.4 when a higher level of the basis set is used. In this case, it is found that the net ORs of a pair of conformers **b–d** or **c–f** are very large (up to 8.5 and −15.6, respectively). However, the net OR values for the two pairs decreased to −7.1.

The absolute OR of −48.3 of **4a** is much larger than the value (−30.0) after Boltzmann statistics. Thus, if only conformer **4a** is used in the OR calculation, the errors will be very large. The OR of −87.4 for **5a** is also much larger than the OR of −47.7 after the Boltzmann sum it could be found that the projection along the direction of C1 to C3 to C5 should be overlapped. Its projection size is the same as a $-C_2H_5$.

It could be expected that the number of pairs of conformations increases quickly when the length of the chain increases. The sum of the OR contribution from the pairs should not be large and its OR sign may be reversed to the one of the lowest conformation in chiral compound with single stereogenic center. The key OR contribution comes from geometries such as **a** in chiral compounds **4** and **5**. However, the weight fraction must be considered to decrease the OR values. The current situation is that not all the most stable conformations with relatively low energy can be found using a single conformational search software. Two or more software packages should be used in the study, and sometimes it may have to be done manually. More powerful methods should be developed.

New methods may comprehensively consider all the factors in OR computations. We have to catch the major contradictions and put the minor contradiction behind. At the same time, energy computations should be avoided.

3.3 Matrix Model

As mentioned above, quantum methods have provided many benefits in AC assignment for relatively rigid chiral compounds. However, when dealing with acyclic chiral compounds, current quantum methods using wave functions are difficult to be used for the AC assignment since there exist many stable conformations with low energy (e.g., from 0 to 2.0 kcal mol^{-1}). The number of conformations is proportional to N^3, where N is the number of single bonds that can rotate freely in an acyclic compound. Obviously, the longer the acyclic chain, the more stable the conformations and, consequently, the much longer the computational time. Indeed, light shows wave–particle duality. Current quantum theory considers wave functions in OR calculations, and has achieved great successes. As mentioned above, new methods without consideration of energy should be developed. The particle function does not consider the molecular energy; this is the new method's characteristic.

The movement of the substituent around its stereogenic center looks like a mass of cloud. For example, the chiral acyclic compound **6** has 189 stable conformations after conformational searches using the MMFF94S force field. If C-1 and C-2 are fixed for superimposing all conformations, it can be found that there is a small area that looks extremely tight in the projections of the $-C_7H_{15}$ group (Figure 3.2, the large white circle section). In contrast, the projection of the Me group looks very fixed (Figure 3.2, small white circle).

The core area exists in all geometries with low energy to high energy based on statistics. It will have the key contribution to the OR values. This is very close to the conclusion from the OR computations of (*R*)-**4** and (*R*)-**5**.

Absolutely, OR must involve the electron's behavior. However, we can think of this behavior from different viewpoints. For example, an atom contains neutron(s) and proton(s), plus electron(s) circling around the core. When we discuss molecular characteristics, we may not consider the electronic behavior. Once we discuss a tissue, we may ignore the molecule's behavior and electronic characteristics. When we face a man's behavior, we may not consider the electrons' characteristics, yet our structure consists of various chemical bonds. The example here shows that, to some degree, even if universe is formed of different kinds of electron combinations (different chemical bond plus H-bond), we may ignore the electron's own characteristics when we face questions at different levels, such as at the electronic level, the molecular level, the tissue level, or the organ level.

On the other hand, when facing the OR phenomenon, we can see that the molecule in a solution is always moving but its movement rate is slow ($0.2–0.5\,cm\,s^{-1}$), and its quantum effect is very small. Thus, we can regard a molecule as a small particle. At the same time, a photon is still a kind of particle. When the both particles meet, it can be found that reflection and refractions can be observed from the chiral molecule. This has been well elaborated. Similarly, a beam of circularly polarized light (CPL) rotates through a definite angle after

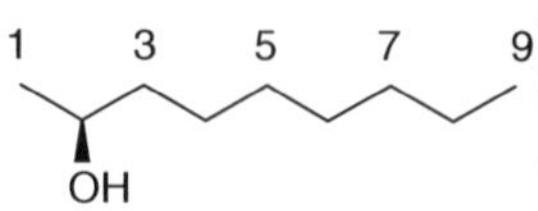

6 (*S*)-nonan-2-ol

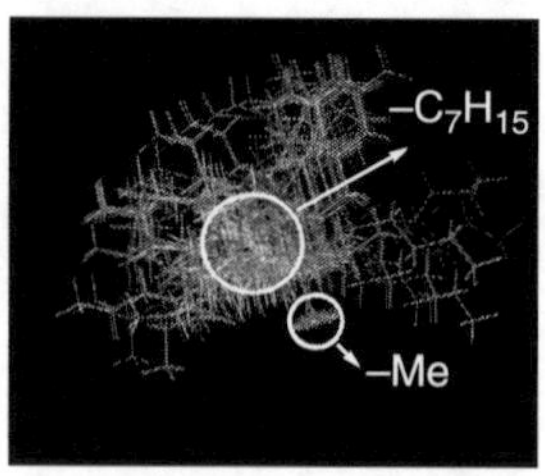

Figure 3.2 Structure of (*S*)-nonan-2-ol and movement projection of all conformations of (*S*)-nonan-2-ol around C1–C2 found using the MMFF94S force field.

Table 3.3 Quantum effect on different particles.

Item	Mass (g)	Velocity, V (cm/s)	Specifying accurate position, Δx (cm)	Uncertainty of velocity, ΔV (cm/s)	$\Delta V/V$ (%)
Electron	9.1×10^{-28}	10^8	10^{-8}	10^9	>1000
Atom (vibration)	1.67×10^{-23}	10^5	10^{-8}	10^5	≈100
Molecule	1.7×10^{-22}	0.2	10^{-3}	4×10^{-2}	≈20
Nanoparticle	10^{-21}	0.1	10^{-3}	10^{-3}	1
Microparticle	10^{-12}	1	10^{-3}	10^{-11}	0

passing a chiral molecule, and this process is possible when we consider two-particle interactions. Quantum effect on this molecule may be ignored, and the loss due to the quantum effect may be compensated in other ways. The different situations are calculated according to the uncertainty principle ($\Delta x \times \Delta p = h$) and summarized in Table 3.3. The molecule's movement is in between the two cases: that is, considering the quantum effect (100%) and not considering it (<1%).

Therefore, it is possible to consider the OR of a chiral molecule in another way. In the new method, it is not necessary to compute molecule's energy. This is the advantages in the OR calculations of acyclic chiral compounds. This conclusion is almost the same as reached from the relationship study for the OR contributions for the chiral compounds **4** and **5** above.

3.3.1 Matrix Basis

A typical flexible chiral molecule 7 was selected that contains one stereogenic center to illustrate the deduction of the matrix model [10]. However, this method can be extended to molecules with two or more stereogenic centers.

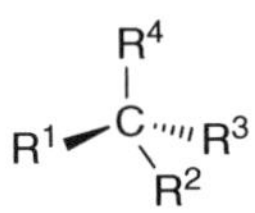

7

The four independent factors that affect the specific OR of 7 are as follows:

1) The comprehensive mass (m) of the substituents (depends on the conformations). The value of m changes with different conformations;
2) The radius (r) of the substituents (depends on conformations);
3) The electronegativity (χ) of the atom connected with the stereogenic center;
4) The symmetry (s) of the substituents. This is a new factor that has not been considered in the previous methods.

This model considers external factors, such as the solvent, temperature, wavelength of light, and internal factors, of a chiral molecule. External factors are those that can be changed by external actions, such as the concentration of solution and the wavelengths and temperatures employed. Their contribution to the OR can be written as a set of functions f_1, f_2, f_3, and f_4 [10].

$$\begin{cases} f_1 = a_1\mathbf{m}_1 + a_2\mathbf{r}_1 + a_3\boldsymbol{\chi}_1 + a_4\mathbf{s}_1 \\ f_2 = a_1\mathbf{m}_2 + a_2\mathbf{r}_2 + a_3\boldsymbol{\chi}_2 + a_4\mathbf{s}_2 \\ f_3 = a_1\mathbf{m}_3 + a_2\mathbf{r}_3 + a_3\boldsymbol{\chi}_3 + a_4\mathbf{s}_3 \\ f_4 = a_1\mathbf{m}_4 + a_2\mathbf{r}_4 + a_3\boldsymbol{\chi}_4 + a_4\mathbf{s}_4 \end{cases}$$

Here, a_1 to a_4 (where $a_1 \neq a_2 \neq a_3 \neq a_4$) are the weighting factors for each of the matrix elements, and represent the magnitudes of the effects exerted by the mass, radius, electronegativity, and symmetry of the substituents, respectively, on the functions f_1–f_4. Each coefficient's value is different and a constant, but they are unknown at the present time.

When photons of light with a frequency (ν) interact with each of the four substituents of the chiral molecule, the function F_1 is generated:

$$F_1 = f(\nu)\begin{bmatrix} f_1 \\ f_2 \\ f_3 \\ f_4 \end{bmatrix} \tag{3.2}$$

This matrix function expresses the interaction of the photons with the four characteristics (elements) of all four substituents simultaneously. Similarly, the functions F_2 for the specific solvent employed, F_3 for the temperature, and others can be written in the same form as Eq. (3.2).

$$F_2 = f(s)\begin{bmatrix} f_1 \\ f_2 \\ f_3 \\ f_4 \end{bmatrix} \text{ and } F_3 = f(t)\begin{bmatrix} f_1 \\ f_2 \\ f_3 \\ f_4 \end{bmatrix}$$

The overall function that describes the interaction of light with the chiral molecule for this conformation can be described as **F**.

$$\mathbf{F} = F_1 + F_2 + F_3 + \cdots$$

where

$$\mathbf{F} = f(\nu)\begin{bmatrix} f_1 \\ f_2 \\ f_3 \\ f_4 \end{bmatrix} + f(s)\begin{bmatrix} f_1 \\ f_2 \\ f_3 \\ f_4 \end{bmatrix} + f(t)\begin{bmatrix} f_1 \\ f_2 \\ f_3 \\ f_4 \end{bmatrix} + \cdots$$

When the solvent, temperature, and other determined conditions are fixed, the functions $f(s)$ and $f(t)$ and others will be constant, and then

$$\mathbf{F} = k_v \begin{bmatrix} f_1 \\ f_2 \\ f_3 \\ f_4 \end{bmatrix} + k_s \begin{bmatrix} f_1 \\ f_2 \\ f_3 \\ f_4 \end{bmatrix} + k_t \begin{bmatrix} f_1 \\ f_2 \\ f_3 \\ f_4 \end{bmatrix} + \cdots$$

where k, k_s, and k_t are constants. Therefore, the expression for F becomes

$$\mathbf{F} = k \begin{bmatrix} \mathbf{m}_1 & r_1 & \chi_1 & \mathbf{s}_1 \\ \mathbf{m}_2 & r_2 & \chi_2 & \mathbf{s}_2 \\ \mathbf{m}_3 & r_3 & \chi_3 & \mathbf{s}_3 \\ \mathbf{m}_4 & r_4 & \chi_4 & \mathbf{s}_4 \end{bmatrix} \begin{bmatrix} a_1 \\ a_2 \\ a_3 \\ a_4 \end{bmatrix}$$

where k is the sum of all the constants (e.g., $k_v + k_s + k_t + \cdots$).

However, F contains a matrix within the function, so it is not a scalar number. Thus, we define $[\alpha] = |\mathbf{F}|$. Also, the sodium D line is used to obtain the OR. Then, one conformation has

$$\begin{aligned} [\alpha]_{\mathrm{D}} &= k \times a_1 \times a_2 \times a_3 \times a_4 \times \det(D) \\ &= k_0 \times \det(D) \end{aligned}$$

where

$$\det(D) = \begin{vmatrix} \mathbf{m}_1 & r_1 & \chi_1 & \mathbf{s}_1 \\ \mathbf{m}_2 & r_2 & \chi_2 & \mathbf{s}_2 \\ \mathbf{m}_3 & r_3 & \chi_3 & \mathbf{s}_3 \\ \mathbf{m}_4 & r_4 & \chi_4 & \mathbf{s}_4 \end{vmatrix}$$

Obviously, $\det(D)$ represents the molecule's characteristics in this matrix model. It has nothing to do with external factors such as the solution concentration, the wavelength of light, the temperature, and so on. This shows that the OR is proportional to the $\det(D)$ value.

A special case needs to be discussed. The groups R^1, R^2, R^3, and R^4 are considered to be *perfect point masses* (e.g., single atoms). The chiral molecule can be placed in a three-dimensional coordinate system with the chiral center at the origin and the smallest group R^4 along the Z-axis. Thus, the corresponding coordinates can be written as: $C_1(x_1, y_1, z_1)$, $C_2(x_2, y_2, z_2)$, $C_3(x_3, y_3, z_3)$, $C_4(x_4, y_4, z_4)$. Next, three new determinants $\det(D_x)$, $\det(D_y)$, and $\det(D_z)$ are formed by projecting the matrix elements from $\det(D)$ onto the X, Y, and Z-axes. From the projected values, $\det(D_x)$, $\det(D_y)$, and $\det(D_z)$ are obtained. Since substituent R^4 is located at the origin ($x_4 = y_4 = 0$), all the projected values for the elements of R^4 on the X and Y axes are zero. This implies that the contributions to the OR can also be represented with $\det(D_z)$, the determinant of the elements of substituents R^1, R^2, R^3, and R^4 which is projected on the Z-axis. Obviously, when there are four different atoms connected to the stereogenic center, this molecule does not have a standard tetrahedral geometry. Thus, $\det(D_z)$ can accurately reflect the characteristics of the molecular OR. However, if the four connected groups are complex substituents, the use of $\det(D_z)$ may result in larger errors than for $\det(D)$ due to the second processing of data.

3.3.2
Explanation of General OR Characteristics

When two or more of the substituents are identical, the corresponding values of **r**, **m**, χ, and **s** will be the same. Then the value of $\det(D)$ must be equal to zero. In this case, the molecule has no dissymmetry. Hence its specific rotation is zero.

$$\det(D) = \begin{vmatrix} m_1 & r_1 & \chi_1 & s_1 \\ m_2 & r_2 & \chi_2 & s_2 \\ m_3 & r_3 & \chi_3 & s_3 \\ m_4 & r_4 & \chi_4 & s_4 \end{vmatrix} = 0$$

When the positions of any two substituents (e.g., R^1 and R^2) are changed, then the following determinant results:

$$\det(D') = \begin{vmatrix} m_2 & r_2 & \chi_2 & s_2 \\ m_1 & r_1 & \chi_1 & s_1 \\ m_3 & r_3 & \chi_3 & s_3 \\ m_4 & r_4 & \chi_4 & s_4 \end{vmatrix} = - \begin{vmatrix} m_1 & r_1 & \chi_1 & s_1 \\ m_2 & r_2 & \chi_2 & s_2 \\ m_3 & r_3 & \chi_3 & s_3 \\ m_4 & r_4 & \chi_4 & s_4 \end{vmatrix} = -\det(D)$$

The specific rotation's direction is now reversed. Thus, in a racemate, the specific rotation will be zero.

When a molecule has two or more stereogenic carbons, the $\det(D)$ values of each different stereogenic center would be calculated. If a compound has n stereogenic centers, then its net OR will be the sum of the OR of each stereogenic center:

$$\det(D_1) + \det(D_2) + \cdots \det(D_n) = \sum \det(D_i)$$

As the bond angles change, the values of the groups R^1, R^2, and R^3 along the Z-axis change. Thus, the determinant $\det(D_z)$ changes when R^1, R^2, R^3, and R^4 are considered to be perfect point masses. That means the magnitude of OR for a chiral molecule would change when the bond angles about the stereogenic atom changes.

3.3.2.1 Sample Calculations

It is crucial to provide accurate values for the variables (elements) to compute OR. If the variables cannot have suitable magnitudes, no matter how reasonable the deduction is, this model will fail. Therefore, reasonable quantitation of the variables and letting the four variables work well in the matrix model becomes an arduous and important task. The definitions for the four variables and their magnitude computations are described below.

Comprehensive Mass *m* The use of mass has a long history in OR computations [18]. Atoms in different positions within the substituents have different degrees of contribution to the specific rotation. If the atom is directly connected to the stereogenic center, its coefficient is b_1. Coefficients of the other atoms that are further removed are b_2, b_3, and so on. Thus, the comprehensive mass of the substituent group becomes:

$$\mathbf{m} = b_1 m_1 + \sum b_2 m_2 + \sum b_3 m_3 + \sum b_4 m_4 + \cdots$$

Here, $\mathbf{m}_1$ is the mass of atom 1, which is directly connected to the stereogenic center, and $\mathbf{m}_2$ is the mass of atom 2, which is directly bound to atom 1. Since more than one atom can be bonded to atom 1, the summation is used to indicate that the contributions of all these atoms must be included. This is also true for each of the substituent atoms that are one step further removed from the stereogenic center. The term b_1 is the coefficient for $\mathbf{m}_1$, b_2 is the coefficient for $\mathbf{m}_2$, and b_n is the coefficient for $\mathbf{m}_n$.

The calculation of each b_i value is now shown for the *n*-propyl group as a simple example (Figure 3.3). In this model, the coefficients of the masses are related to the reciprocal volume. The values of b_i are defined as proportional to the reciprocal of the volume swept out by that specific mass. Atoms attached further from the stereogenic center sweep out larger net volumes about that stereogenic center. Hence, the value of b_i will decrease when an atom is farther away from the stereogenic center since the reciprocal of the volume decreases. C1 is directly bonded to the stereogenic atom (Figure 3.3), and is at the center of the sphere.

The average unit volume of the C1 atom is

$$\frac{1}{v_1} = \frac{1}{(4\pi r_1^3)/3} = \frac{3}{(4 \times 3.14 \times 0.77^3)} = 0.5232\ (\text{Å}^{-3})$$

where r_1 is the radius of an sp^3-hybridized carbon.

C2 and C3 also sweep out volumes around the C1 center during rotations of both the bond connecting C1 to the stereogenic atom and the C1–C2 bond. Their average unit volumes can be calculated, respectively, as

$$\frac{1}{v_2} = \frac{1}{(4\pi r_2^3)/3} = \frac{3}{(4 \times 3.14 \times 1.54^3)} = 0.06538\ (\text{Å}^{-3})$$

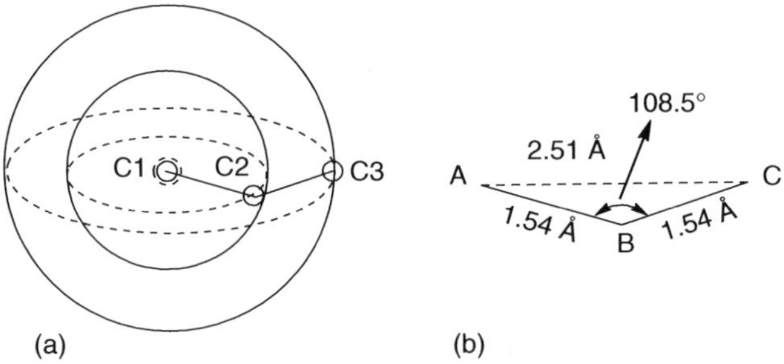

Figure 3.3 *n*-Propyl model used for the calculation of its b_i value. Plot (b) is the amplified and simplified geometry for calculating the distance from C1 to C3 [10].

and

$$\frac{1}{v_3} = \frac{1}{(4\pi r_3^3)/3} = \frac{3}{(4 \times 3.14 \times 2.51^3)} = 0.01516 \quad (\text{Å}^{-3})$$

where r_3 is the distant AC in Figure 3.3.

$$r_3 = \sqrt{(1.54^2 + 1.54^2 - 2 \times 1.54 \times 1.54 \times \cos 108.5)} = 2.51 \ (\text{Å})$$

The b_i values are defined as proportional to the respective calculated reciprocal volumes, that is, $1/v_i$.

$$b_1 : b_2 : b_3 = \frac{1}{v_1} : \frac{1}{v_2} : \frac{1}{v_3} = 0.5232 : 0.06538 : 0.01516 \ = 1 : 0.125 : 0.029$$

Therefore, the mass contributions of C2 and C3 to the specific rotation are 0.125 and 0.029 times that of C1, respectively. This calculation for a carbon bonded to C3 (e.g., C4) results in a value of b_4 that is only 0.008 that of b_1. Different elements have different values of b_i, and the same elements will have different b_i values when they are in different states of hybridization. The values of b_i also change with different conformations. A few example b_i values are calculated and listed in Table 3.3. For example, the comprehensive mass of the CO_2H group is estimated, as shown below, keeping in mind the two oxygen atoms (C=O, C–OH) have to be treated separately. The specific data is summarized in Table 3.4.

$$\mathbf{m} = 12 \times 1 + 0.156 \times 16 \times 2 + 0.114 \times 16 = 18.7$$

Some comprehensive masses of simple groups, calculated in the simple manner described above, are summarized in Table 3.5. There are many ways in which the reciprocal volumes for any given function can be determined. Whatever the method selected, that procedure should be applied consistently to each molecule considered.

Radius *r* The radius is defined as the smallest contact radius of the substituent when that substituent is in a stable conformation. We use the traditional method to obtain their magnitudes. The substituent's *r* value can be estimated by its geometry, or it can be set equal to its Van der Waal's radius. Some example cases are now discussed.

Table 3.4 Sample b_i values calculated for substituents (b_1 = 1.0) [10].

Substituent	b_i	Substituent	b_i
$-n\text{-}C_3H_7$	$b_2 = 0.125$, $b_3 = 0.029$	$-CH_3$	C–H:$b_2 = 0.33$
–Phenyl	$b_2 = 0.125$, $b_3 = 0.024$, $b_4 = 0.0156$	–C(sp^3)–C=C–	$b_2 = 0.152$, $b_3 = 0.0135$
$-CO_2H$	C=O: $b_2 = 0.156$, C–O: $b_2 = 0.114$	–C(sp^3)–O–C(sp^3)	$b_2 = 0.162$, $b_3 = 0.033$
–C(sp^3)–N<	$b_2 = 0.144$ for N	–C(sp^3)–OH	$b_2 = 0.166$
–C(sp^2)=N–	$b_2 = 0.137$ for N	–C(sp^3)–S–	$b_2 = 0.077$

Table 3.5 Values of *m*, *r*, and *s* for some selected substituents [10].

R	*m*	*r* (Å)	χ	*s*
$-CH_3$	13.0	2.0	2.5	0.44
$-C_2H_5$	14.0	2.2	2.5	0
–*n*-Pr	14.6	2.2	2.5	0
–*n*-Bu	14.7	2.2	2.5	0
–*n*-C_5H_{11}	14.9	2.9	2.5	0
c-C_6H_{12}	16.4	2.4	2.5	0
–*i*-Pr	15.6	2.3	2.5	0
$-CMe_3$	16.9	2.8	2.5	0.44
–*i*-Bu	14.9	2.0	2.5	0
–OH	16.3	1.4	3.5	0
–OMe	17.3	2.3	3.5	0
–OAc	19.9	2.9	3.5	0
$-NH_2$	14.7	1.5	3.0	0
–C≡C	16.5	1.5	3.0	0.25
–NCO	17.7	1.5	3.5	0
–CN	17.8	1.45	3.2	0.25
$-CH_2Cl$	15.6	1.8	2.5	0
$-CH_2OH$	15.2	2.1	2.6	0
$-CH_2Br$	17.7	1.95	2.5	0
$-CH_2CN$	18.1	2.9	2.5	0
–NHMe	15.6	2.2	3.0	0
$-NMe_2$	16.6	2.3	3.0	0
$-NH_3^+$	15.0	1.5	3.8	0.44
$-CO_2H$	18.7	1.7	2.7	0
$-CO_2Me$	19.2	2.7	2.8	0
$-CONH_2$	18.5	1.6	2.7	0
2-Furyl	18.1	1.7	2.7	0
–Ph	18.9	1.7	2.7	0.25
4-Cl–Ph	19.0	1.8	2.7	0.25
4-Me–Ph	18.9	1.7	2.7	0
2-Br–Ph	20.1	1.95	2.7	0
1-Naph	20.0	1.7	2.7	0
2-Naph	19.2	1.7	2.7	0
3-Py	18.8	1.7	2.7	0–0.25
4-Py	19.0	1.7	2.7	0.25
$PhCH_2{=}CH-$	15.7	2.0	2.8	0
$PhCH_2CH_2-$	14.7	3.6	2.5	0
$-H_2PO_3$	45.9	2.4	2.1	0
–I	127	2.5	2.5	1
–Br	80.0	1.95	2.8	1
–Cl	35.5	1.8	3.0	1
–F	19.0	1.35	4.0	1
–H	1.0	1.2	2.1	1
–D	2.0	1.2	2.1	1

1) The methyl group's *r* value is taken as that group's Van der Waal's radius (e.g., 2.0 Å) [19]. The radius *r* of an ethyl group is also the group's Van der Waal's radius. This is approximated as the distance from the center of the ethyl group's carbon–carbon single bond to a methyl hydrogen plus an additional 0.8 Å.[1)] For other alkyl groups, the Van der Waal's radii may be estimated by the distance between the group's center of mass and the outer hydrogen plus 0.8 Å [19].
2) The Van der Waal's radii of simple aromatic rings, such as phenyl or furan rings, are approximated as 1.7 Å.
3) If a heavy atom is linked to one or two light atoms, such as in –OH and $-NH_2$, the radius used is simply regarded as the large atom's Van der Waal's radius. For complex substituents, the value of *r* can be calculated by its actual geometry. The radii for selected substituents were calculated and are summarized in Table 3.5.

The effect of a solvent on each substituent is different. Furthermore, the radius of a specific group can change in a different solvent. Chiral molecules, with very small rotation values, may undergo a change in the direction of rotation from positive to negative (or the reverse) as a result of small changes in the value of the substituents' radii. Nevertheless, most chiral compounds have large enough specific rotation values that these small errors are negligible. Thus, it is possible to use either the calculated radii or Van der Waals radii in this matrix model. Polar compounds can have strong intermolecular interactions with each other as their concentration increases, which may induce small changes in the values of the substituents' radii. Thus, the direction of the specific rotation may change if the original value of the rotation was close to zero. In the matrix model, the effects of the solvent, temperature, and concentration are mostly due to changes in the size of the substituents in solution. Holding the solvent and temperature constant eliminates these effects when predicting specific rotations for a series of molecules.

Electronegativity χ The effect of electronegativity on the specific rotation is large. In the matrix model, Pauling electronegativities are used throughout unless stated otherwise. The atom's or group's electronegativity [20] has been used. When the stereogenic atom is carbon, carbon's electronegativity value is used directly. When the stereogenic atom is not carbon, the calculated det(*D*) should be multiplied by the ratio χ_c/χ_y, where χ_c is carbon's electronegativity and χ_y is the stereogenic atom's electronegativity.

Symmetry *s* The symmetry factor *s* represents the space group operation that is widely used to analyze the molecular structure and chirality. Early studies used the term "atomic asymmetry" to calculate and understand OR [1]. Here we quantitate this term "symmetry" and use these values in our model. If one substituent has the

1) Generally, the van der Waal's radius of any atom is about 0.8 Å larger than the same atom's valence radius. We use the average value 0.8 Å as the plus value in the calculation of radius of the substituents.

highest symmetry operation number N, and this symmetric axis passes through the atom that is connected to the stereogenic atom, then the symmetry factor Eq. (3.3) for that substituent is

$$s = \left[\frac{(N-1)}{N}\right]^2 \tag{3.3}$$

The values of s for selected substituents are listed in Table 3.5.

This model needs a chiral molecule having at least one symmetric group for its OR computation, such as a single atom (e.g., H, Cl) or a group (e.g., $-CH_3$, –Ph, $-CCl_3$). If there is no symmetric group connected to the stereogenic center in a chiral molecule, this model would predict its OR values as zero. For example, 5-ethyl 5-propyl-undecane and some glycerides [1] have asymmetric centers. However, their OR values are close to zero. An example may exist in which a chiral molecule has no symmetric group connected to the stereogenic center but has a definite OR value. Therefore, one should be careful to use this model to predict these chiral molecules' ORs.

If the matrix model represents a reasonable method to predict the magnitude and direction of specific rotation, then the value of k_0 should be approximately constant when obtained for different chiral compounds with similar structures in the same solvent (and at the same temperature and light wavelength). This will demonstrate that the value of det(D) is proportional to the magnitude of the specific rotation. The deviation of such k_0 values represents a measure of how well the model works. This premise is tested in Table 3.6, where the calculated values of det(D), the experimental values of specific rotation (using the sodium D line), the solvent, and the calculated k_0 values of series of molecules are summarized.

3.3.2.2 Calculated Values in Same Series of Compounds

Experimental ORs were obtained for eight alcohols (**8–15**) (Table 3.5) and the values of det(D) were calculated. The values of k_0 were determined from the calculated values of det(D) and the experimental values of $[\alpha]_D$. These values (0.59, 0.87, 0.40, 0.35, 0.48, 0.78, 0.64, and 0.76, with the average value of 0.61) are remarkably similar considering the many assumptions used in the model and the use of only four elements (substituent characteristics) as well as the fact that all the experimental $[\alpha]_D$ values were reported by different research groups. The k_0 values of compounds **16–19** measured in methanol are 6.60, 5.73, 5.20, and 6.04, respectively. The k_0 values for the structurally similar series **20–22** are very close to 4.1 (4.20, 4.14, and 3.96). For chiral compounds **23–26** with an extremely small OR, the predicted results are also close to the experimental results except for **23**. The k_0 values for the P-containing chiral compounds **27–29** and dual chiral compounds such as **30–33** are also nearly constants. Among the computed ~100 chiral compounds, compound **25** has a different k_0 value (−0.45) than the others (positive magnitudes). The reason is unknown.

Table 3.6 Calculated values of det(*D*), experimental specific rotation, and the values of k_0 from $[\alpha]_D$/det(*D*) for selected example molecules [10].

Structure number	Structure	det(*D*)	$[\alpha]_D$ (experimental)	k_0
8	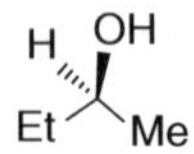	+16.16	+9.57 (neat)	0.59
9	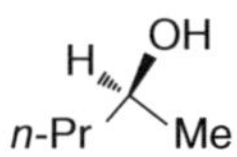	+14.98	+13 (neat)	0.87
10	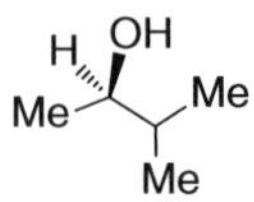	−12.10	−4.9 (neat)	0.40
11	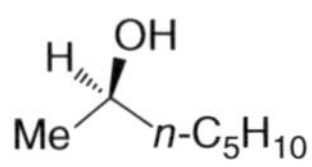	−27.18	−9.5 (neat)	0.35
12	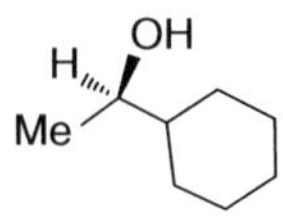	−11.57	−5.6 (neat)	0.48
13	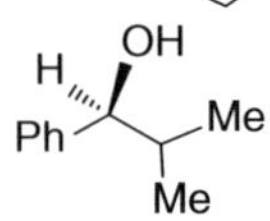	−31.43	−24.6 (neat)	0.78
14	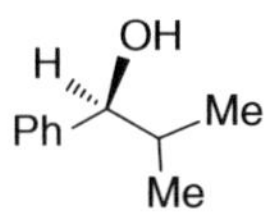	+31.23	+20 (neat)	0.64
15	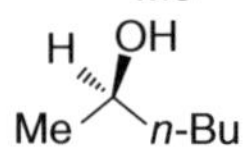	−14.46	−11 (neat)	0.77
16	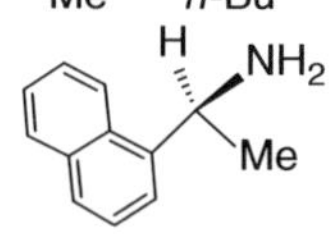	−8.81	−55 (EtOH)	6.24
			−59 (MeOH)	6.60
17	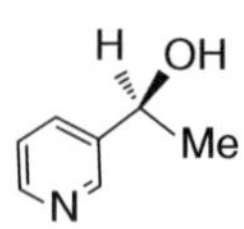	−7.01	−40.2 (MeOH)	5.73
18	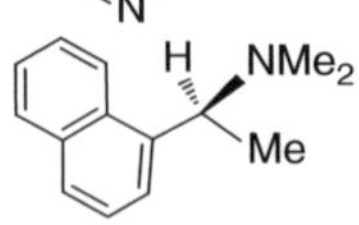	−8.27	−43 (MeOH)	5.20

Table 3.6 (*Continued*)

Structure number	Structure	det(*D*)	$[\alpha]_D$ (experimental)	k_0
19	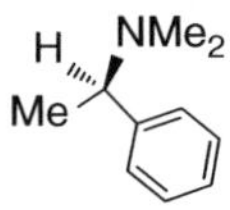	+8.31	+50.2 (MeOH)	6.04
20	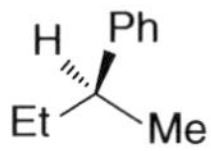	+6.57	+27.6	4.20
21	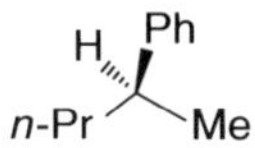	+6.18	+25.6	4.14
22	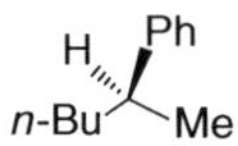	+6.08	+24.1	3.96
23	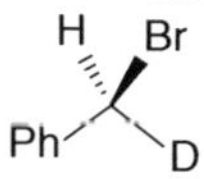	−0.65	+0.0035	−0.05
24	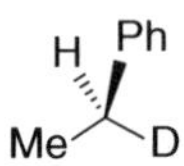	+0.93	+0.0068	0.07
25	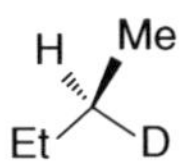	−0.21	+0.0095	−0.45
26	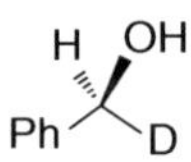	+1.86	+1.58 (c-C_5H_{10})	0.85
27	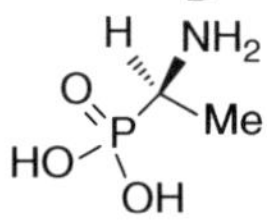	−53.48 ($s_{NH_2} = 0$) −54.51 ($s_{NH_3+} = 0.44$)	−4.8 (H_2O)	0.090 0.088
28	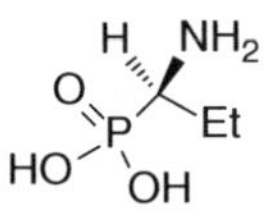	−96.02 ($s_{NH_2} = 0$) −96.67 ($s_{NH_2} = 0.44$)	−16.5 (1 *N* NaOH)	0.17 0.17
29	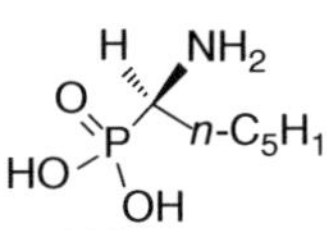	−159.5	−25.0 (NaOH)	0.097
30	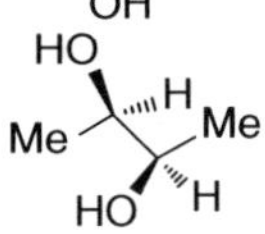	$-7.61 \times 2 = -15.32$	−13 (neat)	0.85

(*continued overleaf*)

Table 3.6 (*Continued*)

Structure number	Structure	det(*D*)	$[\alpha]_D$ (experimental)	k_0
31	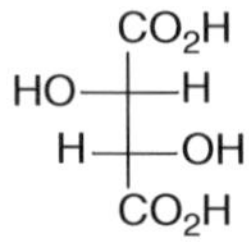	$+7.27 \times 2 = 14.54$	12.4 (H_2O)	0.57
32	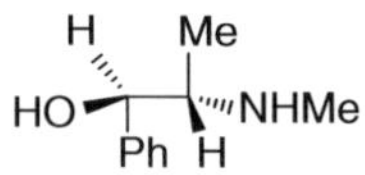 (1*R*,2*S*)	$\det(D_1) = -21.37$ $\det(D_2) = +9.5$ $\Sigma\det(D) = -11.87$	−34.0 (H_2O)	2.86
33	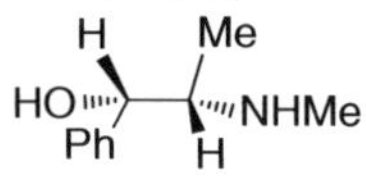 (1*S*,2*S*)	$\det(D_1) = +21.37$ $\det(D_2) = +9.5$ $\Sigma\det(D) = +30.87$	+61.0 (H_2O)	1.98

Some chiral compounds with similar structures have no close k_0 constants in det(*D*) computations. The reason is unknown. However, the range for det(*D*) values is 0.3–7.0. Some det(*D*) magnitudes reach 10 or more. The current structures are of the type >C*H–. For example, when a chiral molecule has a positive det(*D*) value for (*R*)-AC, the determined OR is negative, and the obtained k_0 will be negative. Then it can be judged that the real AC for the specific chiral molecule will be (*S*)-AC.

Once the H atom of >C*H– is replaced by a Me group or a phenyl ring, the rule to use k_0 values to judge AC will be different. If a chiral molecule has a positive det(*D*) value for (*R*)-AC, the determined OR is negative, and the obtained k_0 will be negative. Then, the chiral molecule should be of (*R*)-AC instead of (*S*)-AC. This will be discussed in Section 3.5. These results show that this is a simple and reliable way to make AC assignment for chiral compounds with small OR. This will be discussed later.

This model cannot be used for the AC assignment of a chiral compound that has four groups without any symmetric factor since, once symmetry factor *s* is equal to zero, its det(*D*) equals zero, too. Indeed, if four substituents have no symmetry, the molecular OR is very small, or too small to be detectable. When all conformations are used in the statistics in the matrix, to select one conformation with asymmetric structure is incorrect. However, for a specific conformational OR, the specific data should be used in a matrix; for example, its radius must be a little different from other conformations; and this surely will cause an OR changes.

If there are strong H-bonds in a molecule, it may affect its predictions. This is similar to the other methods. It is still in its infancy, and more tests should be performed.

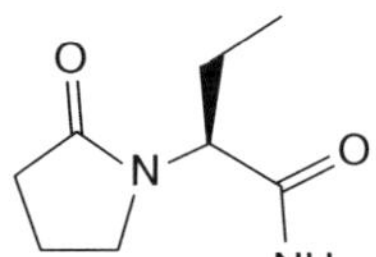

34

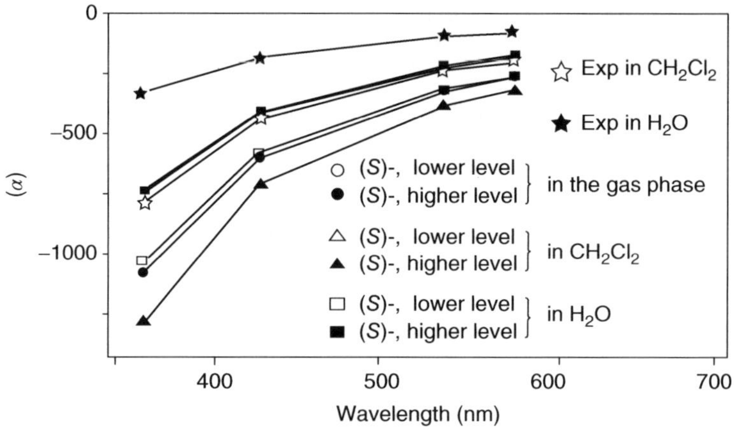

Figure 3.4 Comparison of the calculated ORD of (*S*)-**34** with the experimental ORD [29].

3.4 ORD

As mentioned in Section 3.2, the OR value is the function of the frequency of polarized light. Generally, the wavelength of 589.3 nm of the Na line is used for OR measurements. Determination of a series of OR values across the scanning wavelengths gives the so-called ORD spectrum, which is suitable for those compounds that have no chromophores in the UV–vis region [21, 22]. The advance in quantum mechanical methods in recent years has brought about a renaissance of OR (ORD) in structural elucidation [23–27]. The other usually used wavelengths include 578, 546, 436, and 365 nm. ORD is not as widely used as OR. However, its application is still important in organic stereochemistry.

One example is given here. The chiral levetiracetam **34** [28] obtained marketing authorization from the U.S. Food and Drug Administration (FDA) as an anticonvulsant medication to treat epilepsy and neuropathic pain [29]. Its AC was not determined until a recent ORD study [29]. The predicted ORD was in agreement with experimental data, so the only chiral carbon was assigned as the (*S*) configuration (Figure 3.4).

3.5 Application

Many methods have been developed, such as DFT, CC, CCSD, and others, as mentioned above, and used for OR computations. Some economical methods are introduced here for experimental chemists' use. At the same time, with the development of supercomputer technology, high basis sets can be used for OR computations. Currently, DFT is widely used. The general way to use this method is to optimize the geometries at a relatively high basis sets, such as the 6-311+G(d) level, using different theories, such as B3LYP or B3PW91 and other options set in the Gaussian program. Sometimes, the effect from a solvent may be investigated using the polarizable continuum model (PCM) [29b,c] model or other models such as self-consistent isodensity polarized model (SCI-PCM) or integral equation formalism of polarizable continuum model (IEFPCM). The conformations with lower energy (0–2.0 kcal mol^{-1}) are used for OR calculations at a high level. For example, it could be at the 6-311++G(2d,p) level; or, the 2p orbital should be used rather than just 1p orbital in OR calculations. Solvent effect can also be considered again even if it was investigated in the optimization procedure.

Application of OR in organic stereochemistry has been well investigated. From the small, simple chiral compounds, such as **35** and **36** [30], to the complex natural product **37** [31], it plays an important role. The OR for **37** was recorded as −378, and the calculated OR for (17*S*,18*S*,19*S*) was +339 at the B3LYP/6-311++G(2d,p) level using the B3LYP/6-31+G(d,p)-optimized geometries. Its AC was assigned as (17*R*,18*R*,19*R*).

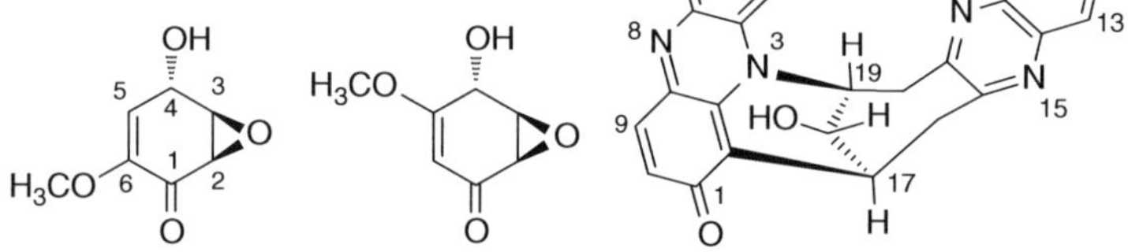

(+)-**35** (−)-**36** (−)-**37**

This method is valid for AC assignment for chiral compounds with single chiral center and a small OR value. For example, a Michal addition product **40** (Eq. (3.4)) [20], had an OR value of only −5.7 (methanol) with ee of 80%. The calculated OR for (*R*)-AC was −9.4 at the B3LYP/6-311++G(2d,2p) level using the IEFPCM model (B3LYP/6-311++G(2d,2p)/IEFPCM) for solvent correction when the M06-2X/6-311++G(2d,2p)/IEFPCM-optimized geometries were used [20]. Its AC was assigned as (*R*).

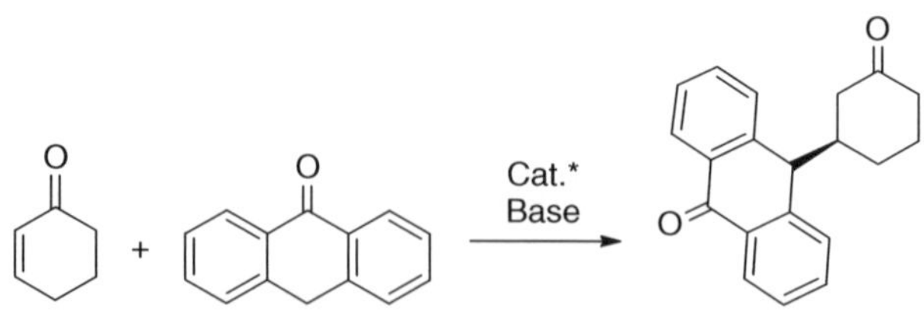

38 **39** **40** (3.4)

3.5.1 AC Assignment for Mono-Stereogenic Center Compounds

Chiral compounds with a single stereogenic center are the simplest. Their AC assignment is simple using computational methods. For example, the chiral concentricolide **41** was assigned as (*S*) [32]. B3LYP/aug-cc-pVDZ//B3LYP/6-31+G(d) (method A) and B3LYP/aug-cc-pVDZ/MP2/6-311+G(d) (method B) were used for it OR computations. Their OR values were −43.7 and −66.1, respectively, using methods A and B. The experimental OR for this structure was −59.2. Similar structures, such as **42, 43** [33], with negative OR values could be predicted as (*S*)-AC, too. The conclusion was further confirmed by B3LYP/6-31+G(d,p)/PCM//B3LYP/6-31+G(d,p) (method C) and PCM/B3LYP/6-31++G(d,p)//B3LYP/6-31+G(d,p) (method D).

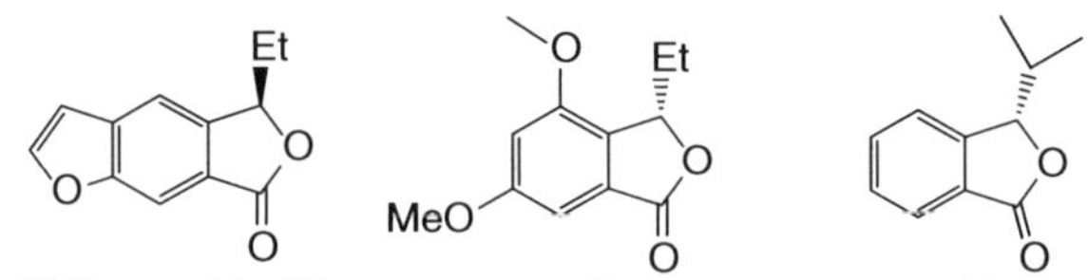

	41 Concentricolide	**42**		**43**
$[\alpha]_{D\ exp}$:	−59.2	−98.8		−64.6
Method A	−43.7	−102.0	Method C	−97.5
Method B	−66.1	−120.8	Method D	−85.6

The AC assignment for the simple cyclic chiral compounds can utilize quantum methods. Only two possible choices exist for this type of compounds, either (*R*) or (*S*). For some cyclic chiral compounds, if the side chain is not long enough, quantum methods are also used for OR computations. For example, (−)-cephalosol (**44**) has an OR of −96.2, and the predicted OR for (*R*) was 78.9 using DFT theory at the B3LYP/aug-cc-pVDZ/B3LYP/6-31G(d) level. Its AC was assigned as (*S*) [34]. Other two spiral chiral examples **45** and **46** were well assigned using DFT theory [35]. The OR for (+)-**45** was +352, and the predicted OR for the current AC was +285 at the B3LYP/6-31G(d) level. OR of −150 for **46** was measured, whereas the calculated OR for the current AC was −327 at the B3LYP/6-31G(d) level or −365 at the B3LYP/aug-cc-pVDZ level.

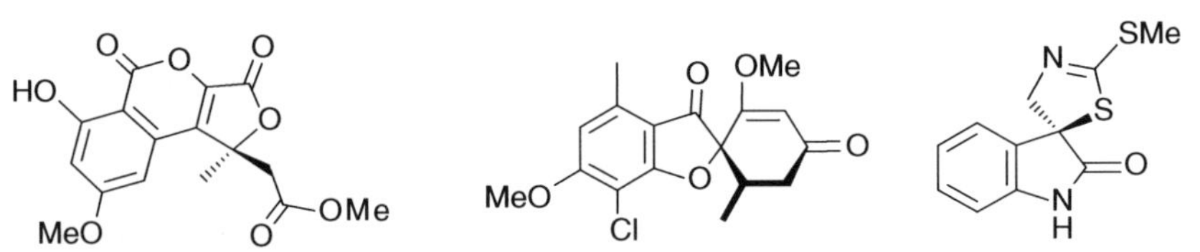

44 (−)-Cephalosol **45** (+)-Griseofulvine **46** (−)-Spirobrassinin

$[\alpha]_{D\ exp}$ −96.2, $[\alpha]_{D\ calcd.}$ −78.9

3.5.2
Matrix Model Application

Acyclic chiral compounds are very important in organic stereochemistry study. As mentioned above, because of the huge number of stable conformations, to predict OR using quantum theory is extremely difficult in reality. Matrix model provides another viewpoint to understand it: for example, two phenylpropanoids, tatarinoids A (47) and B (**48**), and a trinorlignan, tatarinoid C (**49**), from the rhizome of *Acorus tatarinowii* [36]. Their ORs were recorded as −6.85, −4.77, and −10.16, respectively, in $CHCl_3$.

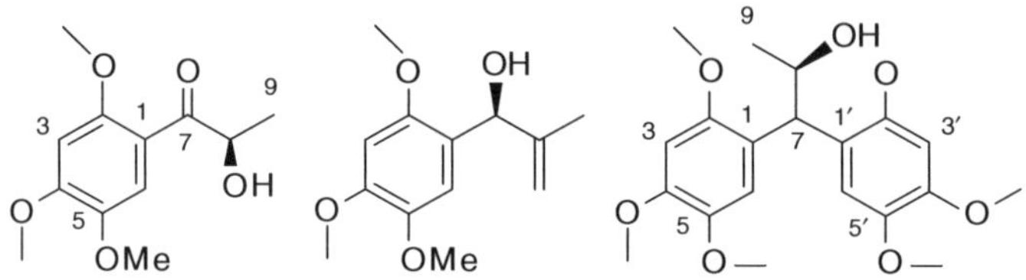

47 **48** **49**

The calculated det(*D*) value for (*R*)-**47** was −8.96. The experimental OR was −6.85 in $CHCl_3$. The calculated k_0 value was +0.76 (= −6.85/(−8.96)), indicating an (*R*)-AC for C-8. (*R*)-**47** was further optimized at the B3LYP/6-31+G(d) level after approximately 40 stable conformations with relative energies from 0 to 2.5 kcal mol^{-1} were found. To reduce the computational time, a total of 22 geometries with relative energy of 0–2.0 kcal mol^{-1} were used for the computation of the specific rotation at the B3LYP/aug-cc-pVDZ level. The predicted specific rotation value was −24.2 when the B3LYP/6-31+G(d)-optimized energy was used. An OR of −23.1 was achieved when single-point energy (SPE) at the B3LYP/aug-ccpVDZ level was used for OR simulations. The sign of the computed OR is negative, which agrees with the sign of there recorded result.

Similarly, the calculated det(*D*) value for (*R*)-**48** was −6.89. The computed k_0 was +0.69, which was close to that of **47**(+0.76), indicating that the AC of **48** was (*R*). The calculated det(*D*) value for the (*R*)-**49** was −20.80, and the measured value was −10.16 in $CHCl_3$. The computed k_0 was +0.50. The k_0 value is also close to the value for **47**(+0.76) and **48**(+0.69).

The atoxic amino acid 2-amino-4-hydroxy-5-hexynoic acid (**50**) was obtained from the mushroom *Trogia venenata* [37]. The AC for **50** was determined as (2*R*,4*S*) by both matrix and OR computations. The calculated det(*D*) for C2(det(D_2)) was −13.32 when it had (*R*) configuration, and that for C4 was +24.21 when it had (*S*) configuration. Thus, the sum of the det(*D*) for (2*R*,4*S*) was −10.89, and k_0 was obtained as 1.21 by division of the observed OR values (−13.2) by the sum of det(*D*). If it had the (2*R*,4*R*) configuration, the sum of det(*D*) would have been −37.53, and k_0 should be 0.35. This k_0 is quite small, although this is located in the reasonable range 0.3–6. However, it is more likely to have (2*R*,4*S*)

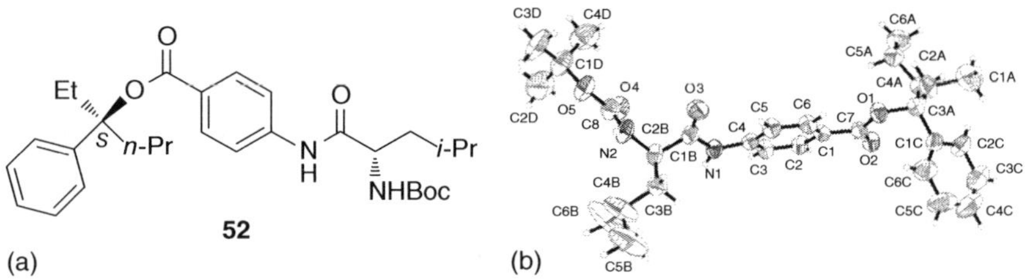

Figure 3.5 (*S*)-**52** (a) and its X-ray structure (b) [38].

AC since its k_0 was 1.21, which lies in a more reasonable range. This result was further confirmed by total synthesis.

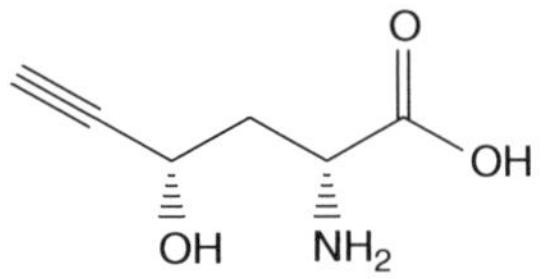

50

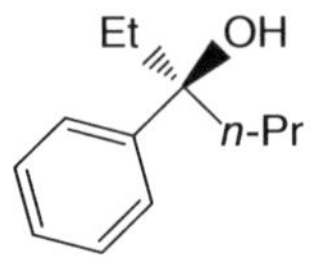

51

When the smallest group is a phenyl ring instead of a proton, like the chiral tertiary alcohol **51** [38], the AC for **51** was well determined by converting **51** to **52** that has a standard stereogenic center (natural chiral amino acid) and can be assigned by X-rays. The determinant det(D) for (S)-**51** was −0.42. The k_0 value was −27.6. The k_0 value is negative. This is different from the rule used for the AC assignment of chiral secondary alcohols or amines. For tertiary alcohols or amines, such as **52** (Figure 3.5), a phenyl ring is used to replace the proton [38]; if the predicted sign of k_0 is positive, the real AC should be the enantiomer's AC. Only when the k_0 is negative, the AC that is used for det(D) calculations should be the same as the real AC for the chiral compound.

When a phenyl ring is replaced by a methyl group, the conclusion is the same: for example, **53**–**55** [39]. The smallest group became –Me. Their ORs were measured as +4.4, +8.2, and +5.8, respectively, in chloroform. The predicted det(D) for (R)-**53**, (R)-**54**, and (R)-**55** were +9.48, +11.21, and +22.97, and the corresponding computed k_0 values were 0.46, 0.65, and 0.47, respectively. As mentioned above, if k_0 value for a tertiary alcohol is positive, the real AC should be the predicted enantiomer's AC. In this case, (R)-AC was used, thus, the real AC for C7 in compounds **53**–**55** should have (S)-AC. Here, it is found that for a similar series of chiral compounds, their k_0 values should be nearly a constant. In this series, it tends to a value of 0.53.

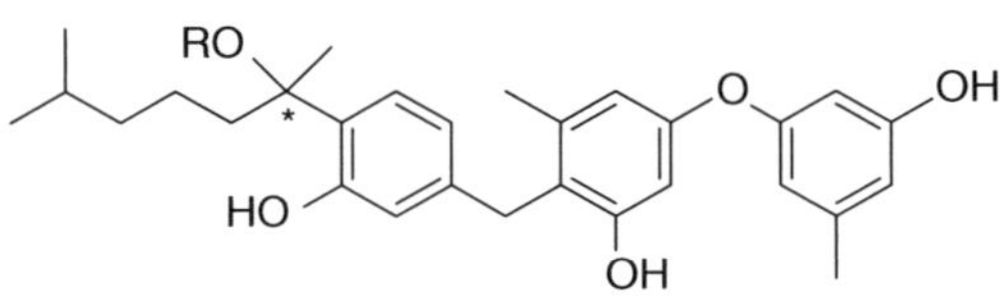

53 R = Me; 54 R = H

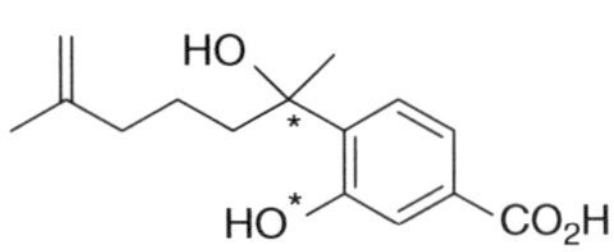

55

3.5.3 AC Assignment for Poly-Stereogenic Center Compounds

In this topic, one case is where the relative configuration (RC) is well established by 2D NMR, or X-ray study via Mo-radiation. Thus, the way to assign its AC is just to compute the OR values for any enantiomer of this structure. When the computed OR has the same sign as the experimental one, one can predict that the AC for the target molecule is the same as that used for OR computations. The OR of griseusin F (**56**) [40] was computed as −145.1 at the B3LYP/6-311++G(2d,p) level. The recorded OR for **56** was +109.0, which is close to the computed value (−145.1). However, the OR signs (negative vs positive) opposite. The proposed structure of **56** must therefore have the AC of (2*R*,2a*S*,8a*S*,9*R*,10*R*,3′*R*,4′*S*,6′*R*).

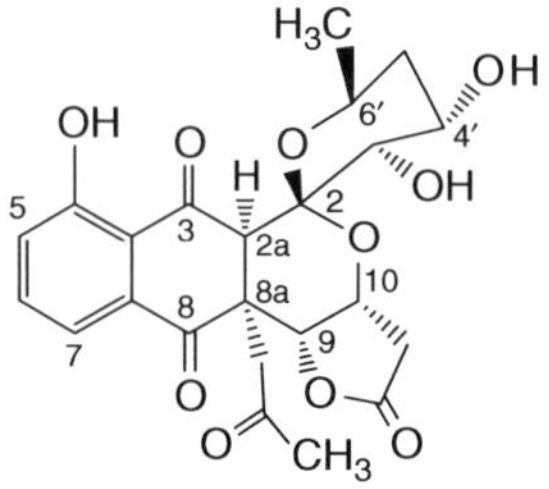

56

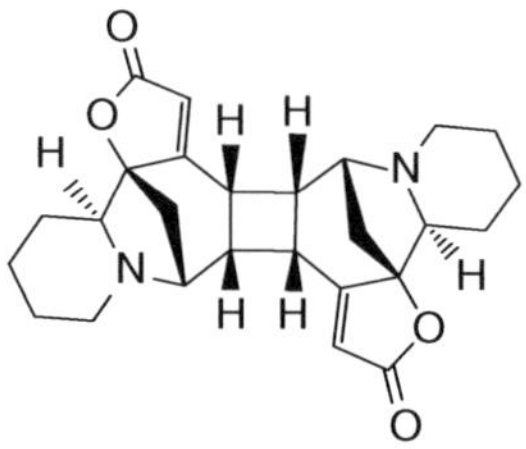

57a

However, not every chiral compound's RC can be well assigned. In this case, different methods should be used for excluding some possible structures. For example, flueggedine (**57**) is a highly symmetrical [2+2] cycloaddition indolizidine alkaloid dimer [41]. NMR experiments showed that it had three possible structures **57a–57c**. Structure **57a** should be excluded because of the great relative chemical shift error (up to 10.2 ppm) at C-15 (C-15′) by computing its ^{13}C NMR. The OR values of **57b** and **57c** were then calculated using the corresponding lower energy conformations at the B3LYP/6-311++G(2d,p) level in the gas phase. The OR value calculated for **55c** was −38.8, which is close to the recorded OR of −33.5. In contrast, the calculated OR magnitude for **57b** was +62.5, which is far away from the experimental result; especially, the opposite OR sign predicted for **57b** allowed the exclusion of this structure. Thus, the structure of **57** was determined as **57c**. Its AC is (2(2′)*S*, 7(7′)*R*, 9(9′)*R*, 14(14′)*S*, and 15(15′)*S*).

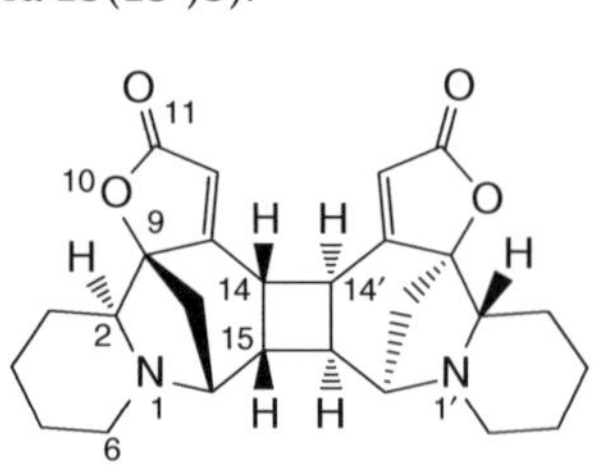

57b

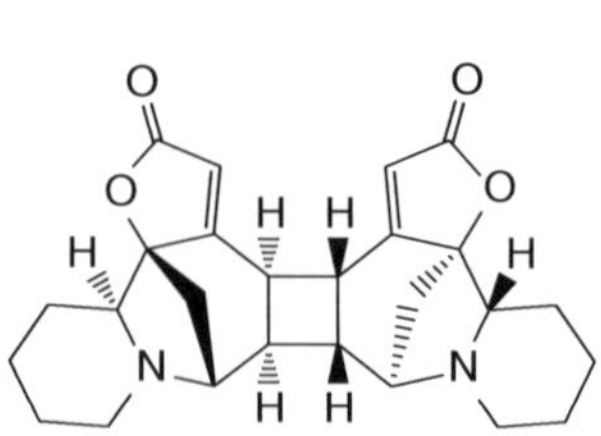

57c

Table 3.7 Computed OR values for (2*R*,7*S*,20*S*,21*S*)-**58** with four wavelengths and their final Boltzmann sum [42].

	$[\alpha]_D$	$[\alpha]_{578}$	$[\alpha]_{546}$	$[\alpha]_{436}$
$[\alpha]_{Dcalcd}$	+42.75	+45.58	+55.41	+137.33
$[\alpha]_{Dexp}$.	+21.63	+23.95	+28.33	+73.40

Table 3.8 Calculated ORD for (4*R*,10*S*)-**60** and (4*R*,10*R*)-**60** using DFT methods [43].

Wavelength (nm)	$[\alpha]_{exp.}$ [a)]	$[\alpha]_{calcd}$ [b)]	
		(4*R*,10*S*)-60	**(4*R*,10*R*)-60**
589	−40.4	−36.3	−260.7
546	−52.0	−50.4	−341.6
435	−227.6	−198.6	−980.9
405	−453.2	−394.0	−1616.6

a) ORs recorded in Et_2O (~0.25 g 100 ml^{-1}).
b) ORs obtained as Boltzmann averages calculated at the B3LYP/aug-cc-pVDZ level employing the Et_2O implicit IEFPCM solvation model.

3.5.4
Using ORD Method

ORD is still a useful method for AC assignment. In some cases, the RCs are known using 2D NMR methods or X-rays (using Mo radiation). For example, the OR magnitudes for schizozygine (**58**) were calculated at the B3LYP/aug-cc-pVDZ//B3PW91/TZ2P at different wavelengths [42]. Obviously, its AC is assigned as (4*R*,10*S*) based on the ORD data. The OR values calculated with different wavelengths are summarized in Table 3.7.

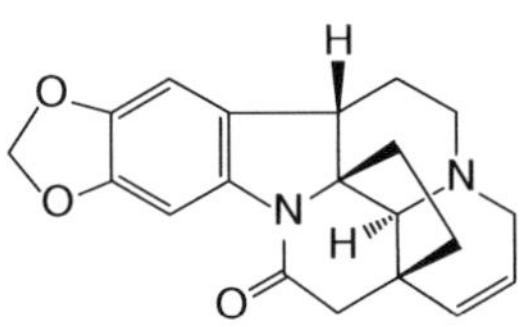

(2*R*,7*S*,20*S*,21*S*)
58 Schizozygine

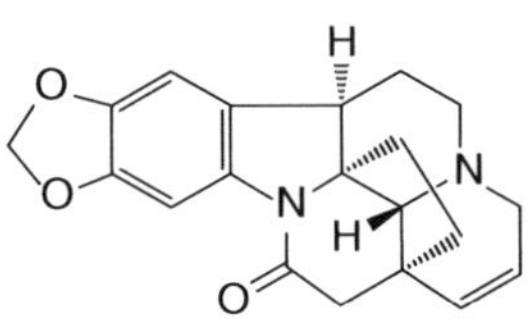

59

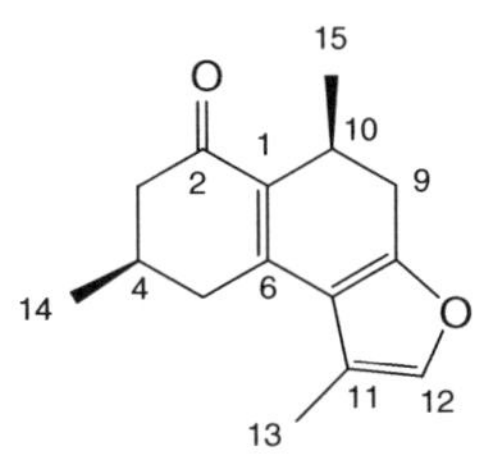

60

Another example is the AC identification of agarsenone (**60**), a new cadinane sesquiterpenoid [43]. Its AC could be (4*R*,10*S*) or (4*R*,10*R*). By ORD calculations using DFT methods, its AC was assigned as (4*R*,10*S*) (Table 3.8).

References

1. (a) Brewster, J.H. (1961) *Tetrahedron*, **13**, 106; (b) Brewster, J.H. (1959) *J. Am. Chem. Soc.*, **81**, 5475; (c) Brewster, J.H. (1959) *J. Am. Chem. Soc.*, **81**, 5483; (d) Brewster, J.H. (1959) *J. Am. Chem. Soc.*, **81**, 5493.
2. (a) Lowry, T.M. (1937) *Optical Rotatory Power*, Dover, New York, pp. 1–24. (b) Frankland, P.P. (1897) *J. Chem. Soc.*, **71**, 683–743; (c) Tinoco, I. Jr., and Woody, R.W. (1963) *J. Chem. Phys.*, **40**, 160; (d) Satyanarayana, B.K. and Stevens, E.S. (1987) *J. Org. Chem.*, **52**, 3170; (e) Gould, R.R. and Hoffmann, R. (1970) *J. Am. Chem. Soc.*, **92**, 1813.
3. Polavarapu, P.L. (1997) *Mol. Phys.*, **91**, 551–554.
4. Yabana, K. and Bertsch, G.F. (1999) *Phys. Rev. A*, **60**, 1271–1279.
5. Cheeseman, J.R., Frisch, M.J., Delvin, F.J., and Stephens, P.J. (2000) *J. Phys. Chem. A*, **104**, 1039.
6. Ruud, K. and Helgaker, T. (2002) *Chem. Phys. Lett.*, **352**, 533–539.
7. Crawford, T.D. and Schaefer, H.F. (2000) in *Review in Computational Chemistry*, vol. 14, Chapter 2 (eds K.B. Lipkowitz and D.B. Boyd), VCH Publisher, New York, pp. 33–136.
8. (a) Kondru, R.K., Wipf, P., and Beratan, D.N. (1998) *J. Am. Chem. Soc.*, **120**, 2204; (b) Grimme, S., Furche, F., and Ahlrichs, R. (2002) *Chem. Phys. Lett.*, **361**, 321; (c) Kongsted, J., Pedersen, T.B., Strange, M., Osted, A., Hansen, A.E., Mikkelsen, K.V., Pawlowski, F., Jorgensen, P., and Hättig, C. (2005) *J. Chem. Phys.*, **401**, 385–392; (d) Giorgio, E., Viglione, R.G., Zanasi, R., and Rosini, C. (2004) *J. Am. Chem. Soc.*, **126**, 12968–12976; (e) Freedman, T.B., Cao, X., Dukor, R.K., and Nafie, L.A. (2003) *Chirality*, **15**, 743; (f) Wiberg, K.B., Wang, Y.G., Vaccaro, P.H., Cheeseman, J.R., and Luderer, M.R. (2005) *J. Phys. Chem. A*, **109**, 3405–3410; (g) Helgaker, T., Ruden, T.A., Jorgensen, P., Olsen, J., and Klopper, W. (2004) *J. Phys. Org. Chem.*, **17**, 913; (h) Mrchesan, D., Coriani, S., Forzato, C., Nitti, P., and Pitacco, G. (2005) *J. Phys. Chem. A*, **109**, 1449–1453; (i) Rinderspacher, B.C. and Schreiner, P.R. (2004) *J. Phys. Chem. A*, **108**, 2867–2870; (j) Krykunov, M. and Autschbach, J. (2005) *J. Chem. Phys.*, **123**, 114103.
9. Gaussian *http://www.gaussian.com* (accessed 27 December 2014).
10. Zhu, H.J., Ren, J., and Pittman, C.U. Jr., (2007) *Tetrahedron*, **63**, 2292–2314.
11. (a) Condon, E.U., Alter, W., and Eyring, H. (1937) *J. Chem. Phys.*, **5**, 753; (b) Rosenfield, L. (1928) *Z. Phys.*, **52**, 161; (c) Inamoto, N. and Masuda, S. (1982) *Chem. Lett.*, 1003.
12. Amos, R.D. (1982) *Chem. Phys. Lett.*, **87**, 23.
13. Helgaker, T., Ruud, K., Bak, K.L., Jorgensen, P., and Olsen, J. (1994) *Faraday Discuss.*, **99**, 165.
14. (a) Condon, E.U. (1937) *Rev. Mod. Phys.*, **9**, 4; (b) Andzelm, J., and Wimmer, E. (1992) *J. Chem. Phys.*, **96**, 1280–1303; (c) Petersson, G., and Al-Laham, M.A. (1991) *J. Chem. Phys.*, **94**, 6081–6090.
15. Stephens, P.J., Mccann, D.M., Cheeseman, J.R., and Frisch, M.J. (2005) *Chirality*, **17**, 52–64.
16. (a) Huang, X., He, J., Niu, X., Menzel, K.D., Dahse, H.M., Grabley, S., Fiedler, H.P., Sattler, I., and Hertweck, C. (2008) *Angew. Chem. Int. Ed.*, **47**, 3995–3998; (b) Pezzuto, J.M., Lea, M.A., and Yang, C.S. (1976) *Cancer Res.*, **36**, 3647–3653.
17. Li, Q.M., Ren, J., Shen, L., Bai, B., Li, Q.M., Liu, X.C., Wen, M.L., and Zhu, H.J. (2013) *Tetrahedron*, **69**, 3067–3074.
18. (a) Brown, C. (1890) *Proc. R. Soc. Edinb.*, **17**, 181; (b) Eliel, E.L., Wilen, S.H., and Mander, L.N. (1994) *Stereochemistry of Organic Compounds*, John Wiley & Sons, Inc., New York, pp. 126–144.
19. Xie, Y.C. and Shao, M.C. (1979) *Structural Chemistry*, Advantage Education Publisher (Book B), p. 205.
20. Wysocki, J., Kwit, M., and Gawronski, J. (2012) *Chirality*, **24**, 833–839.
21. McCann, D.M. and Stephens, P.J. (2004) *J. Org. Chem.*, **69**, 8709.
22. Kong, L.Y. and Wang, P. (2013) *Chin. J. Nat. Med.*, **11**, 193.

23. Mazzeo, G., Giorgio, E., Zanasi, R., Berova, N., and Rosini, C. (2010) *J. Org. Chem.*, **75**, 4600.
24. Lattanzi, A., Scettri, A., and Zanasi, R. (2010) *J. Org. Chem.*, **75**, 2179.
25. Polavarapu, P.L., Scalmani, G., Hawkins, E.K., Rizzo, C., Jeirath, N., Ibnusaud, I., Habel, D., Nair, D.S., and Haleema, S. (2011) *J. Nat. Prod.*, **74**, 321.
26. Li, X.N., Zhang, Y., Cai, X.H., Feng, T., Liu, Y.P., Li, Y., Ren, J., Zhu, H.J., and Luo, X.D. (2011) *Org. Lett.*, **13**, 5896.
27. Fu, G., Doerksen, R.J., and Xu, P. (2011) *J. Mol. Struct.*, **987**, 166.
28. Li, L. and Si, Y.K. (2011) *J. Pharm. Biomed. Anal.*, **56**, 465.
29. (a) Patsalos, P.N. (2000) *Pharmacol. Ther.*, **85**, 77; (b) Tomasi, M., Mennucci, R., and Cammi, R. (2005) *Chem, Rev.*, **105**, 2999–3094; (c) Polavarapu, P.L., Jeirath, N., and Walin, S. (2009) *J. Phys. Chem. A.*, **113**, 5423–5431. Other similar references were cited there.
30. Benedetta, M., Michele, C., Antonio, E., and Carlo, R. (2007) *J. Org. Chem.*, **72**, 6680–6691.
31. Ding, Z.G., Li, M.G., Ren, J., Zhao, J.Y., Huang, R., Wang, Q.Z., Cui, X.L., Zhu, H.J., and Wen, M.L. (2011) *Org. Biomol. Chem.*, **9**, 2771–2776.
32. Ren, J., Jiang, J.X., Li, L.B., Liao, T.G., Tian, R.R., Chen, X.L., Jiang, S.P., Pittman, C.U. Jr., and Zhu, H.J. (2009) *Eur. J. Org. Chem.*, **23**, 3987.
33. Li, L. and Si, Y.K. (2012) *Chirality*, **24**, 987.
34. Huang, W.Y., Chen, J.R., Yan, W.Z., Xie, D.Q., and Tan, R.X. (2008) *Chem. Eur. J.*, **14**, 10670–10674.
35. (a) For **45**; (2001) The Merck Index. 13th ed. Whitehouse Station NJ: Merck and Co Inc; p 290 and 810; For **46**: (b) Suchy, M., Kutschy, P., Monde, K., Goto, H., Harada, N., Takasugi, M., Dzurilla, M., and Balentova, E. (2001) *J. Org. Chem.*, **66**, 3940–3947.
36. Tong, X.G., Wu, G.S., Huang, C.G., Lu, Q., Wang, Y.H., Long, C.L., Luo, H.R., Zhu, H.J., and Cheng, Y.X. (2010) *J. Nat. Prod.*, **73**, 1160–1163.
37. Zhou, Z.Y., Shi, G.Q., Fontaine, R., Wei, K., Feng, T., Wang, F., Wang, G.Q., Qu, Y., Li, Z.H., Dong, Z.J., Dong, Z.J., Yang, Z.L., Zeng, G., and Liu, J.K. (2012) *Angew. Chem. Int. Ed.*, **51**, 2368.
38. Liao, T.G., Ren, J., Fan, H.F., Xie, M.J., and Zhu, H.J. (2008) *Tetrahedron: Asymmetry*, **19**, 808–815.
39. Lu, Z.Y., Zhu, H.J., Fu, P., Wang, Y., Zhang, Z.H., Lin, H.P., Liu, P.P., Zhuang, Y.B., Hong, K., and Zhu, W.M. (2010) *J. Nat. Prod.*, **73**, 911–914.
40. Ding, Z.G., Zhao, J.Y., Li, M.G., Huang, R., Li, Q.M., Cui, X.L., Zhu, H.J., and Wen, M.L. (2012) *J. Nat. Prod.*, **75**, 1994.
41. Zhao, B.X., Wang, Y., Li, C., Wang, G.C., Huang, X.J., Fan, C.L., Li, Q.M., Zhu, H.J., Chen, W.M., and Ye, W.C. (2013) *Tetrahedron Lett.*, **54**, 4708.
42. Stephens, P.J., Pan, J.J., and Devlin, F.J. (2007) *J. Org. Chem.*, **72** (20), 2508–2524.
43. Stefano, S., Stefano, S., Federica, M., Ernesto, S., Ornelio, R., Claudio, S., and Carla Marcotullio, M. (2013) *J. Nat. Prod.*, **76**, 1254–1259.

4
Electronic Circular Dichroism

Theoretically, the absorption intensity of right or left circularly polarized lights (CPLs) by any one chiral compound, like (*R*)- and (*S*)-enantiomer, is different, and this difference ($\Delta\varepsilon$) can be defined as molar electronic circular dichroism (ECD), as shown below:

$$\Delta\varepsilon = \varepsilon_\mathrm{L} - \varepsilon_\mathrm{R} \tag{4.1}$$

where ε_L and ε_R are molar absorption coefficients for left- and right-CPLs, respectively.

In ECD study, the wavelength of CPL used frequently is 190–800 nm, and mostly in organic chemistry, it covers the range of 190–450 nm. This method has been widely used since the first commercial ECD equipment was brought out in about 1960s. In the early days, the use of ECD was mostly focused on comparison of the unknown product's ECD spectrum with that a known compound to assign the absolute configuration (AC) of the unknown product.

Circular dichroism (CD) exciton chirality analysis was also applied for the AC assignment of various chiral compounds [1–4], in particular for the AC assignment of chiral diols. One compound that has a suitable chromophore, such as 4-dimethyl amino benzoic acid, needs to react with a hydroxyl or an amine group of the target molecule if this target chiral molecule has no suitable near the neighbor hydroxyl groups (or amine groups). The UV–vis (or UV) absorption region of the chromophore should be different from that of the target chiral molecule.

After the introduction of the quantum methods, many theories have been applied for the AC assignment for chiral compounds by comparing the experimental ECD with the theoretical value. This has brought modern chemists great benefits in stereochemistry studies.

4.1
Exciton Chirality CD

Exciton chirality CD is based on the UV absorption difference between two aromatic rings (chromophores) in ECD measurement, Nakanishi and others well studied the methods used in AC assignment for chiral organic compounds [1–4]. As a traditional example, this can be explained using 1,2-diol dibenzoate. The

Organic Stereochemistry: Experimental and Computational Methods, First Edition. Hua-Jie Zhu.

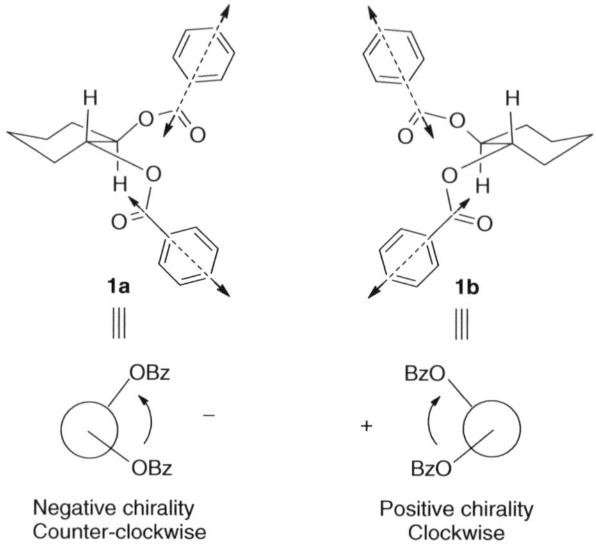

Figure 4.1 The diagrammatic sketch of the chirality of 1,2-diol-dibenzoate [3, 4].

two chromophores (phenyl ring and its conjugated >C=O) that have two strong absorption of $\pi-\pi^*$ are located near the two stereogenic centers. The direction of the electronic transition dipole moment of the two chromophores can be counterclockwise or clockwise. The counterclockwise direction can lead to a negative chirality and the clockwise direction would produce a positive chirality (Figure 4.1).

The interaction between the two chromophores of $\pi-\pi^*$ is called exciton coupling. The energy of the excited states from the two identical chromophores should be different because of the coupling. This is the Davydov (energy) splitting. The difference between the two split energy levels is denoted $\Delta\lambda$ (Figure 4.2), and it leads to two ECD Cotton effects with reversed signs (+ or −).

The strength of optical rotatory dispersion (ORD), ECD, and UV observed should be the sum of the strengths of those variables from the two phenyl rings when a CPL passes through chiral matter. Because of the different directions (+ or −) in ORD and ECD, the sum of the strengths of both ORD and ECD should be different via the wavelength in nanometers (Figure 4.2, up and middle).

There are three signals in the observed ORD curves, whereas generally only two signals exist in ECD corresponding to the two exciton couplings in the system (Figure 4.2 up and middle). This is the direct relationship between the ORD and ECD found in the early studies in stereochemistry.

The relationship between the negative chirality and its Cotton effect in ECD spectra is summarized in Table 4.1 [3]. In positive chirality, the clockwise rotation vector of the electronic transition dipole moment leads to a positive Cotton effect at long wavelengths (the first Cotton effect) and a negative Cotton effect in short wavelengths (the second Cotton effect). The counterclockwise rotation vector of electronic transition dipole moment would cause a positive Cotton effect at short wavelengths and a negative Cotton effect at long wavelengths (Table 4.1). Based on this rule, it is possible to assign the AC of 1,2-diols after they are converted into

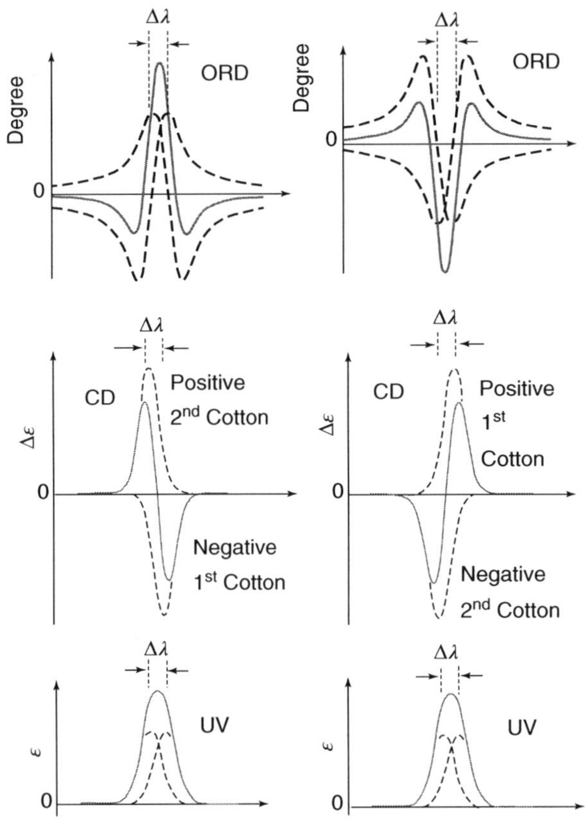

Figure 4.2 The diagrammatic sketch for simulated ORD, ECD, and UV curves for 1,2-diol dibenzoate. The dashed lines are the ORD, ECD, and UV signals generated by the two chromophores (majorly the phenyl rings in this example) interactions (exciton interactions); the grey lines are the sum of both signals from the two chromophores [3].

the corresponding dibenzoates. This method could be used for the AC assignment of other chiral compounds.

Other analogs of benzoic acid, such as 4-dimethyl amino benzoic acid, 4-Cl-benzoic acid, and 4-MeO-benzoic acid, could be used as chromophores in AC assignment for some chiral compounds. Among the analogs, 4-dimethylamino-cinnamic acid (**2**) is used widely [4]. Other aromatic reagents such as **3** and **4** are also developed for AC assignments [4].

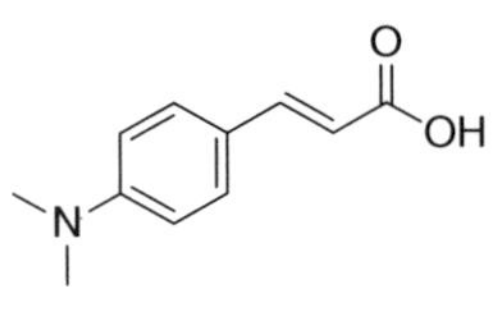

2 UV: 360 nm (31 000)

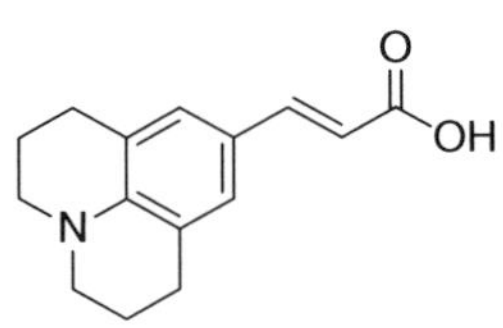

3 UV: 382 nm (27 000)

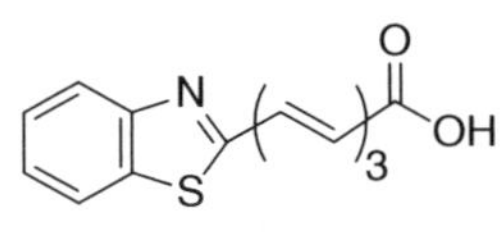

4 UV: 358 nm (58 000)

The key point to correctly use this method for AC assignment is to correctly judge the vector directions of the electronic transition dipole moment of the

Table 4.1 Relationship between positive or negative chirality and their Cotton effects [4].

Chirality	**Rule**	**Long wavelength**	**Short wavelength**
Positive (+)	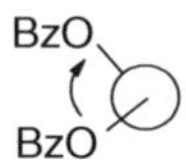	The first positive Cotton effect	The second negative Cotton effect
Negative (−)	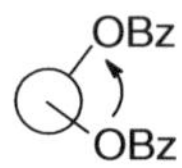	The first negative Cotton effect	The second positive Cotton effect

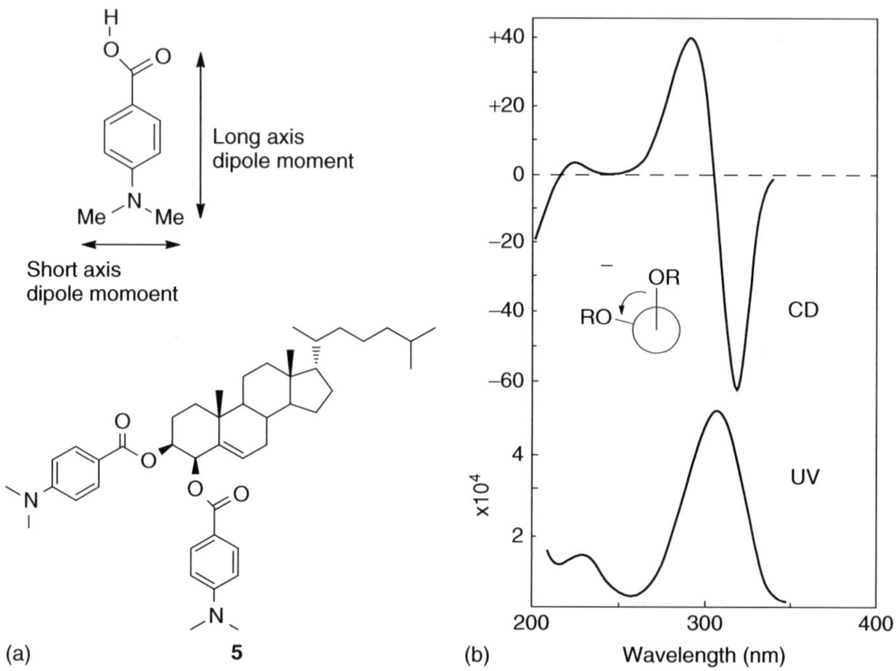

Figure 4.3 The long- and short-axis dipole moments and their vector directions for 4-dimethyl amino benzoic acid (a) and its use as excitons for AC assignment of **5** (b) [4].

two chromophores. For example, there are two transition dipole moments in the derivative of 4-dimethyl amino benzoic acid. The first one is the electronic transition dipole moment in the long axis, and it has a λ_{max} of 309 nm with an ε value of 30 400. The second one is the short-axis moment with a λ_{max} of 229 nm and an ε value of 7200. The first one easily forms exciton coupling between two similar excitons while the second one does not easily. Therefore, the long-axis electronic transition dipole moment should be selected as the vector of the dipole moment in practice (Figure 4.3). For example, the dibenzoate **5** had a negative Cotton effect at about 320 nm and a positive Cotton effect near 290 nm. This agrees with its 3,4-diol configuration. It takes an anticlockwise direction.

Therefore, to determine the long-axis polarized electric transition dipole moment is the key step in the use of exciton chirality method [4]. If this step is wrong, the AC assignment may not be correct. Two frequently used substituents and their long-axis dipole moment directions are listed below.

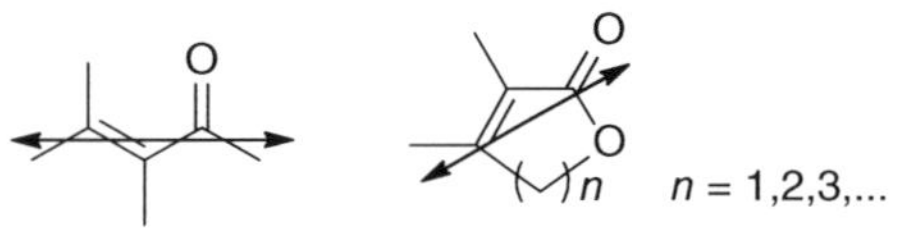

For example, the experimental ECD for the chiral acetate of 7-oxocholest-5-en-3β-ol (**6**) is illustrated in Figure 4.4. After transferring it to the corresponding benzoic acid ester, the examined ECD for **7** is superimposed together.

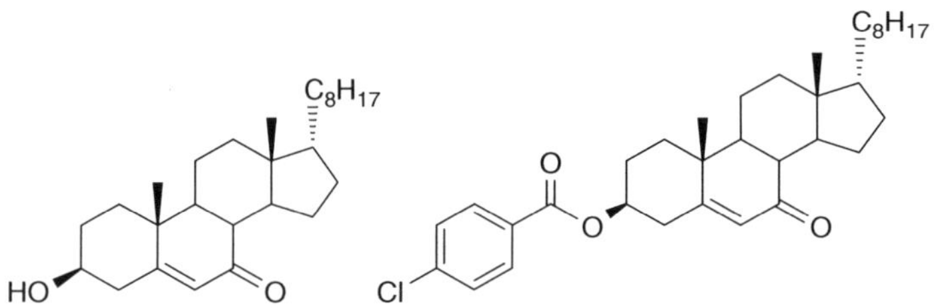

6 **7**

Based on the conclusion listed in Table 4.1, the exciton chirality of C3 is positive (Figure 4.4). The first Cotton effect at long wavelength is positive. Therefore, the prediction agrees well with the experimental results.

Axial chirality can be discussed using exciton chirality. A bisignate ECD Cotton effect couplet centered near 275 nm comes from the exciton coupling between two extended asymmetric aromatic chromophores [5]. They directly correlate with their helical twist. For example, diaporthemin A (**8**) was obtained from *Diaporthe melonis* [6]. Its ECD was measured, which is illustrated in Figure 4.5. It has a

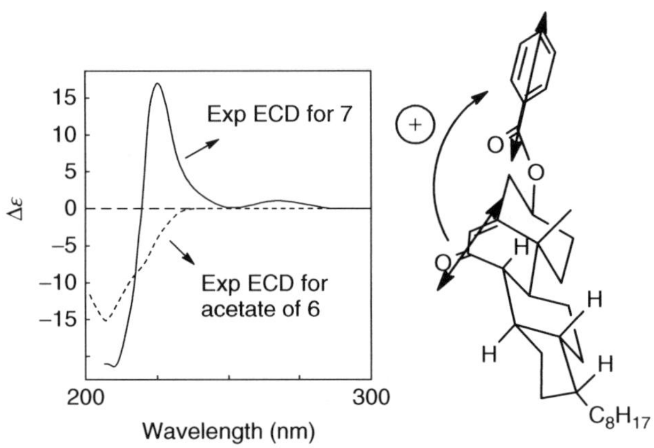

Figure 4.4 Comparison of the ECD spectra of acetate of **6** and its benzoic acid ester **7** including the exciton dipole moment direction [4].

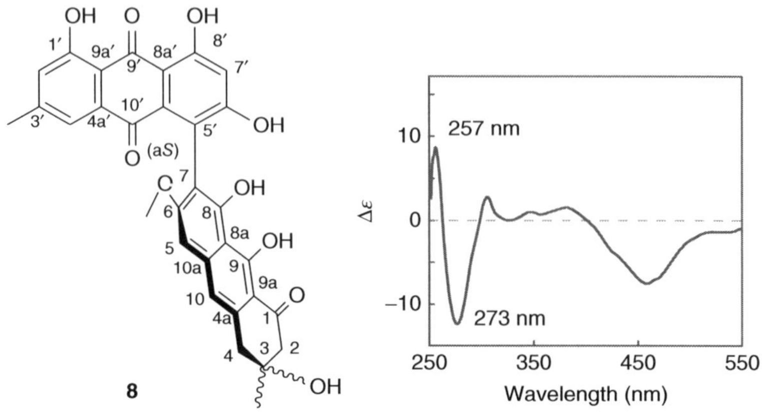

Figure 4.5 Experimental ECD spectra for the axial compound **8** [5].

positive Cotton effect at shorter wavelength (almost 257 nm) and a negative one at around 273 nm. The data suggested an anticlockwise helical twist. Therefore, the axial chirality is assigned as (a*S*).

4.2
ECD Characteristics for Chiral Metallic Compounds

The advantages of using exciton chirality methods are not needed to compute the corresponding ECD. This is useful for AC assignment in coordination chemistry because it is not easy to compute ECD spectra for metallic ion-containing compounds. Research involving metallic ions generally is outside the realm of organic stereochemistry. However, a brief introduction on this important topic is necessary. Readers who are interested in this area can refer to the corresponding textbooks or references from various journals for more details [7].

Different from that of organic chiral compounds, the ECD of chiral metallic compounds contains the contribution from the electron transitions from orbitals of d–d, f–f, including $\pi-\pi^*$, and charge-transfer transitions [8].

Chiral metallic stereogenic compound **9** could be synthesized directly from the (*S*)-2-phenylbutanoic acid (PBA) and 3-(bromomethyl) isoquinoline. ECD values for both (*S*)-and (*R*)-**9** were determined (Figure 4.6) [9]. Based on the general rule mentioned above, this is positive chirality, which means that a positive Cotton effect should appear at long wavelength and a negative Cotton effect at short wavelength.

With the fast developments in solid-state chemistry, chiroptical spectroscopy recorded in the solid state can provide important information for the AC assignment for chiral compounds. In some cases, it can provide evidences that cannot be obtained under measurements in the liquid state with the assistance of other methods recorded in the solid state. The available methods include microcrystalline ECD in KX (X = Cl, Br) disks, liquid suspension ECD in nujol

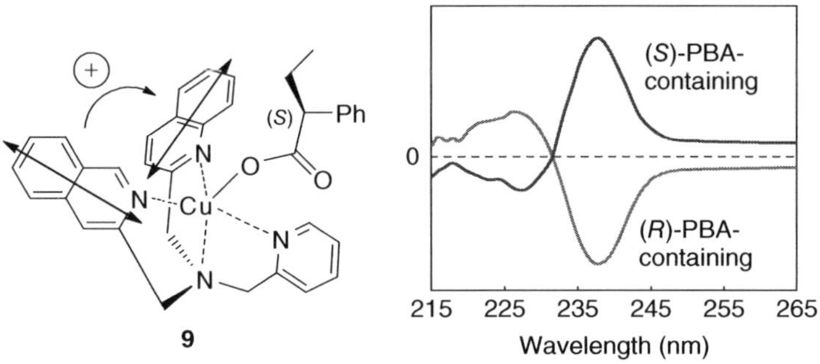

Figure 4.6 The exciton chirality model for the (*S*)-PBA-containing Cu complex and the ECD for the (*R*)-and (*S*)-PBA-containing Cu complex [9].

mull or fluorolube, film ECD on quartz substrates or dispersed in polymer or microfluidic channels, single-crystal ECD of cubic and uniaxial crystals, diffuse reflectance ECD, DRCD (by inserting an integrating sphere in the optical path of the instrument), and so on [10].

Effect of sample concentration on the ECD spectra has been recorded and well studied in solid state ECD investigation (such as microcrystalline disk and film method) [11]. An inappropriate concentration may afford an unreliable ECD spectrum, which finally will lead to a wrong conclusion. Therefore, investigation of sample concentration is necessary, and the reliable ECD results could be achieved based on the concentration gradient measurements for a chiral compound [11].

Racemic metal complexes with high charge could be resolved using some neutral chiral organic compounds as resolving agents. This procedure may produce some new species of *chiral-only-at-metal* coordination compounds. For example, efficient optical resolution were used to separate the racemic $(Me_4N)_4[Fe_4L_6]$ complex using a simple neutral (S)-1,1′-bi-2-naphthol (**11**) in 1:1 water-methanol solution. The separated cage structure exhibited the reversed ECD signals due to different AC of the related cage (Figure 4.7) [12a]. The $\pi-\pi^*$ transitions in the phenyl moiety of the achiral ligands gave rise to an typical negative/positive exciton couplet at 283 nm, from which the *chiral-only-at-metal* configurations can be assigned to the Δ/Λ (or *P/M*) form at each metal center [12b,c,d].

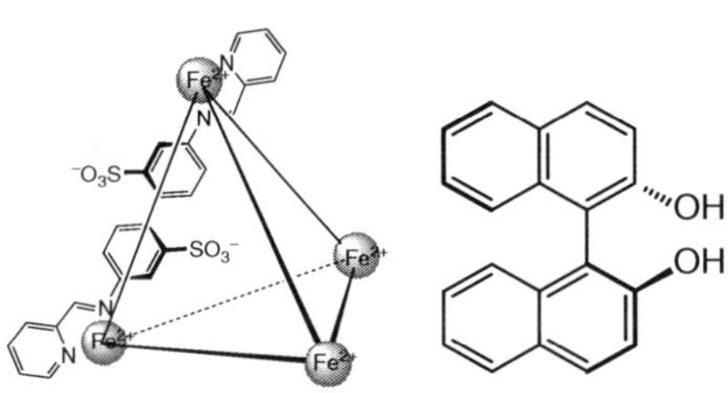

10 T cage **11** (*S*)-BINOL

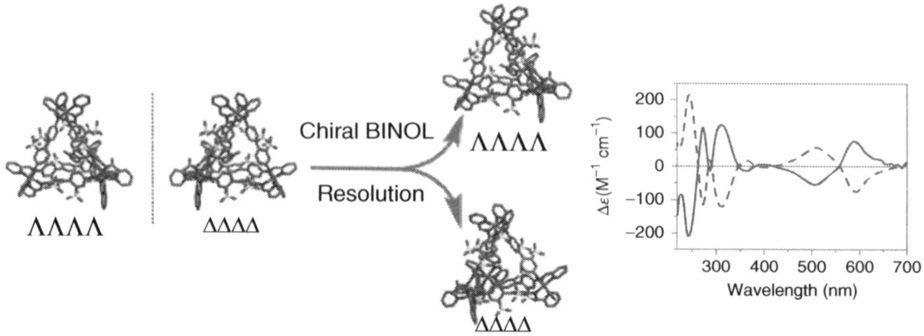

Figure 4.7 CD spectra of enantio-enriched ΔΔΔΔ-**T** (grey line) and ΔΔΔΔ-**T** (dash line) using resolving agent (*S*)-BINOL (**11**) [12a].

> **Notice**
> It must be pointed out that the phenyl ring (or other aromatic group) must be connected more or less rigidly. In most cases, the diol structures or others should be cyclic. Only in this case can the two excitons from the chromophores make correct couplings in ECD experiments. Another important issue involves the directions of the long-axis dipole moment. If the direction used in the procedure is wrong, the final conclusion will be not be correct.

Exciton chirality is very important and has been well used in organic stereochemistry. Experience is important in using exciton chirality to assign the AC for various chiral compounds. However, the requirement of two or more appropriate UV chromophores located at proper orientations limits its application for AC assignment of more chiral compounds.

4.3 Quantum Theory Basis

Although the definition of ECD is simple as pointed out previously (Eq. (4.1)), simulation of an ECD spectrum is not an easy topic. In the viewpoint of quantum chemists, to compute an ECD spectrum one needs to know a molecular parameter β, which is concerned with the frequency-dependent electronic dipole-magnetic polarizability [13]. This is similar to the optical rotation (OR) calculations. The velocity rotatory strength (R_i) is directly related to β and can be computed. Many methods can be used for its computation. In the Gaussian09 package, the function R_i can be calculated as follows:

$$R_i = 2.296 \times 10^{-39} \int \frac{\Delta\varepsilon(\nu)}{\nu} d\nu \tag{4.2}$$

The unit of R_i is 10^{-40} erg·esu·cm per Gauss. This equation is related to each of the specific excitation energies. The obtained curve is not an ECD curve. To simulate the specific ECD curve, a Gaussian function is applied [14]:

$$\Delta\varepsilon(\nu_j) = \Delta\varepsilon_{\max} \exp\left[-\left(\frac{\nu - \nu_i}{\sigma}\right)^2\right] \quad (4.3)$$

where $\Delta\varepsilon(\nu_j)$ is the absorption difference for the jth conformation; σ is the standard deviation (in cm^{-1}), which is defined as the half-width at the peak height of 1/e (some reports call it the half-width, but this is not the width of the half-peak (1/2)); ν_i is the calculated ith excited state wavelength in electronvolts; and ν is the wavelength near ν_i used for the simulation of a Gaussian distribution.

Finally, the ECD simulation function is obtained for the jth conformation at the ith excited state as

$$\Delta\varepsilon(\nu_j) = \frac{R_i \times \nu_i}{2.296 \times 10^{-39}\sqrt{\pi}\sigma} \exp\left[-\left(\frac{\nu - \nu_i}{\sigma}\right)^2\right] \quad (4.4)$$

Theoretically, after the ECD data for all conformations are obtained, the whole ECD simulations can be done using Boltzmann statistics. It plots $\Delta\varepsilon(\nu)$ versus the excitation energies in nanometers, but σ is in electronvolts. The default value for σ should be 0.4 eV in the Gaussian package. However, it depends on each specific case. The smaller the value of σ, the sharper and higher the predicted single ECD peak (absorption difference). Therefore, if the value of 0.4 eV is not suitable, it can be decreased to 0.2 eV or increased to a maximum of 0.8 eV.

Some researchers do report plots with the excitation energies in electronvolts directly. The conversion from energy data in electronvolts to wavelength in nanometers can be effected as

$$\Lambda\ (\text{nm}) = \frac{1239.84}{\text{Val}}\ (\text{eV}) \quad (4.5)$$

where Λ (nm) is the wavelength value in nanometers, which is the corresponding value of the state energy in electronvolts.

For example, if an excited state is located at 3.7514 eV, its corresponding wavelength can be calculated as 330.5 nm.

4.4 Principle Using ECD

ECD is used for chiral compounds with stereogenic centers close to functional groups (chromophore) such as C=O and C=C that have UV–vis absorptions. Therefore, if a chiral molecule has no such UV–vis absorbing functional groups, it is difficult to use ECD data to determine its AC easily. If a stereogenic center is too far away from this double bond (chromophore) in a molecule, it is still difficult to use ECD for AC assignment.

Thus, it cannot be used directly for a chiral compound that has no chromophore. In this case, one possible way is to react a molecule with a suitable chromophore, such as Ph(C=O)–, near the stereogenic center. It is possible to remeasure its ECD to study its AC.

Another way is to convert it to another chiral compound that can use the ECCD method. In this method, the important step is to nicely select a suitable compound having a chromophore whose the UV range does not overlap with the target chiral molecular ECD area, in case the recorded ECD is too complex to be analyzed.

One should be careful that, in some cases, the recorded ECD curves show only small changes when the stereogenic center changes from (*R*) to (*S*). One needs to investigate all possible situations before arriving at a conclusion. If possible, other methods should be used for further confirmation of the AC.

As mentioned in Chapter 1, ECD is very sensitive to the chromophore. Thus, if there is a long aliphatic group, it may be simplified as a short group. This can greatly reduce the computational time.

With the development of novel ECD technologies that can measure the absorption difference from 180 to 210 nm, some structures without any chromophore may be identified using ECD. However, until now, the most used ECD wavelength range is 210–800 nm. Therefore, a chiral compound without any chromophore is difficult to be used in ECD study using an old type of ECD machine that cannot cover the 180–210 nm range well.

Indeed, there are hundreds of chiral compounds' AC are identified using ECD methods. The great achievements greatly encouraged the application of ECD for the AC assignment in various natural products. Some interesting examples (**12** to **16**) [16] are illustrated below.

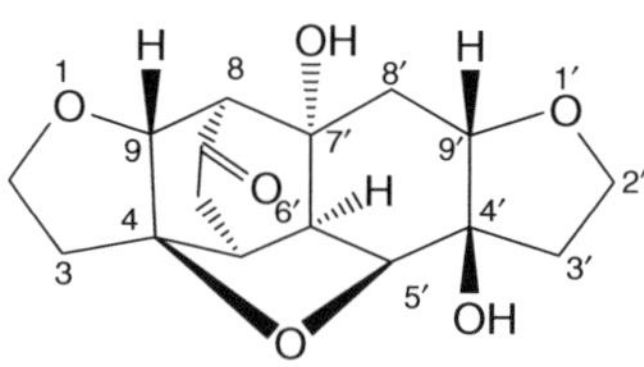

(–)-**12**

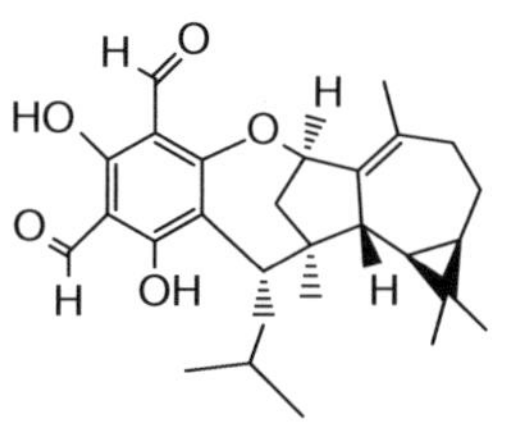

13

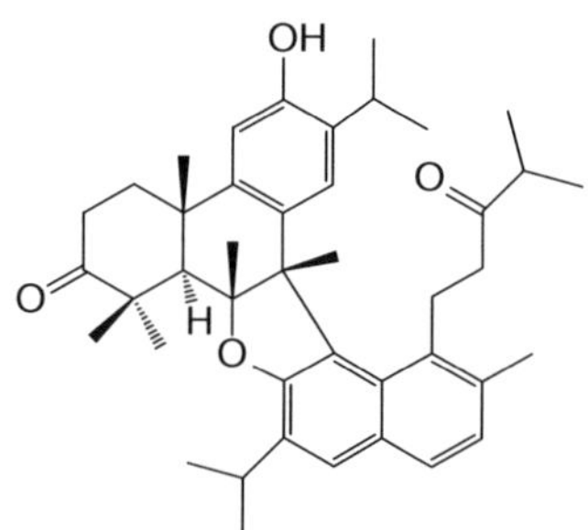

14

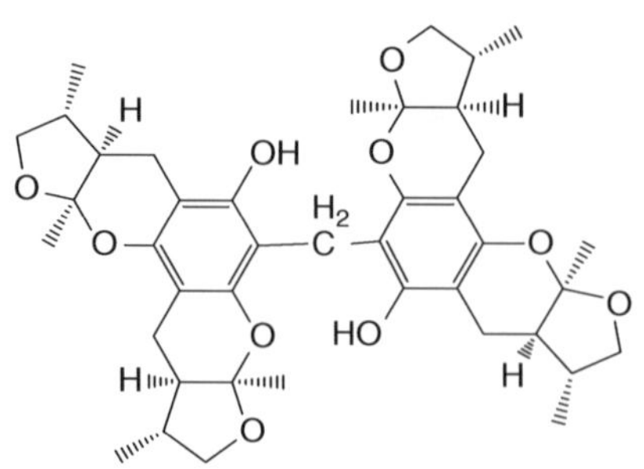

15

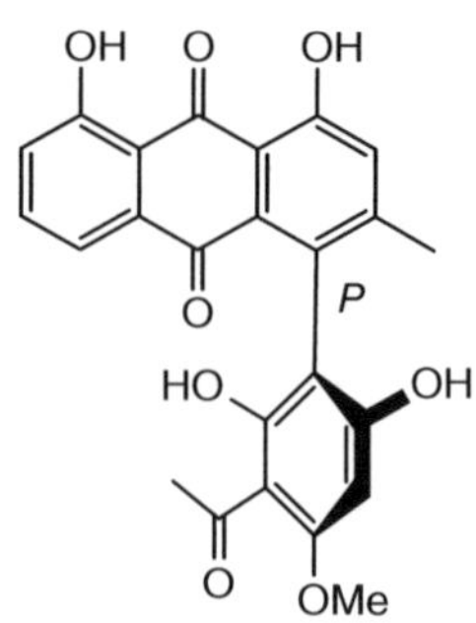

(*P*)-(+)-**16**

4.5 Application

4.5.1 Procedure to Do ECD

As mentioned in Chapters 2 and 3, ^{13}CNMR and OR computations are needed to do a conformational search. To predict an ECD spectrum, conformational searches for any specific chiral compound are also required. After the conformational search, the selected conformations should be further optimized using quantum theory, such as at the B3LYP/6-31G(d) level, at least. The B3LYP/6-31G(d)-optimized conformers can be used for ECD calculations at a higher level such as B3LYP/6-311+G(d) or B3LYP/6-311++G(2d,p).

The B3LYP/6-31G(d)-optimized conformers can be used for further optimization at the B3LYP/6-311+G(d) or higher level again for more accurate geometries; these geometries, in turn, can be used for more accurate ECD spectral predictions. Total electronic energy (TEE), Gibbs free energy (GFE), or the zero-point energy (ZPE) can be used in ECD simulations; this is almost the same as in ^{13}C nuclear magnetic resonance (NMR) and OR computations.

4.5.2 ECD Application

Since the first use of an ECD equipment in the 1960s, there have been many reports involving AC assignment for various chiral compounds, especially for bioactive compounds from nature. For example, the quaternary ammonium salt **17** (betaine) was obtained from the Mediterranean sponge *Axinella polypoides* [16], whose structure elucidated assisted by spectroscopic methods. Its AC was assigned by comparing its computed ECD with the experimental ECD (Figure 4.8).

Diincarvilones (**18**, **19**) were isolated from *Incarvillea arguta* [17]. The biological activity of the substances was screened that exhibited their effects on intracellular Ca^{2+} influx, nitric oxide (NO) production, and human cancer. The concentration was 10 μM (in Hanks Balanced Salt Solution (HBSS) buffer).

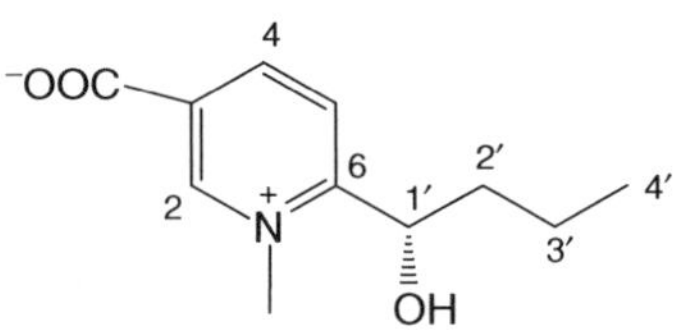

17

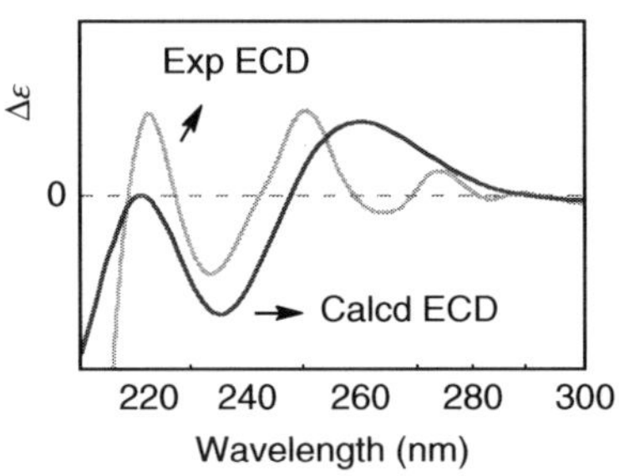

Figure 4.8 Theoretical ECD of (*S*)-**17** versus experimental curve of (−)-**17** [16].

Diincarvilone A caused a persistent increase in cytoplasmic calcium levels in A549 cells. The ECD of (9*S*,10*R*,13*R*,16*R*,19*R*,20*S*)-**18** was computed using the B3LYP/6-311++G(2d,p)//B3LYP/6-31+G(d) method. The comparison of the predicted ECD with the recorded ECD is illustrated below.

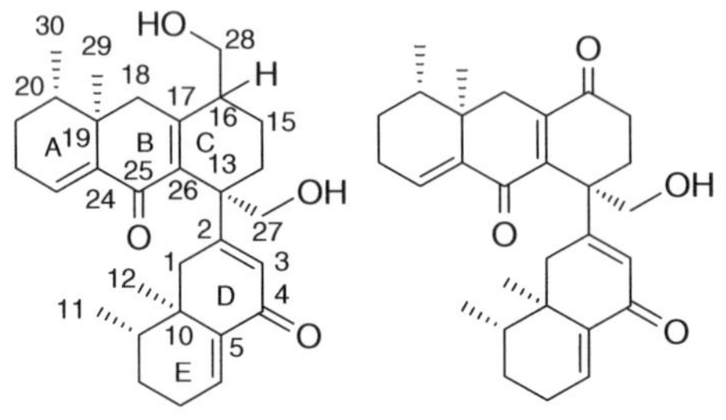

18 **19**

Both predicted and experimental ECDs agreed well except for a red shift in whole ECD for **18** (Figure 4.9) [16]. Therefore, the AC of **18** was assigned as (9*S*,10*R*,13*R*,16*R*,19*R*,20*S*). Similarly, the same computation methods were applied for the ECD calculation of **19**, which had a chirality loss of C16. It was found that both ECD spectra showed strong positive Cotton effect at around 280 nm and negative Cotton effect at 250 nm. Accordingly, the AC of **19** was determined as (9*S*,10*R*,13*R*,19*R*,20*S*).

Spiro curcasone (**20**) was isolated from the root barks of *Jatropha curcas*, a plant extensively cultivated throughout the world [18]. All the chiral centers are close to the C=C bonds and located on the ring. Its stereochemistry could be well characterized by ECD after its relative configuration (RC) was established by 2D NMR. It appears that exciton chirality can be used because there are two α,β-unsaturated ketone units near C10. Indeed, since C8, C9, and C14 are located near C=C, the recorded ECD must be affected by the stereogenic centers. In this case, its computed ECD must be reliable for its AC assignment. The AC was assigned as (8*S*,9*R*,10*S*,14*R*) by

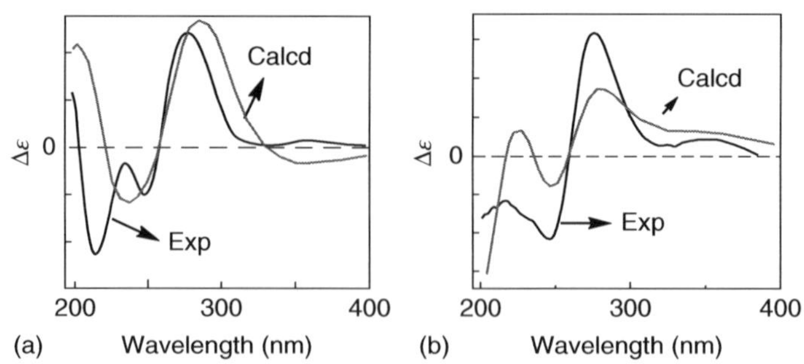

Figure 4.9 The computed and recorded ECDs for **18** (a) and **19** (b) [16].

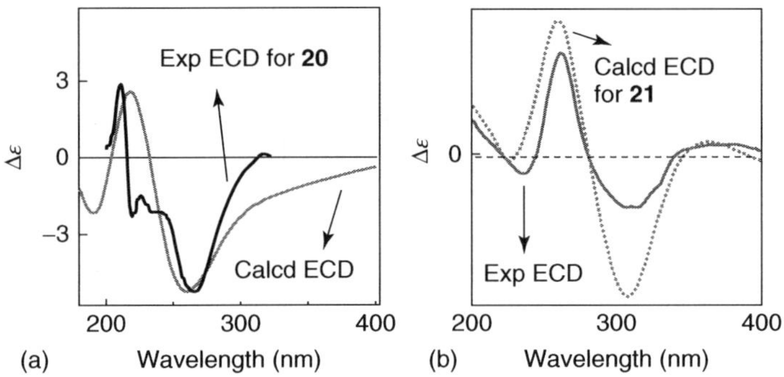

Figure 4.10 (a) Experimental and calculated ECD of spirocurcasone (**20**). (b) Calculated and experimental ECD spectra of (−)-discorhabdin Z (**21**) [18].

comparing the computed ECD of (8*S*,9*R*,10*S*,14*R*)-**20** with experimental results (Figure 4.10a).

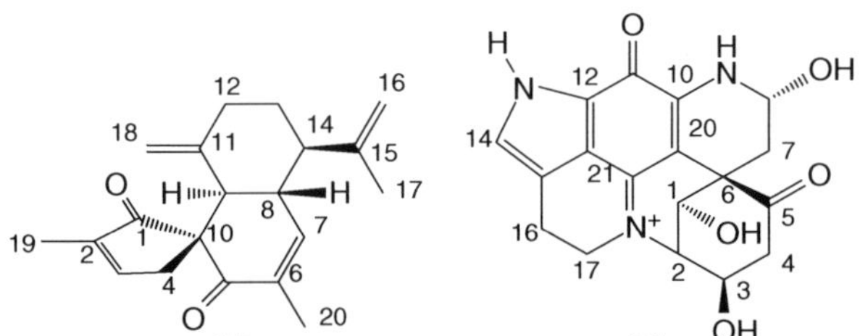

The natural (−)-discorhabdin Z (**21**) possesses an unusual hemiaminal group among the discorhabdin alkaloids [19]. It exhibits moderate to significant cytotoxicity, antibacterial activity, and inhibitory activity against sortase A. The ECD of (1*R*,2*R*,3*R*,6*R*,8*R*)-**21** was calculated at the B3LYP/6-31G(d,p) level in the gas phase. The predicted and experimental ECD spectra matched well, and the results are illustrated in Figure 4.10b.

A trimeric pyrroloindoline derivative, psychotripine (**22**), has an unprecedented hendecacyclic system that bears a hexahydro-1,3,5-triazine unit [20]. The structure was elucidated on the basis of spectroscopic measurements and quantum theory. The ECD for (3a*R*,8a*R*,3a′*R*,8a′*R*,3a″*S*,8a″*R*) was investigated at the B3LYP/6-311++G(2d,p) level using B3LYP/6-31+G(d)-optimized geometries. The half-width of 0.2 eV was used in the ECD simulations. Both ECDs had good agreement (Figure 4.11).

Compared to rigid chiral compounds, for a linear chiral compound it is difficult to do an AC assignment due to a large number of stable conformations that are needed to be computed. If the chiral center is close to a chromophore, ECD can be used to assign its AC, but this may require more computational time. Mandassion A (**23**) and mandassion B (**24**) have two stereogenic centers, and one of them is on a long side chain [21]. Both centers are close to an α,β-unsaturated ketone,

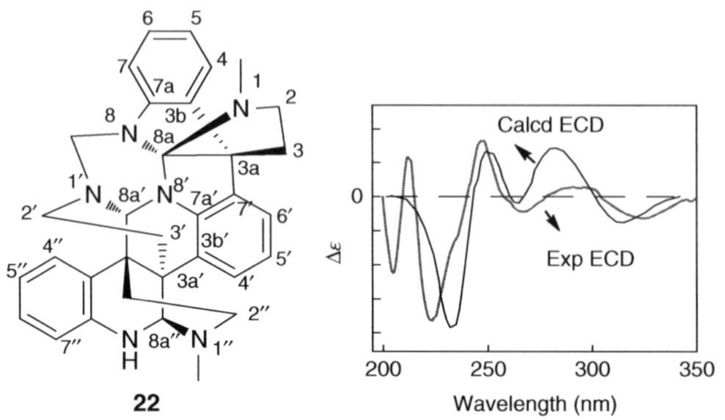

Figure 4.11 Experimental and computed ECD for **22** [20].

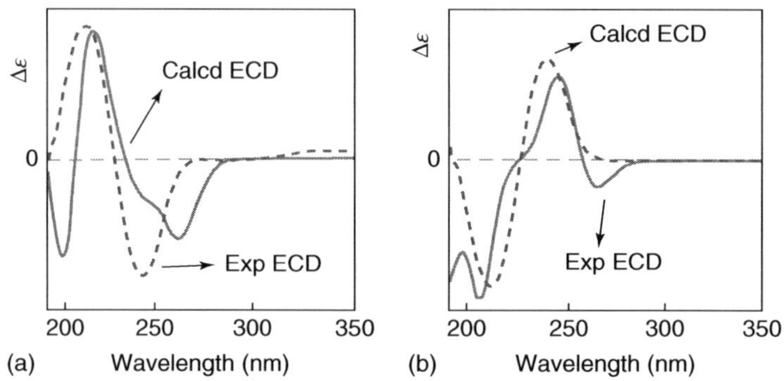

Figure 4.12 Experimental and calculated ECD of (a) mandassion A (**23**) and (b) mandassion B (**24**) [21].

and both compounds' ACs could be determined by ECD. Structures (6*Z*,4*S*,7*S*)-**23** and (6*Z*,4*S*,7*R*)-**24** were used for ECD calculations, and the predicted ECD for the two geometries **23** and **24** agreed well with the experimental results. Clearly, (6*Z*,4*S*,7*S*)-**23** and (6*Z*,4*S*,7*R*)-**24** have the desired stereochemistry (Figure 4.12).

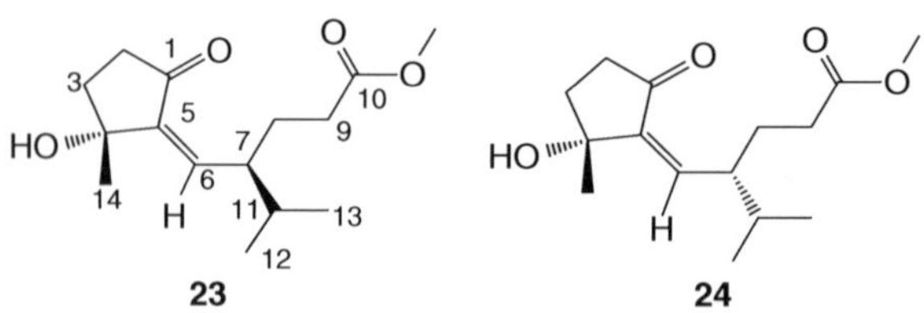

The example above is one case where the C=C is out of the ring. This double bond can be on the ring. In this case, it does not affect the results. For example, in the chiral compound **25**, the side chain has a chirality on C11 [17]. The (*R*) or (*S*) AC can affect the observed ECD (Figure 4.13), and based on their matching degrees, it could be predicted that C11 has (*S*) AC.

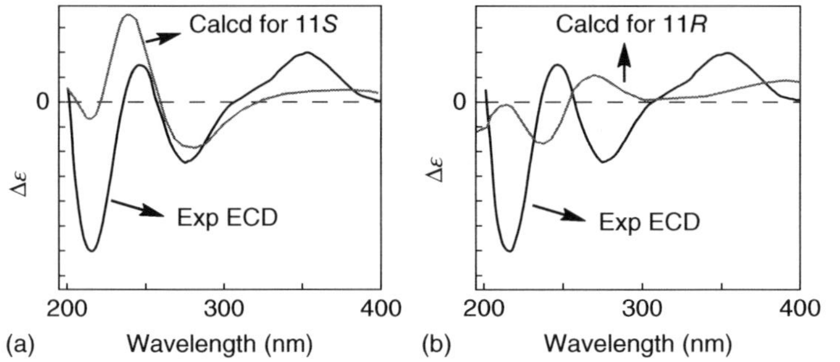

Figure 4.13 Computed and recorded ECDs for **25** with (11*S*) (a) and (11*R*) (b) configuration [17].

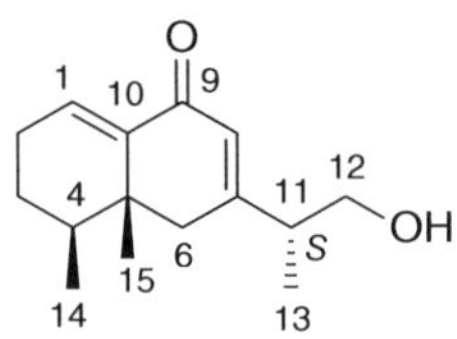

25

Streptocarbazole A (**26**) was isolated from the marine-derived actinomycetes strain *Streptomyces* sp. FMA [22]. Its RC structure was established by spectroscopic methods. It was cytotoxic to HL-60 and A-549 cell lines since it could arrest the cell cycle of HeLa cells at the G2/M phase. Its experimental ECD and predicted ECD were compared (Figure 4.14), and the AC was identified as (1′*R*,3′*S*).

Bioassay-guided study is a useful method to find bioactive compounds. This procedure is not complex. If a rough extraction exhibits some bioactivity on a certain model, such as the anti-HL-60 cell model, then it can be separated into several new fractions (e.g., 10 fractions). Then the same bioactivity method is used for the study of the 10 fractions. Possibly, similar or higher activity may be found in two or three fractions, and this is a good sign. After combining two or three fractions together,

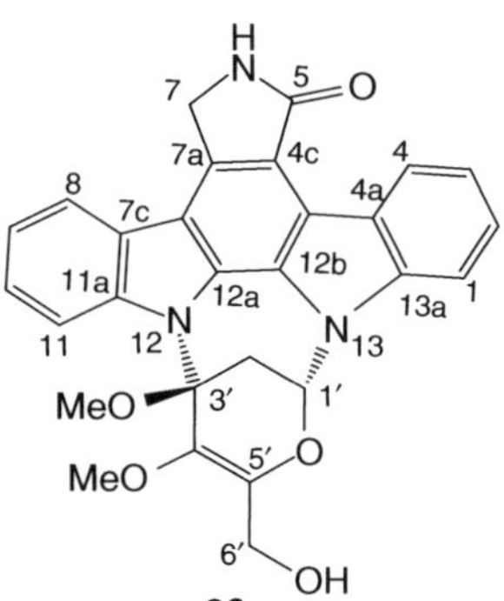

26

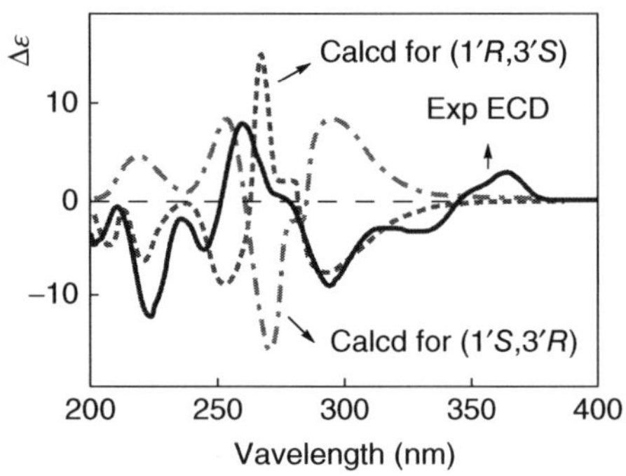

Figure 4.14 Measured and calculated ECD spectra for **26** [22].

the combined new fraction is used for further separation and affords some new fractions again. Bioactivity scanning for these fractions can be performed again to select the new bioactive fraction, and these fractions can be used for next cycle study. The procedure is repeated until the bioactive compounds are found.

For example, based on the above principle, natural streptosetin A (**27**) was obtained from one of the active strains identified through yeast assay [23]. After its planar structure was identified using 1D and 2D NMR experiments, and its RC was determined based on X-ray analysis. The simulation of ECD spectra using density functional theory (DFT) methods confirmed its AC as (4*R*,5*R*,9*R*,10*S*,13*R*) (Figure 4.15).

> **Notice**
> The same scale should be used at the *X*-axis when comparing experimental and theoretical ECD's. The best way is to place the experimental and theoretical ECD in the same plot. The experimental ECD spectra here covered the range of 220–400 nm instead of the computed range of 170–350 nm; this is not frequently used in reports.

4.5.3
UV Correction

The examples mentioned above, especially example **27**, bring another academic question: the calculated spectra curve may appear red-shifted or UV-shifted in practice. If the signal with the largest Δε value moves to the long-wave direction, it is then called the ECD red shift. If it moves to shorter values, this is a UV shift. The shift range may be from 10 to 30 nm in most cases. Because the accuracy is not high enough in computations using DFT methods, the energy of the excited state may not match with the experimental results well. In example **27**, the predicted ECD has a strong UV shift of almost 40 nm with a negative Cotton effect in experiments. It reduces to about 30 nm with a positive Cotton effect near 310 nm. This could be compensated using the UV correction method [24]. The procedure for UV correction is illustrated below.

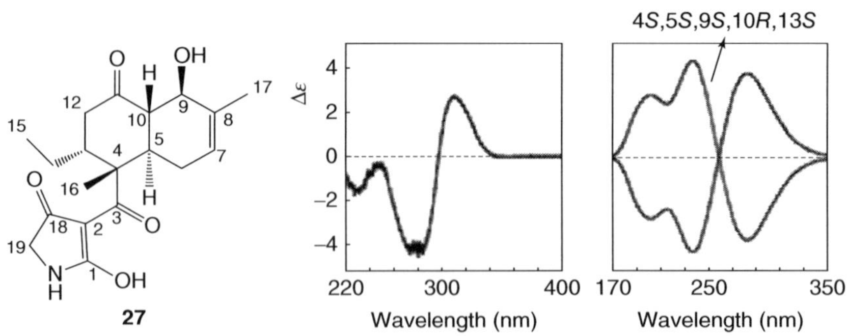

27

Figure 4.15 Experimental and calculated ECD spectra for **27** [24a].

1) UV spectral simulation is required while simulating of its ECD spectra. The UV absorption spectra from oscillator or dipole strength can be simulated using the formula

$$\varepsilon(\nu_j) = \frac{D_i \times \nu_i}{4 \times 2.296 \times 10^{-39}\sqrt{\pi}\sigma} \exp\left[-\left(\frac{\nu - \nu_i}{\sigma}\right)^2\right] \tag{4.6}$$

where D is the dipole strength of the ith excited state of the conformer, and σ is the half-width at ε_{max}/e. The default is 0.4 eV in the Gaussview Program.

In the Gaussian output file, the computed data are the oscillator strengths instead of dipole strengths. Thus, the following conversion formula is used:

$$f_i = \frac{8\pi^2{}_i mc}{3he^2} D_i \tag{4.7}$$

or

$$D_i = \frac{f_i}{4.701755608 \times 10^{29} \times \nu_i} \tag{4.8}$$

If ν_i is in electronvolts, the predicted D_i should be

$$D_i = \frac{f_i}{3.79222203232 \times 10^{33} \times \nu_i} \tag{4.9}$$

If ν_i is in nanometers, the predicted D_i should be

$$D_i = \frac{f_i \times \nu_i}{4.701755608 \times 10^{36}} \tag{4.10}$$

where f_i is the oscillator strength and ν_i is the value of the ith excited state.

Similar to the simulation of ECD introduced above, the simulation can be performed under the same methods. After the UV simulation, the following steps is carried out:

1) Measure the difference of the extremum value of UV between the two experimental and theoretical UV spectra, for example, this is $\Delta\lambda$ ($\Delta\lambda = \lambda_{exp} - \lambda_{calcd}$); it could be in nanometers or in electronvolts.
2) Convert $\Delta\lambda$ in nanometers to the corresponding value ($\Delta\nu$) in electronvolts if nanometers was used in the UV spectra.
3) Insert this value (in electronvolts) into the calculation equation of ECD ($\Delta\varepsilon$) below.

$$\Delta\varepsilon(\nu_j) = \frac{R_i \times (\nu_i + \Delta\nu)}{2.296 \times 10^{-39}\sqrt{\pi}\sigma} \exp\left[-\left(\frac{\nu - (\nu_i + \Delta\nu)}{\sigma}\right)^2\right] \tag{4.11}$$

The value of $\Delta\nu$ could be positive or negative. Thus, when it is added to the formula of ECD ($\Delta\varepsilon$), the new ECD curve should move to the corresponding direction.

4) Simulate the corresponding ECD curves using the Boltzmann sum.

This procedure is complex. In practice, there is another simple way to perform UV corrections. The first two steps are the same.

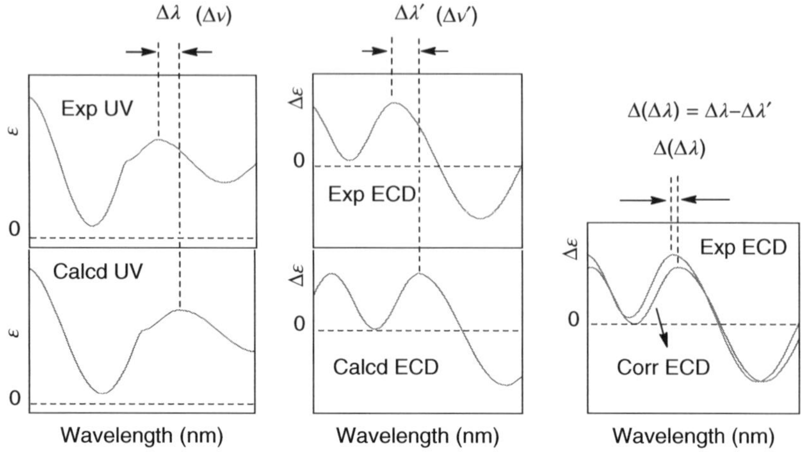

Figure 4.16 A diagrammatic sketch for UV corrections used in the simulation of ECD. In this proposed case, this UV appears red-shifted. After moving the ECD to the short wavelength direction (UV-shift direction) with a value of $\Delta\lambda$, the new ECD curve (the lower line labelled as 'corr ECD') can match the experimental ECD well generally. The difference of $\Delta(\Delta\lambda)$ is unavoidable, and, sometimes, this is still a large value.

1) Simulate the UV spectra.
2) Measure the difference of the extremum value of the Cotton effect ($\Delta\lambda = \lambda_{exp} - \lambda_{calcd}$), namely, the point with largest ε value between the predicted UV and experimental UV.
3) Translate the predicted ECD with a step of $\Delta\lambda$; the new ECD should closer to the experimental ECD.

A diagrammatic sketch is given in Figure 4.16. Indeed, even if the new ECD curves are translated with a value of $\Delta\lambda$ to the opposite direction, the corrected ECD has still a small difference $\Delta(\Delta\lambda)$. This is normal. In many cases, this value is acceptable.

For example, the derivative **28** is the analog **29**, which has strong bioactivity [25]. The total synthesis of **28** provides a chance to study its stereochemistry [26]. The ECD prediction was performed at the B3LYP/6-311++G(2d,p) level by using the conformations obtained at the B3LYP/6-311++G(2d,p) level. GFE and ZPE were used in ECD simulations.

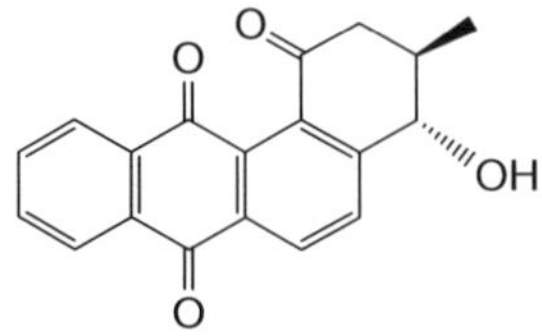

(2*R*,3*S*)-**28**

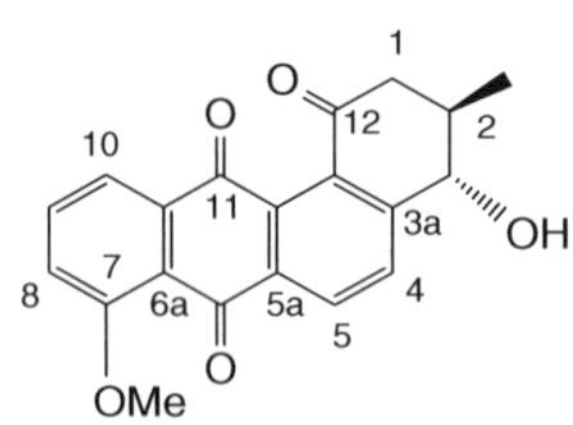

29 Rubiginone A$_2$

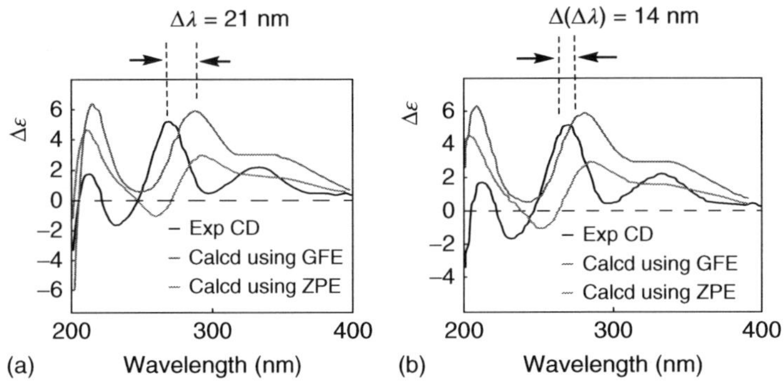

Figure 4.17 The simulated ECD for (2*R*,3*S*)-**28** using zero-point energy and Gibbs free energy without UV corrections (a) and with UV corrections (b) [26].

The predicted ECD had a large red shift compared to the experimental ECD (Figure 4.17a). In UV simulations, it was found that the computed UV had about 7 nm red shift. Both methods were used for ECD simulations again. It was found that both methods gave nearly the same conclusion: the corrected ECD needed to move to shorter wavelength by about 7 nm (Figure 4.17b). The ECD curves before and after UV corrections are illustrated below.

Obviously, after the UV correction, the new simulated ECD can be matched with the experimental ECD better. It is worth carrying out this step. However, in many cases, it is usual to compare the ECD shapes between the predicted ECD and calculated ECD without UV correction; this is absolutely acceptable.

References

1. (a) Gaffield, W. (1993) *Bioactive Natural Products*, Chapter 7, CRC Press Inc, New York, p. 147; (b) Harada, N., Chen, S.L., and Nakanishi, K. (1975) *J. Am. Chem. Soc.*, **97**, 5435–5452; (c) Tinoco, I. Jr., Woody, R.W., and Bradley, D.F. (1963) *J. Chem. Phys.*, **38**, 1317.
2. (a) Yu, D.Q. (1989) *Huaxue Tongbao (Chin)*, **4**, 5–12; (b) Yin, B.P., Qin, G.W., and Xu, R.S. (1987) *Youji Huaxue*, **3**, 165–173. Some examples were cited there.
3. Berova, N., Di Bari, L., and Pescitelli, G. (2007) *Chem. Soc. Rev.*, **36**, 914–931.
4. (a) Chen, Y.G., Qing, G.W., and Xie, Y.Y. (2000) *J. Nat. Prod. Res. Dev. (Chin.)*, **3** (1), 64–68. Some examples were cited there; (b) Harada, N. and Nakanishi, K. (1983) *Circular Dichroic Spectroscopy-Exciton Coupling in Organic Stereochemistry*, Oxford University Press; (c) Cai, G., Bozhkova, N., Odingo, J., Berova, N., and Nakanishi, K. (1993) *J. Am. Chem. Soc.*, **115**, 7192–7198; (d) Verdine, G.L., and Nakanishi, K. (1985) *J. Chem. Soc., Chem. Commun.*, 1093–1095.
5. (a) Bara, R., Zerfass, I., Aly, A.H., Goldbach-Gecke, H., Raghavan, V., Sass, P., Mandi, A., Wray, V., Polavarapu, P.L., Pretsch, A., Lin, W., Kurtan, T., Debbab, A., Broetz-Oesterhelt, H., and Proksch, P. (2013) *J. Med. Chem.*, **56**, 3257–3272; (b) Buchanan, M.S., Gill, M., Millar, P., Phonh-Axa, S., Raudies, E., and Yu, J. (1999) *J. Chem. Soc., Perkin Trans. 1*, 795–802; (c) Mueller, M., Lamottke, K., Steglich, W., Busemann, S., Reichert, M., Bringmann, G., and Spiteller, P. (2004) *Eur. J. Org. Chem.*, **2004**, 4850–4855.
6. Ola, A.R.B., Debbab, A., Kurtán, T., Bröz, O.H., Aly, A.H., and Proksch, P. (2014) *Tetrahedron Lett.*, **55**, 3133–3136.

7. Canary, J.W. (2009) *Chem. Soc. Rev.*, **38**, 747–756.
8. Zhang, H. and Zhang, L.R. (2011) *Daxue Huaxue*, **25**, 8.
9. Joyce, L.A., Maynor, M.S., Dragna, J.M., da Cruz, G.M., Lynch, V.M., Canary, J.W., and Anslyn, E.V. (2011) *J. Am. Chem. Soc.*, **133**, 13746–13752.
10. Zhang, H., Chen, Y.C., Wang, F., Qiu, X.M., Li, L., and Chen, J.G. (2006) *Acta Phys.-Chim. Sin.*, **26** (6), 666–671.
11. Zhang, H., Yan, J.X., Wu, X.T., Li, D., Ding, L., and Lin, L.R. (2013) *Acta Phys. -Chim. Sin.*, **29**, 2481–2497.
12. (a) Wan, S., Lin, L.-R., Zeng, L., Lin, Y., and Zhang, H. (2014) *Chem. Commun.*, **50**, 15301–15304; (b) Kerckhoffs, J.M.C.A., Peberdy, J.C., Meistermann, I., Childs, L.J., Isaac, C.J., Pearmund, C.R, Reudegger, V., Khalid, S., Alcock, N.W., Hannon, M.J. and Rodger, A. *Dalton Trans.*, 2007, 734–742; (c) Khalid, P. M. Rodger, and Rodger, A. (2005) *J. Liq. Chromatogr. Relat. Technol.*, **28**, 2995–3003; (d) Kerckhoffs J.M.C.A., Peberdy, J.C., Meistermann, L.J. Childs, C.J., Isaac, C.R., Pearmund, V., Reudegger, S., Khalid, N.W., Alcock, M.J.Hannon, and Rodger, A. (2007) *Dalton Trans.*, 734–742.
13. (a) Berova, N., Polavarapu, P., Nakanishi, K., and Woody, R.W. (2012) *Comprehensive Chiroptical Spectroscopy*, vol. 1, John Wiley & Sons, Inc., Hoboken, NJ; (b)From the materials provided by Gaussian Company *www. Gaussian.com* (accessed 28 December 2014).
14. Harada, N. and Nakanishi, K. (1983) *Circular Dichroism Spectroscopy-Exciton Coupling in Organic Stereochemistry*, University Science Books, Mill Valley, CA.
15. (a) for 12, see: Gao, Y.P., Shen, Y.S., Zhang, S.D., Tian, J.M., Zeng, H.W., Ye, J., Li, H.L., Shan, L., and Zhang, W.D. (2012) *Org. Lett.*, **14**, 1954; (b) for 13, see: Wang, J., Zhai, W.Z., Zou, Y.K., Zhu, J.J., Xiong, J., Zhao, Y., Yang, G.X., Fan, H., Hamann, M.T., Xia, G., and Hu, J.F. (2012) *Tetrahedron Letters*, **53**, 2654; (c) for 14, see: Hou, X.F., Yao, S., Mandi, A., Kurtan, T., Tang, C.P., Ke, C.Q, Li, X.Q., and Ye, Y. (2012) *Org. Lett.*, **14**, 460; (d) Wu, X.Y., Liu, X.H., Lin, Y.C., Luo, J.H., She, Z.G., Jin, L.H., Chan, W.L., Antus, S., Kurtán, T., Elsässer, B., and Krohn, K. (2005) *Eur. J. Org. Chem.*, 4061–4064; (e) for 16: Bringmann, G., Maksimenka, K., Comar, J.M., Knauer, M., and Bruhn, T. (2007) *Tetrahedron*, **63**, 9810–9824.
16. Menna, M., Aiello, A., Aniello, F.D., Fattorusso, E., Imperatore, C., Luciano, P., and Vitalone, R. (2012) *Mar. Drugs*, **10**, 2509–2518.
17. Yan, Y.M., Wu, G.S., Dong, X.P., Shen, L., Li, Y., Su, J., Luo, H.R., Zhu, H.J., and Cheng, Y.X. (2012) *J. Nat. Prod.*, **75**, 1025–1029.
18. Chianese, G., Fattorusso, E., Aiyelaagbe, O.O., Luciano, P., Schrőder, H.C., Müller, W.E.G., and Taglialatela-Scafati, O. (2011) *Org. Lett.*, **13**, 316.
19. Jeon, J., Na, Z., Jung, M., Lee, H.S., Sim, C.J., Nahm, K., Oh, K.B., and Shin, J. (2010) *J. Nat. Prod.*, **73**, 258–262.
20. Li, X.N., Zhang, Y., Cai, X.H., Feng, T., Liu, Y.P., Li, Y., Ren, J., Zhu, H.J., and Luo, X.D. (2011) *Org. Lett.*, **13**, 5896.
21. Luo, J.G., Lv, X.Q., Wang, X.B., and Kong, L.Y. (2012) *Phytochem. Lett.*, **5**, 134.
22. Leverrier, A., Dau, M.E.T.H., Retailleau, P., Awang, K., Gueritte, F., and Litaudon, M. (2010) *Org. Lett.*, **12**, 3638–3641.
23. Amagata, T., Xiao, J., Chen, Y.P., Holsopple, N., Oliver, A.G., Gokey, T., Guliaev, A.B., and Minoura, K. (2012) *J. Nat. Prod.*, **75**, 2193–2199.
24. (a) Amagata, T., Xiao, J., Chen, Y.-P., Holsopple, N., Oliver, A.G., Gokey, T., Guliaev, A.B., and Minoura, K. (2012) *J. Nat. Prod.*, **75**, 2193–2199; (b) Stephens, P.J. and Harada, N. (2010) *Chirality*, **22**, 229–233.
25. (a) Huang, X., He, J., Niu, X., Menzel, K.D., Dahse, H.M., Grabley, S., Fiedler, H.P., Sattler, I., and Hertweck, C. (2008) *Angew. Chem. Int. Ed.*, **47**, 3995–3998; (b) Oka, M., Kamei, H., Hamagishi, Y., Tomita, K., Miyaki, T., Konishi, M., and Oki, T. (1990) *J. Antibiot.*, **43**, 967–976.
26. Zhou, B.D., Ren, J., Liu, X.C., and Zhu, H.J. (2013) *Tetrahedron*, **69**, 1189–1194.

5
Vibrational Circular Dichroism and Raman Optical Activity

Optical rotation (OR), optical rotatory dispersion (ORD), and electronic circular dichroism (ECD) have been widely used in absolute configuration (AC) assignment for chiral compounds, while vibrational circular dichroism (VCD) was still under application promotion in the year 2000. Now, its application has become popular not only for obtaining abundant structural information on various bond vibrations in a chiral molecule itself but also information on the interaction between two or more chiral compounds. The vibration caused by atom oscillators forms the fundamental basis of AC assignment.

Different absorption of the left- and right-handed circularly polarized infrared (IR) radiation forms the basis of VCD. Its definition is the same as that of ECD. However, the range of wavelengths involved is different. In ECD spectral studies, UV and the corresponding circularly polarized UV in the range 200–800 nm is used in most cases. Because the energy of UV radiation is very large, the time taken for ECD determination is short. In VCD, IR and polarized IR in the range 200–8000 cm^{-1} is generally used, but in particular from 500 to 4000 cm^{-1}, or just from 1000 to 2000 cm^{-1} for many chiral organic compounds. Since the IR energy is much lower than that of UV, the determination time of VCD is much longer than for ECD measurements. Normally, it may need 10 h or more.

Similarly, Raman scattering spectra and the corresponding Raman optical activity (ROA) provide other information for different structures. ROA involves a physical variable: that is, the scattering cross section when polarized light passes through a chiral sample in solution or in solid [1]. It is also the intensity differences between left- and right- circularly polarized radiation. The form of the formula is very similar to the ECD formula:

$$\Delta A = A_L - A_R \tag{5.1}$$

where A is the scattering cross section and A_L and A_R are the scattering cross sections to the corresponding left- and right-handed circularly polarized light (CPL), respectively.

The wavelength range in ROA is mostly from 200 to 4000 cm^{-1} when used in organic stereochemistry. This also depends on the requirement of measuring a chiral sample. In some cases, the window just covers 200–2000 cm^{-1}.

Organic Stereochemistry: Experimental and Computational Methods, First Edition. Hua-Jie Zhu.

5.1
Exciton Chirality

Exciton chirality can be used for AC assignment using VCD. This is tentatively called the VCD exciton chirality method. It relies on bisignate VCD couplets that have a sign that reflects the relative spatial orientations of two IR chromophores (vibrational frequencies). It is the same as that appeared in the ECD study; the positive dihedral angle of two chromophores, such as carbonyl groups, generates a positive–negative couplet from lower to higher frequencies, and the negative angle gives a negative–positive couplet. Until now, the method has been tested using known chiral compounds. In this case, the chiral compounds must have two >C=O groups because of their strong, sharp, and isolated absorption around 1650–1800 cm^{-1}. On the other hand, the >C=O group can be introduced into a molecule by esterification or other reactions, which is convenient when using VCD exciton couplet methods in AC assignments.

For example, (*S*)-3-hydroxydihydrofuran-2(3*H*)-one (**1**) and its derivatives (*S*)-**2** have very weak VCD signals from 1720 to 1820 cm^{-1}. However, if they are converted to the acetyl ester (*S*)-**3** [2], strong VCD signals in this area are obtained as a result of the exciton coupling from the two >C=O groups: a positive signal at 1790 cm^{-1} and a negative signal near 1750 cm^{-1} (grey line labelled as 'Exp VCD for (*S*)-**3**' in Figure 5.1). It looks like they are negative–positive couplets from the low wavelength to the high wavelength, namely, from the high frequency to the low frequency direction. Once (*R*)-**3** was used, almost a mirror image of the VCD curve was obtained from low wavelength to high wavelength (dash line labelled as 'Exp VCD for (*R*)-**3**' in Figure 5.1). The VCD signal exhibits the positive–negative coupling signals.

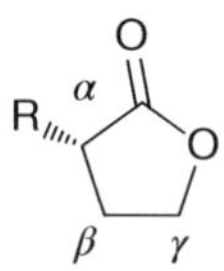

1 R = OH
2 R = OMe
3 R = OAc

It is different from the exciton chirality method in ECD study, because one needs to do conformational searches to obtain the most stable conformation for the relative orientation of the two >C=O groups. This procedure is the same as what we did for conformational searches in Chapter 3. By using the MMFF94S force field or any other molecular mechanics force field, a series of low-energy conformations can be recorded for further optimization at the B3LYP/6-311+G(d,p) level. The DFT (density functional theory)-level optimized geometry with the lowest energy is used for the determination of the dihedral angle of the two >C=O groups. If the angle is positive ($0° < \theta < +180°$), it hints of a clockwise twist, and it will have a positive–negative VCD signal. When it is a negative angle (counterclockwise, $-180° < \theta < 0°$), a negative–positive signal should be predicted.

In the VCD exciton chirality study, a conformational search is necessary. However, it reduces the steps of VCD computations. This has the advantage

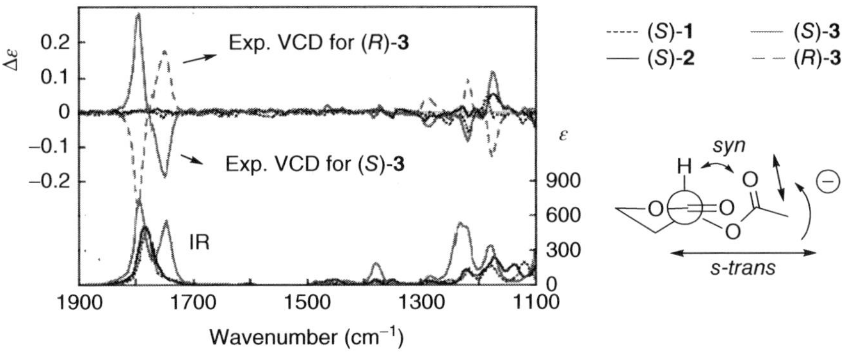

Figure 5.1 The VCD spectra for (*S*)-**1**–(*S*)-**3** and (*R*)-**3**, including the dihedral angle "direction" of the two C=O groups based on the DFT-optimized geometry (the VCD signals for (*S*)-**1** and (*S*)-**2** are very weak) [2].

of economy. However, it must have the corresponding C=O groups at suitable positions, or introduced acetyl groups in the chiral molecule by esterification. That means this method has limited uses in AC determinations, although it has some advantages when a C=O group can be introduced for increasing the VCD signal strength near $1650-1850\,cm^{-1}$. If either of these conditions cannot be satisfied, this method cannot be used in AC assignment.

5.2 Quantum Theory Basis

5.2.1 VCD and IR

VCD and ROA are the two fundamental forms of vibrational optical activity (VOA). Both involve vibrational transitions in molecules involving internal normal modes of vibrational motion, such as the stretching and angle-bending of bonds. The quantum theory of VCD can be found in many specific references, including comprehensive descriptions of all aspects of VOA [1, 3]. The discoveries of VCD and ROA have been reported in the many pioneering works by, for example, Stephens *et al.*, [4] Barron *et al.*, [5] Holzwarth and Chabay, [6] and Nafie *et al.* [7]. These study have opened the field of VOA and made possible today's technology for the assignment of AC in chiral molecules.

VCD is a measure of the difference in left versus right circularly polarized infrared (IR) radiation. An important advance for VCD was the development of Fourier transform (FT) VCD measurement [7d] and its subsequent commercial availability by BioTools in 1997 that made VCD widely accessible to for example organic chemists. The intensity of VCD is proportional to the vibrational rotational strength, and this is related to vibrational (1) electric and (2) magnetic dipole transition moments. The former two are related to the harmonic force field and atomic polar tensor, and the latter is concerned with the harmonic force field and the atomic axial tensor. Currently, DFT methods can give sufficiently good prediction for IR vibrational absorption intensities, as the dipole strength, *D*, and for VCD intensities as the rotational strength, *R*.

In modern calculations, the computations can be performed on either personal computers with sufficient memory and speed or on larger mainframes such as a super-computer-system. From the viewpoint of experimental chemists, the important aspect of a VCD calculation is how to treat the obtained numerical VCD and IR intensities for each normal mode of vibration. In Gaussian 09 software, once the rotational strength and dipole moment values are obtained for each vibrational frequency in a particular optimized molecular conformation, the following formulas can be used for IR and VCD simulation with the frequency as X-axis [8a]. For the ith frequency, it gives

for IR :

$$\varepsilon(\nu) = 3.4651907 \times 10^{-3} \cdot \nu \cdot D_i \frac{\gamma}{(\nu - \nu_i)^2 + \gamma^2} \tag{5.2}$$

for VCD :

$$\Delta\varepsilon(\nu) = 4 \times 3.4651907 \times 10^{-3} \cdot \nu \cdot R_i \frac{\gamma}{(\nu - \nu_i)^2 + \gamma^2} \tag{5.3}$$

where D_i is the dipole strength for the *i*th frequency with the unit of $10^{-40}\,\text{esu}^2\,\text{cm}^2$, R_i is the rotational strength in $10^{-44}\,\text{esu}^2\,\text{cm}^2$, and γ is the half-width of the Lorentzian band at one-half of the peak height (in cm^{-1}) ($4\,\text{cm}^{-1}$ is used as default). All of them are used in the Gaussian software. For more details of specific deduction procedure for calculations of D_i and R_i, one should read the corresponding materials from Gaussian [8b].

After the IR and VCD intensities for all vibrational frequencies are calculated using Eqs. (5.2) and (5.3), the complete IR and VCD spectra can be simulated by adding all calculated intensities, each centered at its vibrational transition frequency for a particular optimized molecular conformation. Once all sufficient low-energy conformations are investigated, the conformer population-weighted IR and VCD spectra can be simulated using Boltzmann statistics. The final Boltzmann-averaged IR and VCD spectra are compared with the corresponding experimentally measured IR and VCD spectra. Comparison of the signs of the measured and calculated VCD spectra allow for the assignment of AC of the molecule. If the signs agree, the AC of the calculated molecule is the same as the AC of the measured molecule, and if they are opposite in sign, then the AC of measured sample molecule is opposite to the one chosen for the calculation.

5.2.2 ROA and Raman Scattering

ROA studies the behavior of with circularly polarized Raman scattering light in chiral matter [1, 3]. Prior to the discovery of ROA, it was speculated that the principles optical activity in the UV and IR could be extended to Raman scattering and that ROA could be determined under suitable experimental conditions. This assumption was confirmed by Barron and coworkers in 1973 by using α-phenylethanol and other chiral compounds [9]. The commercial development of ROA equipment by BioTools in 2004 provided the opportunity for organic chemists to complement the assignment of AC by VCD and thereby further extend the determination of AC assignment for all classes of chiral compounds.

ROA is a kind of scattering phenomenon, with three different geometries, namely right-angle scattering (90°), backscattering (180°), and forward scattering (0°). The intensity of ROA is determined by the three generalized polarizabilities. The major conclusions will be presented here.

The definition of ROA in computations was developed in 1971 by Barron and Buckingham [10], who introduced the dimensionless circular intensity difference (CID) (Δ).

$$\Delta = \frac{(I^{\mathrm{R}} - I^{\mathrm{L}})}{(I^{\mathrm{R}} + I^{\mathrm{L}})} \tag{5.4}$$

where I^{R} and I^{L} are the scattered intensities in for right- and left-circularly polarized incident light, respectively. The specific relationship can be written as shown below [1–4, 11]:

$$I^{\mathrm{R}} + I^{\mathrm{L}} \propto \frac{(\nu_{in} - \nu_i)^4}{\left[1 - \exp\left(-\frac{hc\nu}{RT}\right)\right]}[45\alpha_i^2 + 7\beta_i^2] \tag{5.5}$$

$$I^{\mathrm{R}} - I^{\mathrm{L}} \propto \frac{(\nu_{in} - \nu_i)^4}{\left[1 - \exp\left(-\frac{hc\nu}{RT}\right)\right]}[12\beta_G^2 + 4\beta_A^2] \tag{5.6}$$

where α_i^2 and β_i^2 are the Raman invariants, and β_G^2 and β_A^2 are the ROA invariants. They are all related to the electric dipole–electric dipole molecular polarizability tensor ($\alpha_{\alpha\beta}$) and the electric dipole–magnetic dipole and electric dipole–electric quadrupole optical activity tensors ($G'_{\alpha\beta}$) and ($A_{\alpha\beta\gamma}$). These polarizablity tensors can be computed using a program such as the Gaussian package. ν_{in} and ν_i are the incident light frequency (in cm^{-1}) and vibrational frequency mode (in cm^{-1}), respectively, for the ith conformation.

In the Gaussian computation program, the complex tensors are simplified using *Raman* X_i and *ROA* X_i, which refer to the intensity values (sum or difference) for each normal mode i for three different types of experiments, and *CID* X_i is the quotient *ROA* X_i/*Raman* X_i, which is also reported in the Gaussian output (note that the ROA intensity values in the Gaussian output are already multiplied by 10^4). It can simulate the corresponding ROA, Raman, and CID spectra, respectively, for the ith mode (frequency in cm^{-1}) by using the following Lorentzian line shape function:

Raman spectra :

$$(I_{\mathrm{R}} + I_{\mathrm{L}})(\nu) = (\nu_{in} - \nu_i)^4 \ \ \mathrm{Raman\,X}_i \frac{1}{\pi} \frac{\gamma}{(\nu - \nu_i)^2 + \gamma^2} \tag{5.7}$$

$$\text{ROA spectra}:$$
$$(I_R - I_L)(\nu) \times 10^4 = (\nu_{in} - \nu_i)^4 \ \text{ROA}\ X_i \frac{1}{\pi} \frac{\gamma}{(\nu - \nu_i)^2 + \gamma^2} \tag{5.8}$$

$$\text{CID spectra}:$$
$$\frac{(I_R - I_L)(\nu) \times 10^4}{(I_R + I_L)(\nu)} = \text{CID}\, X_i \frac{1}{\pi} \frac{\gamma}{(\nu - \nu_i)^2 + \gamma^2} \tag{5.9}$$

where ν_{in} is the incident (laser) frequency and γ is the half-width of the Lorentzian band at one-half of the peak's height (in cm^{-1}). The default value is $4\,cm^{-1}$ in the Gaussian program.

For each frequency mode, the intensity can be simulated to a Gaussian distribution (normal distribution), if desired, using the formulas above by changing the ν values. After summing all the modes' Raman, ROA, or CID, it will give the corresponding Raman, ROA, and CID spectra, respectively, for the *i*th conformer. By Boltzmann statistics using different energy data, such as TEE (total electronic energy), GFE (Gibbs free energy), or ZPE (zero-point energy), whole Raman, ROA, or CID spectra can be obtained using the vibrational frequency mode as the X-axis (in cm^{-1}).

The relationship between the *Raman* X_i and *ROA* X_i is listed in Table 5.1 [1]. In the far-from resonance (FFR) approximation, the incident circular polarization (ICP) and scattered circular polarization (SCP) forms lead to the same result; that is why the boxes contain "ICP = SCP."

"DCP_I" (dural circular polarization) is the in-phase dual circular polarization form. In the FFR approximation

DCP_I-ROA (180°) = ICP_U-ROA (180°) = SCP_U-ROA (180°), but
DCP_I-Raman (180°) < ICP_U-Raman (180°) = SCP_U-Raman (180°).

The angles in parentheses indicate whether they are the intensities for backscattering (180°) or right-angle scattering (90°) experiments.

The wavelength for the static limit is 532 nm used in the Gaussian package (which is the laser frequency of the BioTools instrument), so the frequency ν_{in} is $18\,796.99\,cm^{-1}$ in this case. The ROA instrument that have been constructed include the right-angle, forward scattering, and backscattering ICP ROA setups as well as ICP, SCP and DCPI and DCPII backscattering ROA, and,

Table 5.1 Relationship between the *Raman* X_i and *ROA* X_i [1].

i of X_i	*Raman X*	*ROA X*
1	ICP_U(180°) = SCP_U(180°)	ICP_U(180°) = SCP_U(180°)
2	ICP_D(90°) = SCP_D(90°)	ICP_D(90°) = SCP_D(90°)
3	DCP_I(180°)	DCP_I(180°)

The subscripts "U" and "D" refer to use of unpolarized incident radiation or depolarized ROA (a polarization analyzer is placed in the scattered beam parallel to the scattering plane), respectively.

currently, available commercial ROA instruments from BioTool include the SCP backscattering and simultaeous backscatering for all four forms of ROA [12].

Currently, it is difficult to measure the absolute Raman/ROA cross section, which could be related to the calculated intensities on a per molecule (mole) basis just like IR/VCD for epsilon and delta epsilon. So, the Raman/ROA counts or Raman/ROA intensity is plotted on an arbitrary scale for the comparison of measured and calculated Raman/ROA spectra, respectively.

5.3 Principles Using VCD and ROA

Computation of VCD just needs the investigation of molecular geometries at the ground state instead of excited state like ECD; thus, the computational basis sets have no big effect on the VCD spectra. The basis set at B3LYP/6-31G(d) in many cases is reasonable for the prediction of a simple chiral molecular VCD spectra. However, for a relatively complex chiral compound, to obtain a reasonable geometry that matches a real case, a 6-311+G(d) set may be perfect, especially in energy data predictions [13].

Because of the existence of the C–H bond in all organic compounds, to reduce the interfering signals of C–H from the solvent, solvents such as $CDCl_3$, CD_3OD, and CCl_4 must be used in VCD measurements. The effect of the solvent on VCD needs to be removed from the whole spectra. After the deduction, it can give important information on the C–H bond or other H– connected bonds.

As mentioned above, the VCD signals are very sensitive to the structure, including the relative positions between different chiral molecules. Also, because of this characteristic, sometimes the predicted VCD will not agree well with the experimental VCD when a chiral molecule has two or two more hydroxyl groups inside (other groups like $-NH_2$ have the same effects), especially when the two –OH groups are close enough.

A VCD study to determine the orientations of four axial compounds **4** in solution was performed [14]. Compound **4** has two ways in which to connect the H-bond, **a** and **b**, respectively (Figure 5.2). The energy difference between the two H-bond schemes is not high, as seen from the potential energy scan (PES) searches. The VCD results confirmed that the full connection is **aaab**, as illustrated below. Their orientations were well predicted. This is an interesting and successful example in the orientation study of four chiral molecules.

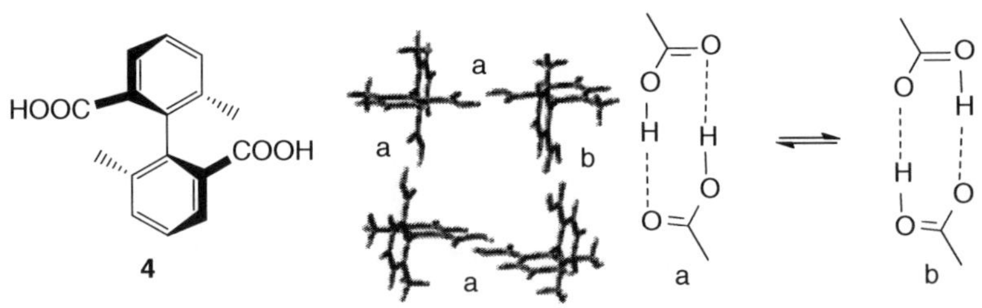

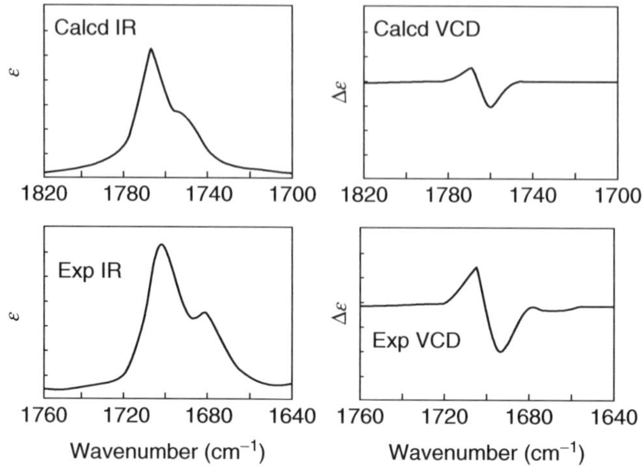

Figure 5.2 Experimental and theoretical IR and VCD spectra for **4** [14].

This example suggests that the interaction between the different molecules should be frequent. This not only provides us with much information on the structures but also points the difficulty in structural study. For instance, the example listed above can be estimated on the basis of the X-ray structures of their (*R*)-isomer [14]. However, if there is no crystal structure to be used as a reference, then arriving at a conclusion on the number of molecules in computation becomes a challenge. If the number is more than three, how do we construct their relative orientations (relative positions) in optimizations? These are the fundamental questions in VCD study.

So, the best candidate molecule used in VCD study should contain either the –OH, $-NH_2$, or $-CO_2H$ group. If there are two or two more these groups, they should not connect closely, like 1,2-diol structures. If one of the groups appears in a chiral molecule, the VCD signals may not be good enough for comparison of the theoretical result with the experimental one. Of course, if the polar groups do not interact strongly among different molecules in solution or in solid, it is still a useful way to use VCD for AC study. The cases discussed for VCD study may exist in experiments with ROA as well.

Contrary to the cases mentioned above, if a chiral molecule has no –OH or other active-proton-free groups, then VCD spectra of generally high resolution and quality can be obtained. They generally agree well with the predictions.

When a chiral molecule has no UV absorption, it is impossible to use ECD for AC determinations; or a chiral molecule may have a small OR value that cannot be well predicted using quantum methods. In the both cases, it is a good choice to use VCD for AC study.

In VCD study, the corresponding IR spectra must be measured simultaneously. It should provide reliable results after comparing the predicted VCD and IR with the experimental VCD and IR. This point must be emphasized.

In ROA studies, it is found that the Raman spectral intensities are sensitive to the fuse augmentation of the basis set. For example, the minimum of 3-21++G basis set has been recommended [15].

As expected, because of the characteristics of the scattering of light, the "shape" of a molecular geometry has an effect on its scattering spectra. It appears that because of the strong interaction between two chiral molecules, or among several different chiral molecules, the "shape" of a "dimer" or "trimer" may be different from its monomer structure, so their Raman, ROA, and CID spectra may be different from the corresponding monomer spectra. This is similar to the VCD spectra.

It is different from the IR/VCD spectra, which give the absolute values for the absorption difference when right- and left-CPL pass through the chiral sample. It is possible, with careful calibration, to measure the absolute Raman/ROA cross section, which could be related to the calculated intensities on a per molecule (mole) basis just like IR/VCD for epsilon and delta epsilon. However, it is really very hard to achieve experimentally, so the Raman/ROA counts or Raman/ROA intensity is plotted on an arbitrary scale.

5.4 Application

5.4.1 VCD Application

VCD can be widely used in AC determinations [1–3]. The use of different basis sets will clearly affect the VCD spectra in many cases. Discussion of the effect is useful. As an example, α,β-unsaturated ketone represents a typical structure existing widely in synthetic and natural products chemistry, like **5**. Because only one >C=O exists in molecule **5**, one expects no VCD signals near 1700–1800 cm^{-1}. The major VCD signals will appear from 1000 to 1400 cm^{-1} caused by various stretching vibrations. Thus, the VCD of monoterpene (1*R*)-**5** was studied using B3LYP theory with the basis sets 6-31G(d,p), 6-31+G(d,p), 6-311+G(d,p). DGDZVP, and B3PW91/DGTZVP, respectively [16].

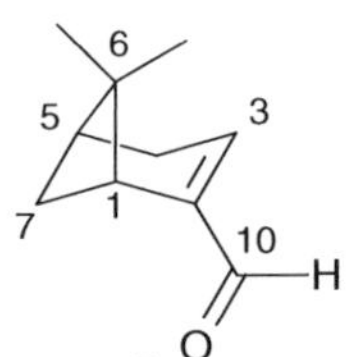

5

All calculation results provided a good balance between computation cost and VCD spectral accuracy. The DGDZVP basis set just needed about a quarter of the time taken by the 6-311+G(d,p) basis set. (−)-Myrtenal **5** has AC of (1*R*) by comparison of its calculated VCD and experimental VCD (Figure 5.3).

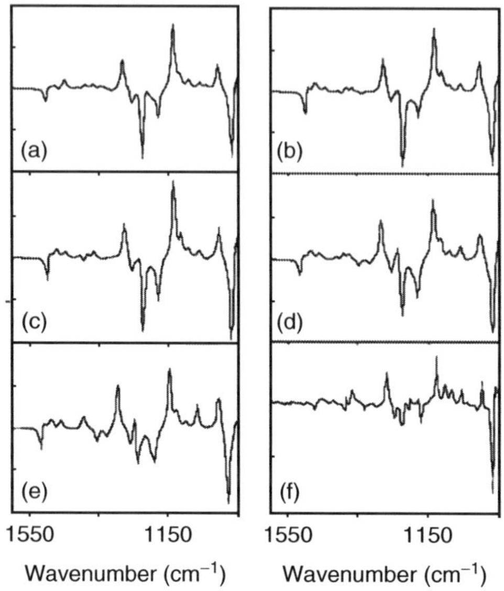

Figure 5.3 Calculated VCD spectra at the (a) 6-31G(d), (b) 6-31+G(d,p), (c) 6-311+G(d,p), (d) DGDZVP, and (e) B3PW91/DGTZVP level. Experimental VCD (f) of (−)-myrtenal (**5**) is also illustrated. The IR spectra are omitted here [16].

It was found that the higher basis sets in VCD calculation have no big effect on the VCD spectra. This is highly advantageous in AC assignment. A reasonable basis set, such as 6-31+G(d), may be good enough for VCD predictions for many chiral compounds. This is highly beneficial for researchers who do not have enough computational sources in their hands, because in this case a general PC can do the VCD calculations using the recommended basis sets.

An atropisomer is an important kind of chiral compound, such as **6** [17] isolated from the marine sponge *Didiscusaceratus,* whose AC was assigned as the (*S*)-(+)-curcuphenol dimer using the VCD method (Figure 5.4).

Chiral sulfoxide is a valid intermediate in organic and pharmaceutical chemistry, and can be synthesized quickly and measured using VCD. Perdeuteriophenyl-phenyl-sulfoxide (7) was used for VCD and IR predictions using B3LYP/TZ2P and B3PW91/TZ2P methods, respectively [18]. For this simple chiral compound, only one dominant conformer was found. The VCD and IR spectra simulated using B3LYP/TZ2P are more accurate than those using B3PW91/TZ2P methods (Figure 5.5). The predicted and experimental VCD and IR matched well except for the frequency difference near 1450 – 1500 cm^{-1}.

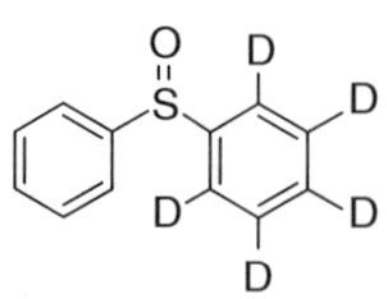

7

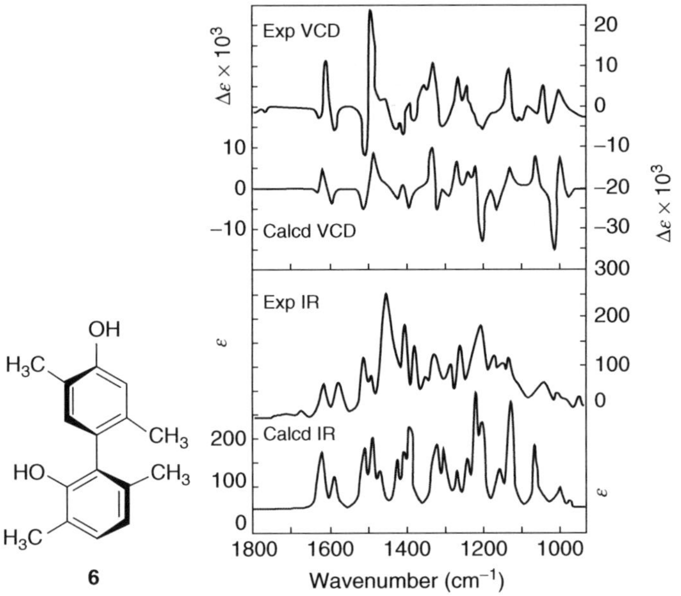

Figure 5.4 The observed and calculated IR/VCD [18].

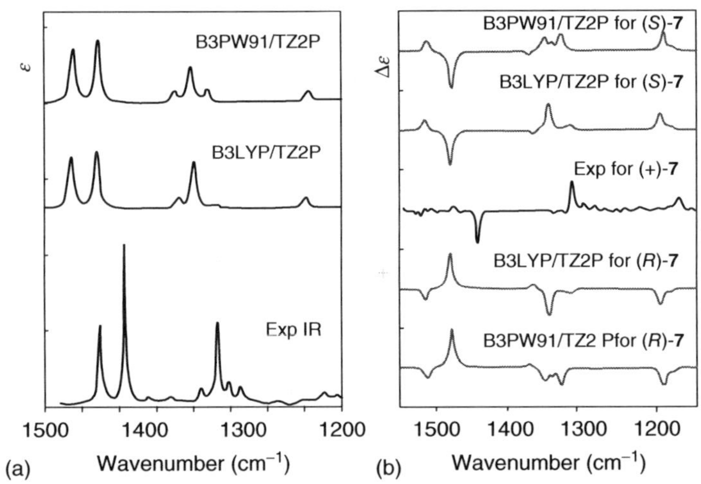

Figure 5.5 (a) Comparison of the IR spectra of **7** at the B3LYP/TZ2P and B3PW91/TZ2P level to the experimental IR spectrum. (b) Comparison of the VCD spectra of (*R*)-**7** and (*S*)-**7** at the B3LYP/TZ2P and B3PW91/TZ2P level to the experimental VCD spectrum of (+)-**7** [18].

Usually, for sulfoxide-containing molecules, the agreement between the experimental and calculated VCD spectra is not very good when 6-31G(d) basis sets are used because of a lower accuracy frequency simulation for the S=O stretching mode. Among the larger basis sets, it was found that B3LYP/TZ2P is

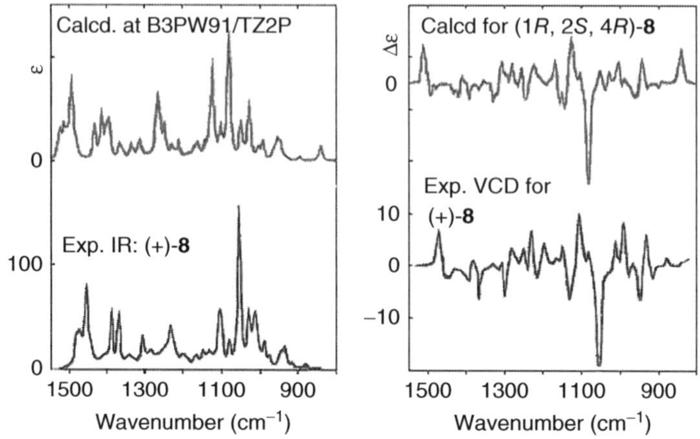

Figure 5.6 Experimental and calculated IR and VCD for **8** [19].

a better method for calculating the geometry and VCD for sulfoxide-containing molecules [18].

The hydroxyl-containing *endo*-borneol **8** is conformationally rigid [19]. Three thermally accessible stable conformations were found for this molecule. These geometries were used for VCD and IR analysis using DFT methods. Comparison of IR and VCD spectra to the experimental spectra unambiguously confirmed its AC as listed in the structure was correct (Figure 5.6).

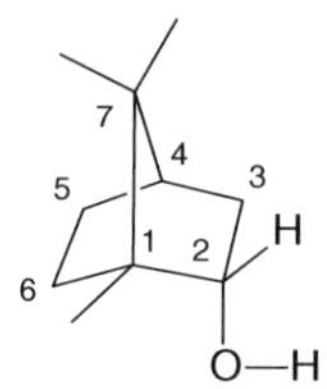

(1*R*,2*S*,4*R*)-**8**

The phenolic compound **9** is a very potent aromatase inhibitor with IC_{50} 0.6 nM, comparable to letrozole. This structure contains the groups –OH, –CN, and triazole. It is a polar chiral acyclic compound. This structure does not easily form a dimer, trimer, or tetramer in solution. The (*R*) AC was chosen for calculations of VCD at the B3LYP/6-31G(d) level [20]. Eight conformations were optimized, and the GFE values were utilized in Boltzmann statistics to simulate the corresponding VCD spectra (Figure 5.7).

It is valuable to compare the differences of VCD and IR for the same series of chiral natural products. This offers a chance for us to see the effect of different substituents on vibrational modes (VCD). The VCD curve is extremely sensitive to the changes of substituents. For example, three peperomin-type secolignans (**10–12**) from *Peperomiablanda* (Piperaceae) were used for VCD study using B3LYP/6-31G(d) methods [21]. It was found that the three VCD spectra

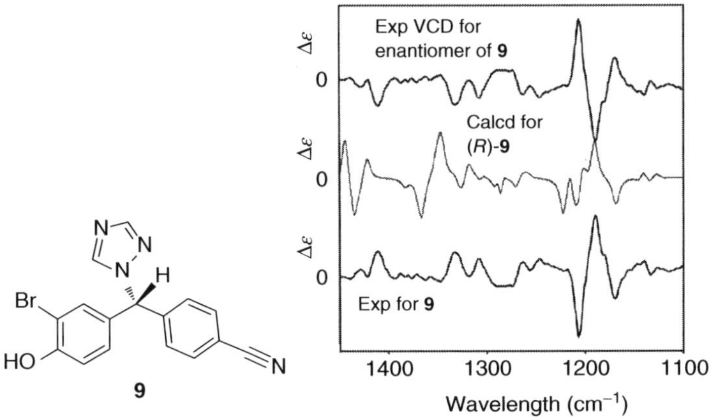

Figure 5.7 VCD and IR spectra for the phenolic compound **9** [20].

were totally different. This demonstrates the use of VCD obtained through computations for AC assignment (Figure 5.8). This is different from the ECD method. In ECD study, if the substituents located far away from stereogenic center change, the ECD curve may change only very little.

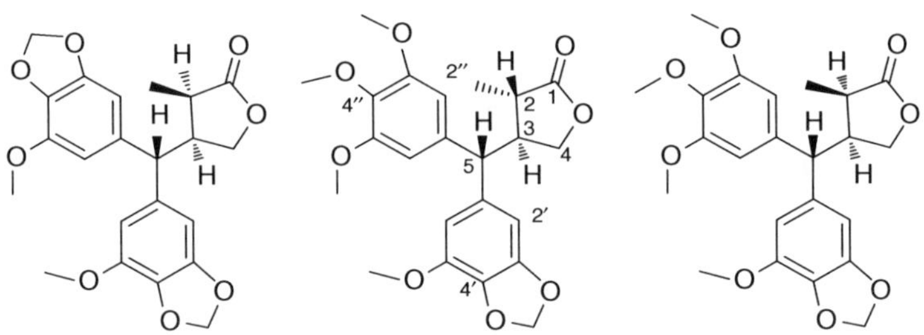

A chemical conversion to seco-oxacassane diterpenes **13** was obtained with conformational restriction [22]. It was assigned an AC of (5*S*,8*R*,9*R*,10*S*), which was studied by VCD at the B3PW91/DGDZVP2 level. Again, the active-proton-free compound showed good agreement between the experimental and theoretical VCD spectra (Figure 5.9).

Preussidone (**14**) [23], an uncommon 2,5-diarylcyclopentenone compound, was obtained from *Preussia opharum*. This structure can be converted to an anion structure **15** below a pH value of 12. This is accompanied by a fresh red color formation (it is dark grey color in the b/w figure). Its RA structure was established by NMR, while the AC of **14** was assigned by the VCD technique (Figure 5.10).

The agreement between experimental and theoretical VCD clearly shows that VCD is a good tool to investigate real cases in the microscopic world. This also allows theoretical chemists to study the stereochemistry of chiral compounds just by means of the VCD method, for example, the stereochemistry of the helical compounds **16–23** [24]. The relative energies show that the *M* atropisomer is more

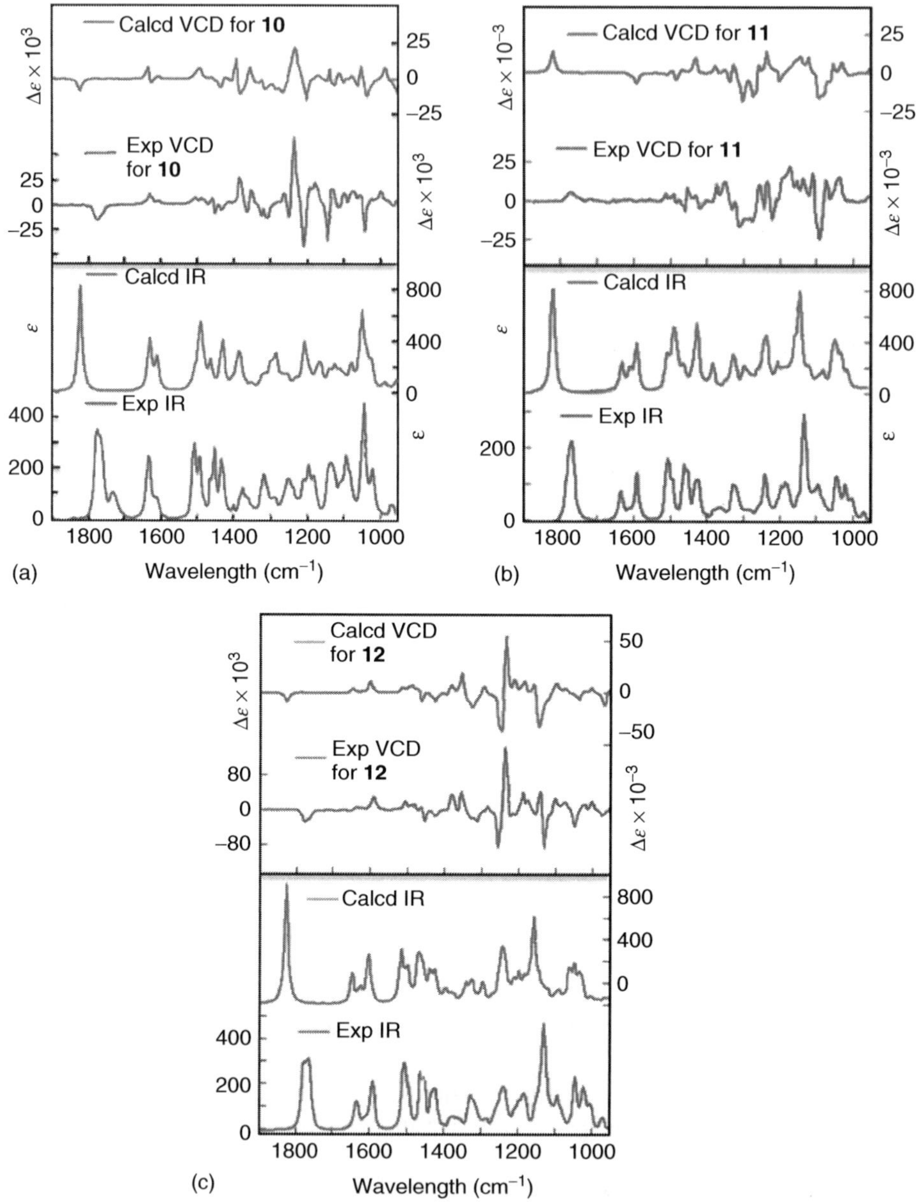

Figure 5.8 Comparison of the VCD and IR spectra of the measured (−)-**10** with the predicted (2*R*,3*S*)-**10**, (+)-**11** with (2*S*,3*S*,5*S*)-**11**, and (−)-**12** with (2*R*,3*S*,5*S*)-**12**. The B3LYP/6-31G(d) method was used for VCD and IR calculations for (2*R*,3*S*)-**10** and (2*S*,3*S*,5*S*)-**11**, and B3PW91/TZVP was used for the VCD and IR calculations for (2*R*,3*S*,5*S*)-**12** [21].

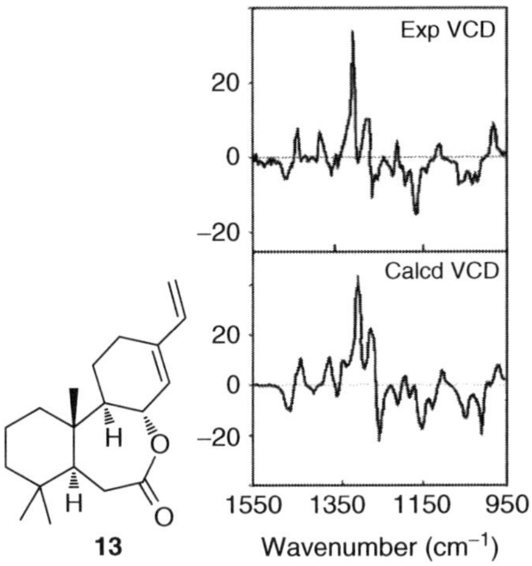

13

Figure 5.9 Experimental and calculated VCD spectra at the B3PW91/DGDZVP2 level [22].

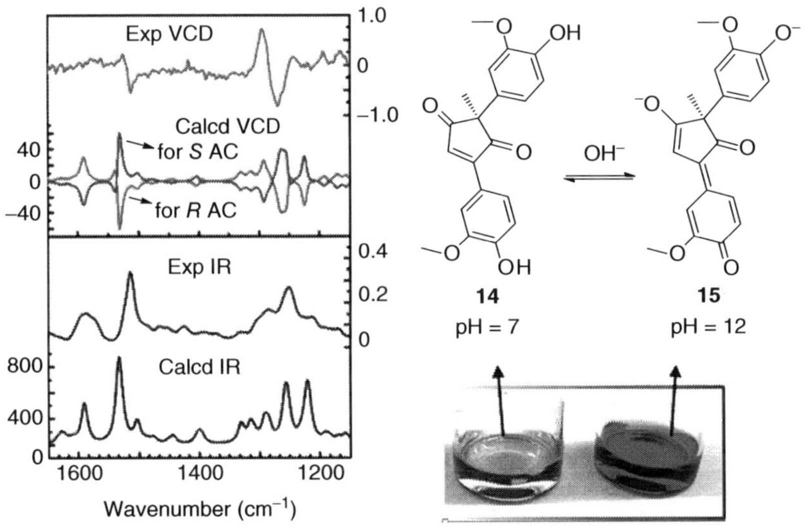

Figure 5.10 Comparison of the calculated VCD for **14** and the experimental VCD [23].

stable than the *P* atropisomer for both the helicenes, with or without the camphanate group. The order of the relative energy for different isomers (in kcal/mol) is *M*-**16** = *P*-**20**, *M*-**17** (0.00) < *P*-**21** (0.02), *M*-**18** (0.00) < *P*-**22** (1.67), and *M*-**19** (0.00) < *P*-**22** (1.38). As expected, *M*-**16** and *P*-**20**, and *M*-**17** and *P*-**21** are enantiomers, so their VCD almost exhibited mirror-image-like curves. *M*-**18** and *P*-**22**, and *M*-**19** and *P*-**23** are not enantiomers since there is a chiral camphanate group, so their VCD spectra showed large differences (Figure 5.11) [24].

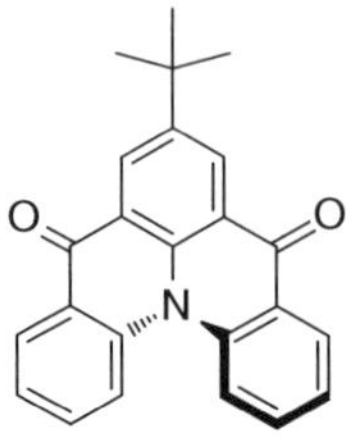

M-**16**

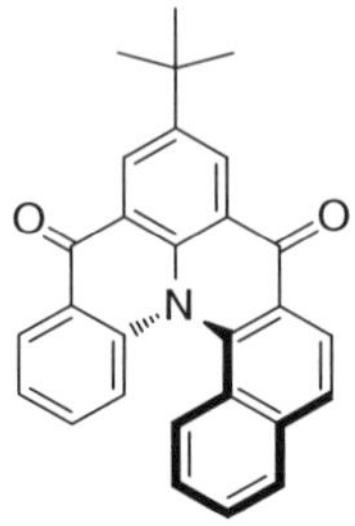

M-**17**

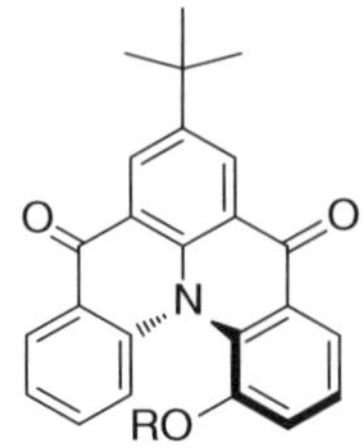

M-**18**

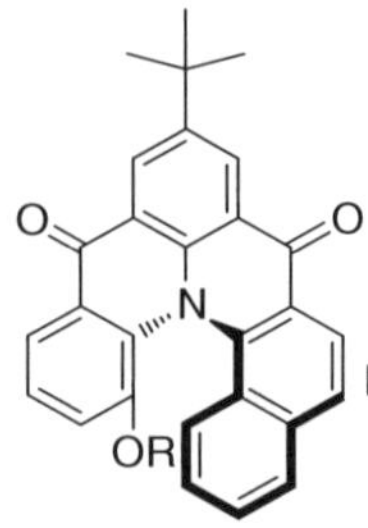

M-**19**

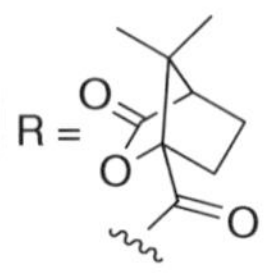

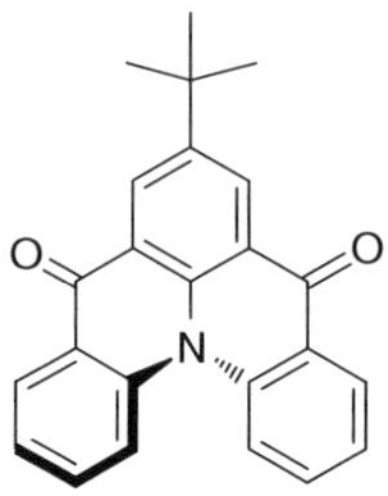

P-**20**

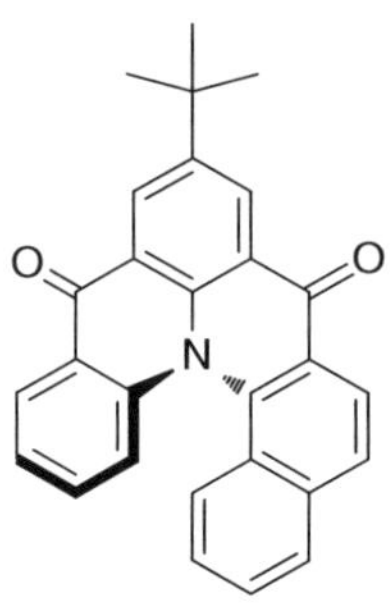

P-**21**

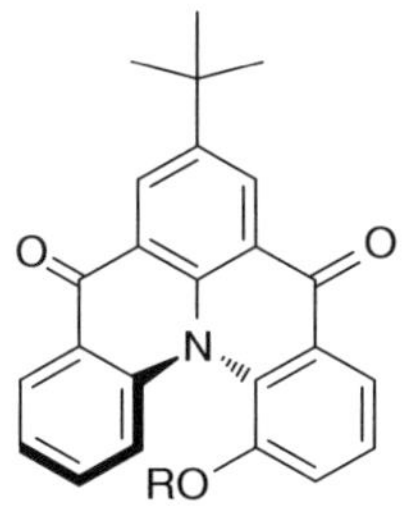

P-**22**

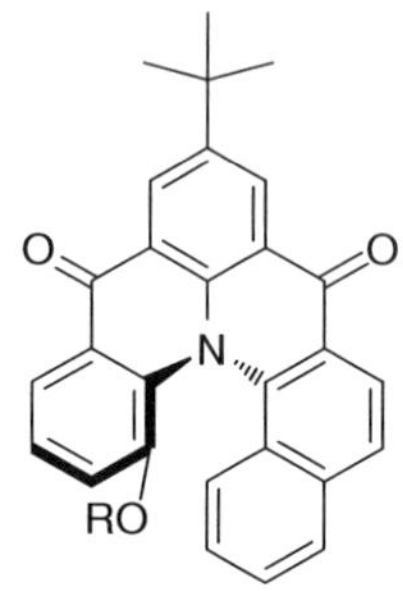

P-**23**

It is understandable that *M*-**16** and *P*-**20** should have almost the same energy since they are enantiomers. However, *M*-**17** has 0.02 kcal mol^{-1} lower energy than *P*-**17**. On removing the *tert*-butyl group in calculations, both *M* and *P* structures exhibited the same energy. It may hint that *M*-**17** and *P*-**21** are not mirror images in energy computations as in some original reports. For example, the tertiary butyl group should have two stable conformations in *M* and *P* structures, respectively, as illustrated below. For *M*-**17**, the geometry *M*-**17a** was lower in energy by 0.0196 kcal mol^{-1} than *M*-**17b** at the B3LYP/6-31G(d) level in the gas phase. When the total mirror-image-like geometries *M*-**17a** and *P*-**21a** were used in the calculations, they both exhibited almost the same energy ($<10^{-6}$ kcal mol^{-1}). To conclude that *M*-**17** has a lower energy of 0.02 kcal mol^{-1} than *P*-**21** may use the energetics of *M*-**17a** and *P*-**21b** in the calculations. The small energy difference did not affect the VCD simulation greatly. However, to conclude that two enantiomers do not have the same energy may mislead other researcher's conclusion. For example, in study of the origin of life, this 0.02 kcal mol^{-1} energy difference is absolutely a big data, which would lead the enantiomeric excess values to reach 1.6% in almost racemic compound formation. If this *M* structure were more stable than the *P* structure, it means it would have a natural 1.6% ee. This natural 1.6% ee may play a key role in the evolution process in the origin of life. This is absolutely a big discovery that may change the theory of the origin of life. Owing to some asymmetric force effect, the formation of enantiomers might exist with a tiny enantiomeric excess, such as $10^{-5}-10^{-6}$%. Therefore, the 1.6% ee is absolutely a big value in nature. This clearly hints that, when the relative

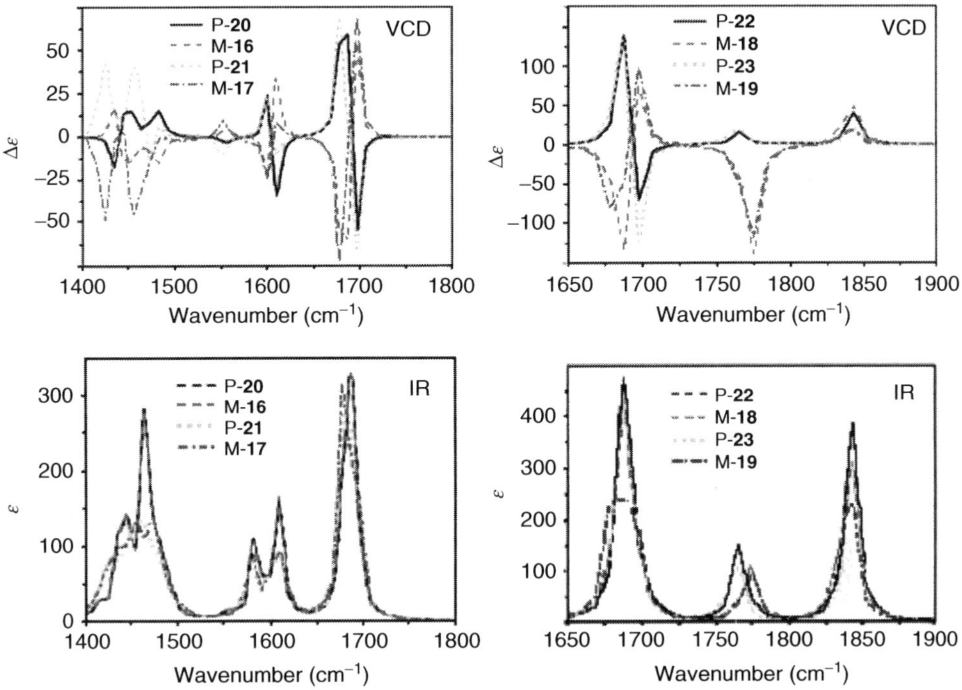

Figure 5.11 Comparison of VCD and IR for four pairs of helical compounds **15–22** [24].

energies are to be compared, the most basic and fundamental step.is to use the correct geometry.

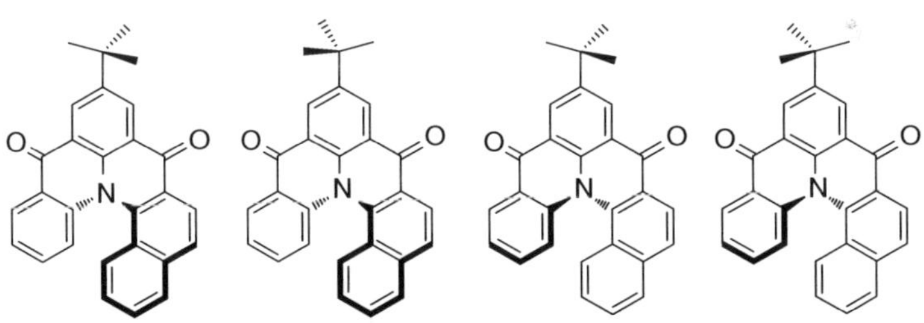

	M-**17a**	M-**17b**	P-**21a**	P-**21b**
AM1	0.0000(00)	0.0120(54)	0.0000(00)	0.0120(16)
B3LYP/ 6-31G(d)	0.0000(00)	0.0196(09)	0.0000(00)	0.0196(09)

The organocatalytic Michael addition is an important reaction to form a new C–C bond with various enantioselectivities. The addition of malonate (**24**) to the symmetric unsaturated 1,4-diketone (**25**) afforded the product **26** with up to 93% ee catalyzed by thiourea and squaramide derivatives with *Cinchona* alkaloids [25]. The AC of the product was suggested from a comparison of the experimental and calculated VCD spectra of the reaction product **26**. DFT calculations at the

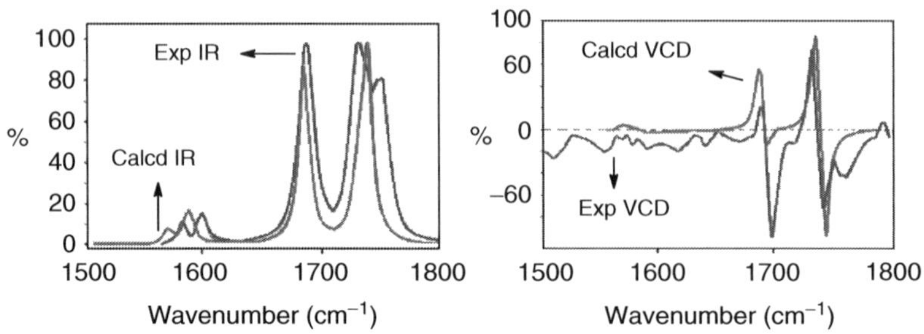

Figure 5.12 Calculated and experimental IR and VCD spectra of compound (*R*)-**25** [25].

B3PW91/6-311G(d,p) level were performed for (*R*)-**26**. VCD was simulated by Boltzmann distribution using GFE (Figure 5.12). Good agreement between the calculated and experimental spectra directly allowed the assignment of the AC of **26** as the (*R*)-enantiomer. By the way, in this example, the exciton coupling phenomenon from the neighboring C=O groups was observed at the vibrational frequency near 1750 cm^{-1}. Readers who are interested in analyzing the relationship can do the conformational searches first and then to conclude the AC assignment.

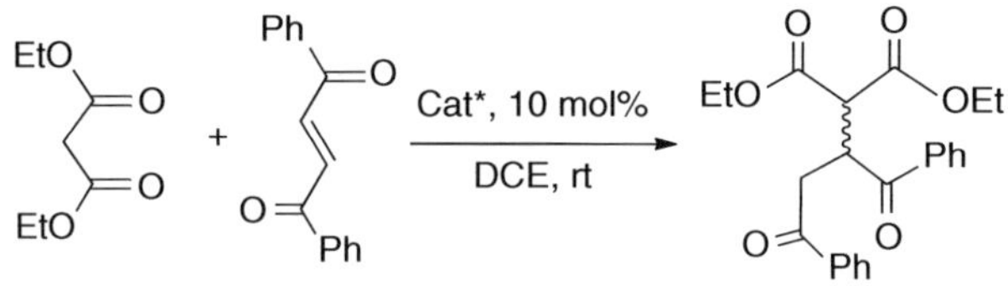

24 **25** **26**

5.4.2 ROA Application

After the first ROA spectrum was reported in 1973 [4, 9], it was soon established as a valid tool in AC study for chiral molecules. Apart from its achievements, the stereochemistry of biological systems, such as proteins [26] and nucleic acids [27], can be conveniently measured in an aqueous environment. Because of the useful information recorded in ROA signals, its usage is becoming wider and wider in AC assignment for small chiral compounds. For example, two enantiomers of methyloxirane (**27**) were used for an ROA study [28]. Methyloxirane has a very low boiling point (33 °C), which allowed the ROA measurement to be performed relatively easily. However, it took much time for accumulation. In contrast, the larger concentration in the neat liquid (14.3 M) provided the spectra much faster (minutes) and with smaller noise (Figure 5.13).

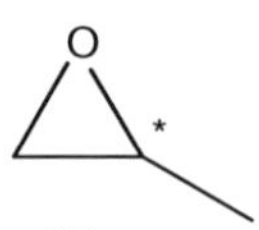

27

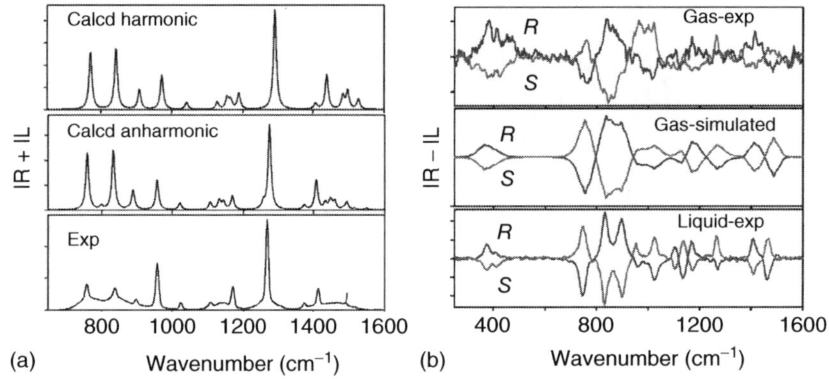

Figure 5.13 (a) Raman spectra (from top to bottom) calculated (B3LYP/aug-cc-pVTZ, Lorentzian profiles used only) harmonic and vibrational configuration interaction (VCI) anharmonic Raman spectra and the experimental spectrum of (*S*)-methyloxirane vapor. The calculated intensities were scaled arbitrarily by a common factor according to the experiment. (b) ROA spectra of the methyloxirane enantiomers, measured with vapor (top), simulated with the scaled B3LYP/aug-cc-pVTZ calculation (middle, frequencies scaled), and neat liquid (bottom). The *y*-axis is arbitrary [28].

ROA calculations were performed at different levels, and the backscattered intensities were generated by convolution with Lorentzian bands of 5 cm^{-1} width. The predicted ROA spectra agreed well with the experimental ones (Figure 5.13).

Traditionally, in protein structure study, an important assumption is that relatively simple general principles determine the shapes of all peptides and proteins by the preferred and forbidden values of the torsional angles φ and ψ [29]. Now, the ROA technique can be used to study the conformation for a chiral compound. For example, a dipeptide **28**, because of the rotation of the dihedral angles ψ and φ, theoretically, can form several stable conformations [30]. By PES via changing ψ and φ using theoretical free-energy surfaces at the PMF/ff03 and *m*PWPLYPD/6-311++G(d,p)/SMD, respectively, two stable conformers (**28a** and **28b**) were found.

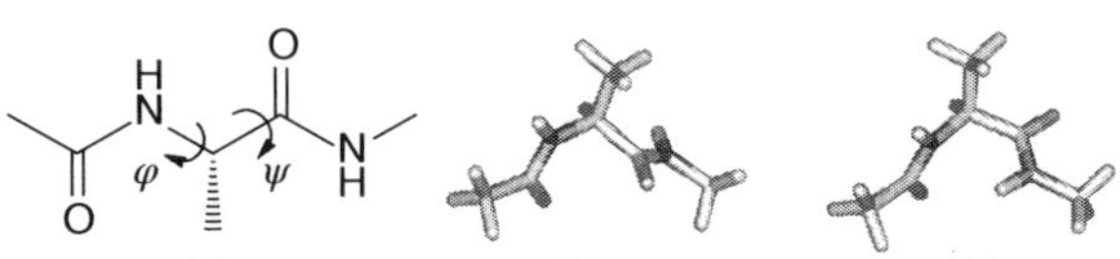

28 **28a** **28b**

Comparison of the theoretical and experimental spectra is illustrated in Figure 5.14. Also, spectra of the two highest populated conformers, **28a** and **28b**, are compared with the whole grid fit and experiments. It could confirm the multicomponent (multiconformational) character of the dipeptide's ROA response instead of the two major conformers. For example, more conformers are needed to reproduce the double-negative ROA signal at 348/399 cm^{-1} or the couplet centered around 1300 cm^{-1}.

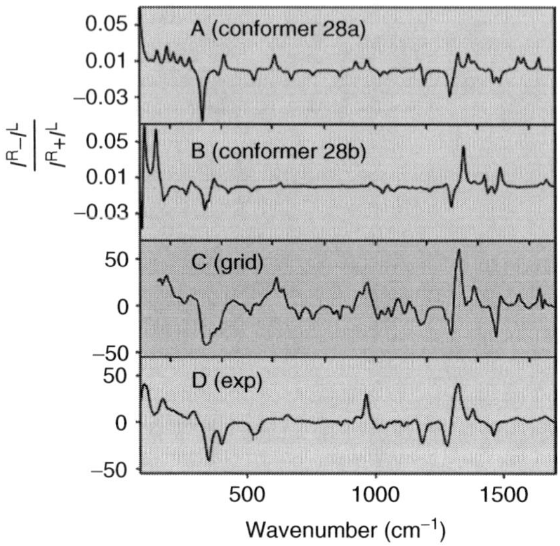

Figure 5.14 ROA spectra for **28** calculated for (φ, ψ) = (−59°, 144°) (A, conformer **28a**) and (−77°, −15°) (B, conformer **28b**), the decomposition fit using all grid spectra (C), and the experiment (D). Selected peak wavenumbers are indicated. The intensity scale is in atomic units (A, B) and arbitrary (C, D) for normalized spectra) units [29].

References

1. (a) Berova, N., Polavarapu, P., Nakanishi, K., and Woody, R.W. (2012) *Comprehensive Chiroptical Spectroscopy*, John Wiley & Sons, Inc., Hoboken, NJ; (b) Cheeseman, J.R., Frisch, M.J., Devlin, F.J., and Stephens, P.J. (1996) *Chem. Phys. Lett.*, **252**, 211–220.
2. Taniguchi, T. and Monde, K. (2012) *J. Am. Chem. Soc.*, **134**, 3695–3698.
3. (a) Nafie, L.A. (2011) *Vibrational Optical Activity: Principles and Applications*, John Wiley & Sons, Ltd., Chichester; (b) Stephens, P.J., Devlin, F.J., and Cheeseman, J.R. (2012) *VCD Spectroscopy for Organic Chemists*, CRC Press; (c) Barron, L.D. (2004) *Molecular Light Scattering and Optical Acitivity*, 2nd ed., Oxford University Press.
4. Oshorne, G.A., Cheng, J.C., and Stephens, P.J. (1973) *Rev. Sci. Instrum.*, **44**, 10–15.
5. Barron, L.D., Bogaard, M.P., and Buckingham, A.D. (1973) *J. Am. Chem. Soc.*, **95**, 603; (b) Nafie, L.A., Keiderling, T.A., Stephens, P.J. (1976) *J. Am. Chem. Soc.*, **98**, 2715–2723; (c) Nafie, L.A., Diem, M., Vidrine, D.W. (1979) *J. Am. Chem. Soc.*, **101**, 496–498.
6. Chabay, I. and Holzwarth, G. (1975) *Appl. Opt.*, **14**, 454–459.
7. Nafie, L.A., Cheng, J.C., and Stephens, P.J. (1975) *J. Am. Chem. Soc.*, **97**, 3842. (b) Nafie, L. A., Keiderling, T. A., and Stephens, P. J. (1976) *J. Am. Chem. Soc.*, **98**, 2715–2723. (c) Nafie, L. A., Diem, M., and Vidrine, D. W. (1979) *J. Am. Chem. Soc.*, **101**, 496–498.
8. (a) Kwiecki, R.W., Devlin, F., Stephens, P.J., Amos, R.D., and Handy, N.C. (1998) *Chem. Phys. Lett.*, **145**, 411–417; (b) Gaussian, Inc. *http://www.gaussian.com* (accessed 18 July 2016)
9. (a) Barron, L.D., Bogaard, M.P., and Buckingham, A.D. (1973) *Nature*, **241**, 113; (b) Ruud, K., Helgaker, T., and Bour, P. (2002) *J. Phys. Chem. A*, **106**, 7448–7455.
10. Barron, L.D. and Buckingham, A.D. (1971) *Mol. Phys.*, **20**, 1111.
11. (a) Cheeseman, J.R. and Frisch, M.J. (2011) *J. Chem. Theory Comput.*, 7(10), 3323–3334; (b) Berova, N.,

Polavarapu, P., Nakanishi, K., Woody, R. W. *Comprehensive Chiroptical Spectroscopy*, John Wiley & Sons, Inc., Hoboken, NJ, 2012, pp. 147-153.

12. (a) Li, H., Nafie, L.A. (2012) *J. Raman Spectrosc.*, **43**, 89–94; (b) Barron, L.D. and Buckingham, A.D. (2010) *Chem. Phys. Lett.*, **492**, 199–213.
13. Li, Q.M., Ren, J., Shen, L., Bai, B., Liu, X.C., Wen, M.L., and Zhu, H.J. (2013) *Tetrahedron*, **69**, 3067–3074.
14. Urbanová, M., Setnika, V., Devlin, F.J., and Stephens, P.J. (2005) *J. Am. Chem. Soc.*, **127**, 6700–6711.
15. (a) Furche, F. (2007) *J. Chem. Phys.*, **126**, 201104; (b) Zuber, G. and Hug, W. (2004) *J. Phys. Chem. A*, **108**, 2108–2119.
16. Burgueño-Tapia, E., Zepeda, L.G., and Joseph-Nathan, P. (2010) *Phytochemistry*, **71**, 1158–1161.
17. CichewiczR, H., Clifford, L.J., Lassen, P.R., Cao, X.L., Freedman, T.B., Nafie, L.A., Deschamps, J.D., Kenyon, V.A., Flanary, J.R., Holman, T.R., and Crews, P.B. (2005) *Med. Chem.*, **13**, 5600–5612.
18. Drabowicz, J., Zajac, A., Lyzwa, P., Stephens, P.J., Pan, J.J., and Devlin, P.J. (2008) *Tetrahedron: Asymmetry*, **19**, 288.
19. Devlin, F.J., Stephens, P.J., and Besse, P. (2005) *J. Org. Chem.*, **70**, 2980–2993.
20. Wood, P.M., Woo, L.W.L., Labrosse, J.R., Trusselle, M., Abbate, S., Longhi, G., Castiglioni, E., Lebon, F., Purohit, A., Reed, M.J., and Potter, B.V.L. (2008) *J. Med. Chem.*, **51**, 4226–4238.
21. Felippe, L.G., Batista, J.M. Jr.,, Baldoqui, D.C., Nascimento, I.R., Kato, M.J., He, Y., Nafie, L.A., and Furlan, M. (2012) *Org. Biomol. Chem.*, **10**, 4208–4214.
22. Manriquez-Torres, J.J., Torres-Valencia, J.M., Gomez-Hurtado, M.A., Motilva, V., Garcia-Maurino, S., Avila, J., Talero, E., Cerda-Garcia-Rojas, C.M., and Joseph-Nathan, P. (2012) *J. Nat. Prod.*, **74**, 1946–1951.
23. Du, L., King, J.B., Morrow, B.H., Shen, J.K., Miller, A.N., and Cichewicz, R.H. (2012) *J. Nat. Prod.*, **75**, 1819–1823.
24. Pandith, A., Islam, N., Syed, Z.F., Suhail-ul, R., Bandaru, S., and Anoop, A. (2011) *Chem. Phys. Lett.*, **516**, 199–203.
25. Žari, S., Kailas, T., Kudrjashova, M., Öeren, M., Järving, I., Tamm, T., Lopp, M., and Kanger, T. (2012) *Beilstein J. Org. Chem.*, **8**, 1452–1457.
26. (a) Barron, L.D., Blanch, E.W., and Hecht, L. (2002) *Adv. Protein Chem.*, **62**, 51–90; (b) Syme, C.D., Blanch, E.W., Holt, C., Jakes, R., Goedert, M., Hecht, L., and Barron, L.D. (2002) *Eur. J. Biochem.*, **269**, 148–156.
27. Barron, L.D., Hecht, L., Blanch, E.W., and Bell, A.F. (2000) *Prog. Biophys. Mol. Biol.*, **73**, 1–49.
28. Jaroslav, S. and Petr, B. (2011) *J. Phys. Lett.*, **2**, 498–502.
29. Ramachandran, G. N., Ramakrishnan, C., Sasisekharan, V. *J. Mol. Biol.* 1963, **7**, 95–99.
30. Parchaňský, V., Kapitán, J., Kaminský, J., Šebestík, J., Bouř, P. *J. Phys. Chem. Lett.* 2013, **4**, 2763–2768.

6
Combinational Use of Different Methods

For complex chiral compounds, it is not easy to identify their absolute configuration (AC) in a short time experimentally and theoretically. Earlier, for example, before the 1960s, most researchers identified the decomposed fractions from the target molecules and then reconnected the decomposed molecules and recovered the target molecules' structure including their stereochemistry. With the application of the X-ray method, and some empirical methods, such as optical rotation (OR) studies in which an unknown chiral compound's OR sign is compared with that of a similar known chiral compound, more and more complex ACs of chiral compounds have been identified. However, with the development of various separation methods, such as high-performance liquid chromatography (HPLC), many new compounds are obtained in very limited quantities, such as 1–4 mg, and some of them cannot be either crystallized for X-ray studies or converted into Mosher esters. These difficulties lead to a big puzzle in AC assignment.

With the development of computational technology, especially supercomputer technology, quantum theoretical studies go before experimental work, after the former stayed behind the scene for a long time. The maturing of different software packages, such as the Gaussian package, has brought big benefits not only for theoretical but also for experimental chemists. Currently, computation of ^{13}C nuclear magnetic resonance (NMR), OR, electronic circular dichroism (ECD), vibrational circular dichroism (VCD), and Raman optical activity (ROA) have all been well applied in AC identification of complex chiral compounds involved in natural products chemistry and organic synthetic chemistry, as well as extended to polymer and other corresponding researches.

At the same time, because of the fact that every method has its usage range or drawbacks, the combined use two or two more methods should be encouraged in AC assignment for complex chiral compounds.

6.1
Tactics to Select Methods

Before a method is selected for AC or RC (relative configuration) assignment, it is necessary to know the usage range of that method. If we work outside of this

Organic Stereochemistry: Experimental and Computational Methods, First Edition. Hua-Jie Zhu.

usage window, it may lead to wrong prediction. Here, the usage window of different methods and their characteristics are summarized.

6.1.1
^{13}C NMR Methods

A comparison of the computed ^{13}C NMR spectrum of a compound with the experimental data is a good way to test whether its structure is correct or not in many cases [1]. The predicted configurations are relative instead of absolute by the use of ^{13}C NMR. However, if one stereocenter is known, other chiral centers' AC can be identified by comparison with the stereogenic centers with known ACs.

In many reports, the coefficients are used to judge whether the proposed structure is correct or not. This is generally correct in many cases. Indeed, some cases need to be discussed here. First, assume there are two structures whose coefficients have different values. If the maximum error is >8.0 ppm, both structures should not be reliable no matter what the coefficients are. Second, if the maximum error is <8.0 ppm, the structure may be corrected by comparing the magnitudes of the coefficients, but further confirmation should be obtained using more experimental or theoretical evidences. The third case is when the difference between two coefficients is very small. It can lead to a dangerous conclusion if the structure with the larger coefficient is selected. An example was discussed in Chapter 2, but it is repeated here for easy recapitulation. The predicted coefficients using ^{13}C NMR, maximum error, and OR values are summarized in Table 6.1 for the four structures **1–4** [1c]. As discussed above, the correct natural structure is **3**.

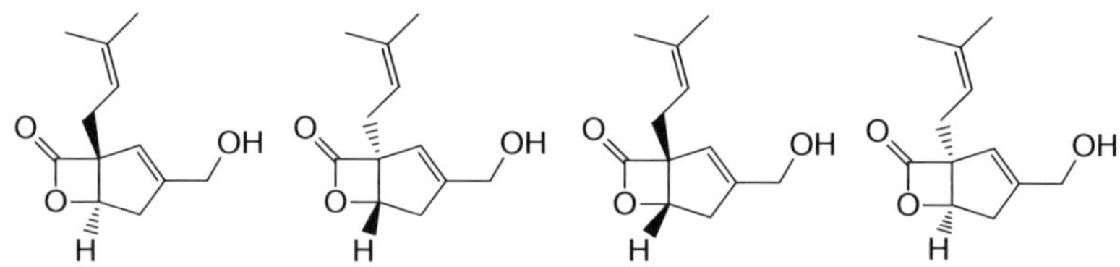

1 **2** **3** **4**

This is a typical example, which clearly shows that the coefficient and maximum error control used as principles are necessary for the prediction of RC or AC in any study.

Table 6.1 Summary of coefficient constant, maximum errors, and ORs for **1–4** [1c].

	1	2	3	4
Coefficient constant	0.99468	0.99402	0.99932	0.999358
Maximum error (ppm)	−11.2	−11.4	−2.8	2.4
OR (°)	+110.5	−110.5	−127.4	+127.4

When aromatic compounds are used in ^{13}C NMR computations, the maximum of 8.0 ppm is too large to be used. The recommended maximum generally is 4.0 ppm.

For a linear chiral compound, because of the fact that a single bond, such as C–C, can rotate freely, the computed ^{13}C NMR shift errors may be less than 8.0 ppm [1d]. If the largest coefficient constant is used for structure determination, a wrong conclusion may result. It is suggested not to use the computed ^{13}C NMR of linear chiral compounds for structure identification.

6.1.2
OR and ORD

OR and optical rotatory dispersion (ORD) are important characteristics of chiral compounds. In most cases, chiral compounds have large OR or ORD values, which can be used for their AC assignment.

It was reported that, only when the OR value is above 57°, the predicted OR may be reliable [2]. Once the experimental OR magnitudes are less than this value, the computed OR may not have a high reliability. However, if there is a similar chiral analog used as a standard, the same methods may be applied for the unknown structure's OR calculations even if its OR is not large enough, such as ~10°.

Notice

In natural products chemistry study, one may find that a recorded OR value for a pure chiral compound is much smaller than the predicted value. In this case, do not come to the conclusion that theoretical methods are not suitable for OR calculation for this chiral compound. The reason may be that the obtained chiral compound is a mixture of two enantiomers with unequal amounts. A plant body (or any other living body) is a "chiral pool," which may lead to the formation of two enantiomers without an equal molar ratio. When this happens, one should be careful to select a chiral column for the enantiomer separation. Then one can obtain two chiral compounds and carefully study their ACs.

It is clear that the larger the OR values, the more reliable will the predictions be using quantum theory. Thus, helical chiral compounds and axial chiral compounds are good examples to be used for AC assignment since they generally have large OR magnitudes.

It should be noted that, when two chiral molecules form a very stable "dimer" in a solution via strong H-bonds, this may cause a change in the OR values. There are many examples reported. This case will be discussed later.

Linear chiral compounds generally have small OR values, mostly 1–50, in different solvents. At the same time, they have many stable conformations with low energy. It is difficult to predict their AC by comparing their experimental ORs with the computed values.

6.1.3
Matrix

This model is constructed for acyclic chiral compounds' AC determination. It is suitable for the AC assignment of chiral compounds with small OR values. For example, in the case of (*S*)-(+)-**5** and (*S*)-(+)-**6**, their OR values are +0.0068 and +0.0095, respectively [3a]. Their determinants are computed and illustrated below. The computed k_0 values are 0.07 and 0.095, respectively [3b], which are very close.

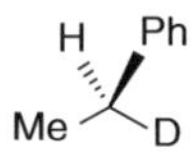

5

$$\det(D) = \begin{vmatrix} 18.9 & 1.7 & 2.5 & 0.25 \\ 2.0 & 1.2 & 2.1 & 1.0 \\ 13.0 & 2.0 & 2.5 & 0.25 \\ 1.0 & 1.2 & 2.1 & 1.0 \end{vmatrix} = +0.93 \qquad k_0 = [\alpha]_D/\det(D) = 0.07$$

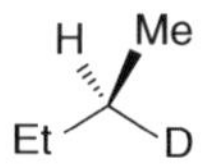

6

$$\det(D) = \begin{vmatrix} 13.0 & 2.0 & 2.5 & 0.25 \\ 2.0 & 1.2 & 2.1 & 1.0 \\ 14.1 & 2.2 & 2.5 & 0.0 \\ 1.0 & 1.2 & 2.1 & 1.0 \end{vmatrix} = +0.10 \qquad k_0 = [\alpha]_D/\det(D) = 0.095$$

It is useful for the AC determination of long-chain chiral compounds, as well as for linear chiral compounds with two stereogenic centers when the substituents on stereogenic centers do not interact with each other. However, if a chiral compound has two or more stereogenic centers with one located on a ring, it is difficult to use it now.

6.1.4
ECD

This method can be widely used for chiral compounds with strong chromophore groups. Even if there is no such large chromophore group, it is possible to introduce a UV–vis group, such as 4-NO_2PhCO–, in which case exciton chirality methods could be used.

Notice
In some cases, the chiral molecular RC is well assigned using 2D NMR. When one or two stereogenic centers are located near the C=C or other chromophore groups, use of ECD is possible for AC determination. This requires that the RC be correct! If one of the stereogenic centers was wrongly assigned, the AC assignment would be wrong when using the ECD method. If a chiral center is far away from a chromophore, even if (*R*) or (*S*) configuration was used in ECD computations, the ECD spectra will not show clear differences that can be used for AC assignment. Thus, using ECD for these chiral compounds' AC determination may be dangerous, especially for complex natural products since their RC is not easy to be identified correctly using 2D NMR because of serious overlap of the ^1H NMR signals.

In some structures, the changes of configuration (*R*) to (*S*) cannot cause a large change in the ECD spectra. In this case, other evidences, such as ^{13}C NMR, for RC assignment are helpful for AC assignment economically. Other methods like OR may also be valid if the recorded OR is over at least 20°.

6.1.5
VCD Method

VCD is a very powerful method used in AC studies; it could be also used to study the interaction between two chiral molecules or the relationship between three or more chiral compounds since it records all signals of the circularly polarized infrared (IR) light passing through chiral compounds. For example, strong H-bonds will affect the vibrational spectrum, and especially when H-bonds are located near the stereogenic centers, this may affect the VCD spectral structure also. Thus, if a chiral molecule such as a ketone, an ester, or a lactone with α-hydroxyl or amido group is used in a VCD study, one should be careful while handling their VCD spectra. Since the structure easily forms a "dimer," "trimer," or even "tetramer" via intermolecular H-bonds, these different structures may produce different VCD signals.

Therefore, the best candidates for VCD measurement should have no –OH, $-NH_2$, or similar groups, or just contain any one of such polar groups, or two or more polar groups that are far away from each other. Chiral compounds without any polar group such as –OH can avoid the intermolecular interaction and produce good VCD spectra in many cases. The use of $CDCl_3$ to dissolve the sample can provide much better VCD spectra than when CD_3OD, DMSO-d_6, or D_2O is used.

6.2
Examples and Discussion

Terpenes are widely distributed in nature. Because of their "head-to-head," "head-to-tail," or "tail-to-tail" connections, their planar structure exhibits regular changes. A dihydronaphthalene-2,6-dione derivative from *Phoma* sp. LN-16, an endophytic fungus, was identified by spectroscopic data as a previously reported compound, botryosphaeridione [3c]. Theoretically, it has two pairs of enantiomers (four isomers). Thus, only two isomers (5*R*,6*S*)-**7** and (5*S*,6*S*)-**8** were analyzed. The real AC could be either **7** or **8**, or one of their two enantiomers. Its AC was assigned by using ECD spectra at the B3LYP/aug-cc-pVDZ level. The predicted ECD for (5*R*,6*S*)-**7** was close to the experimental result, and this clearly showed that its AC should be (5*R*,6*S*) (Figure 6.1). It was also established by means of X-ray diffraction.

Other chiral terpenoids, such as *endo*-borneol **9** [4], perilladehyde **10** [5], and carvone **11** [6], were well established for their ACs using VCD methods. As a powerful method, it should be carefully introduced. The details will not be presented here.

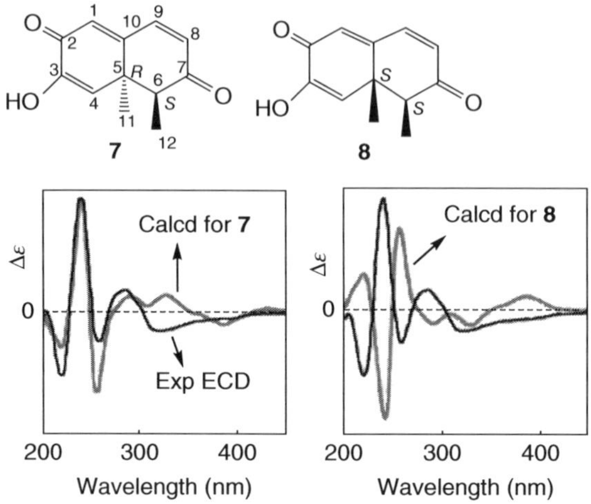

Figure 6.1 Experimental and calculated CD spectra for **7** and **8** [3c].

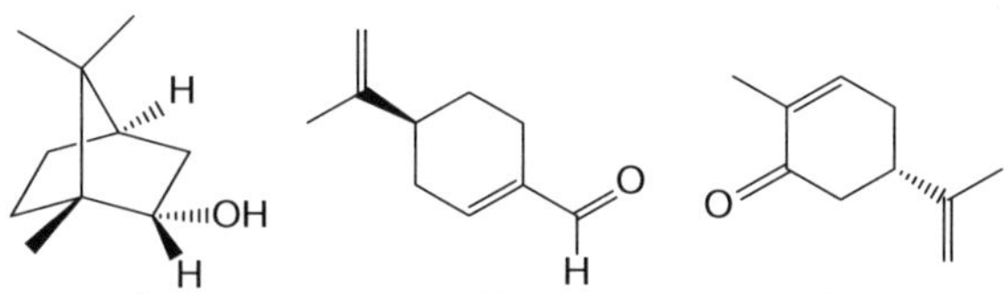

9 **10** **11**

Density functional theory (DFT) calculations of VCD, ECD, and ORD to determine ACs of the isoschizozygane alkaloids, isoschizogaline and isochizogamine [2], whose ACs had not been determined previously, have been performed (Figure 6.2–6.4). The ACs of naturally occurring (2)-isoschizogaline (**12**) was determined as (2*R*, 7*R*, 20*S*, 21*S*).

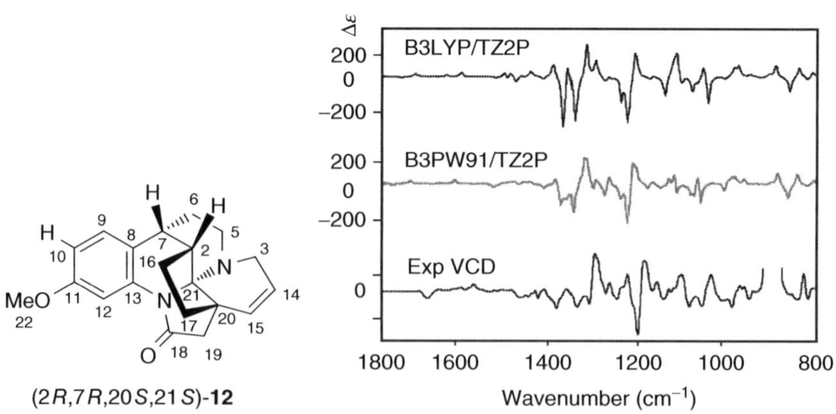

Figure 6.2 Comparison of the experimental and theoretical VCD using B3LYP/TZ2P and B3PW91/TZ2P methods for isoschizogaline **12** [2].

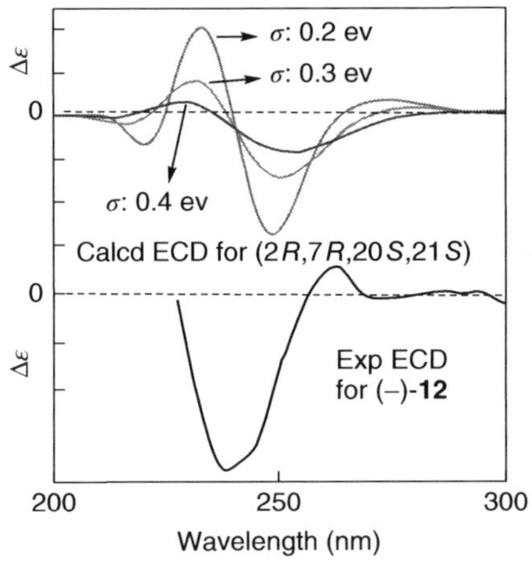

Figure 6.3 Comparison of the experimental and predicted ECD spectra [2].

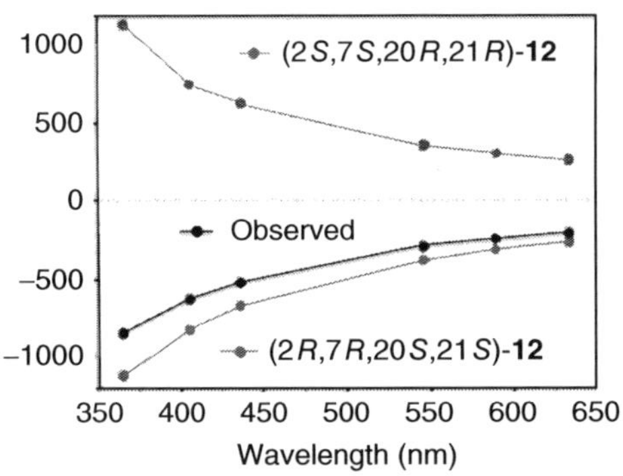

Figure 6.4 Comparison of experimental and calculated specific rotations of isoschizogaline **12** [2].

As mentioned in Chapter 4, the AC of the trimeric pyrroloindoline derivative psychotripine (**13**) was assigned based on ECD [7]. Indeed, its RC was identified by NMR and then tested by comparing its ^{13}C NMR obtained at the B3LYP/6-311++G (2d,p)//B3LYP/6-31+G(d) level with the experimental data. The relative errors between the predicted ^{13}C NMR and experiments were computed. The small relative error (<4.0 ppm) supported the RC deduced by rotating-frame Overhauser enhancement spectroscopy (ROSEY) experiments (Figure 6.5). Its OR was also computed at the same level. The predicted OR for (3a*R*,8a*R*,3a′*R*,8a′*R*,3a″*S*,8a″*R*)-**13** was −97.2, which is very close to the experimental OR value of −84.2. Clearly, its AC was well determined on the basis of the three major chiroptical spectroscopies.

Discussion of the solution behavior of compounds **14** and **15** is valid. The OR signs of **14** and **15** were not the same though both have the same AC. The OR for

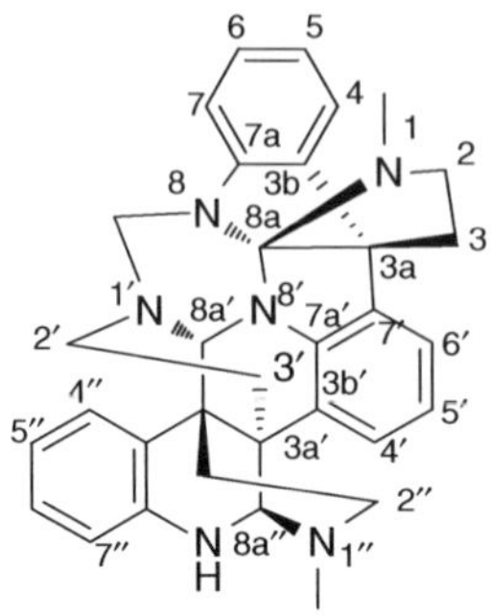

13

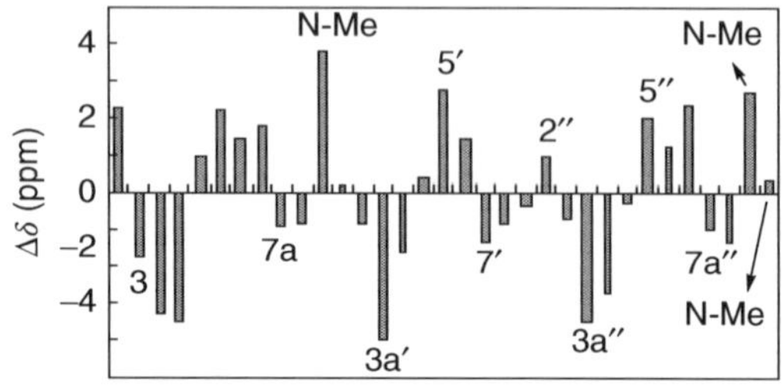

Figure 6.5 Δδ values obtained by experimental and theoretical ^{13}C NMR [8].

(*R*)-**14** is −21, while that of (*R*)-**15** is +36. The AC for **14** and **15** are well established by VCD [8]. The VCD spectra for (*R*)-**14** and (*R*)-**15** were almost the same as the experimental results. All evidences showed that the traditional empirical method, which compares the OR sign to assign AC for a series of chiral analogs, is not suitable for the AC assignment for the two examples.

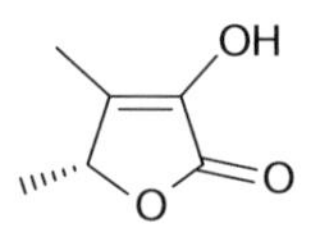

14

$[\alpha]_D = -21$

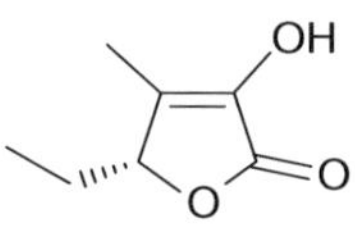

15

$[\alpha]_D = +36$

Why do the two chiral compounds have different OR signs? Can quantum theory give the answer?

Unfortunately, the predicted OR for (*R*)-**14** was +0.89 by the B3LYP/6-31+G(d) method in the gas phase. There is a large difference from the reported OR and the OR sign. If the 6-311++G(2d,p)-optimized geometries were used in OR calculations, the predicted OR was −1.84 at the 6-311++G(2d,p) level. The OR signs are different while using different methods. However, the OR calculated at a high level has the same sign (−) as the experimental OR. The calculated OR for (*R*)-**15** was +73.4 at the B3LYP/6-31+G(d)//B3LYP/6-31+G(d) level in the gas phase, whereas its OR became +81.78 at the 6-311++G(2d,p) level. The OR values for (*R*)-**15** are larger than the experimental OR, but the OR signs are the same (H.J. Zhu, unpublished results).

Possibly, chiral compounds **14** and **15** may form "dimers" **16** and **17**, respectively, after overcoming the intramolecular H-bond barrier by the formation of two new strong H-bonds in solution, as illustrated below:

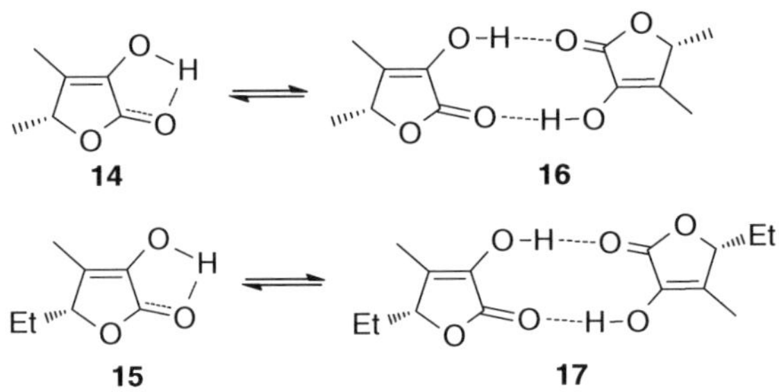

The computed OR for "dimer" **16** was −34.7 at the B3LYP/6-31+G(d)//B3LYP/6-31+G(d) level in the gas phase, while it is −42.7 at the B3LYP/6-311++G(2d,p)//B3LYP/6-311++G(2d,p) level. It was initially thought that the predicted OR was close to the experimental OR of −21. It at least showed the formation of some "dimer" **16**, and the "dimer" had a major contribution to the observed OR values and OR sign for (*R*)-**14**.

The VCD calculation for (*R*)-**16** was done using the B3LYP/6-311++G(2d,p)//B3LYP/6-311++G(2d,p) methods. The predicted VCDs were almost the same as those while using (*R*)-**14** in the VCD calculations (Figure 6.6). Especially, the details near 1050 cm^{-1} were closer to the experimental results than the

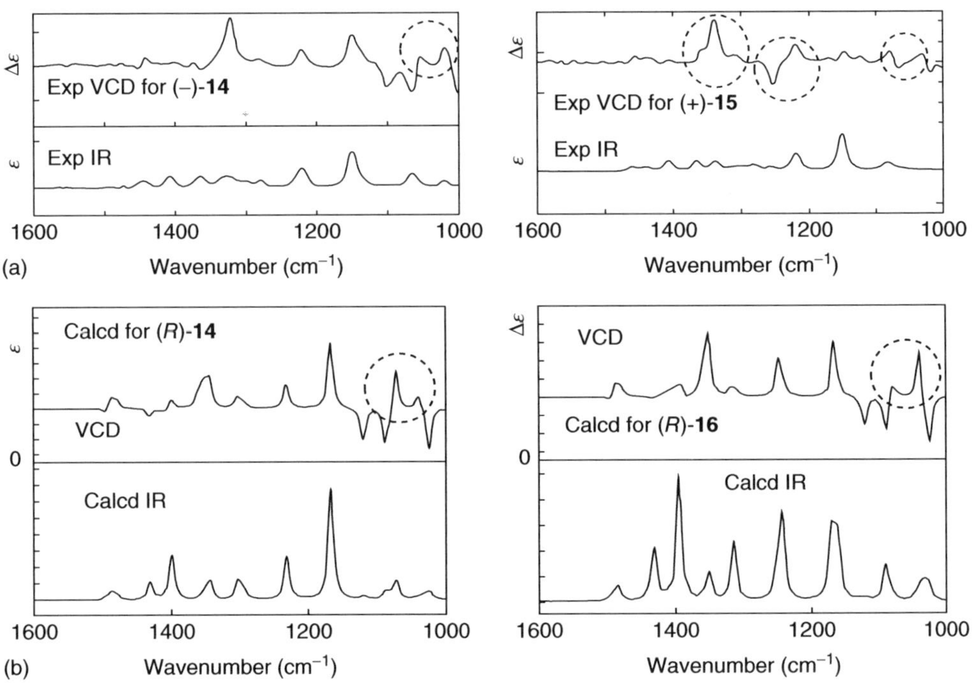

Figure 6.6 (a) Experimental VCD for (−)-**14** and (+)-**15**. (b) Predicted VCD for (*R*)-**14** and its "dimer" (*R*)-**16** [9].

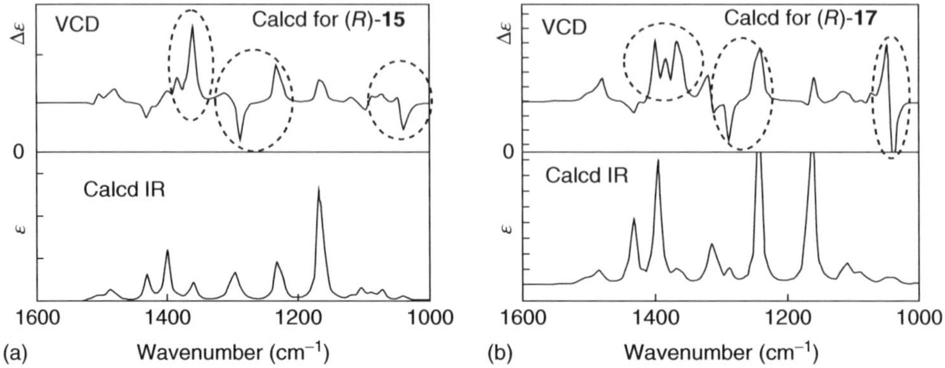

Figure 6.7 (a) Predicted VCD spectra for (*R*)-**15** and (b) its "dimer" (*R*)-**17**.

monomer's VCD. Both the OR and VCD results clearly showed that the partial "dimer" **16** was formed in the solution.

Similarly, the same methods were used for VCD calculations for (*R*)-**15**. The predicted VCD was also almost the same as that of (*R*)-**15**. The VCD spectra are illustrated in Figure 6.7.

Similarly, (*R*)-**17** was investigated using the same methods under optimization. The calculated OR for (*R*)-**17** was +95.1 at the B3LYP/6-31+G(d)//B3LYP/6-31+G(d) level in the gas phase. This value decreased to +79.3 at the B3LYP/6-311++G(2d,p)//B3LYP/6-311++G(2d,p) level. Until now, the predicted OR data are much larger than the experimental OR data. More importantly, the predicted VCD spectrum had obvious differences from the experimental spectrum (Figure 6.7), while the calculated VCD of monomer (*R*)-**15** was in good agreement with the experimental results (dash-line circle sections). Both results prove that the "dimer" formation from (*R*)-**15** is not favorable.

Based on the OR data and VCD spectra, it could be found that some "dimer" **16** was formed from (*R*)-**14,** and relatively very little "dimer" **17** was formed from (*R*)-**15**. This shows that VCD should be carefully used when similar structures are involved.

6.3 Revised Structures

Because of various limitations of experimental and theoretical methods used in AC assignment, the wrong identification of chiral compounds is absolutely unavoidable, especially in the identification of natural products. This is not the duty of this researcher. As mentioned by Nicolaou in 2005, "over the course of the past half century, the structural elucidation of unknown natural products has undergone a tremendous revolution. Before World War II, a chemist would have relied almost exclusively on the art of chemical synthesis, primarily in the form of degradation and derivatization reactions, to develop and test structural hypotheses in a process that often took years to complete when grams of material were available. Today, a battery of advanced spectroscopic methods, such as multidimensional NMR spectroscopy and high-resolution mass spectrometry, not to mention X-ray crystallography, exist for the expeditious assignment of structures to highly complex molecules isolated from nature in milligram or sub-milligram quantities. In fact, it could be argued that the characterization of natural products has become a routine task, one which no longer even requires a reaction flask!" [10].

Absolutely, "imaginative detective work and chemical synthesis still have important roles to play in the process of solving nature's most intriguing molecular puzzles" [10]. However, the cost of revising a structure is also expensive using synthesis. A total 51 structures have wrong planar structures including wrong AC determination, and 10 chiral compounds have wrong AC assignment. The revised structures from the wrong AC assignment selected from the review are illustrated in Table 6.2.

Theoretical methods are developing very fast now, and they have provided many benefits for structural evaluations, including AC assignment. The following examples are selected from published reports. The possible reasons that led to the wrong assignment are analyzed for some examples.

6.3.1
ORD Method

The ORD method is valid and used in AC assignment. For example, epimer (**39**) was assigned as (4*S*,5*S*) [11]. OR computations for **39** with consideration of solvent effect in acetonitrile using the polarizable continuum model (PCM) model were carried out. The ORs were then compared with the experimental OR data taken at four different wavelengths (Figure 6.8). The comparison clearly showed that the previously proposed AC for **39** was incorrect and that the correct AC was (4*R*,5*R*) [12].

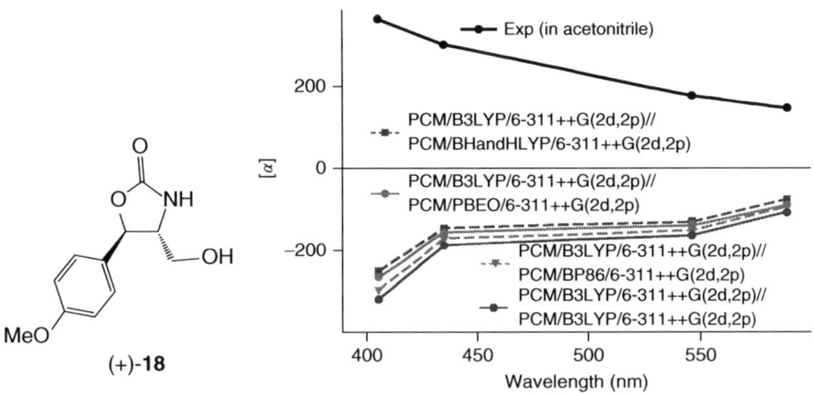

Figure 6.8 Experimental and theoretical ORD spectra of (+)-**18** [12].

Table 6.2 The wrong structures and their corrected structures [10].

Entry	Wrong structure	Correct structure
1	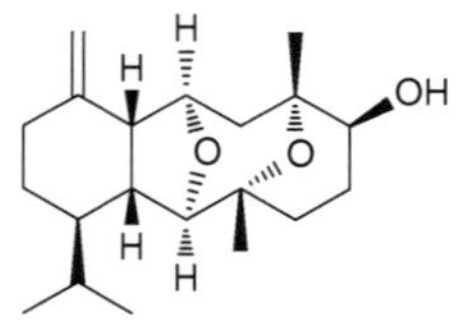 **19** Sclerophytine A	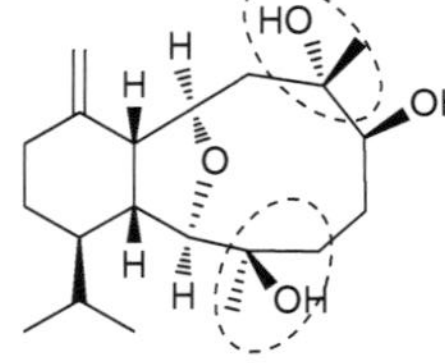 **20**
2	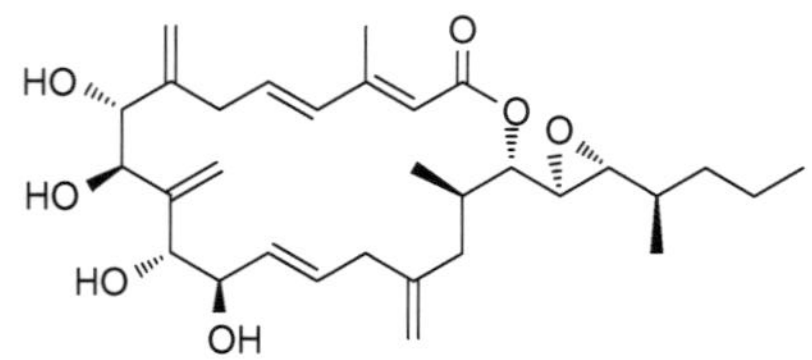 **21** (+)-Amphidinolide A	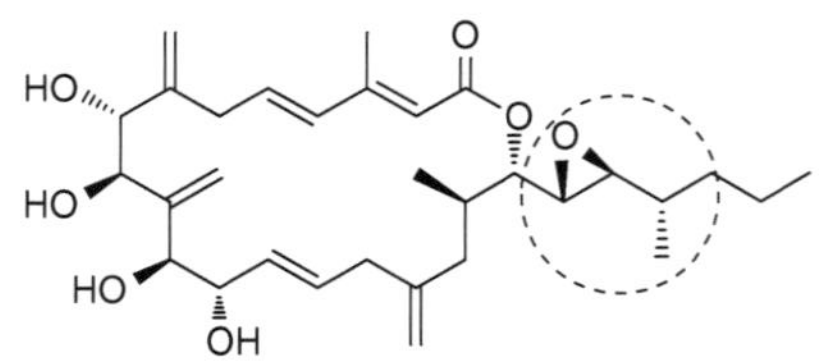 **22**
3	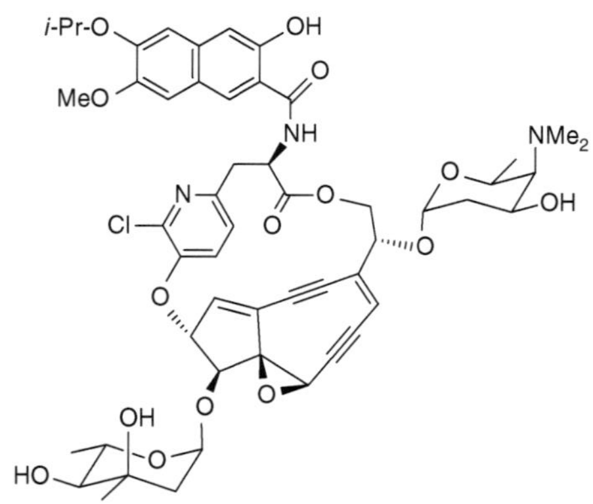 **23** Kedarcidin chromophore	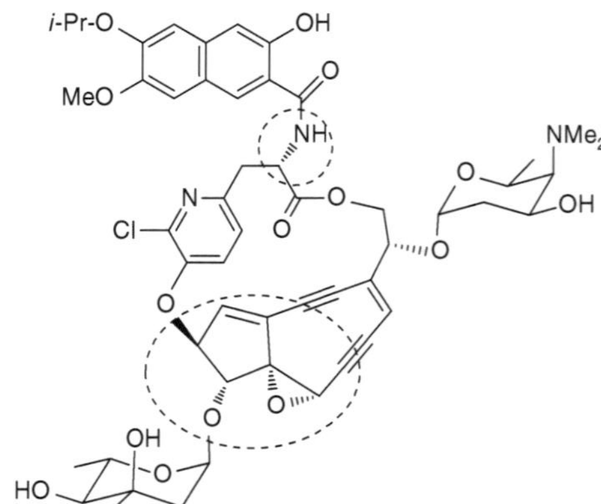 **24**

4	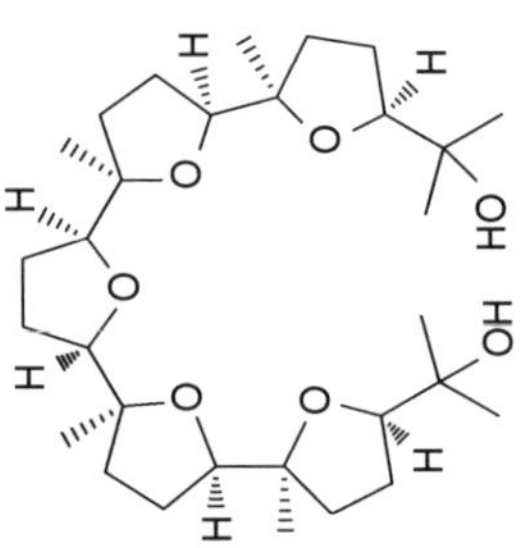 25 Glabrescol	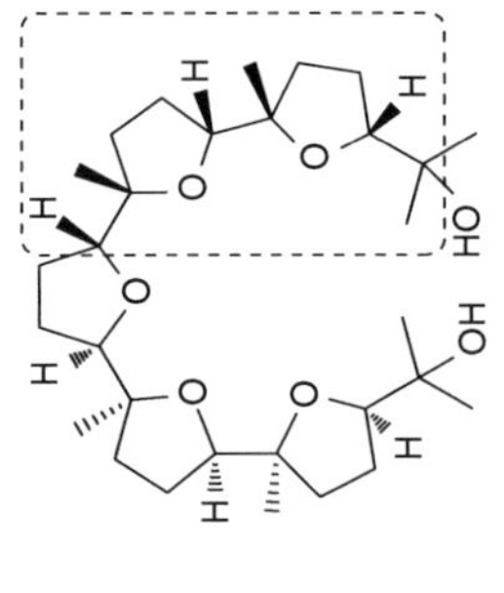 26
5	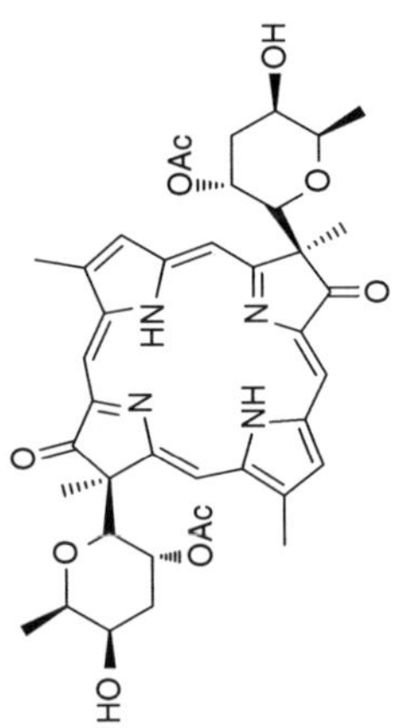 27 (+)-Tolyporphin A	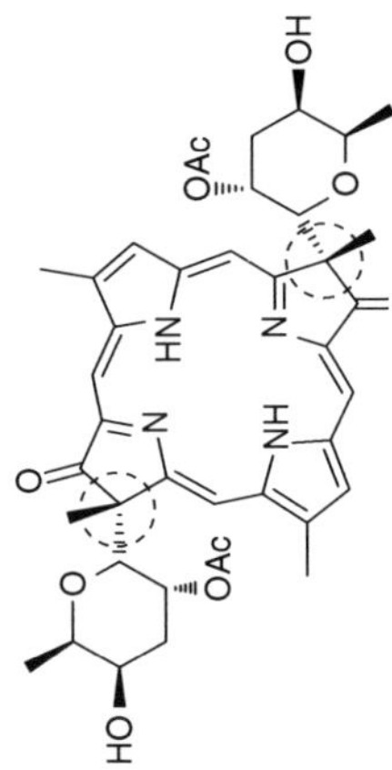 28
6	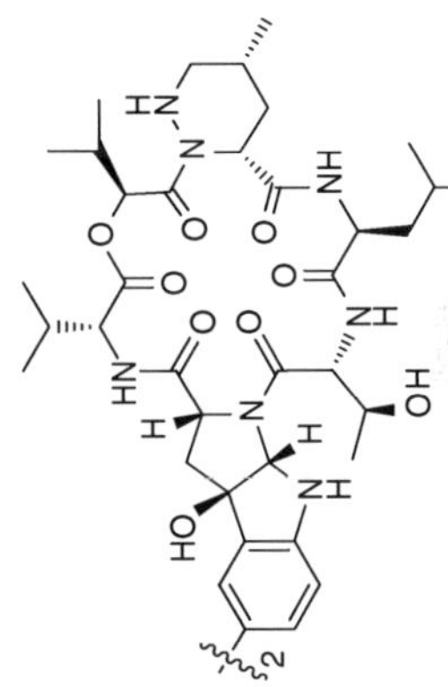 29 Himastatin	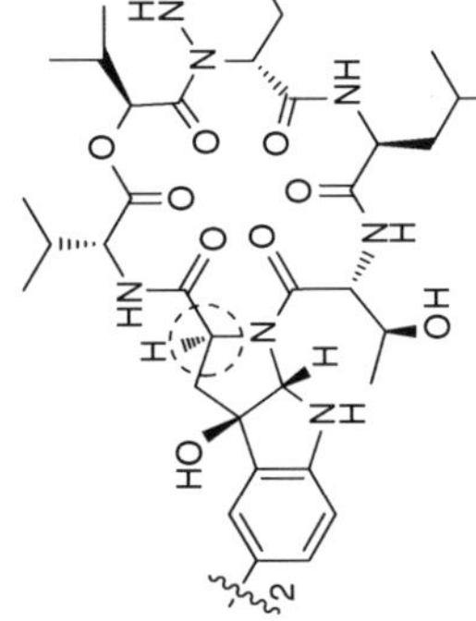 30

(continued overleaf)

Table 6.2 (Continued)

Entry	Wrong structure	Correct structure
7	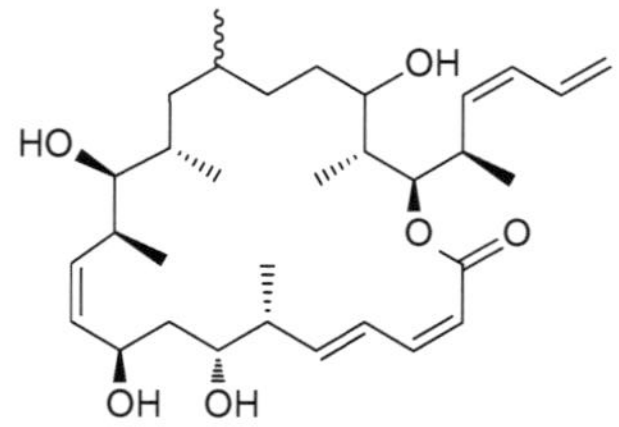 **31** Dictyostatin	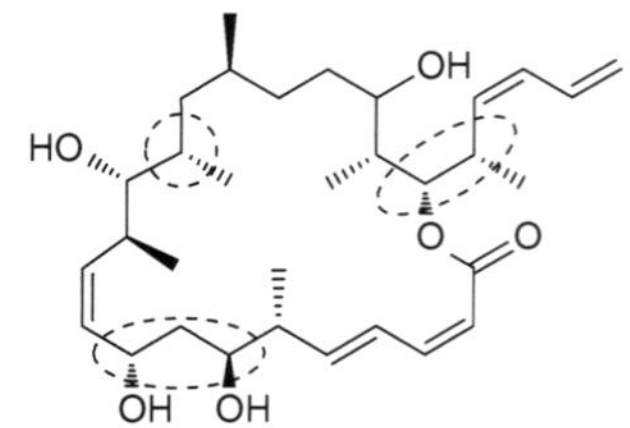 **32**
8	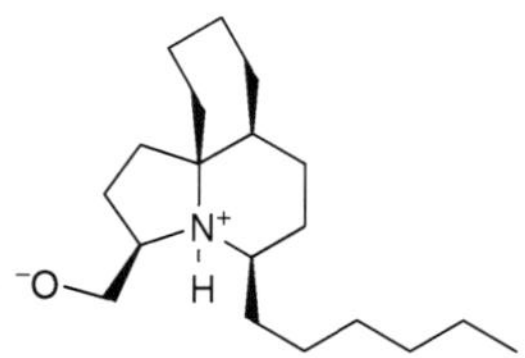 **33** Lepadiformine	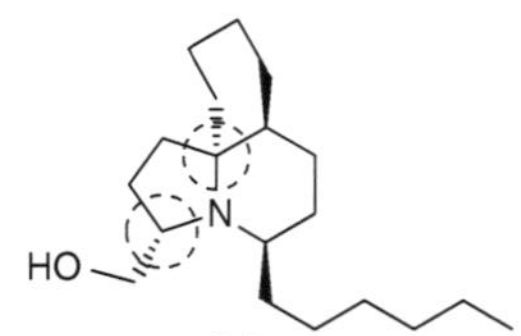 **34**

9	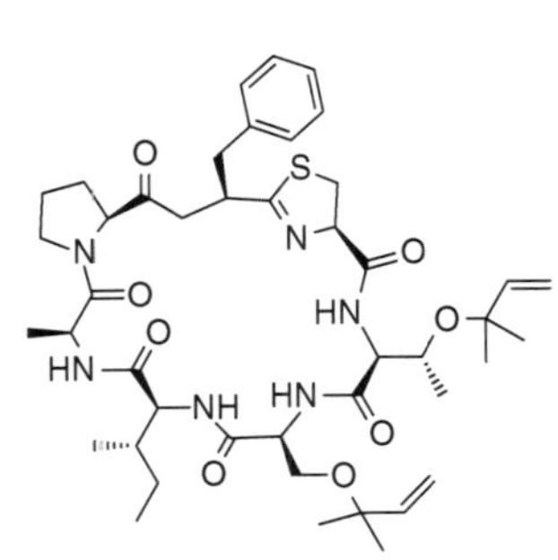 **35** Trunkamide A	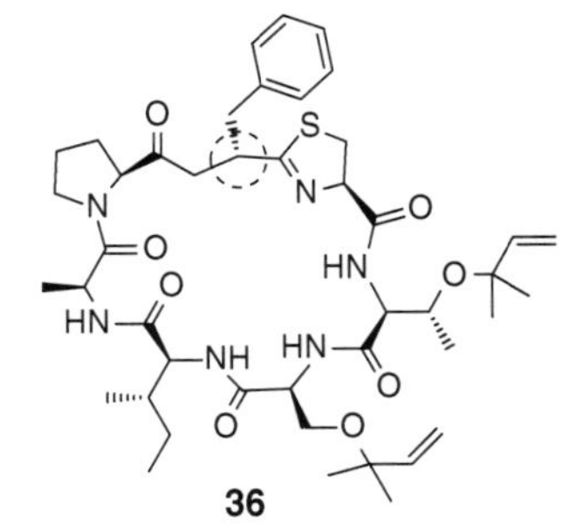 **36**
10	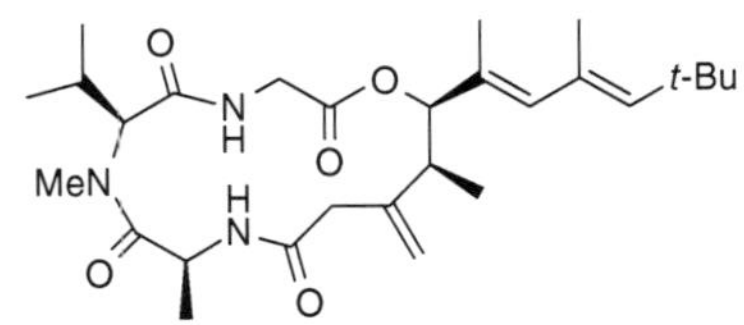 **37** Antillatoxin	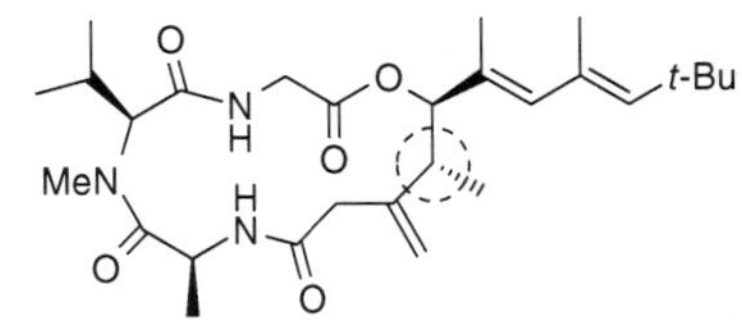 **38**

6.3.2
Combinational Use of OR and ECD

This could be a powerful tool in the assignment for complex natural products. It is found that some errors happen by the misuse of empirical methods in AC determinations. For example, one of the widely used methods is to compare the ORs two similar compounds to figure out the AC of one of them. Another is the Mosher method, which can lead to wrong conclusions by using it on a wrong molecule. For example, **40** was assigned as having the same AC of (2*R*,3*S*) as the reported analogs of **41** based on the fact that **40** has +38 in $CHCl_3$ and **41** has +50 in chloroform [13].

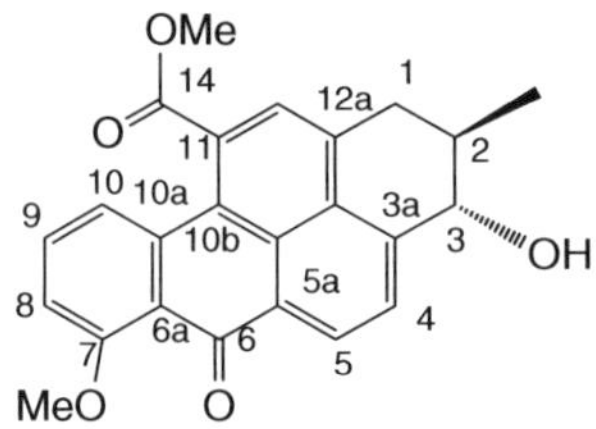

40 Bezopyrenomycin

Exp $[\alpha]_D$: +38 in $CHCl_3$

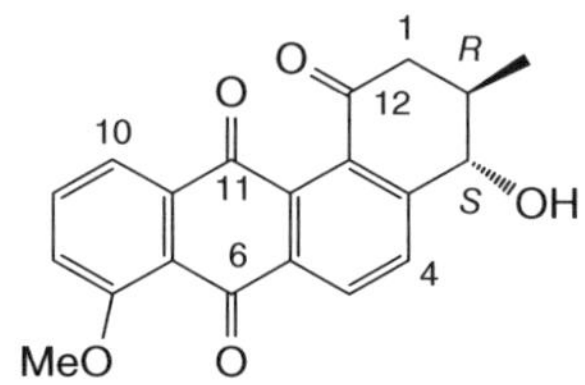

41 Rubiginone A_2

Exp $[\alpha]_D$: +50 in $CHCl_3$

Benzopyrenomycin (**40**) has relatively strong cytotoxic activity. For instance, benzopyrenomycin exhibited a GI_{50} of 3.2 and 4.2 μg ml^{-1} for L-929 and K562 cell lines, respectively. Meanwhile, the AC of **40** was the same as that of **41** based on the ^1H NMR shift differences of the synthesized (*R*)- or (*S*)-Mosher ester of **42**, whose structure was eventually converted to **40** [14]. However, when using Mosher ester to determine the AC of **42** analogs, wrong predictions could be made because it has a bulky group near the C3 atom [15a]. An example is **43**, which has a bulky group near the chiral center, and the wrong sign of the Δδ values was recorded, and the data are illustrated in its structure [15b].

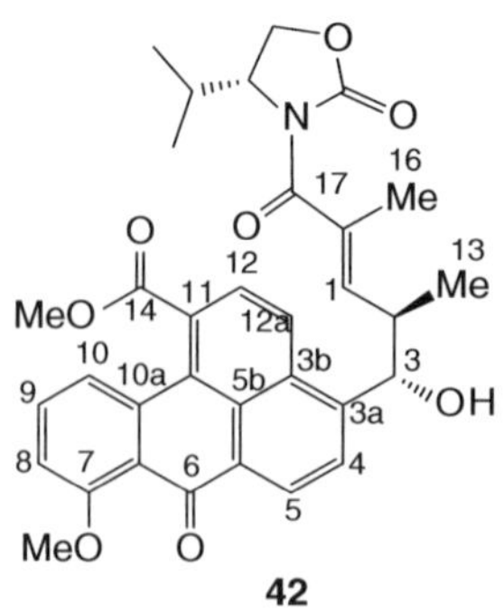

42

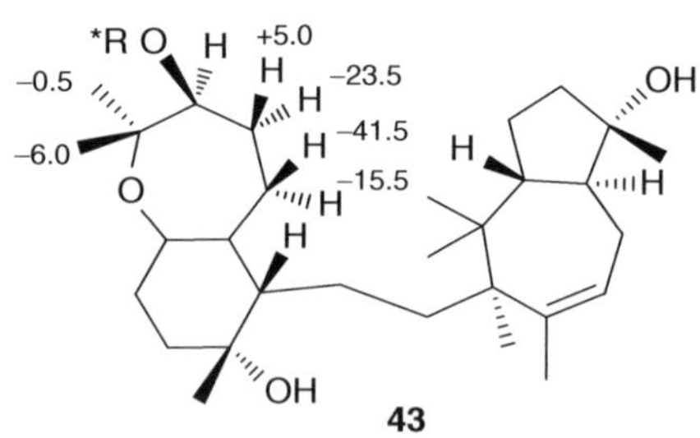

43

R* = (*R*)- or (*S*)-Mosher ester moiety, the number labelled near atom is the Δδ value.

Different methods were used for OR calculation for **40** and **41** [15c]. Compound (2*R*,3*S*)-**40** has OR values between −28° and −86°, mostly ranging from −28° to −50°, using total electronic energy (TEE). The OR values are also negative and most of them in the range −16° to −32° using zero-point energy (ZPE).

Table 6.3 Computed OR values using different methods [15c].

	(2*R*,3*S*)-**40**	(2*R*,3*S*)-**41**
$[\alpha]_{D\ exp}$	+38	+50
Method 1[a)]	−50.4/−16.2[b)] /+9.5[c)]	+48.7/+47.7[b)] /+87.2[c)]
Method 2[d)]	−44.1	+18.7
Method 3[e)]	−27.8/−26.8[b)] /−28.4[c)]	+33.4/+72.6[b)] /+104.7[c)]
Method 4[d)]	−40.5	+34.1
Method 5[f)]	−28.9	+32.4
Method 6[g)]	−86.1	+2.12
Method 7[h)]	−49.2/−28.6[b)] /−14.0[c)] ($CHCl_3$)	+38.4/+61.4[b)] /+87.8[c)]
Method 8[d)]	−12.2	+13.5

a) B3LYP/6-311++G(2d,p)//B3LYP/6-31G(d), total electronic energetics were used in OR computations.
b) Vibration corrections were performed in OR computations.
c) Free energy data were used in OR computations.
d) Single-point energy at the B3LYP/aug-cc-pVDZ level in chloroform via PCM model was used in OR computations.
e) B3LYP/6-311++G(2d,p)//B3LYP/6-31+G(d,p).
f) B3LYP/6-311++G(2d,p)//PCM($CHCl_3$)/B3LYP/6-311+G(d).
g) PCM($CHCl_3$)/B3LYP/6-311++G(2d,p)//PCM($CHCl_3$)/B3LYP/6-311+G(d).
h) B3LYP/6-311++G(2d,p)//B3LYP/6-311++G(2d,p).

Meanwhile, the predicted OR values for (2*R*,3*S*)-**41** are mostly in the range +2.1° to +104° using TEE, ZPE, and GFE (Gibbs free energy) data, respectively, mostly from +32° to +87°, which are in good agreement with the experimental results (+50). Relatively, it predicted low OR values by using the single-point energy (SPE) in solutions (method 8, Table 6.3).

Similarly, the predicted ECD simulations for (2*R*,3*S*)-**40** and (2*R*,3*S*)-**41** were computed and simulated using TEE, ZPE, and GFE data in Boltzmann statistics. ECDs for both **40** and **41** looked as though they were reversed stereostructures (Figure 6.9a–c). Especially, the ECD simulated using GFE had large differences between **40** and **41**. The ECD spectra indicate that the two compounds have different ACs at C2 and C3. This independent evidence supported the conclusion from OR data.

Based on the calculated ECD and OR results and the fact that the RC of **40** was well established, it was proposed that the AC of (+)-**41** is (2*S*,3*R*), namely, (+)-(2*S*,3*R*)-**41**.

Indeed, **40** and **41** can form helical structures, which generally produce large OR values. Thus, the sign of OR (positive or negative) depends on the OR magnitude of the most stable helical conformation instead of the chirality of the carbon tetrahedron. As expected, the two helical geometries that are the most and second most stable conformations have shown opposite OR signs. For example, two pairs of B3LYP/6-311++G(2d,p)-optimized helical conformations have almost the same absolute OR values but opposite signs for **40,** and they make over 70% contribution to the total OR values. Similarly, four conformations make about 80% contributions to the total OR of **41.** The results are summarized in Figure 6.10.

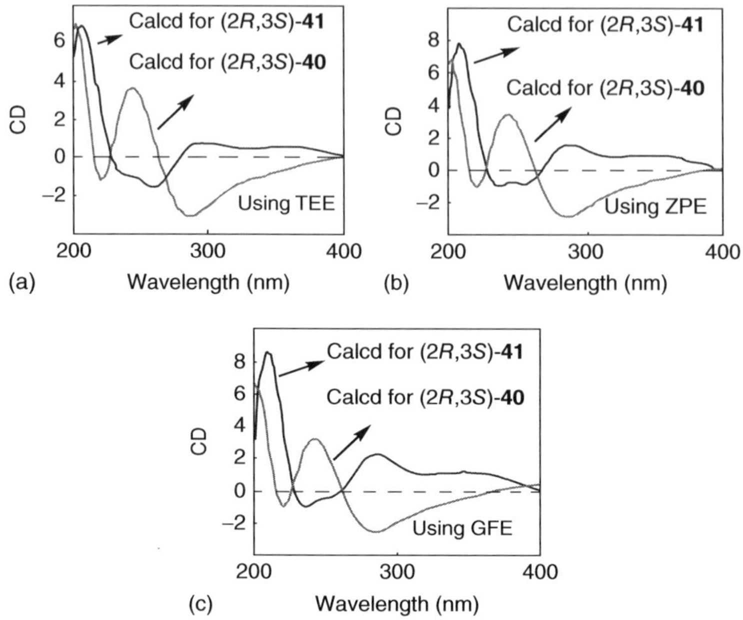

Figure 6.9 Comparisons of the computed ECD spectra for (2*R*,3*S*)-**40** and (2*R*,3*S*)-**41** simulated using (a) TEE, (b) ZPE, and (c) GFE [15c].

6.3.3
VCD and ECD

VCD and ECD are powerful tools used in AC assignments. For example, the AC of the bioactive natural product (+)-schizandrin was assigned as (7*S*,8*S*) using NMR and was also confirmed by experiments [15d]. However, its AC is wrong. Conformational searches for (7*S*,8*S*)-**44** were performed first using the MMFF94S force field. Totally 63 conformations were found. All B3LYP/6-31G(d)-optimized geometries were used for VCD computations at the same level using the Gaussian 09 package. The predicted VCD and IR spectra for (7*S*,8*S*)-**44** are illustrated in Figure 6.11. The VCD was measured in $CDCl_3$ at room temperature.

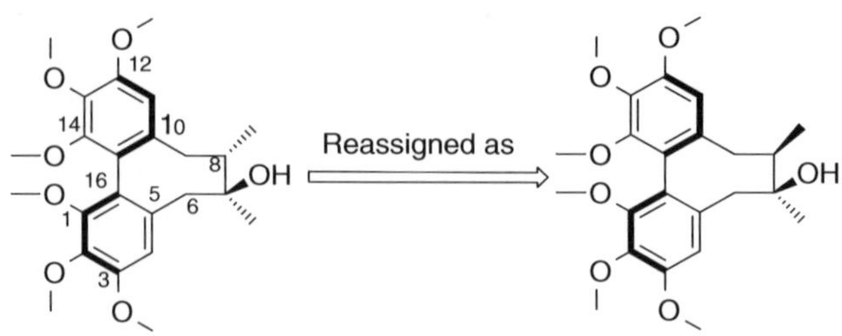

44 (+)-Schizandrin
[α]$_D$ +86 to +92

45 (+)-Schizandrin
[α]$_D$ +81 to +91

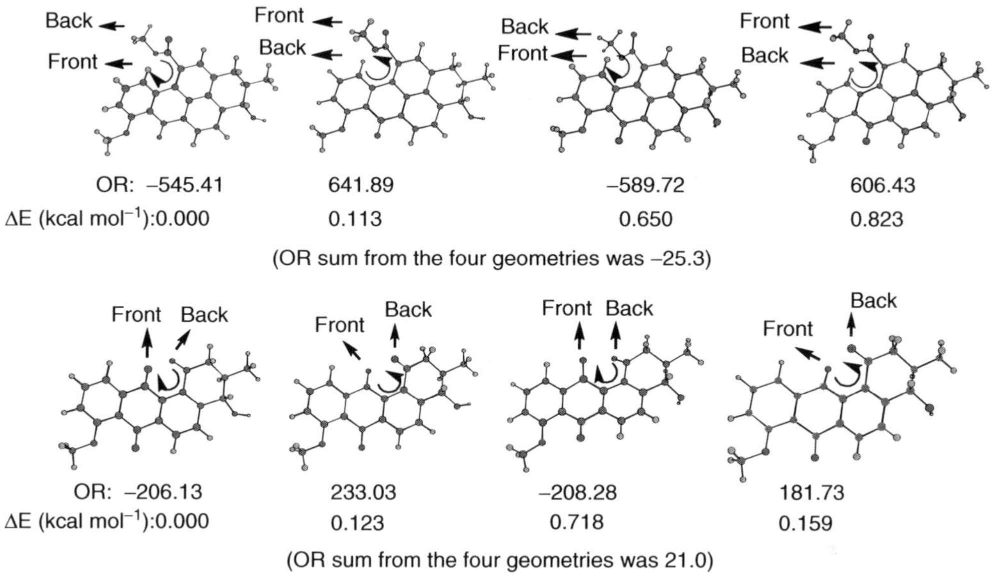

Figure 6.10 OR, relative energy, and contribution to the total OR of two pairs of stable conformation [15d].

From the bond vibrational modes (e.g., stretching and swing), the corresponding window viewer can be seen, for example, the Gaussview from the frequency computation results. It was found that the frequency near 1400–1500 cm^{-1} in VCD mainly corresponded to the different stretching vibrations of the bonds near the stereogenic centers C7 and C8. The different VCD structures in this range may hint at different ACs for (+)-**44**. Therefore, compound **45** with (7*S*,8*R*) AC was considered for further VCD study.

The predicted VCD and IR spectra for (7*S*,8*R*)-**45** look extremely similar to the experimental results. They are superimposed in Figure 6.12. It clearly shows that the earlier reported structure (+)-(7*S*,8*S*)-schizandrin **44** has the wrong AC. The correct AC structure should be (7*S*,8*R*)-**45**.

Compared to the VCD results, those of ECD could not arrive at the same conclusion. For example, (7*S*,8*S*)-**44** and (7*S*,8*R*)-**45** were used for ECD calculations. From plot A in Figure 6.13, one cannot conclude that the real AC is (7*R*,8*R*) since both look like mirror images. However, once the (7*S*,8*R*) ECD was computed and simulated, it was found that its AC should be (7*S*,8*R*). Therefore, it was not possible to arrive at a decision just by ECD results alone for this example.

6.3.4
Comprehensive Use of OR, ECD, and VCD

The combined use of OR, ECD, and VCD constitutes a very powerful method in AC assignments. Recently, a novel bioactive alkaloid, brevianamide M (**46**), was reported [16a]. The RC of **46** was established as (2*R*,13*R*) or (2*S*,13*S*) by X-ray

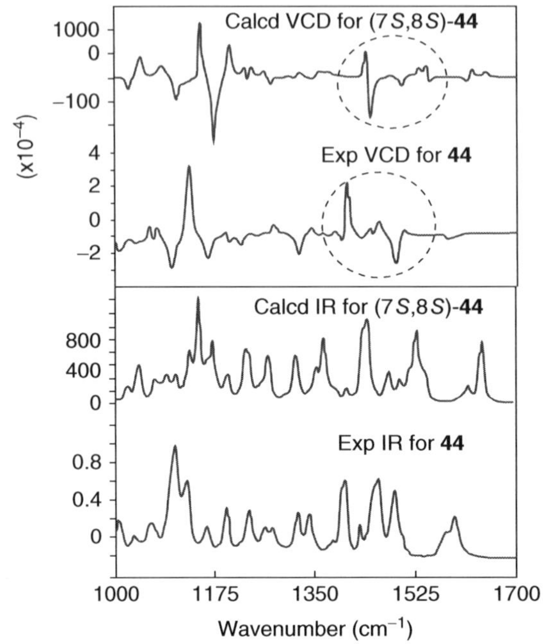

Figure 6.11 Calculated IR and VCD for (7*S*,8*S*)-**44** and comparison of both VCDs. The measurement of VCD was done at room temperature with a concentration of 50 mg ml^{-1} [15d].

crystallography using Mo radiation. This compound can be hydrolyzed in acidic aqueous solution at 100 °C to afford L-phenylalanine, which strongly implied that C13 should have an (*S*) AC. Therefore, the AC of **46** was assigned as (2*S*,13*S*). This procedure for AC assignment is normally definitive and cogent. However, this misassignment of AC is the most extreme case uncovered to date in our theoretical studies [16b]. It demonstrates that, given the power of current computational chemistry using DFT, chemists should now reexamine AC assignments based in part on indirect chemical evidence, such as the acid hydrolysis mentioned above.

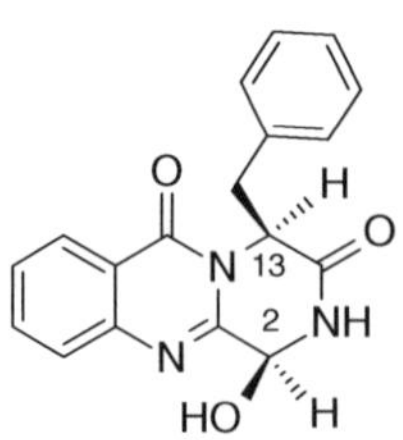

46 Brevianamide M

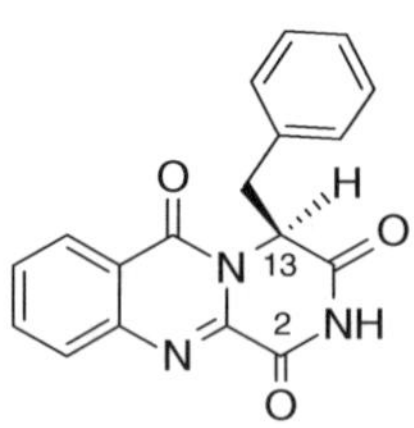

47

The first evidence is from its OR data. The experimental OR is −147 in $CHCl_3$ [16a]. However, (2*S*,13*S*)-**46** had positive OR values (Table 6.4) both in the gas phase and in chloroform using DFT methods. All OR values are summarized in Table 6.4.

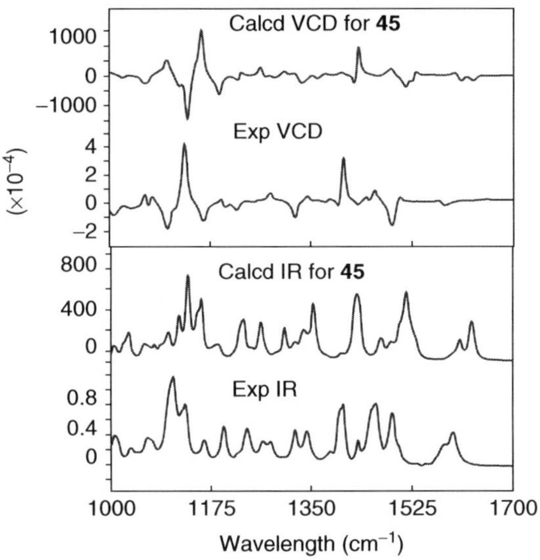

Figure 6.12 Comparison of experimental and computed VCDs and IRs for **44** and **45** [16b].

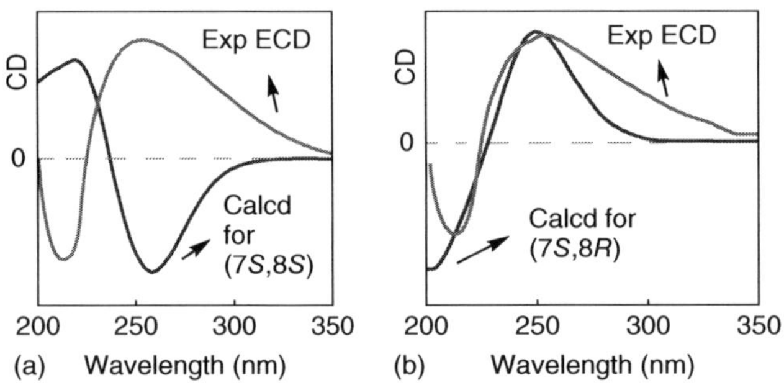

Figure 6.13 Predicted ECD for (a) (7*S*,8*S*)-**44**, (b) (7*S*,8*R*)-**45**, and their comparison with experimental ECD [16b].

All OR predictions for (2*S*,13*S*)-**46** are positive, ranging from about +210 to +600. The absolute values of OR are much larger than the experimental OR (−147), which is typical for DFT-level calculations. Since the RC was well established using X-ray, the positive OR values from +210 to +600 suggest that (−)-**46** should have the AC of (2*R*,13*R*).

Considering that the computation of OR values is based on the determination of electronic transitions in the inaccessible UV region, it is not easy to calculate it with definite reliability. By contrast, the RC of the structure of (2*R*,13*R*) or (2*S*,13*S*)-**46** was well established by X-ray, and its AC was confirmed by the hydrolysis of **46** to afford the L-amino acid. Thus, are the DFT methods not suitable for prediction of OR for this example?

The second evidence is the ECD of (2*R*,13*R*)-**46**. The simulated ECD using GFE for (2*R*,13*R*)-**46** was in good agreement with the experimental value (Figure 6.14).

Table 6.4 OR values for (2*S*,13*S*)-**46** from four methods [16b].

	Method 1[a)]	Method 2[b)]	Method 3[c)]	Method 4[d)]
$[\alpha]_D$	+510.5	+532.9	+519.5	+600.7
$[\alpha]_D$[e)]	+209.8	+222.4	+217.0	+354.0

a) B3LYP/6-311++G(2d,p)//B3LYP/6-31G(d).
b) B3LYP/6-311++G(2d,p)//B3LYP/6-31+G(d,p).
c) B3LYP/6-311++G(2d,p)//B3LYP/6-311++G(2d,p).
d) B3LYP/6-311++G(2d,p)//PCM/B3LYP/6-311++G(2d,p).
e) Single-point energy at the B3LYP/aug-cc-pVDZ level in chloroform using PCM model was used in OR computations.

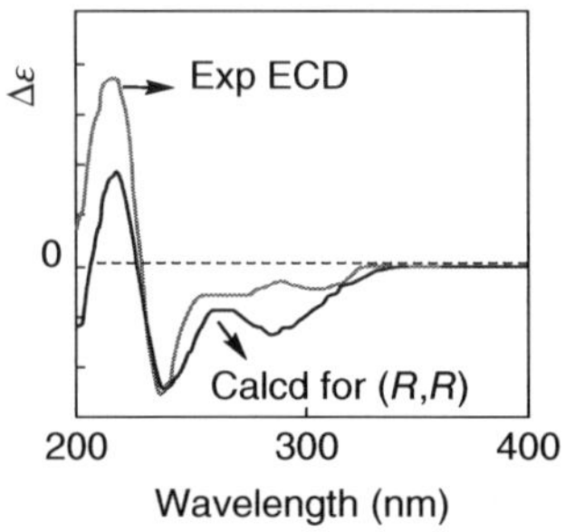

Figure 6.14 Comparison of the computed ECD with the experimental ECD [16b].

The theoretical prediction of positive and negative Cotton effect matched very well with the experiments. Both OR and ECD provided evidences that the previously assigned AC (−)-(2*S*,13*S*)-**46** may not be correct.

Why did the early experiments provide the wrong predictions? The major reason must be the low barriers of isomerization from (2*R*,13*R*)-**46** to (2*R*,13*S*)-**47**. Theoretically, four transition state (TS) structures were plausible for the formation of (2*R*,13*S*)-**46**. However, only **TS-1, TS-2,** and **TS-4** were found in the calculations; **TS-3** was not found using the reported methods [17].

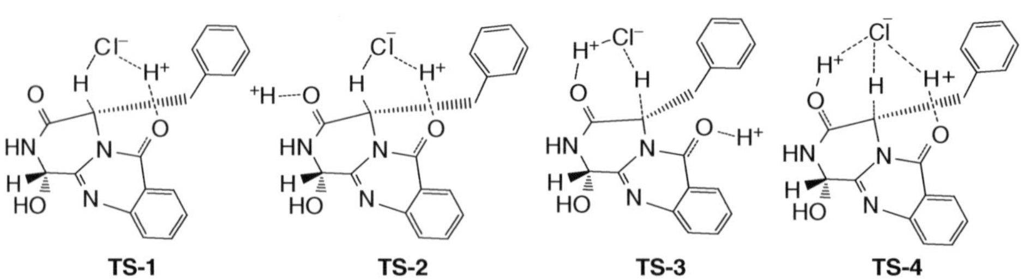

In the absence of H^+ catalysis (**TS-1**), a 33.4 kcal mol^{-1} barrier was obtained at the B3LYP/6-311++G(2d,p) level in the gas phase using TEE. It was 32.9 kcal mol^{-1} using GFE and 31.5 kcal mol^{-1} using ZPE. Here, all the barriers are quite high, and it is difficult to get conversion via **TS-1.** The energy barrier

Table 6.5 TS activation energy magnitudes for the four different procedures computed at the B3LYP/6-311++G(2d,p) level in the gas phase or in water [16b].

	ΔE_0 a)	ΔG b)	ΔE c)	ΔE_l d)
TS-1	31.5	32.9	33.4	25.1
TS-2	26.8	28.6	28.2	16.5
TS-4	21.5	23.3	23.1	17.9
TS-5	25.4	26.7	27.7	21.3

a) Using zero-point energy correction.
b) Using Gibbs free energy.
c) Using total electronic energy.
d) Using single-point energy obtained in water via PCM model.

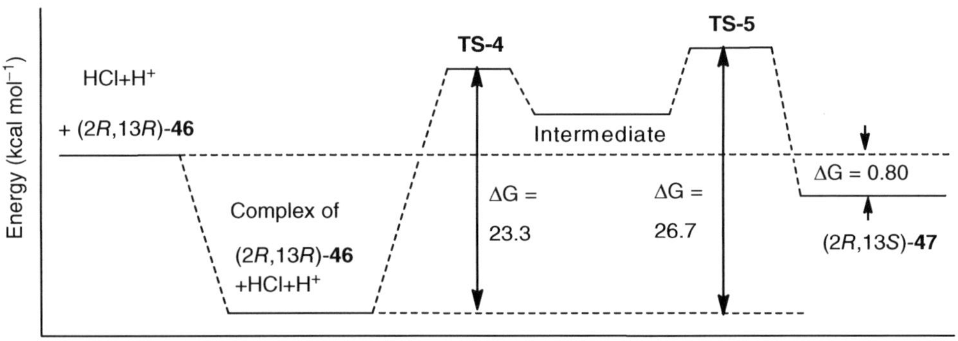

Figure 6.15 3D structures of **TS-4** and **TS-5**, and the coordinates for the conversion from (2*R*,3*R*)-**46** to (2*R*,3*S*)-**47** [16b].

decreased to 25.1 kcal mol^{-1} at the B3LYP/6-311++G(2d,p) level in water with the PCM model using SPE data. However, in the presence of another H^+, for example, **TS-2** and **TS-4**, the conversion benefits from a lower TS barrier of 23.1 kcal mol^{-1} (**TS-4**) in the gas phase in TEE or 17.9 kcal mol^{-1} in SPE in water (Table 6.5). If the reaction involves **TS-2**, the barriers in the gas phase would be higher by 5.1 – 5.3 kcal mol^{-1} than those in **TS-4**. However, the barrier via **TS-2** decreased to 16.5 kcal mol^{-1} in water and was about 1.4 kcal mol^{-1} lower than **TS-4**. Both barriers are low enough for isomerization. The procedure involved HCl again. This energy barrier is only 25.4 kcal mol^{-1} in ZPE, 26.7 kcal mol^{-1} in GFE, and 27.7 kcal mol^{-1} in TEE. The barrier decreased to 21.3 kcal mol^{-1} in water using the PCM model. The reaction coordinate for the whole procedure is illustrated in Figure 6.15. Compound (2*R*,13*S*)-**46** finally decomposed into L-phenylalanine, as reported previously.

Finally, X-ray diffraction data was obtained using a Cu-radiation source. The X-ray experiments confirmed that the theoretical predictions were correct. The structure (−)-**46** should be (2*R*,13*R*).

Notice

Experimental results are always the final proof for different theories. However, if the interpretation of the experimental results is not correct, the conclusions reached will be affected. With this idea in mind, the example analyzed here carries a high level of significance for both experimental and theoretical chemists. All sources of evidence need to be considered carefully, and computational methods should be used, where possible, to confirm conclusions that were arrived at based only on chemical reasoning.

There is one other example that used VCD and DFT calculations for AC reassignment of two chromanes from *Peperomia obtusifolia* (Piperaceae) [9]. Previous analysis of the ECD spectra of two prenylated benzopyrans by means of the helicity rule for the chromane chromophore resulted in the incorrect assignment of their AC (**48**) instead of (*R*) for (+)-enantiomers [9]. After the application of VCD study using DFT methods (B3LYP/6-31G(d)), and comparison of the experimental and calculated VCD and IR spectra, AC of (+)-**48** and (+)-**49** was reassigned directly in solution as (*R*). Moreover, time-dependent (TD)-DFT (B3LYP/6-311++G(2d,2p)//B3LYP/6-31G(d)) calculations of the ECD spectra confirmed this conclusion (Figure 6.16).

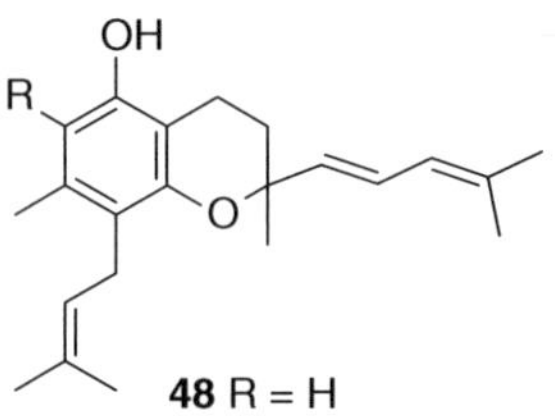

48 R = H
49 R = CO_2H

As the final example, we discuss the AC for chiral compounds from enantioselective hydrosilylation of C=N with $HSiCl_3$, which has been proven to be a powerful method for producing chiral amines [18]. The addition product of (−)-*N*-(1-phenylethyl) aniline (**52**) from **50** was assigned as (*R*) in recent reports [19].

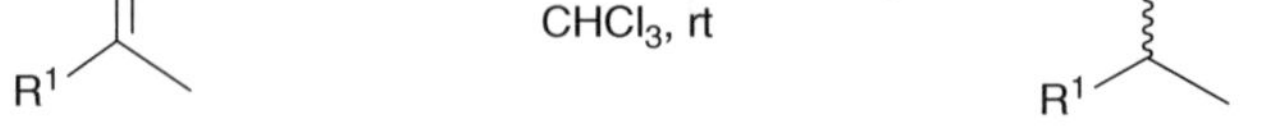

50 R^1 = Ph, R^2 = Ph
51 R^1 = 4-NO_2-Ph, R^2 = Ph

52 R^1 = Ph, R^2 = Ph
53 R^1 = 4-NO_2-Ph, R^2 = Ph (6.1)

Compound **52** has an OR of −13.5 in methanol, while **53** has OR values of +16.5 obtained in the additions. Therefore, their AC should be (*R*) and (*S*), respectively, based on recent reports. Because an asymmetric-axis-supported chiral catalyst **54** was used and high enentiometric excess values were achieved

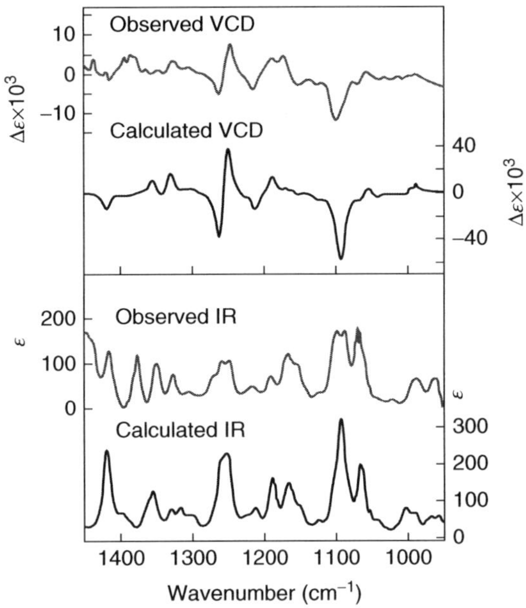

Figure 6.16 Comparison of the VCD and IR spectra of the measured (+)-**49** with the calculated VCD and IR spectra of the Boltzmann average of the four lowest energy conformers of the corresponding (*R*)-**49** [9].

in the reaction, the study of its mechanism using HF (Hatree–Fock) and DFT theory should be valid for examination. However, it showed that **52** and **53** should have (*S*) and (*R*) AC, respectively, instead of (*R*) and (*S*) based on the TS barrier magnitudes [20].

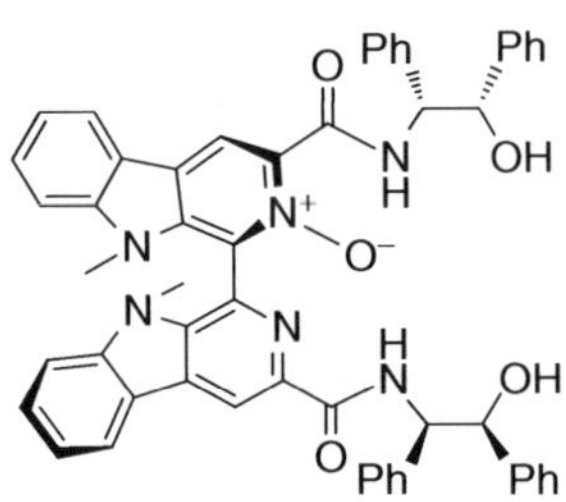

54

The AC of **52** was investigated using the matrix method since it is an acyclic chiral compound [20]. The predicted det(D) for (*S*)-**52** was −9.76 and the recorded OR was −13.5 in CH_3OH. The k_0 value was +1.38 (=−13.5/−9.76). This proves that its AC should be (*S*) instead of (*R*).

The ECD for (*S*)-**52** was computed at the B3LYP/6-311++G(2d,p)//B3LYP/6-311++G(2d,p) level and simulated using GFE (Figure 6.17). After UV corrections, the experimental and predicted ECDs were superimposed, and both ECD curves

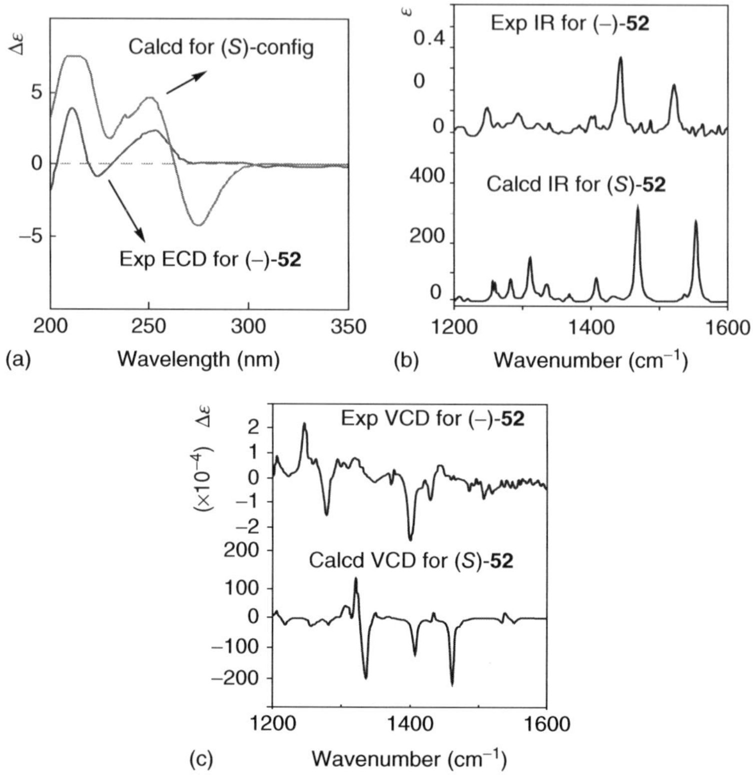

Figure 6.17 Predicted and experimental (a) ECD, (b) IR, and (c) VCD for **52** [20].

looked similar. At the same time, the VCD for (*S*)-**52** was performed at the B3LYP/6-311++G(2d,p)//B3LYP/6-311++G(2d,p) level, and the predicted VCD for (*S*)-**52** looked almost the same as the experimental VCD. Therefore, (−)-**52** should have (*S*) AC.

Compound **53** has positive OR values [19d, 20]. Logically, it should have (*R*) AC. Furthermore, the product (*S*)-**53** was selected for det(D) and ECD computations first. The AC for (+)-**53** was predicted to be (*R*) using the matrix method. The predicted ECD for (*S*)-**53** at the B3LYP/6-311++G(2d,p)//B3LYP/6-311++G(2d,p) level looked like a mirror image of the experimental ECD of (+)-**53** after UV corrections (Figure 6.18). It confirmed that (+)-**53** should be (*R*) AC too. The predicted VCD for (*S*)-**53** at the B3LYP/6-311++G(2d,p)//B3LYP/6-311++G(2d,p) level looked like a mirror image of the experimental VCD of (+)-**53**. This confirmed again that (+)-**53** should be (*R*)-AC.

Amazingly, some earlier reports of this compound had found that the AC of (−)-**52** was (*S*) [21]. Landor and coworkers confirmed that phenylation of (*S*)-(−)-1-phenylethylamine afforded (−)-*N*-(1-phenylethyl) aniline (−)-**52**. It is

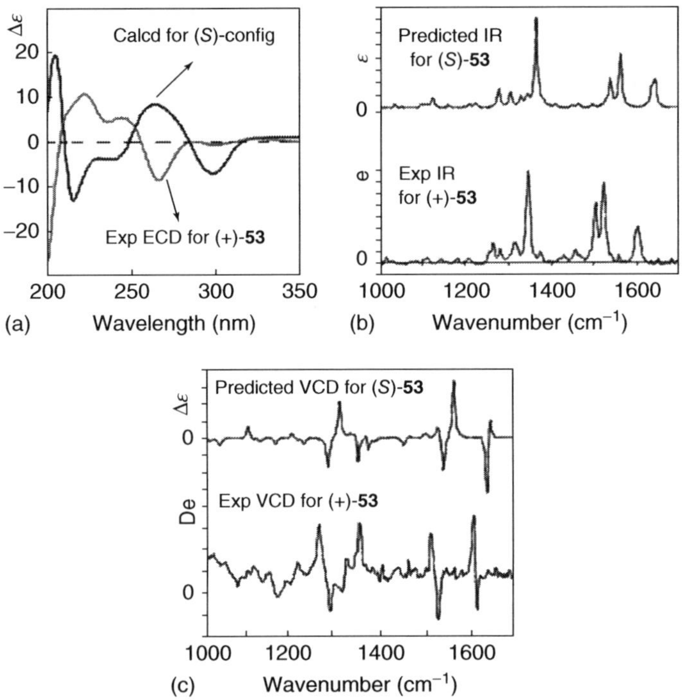

Figure 6.18 Predicted and experimental (a) ECD, (b) IR, and VCD for **53** [20].

unclear why the reports that appeared near 1997 or earlier assigned the (*R*)-AC for (−)-**52**.

Other structures reassigned using chiroptical spectroscopies and NMR methods are given in Table 6.6. The procedure for the reassignment is not analyzed here. Readers may consult the corresponding literature for details.

Table 6.6 Reassigned structures.

Entry	Original structure and methods used for AC identification	Corrected structure and methods for assignment of AC
1	Chromanes (**48, 49**) from *Peperomia obtusifolia* [22, 23], empirical ECD 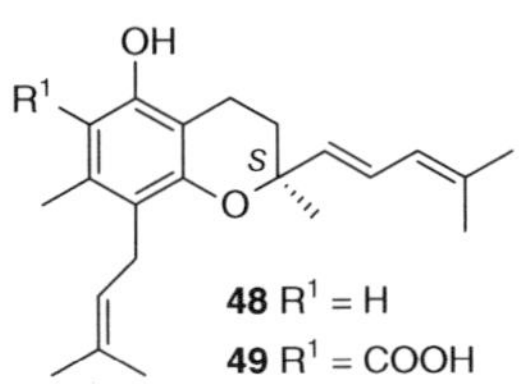 **48** R^1 = H **49** R^1 = COOH	Using the following models **55** and **56** in VCD study [9]. Reassigned AC as (*R*) 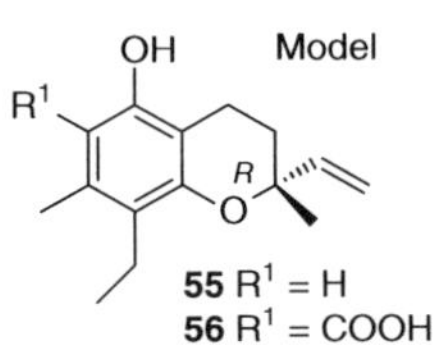 **55** R^1 = H **56** R^1 = COOH

(*continued overleaf*)

Table 6.6 (*Continued*)

Entry	Original structure and methods used for AC identification	Corrected structure and methods for assignment of AC
2	Isoepitaondiol from *Stypopodium flabelliforme* [24], NMR 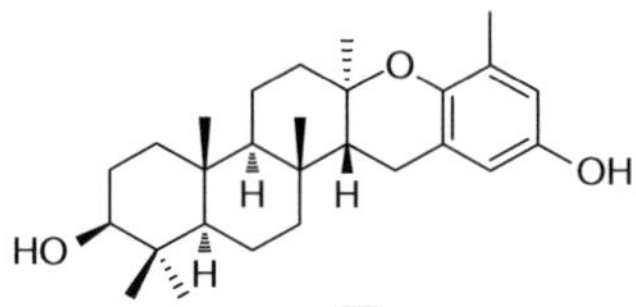 **57**	X-ray and VCD [25] 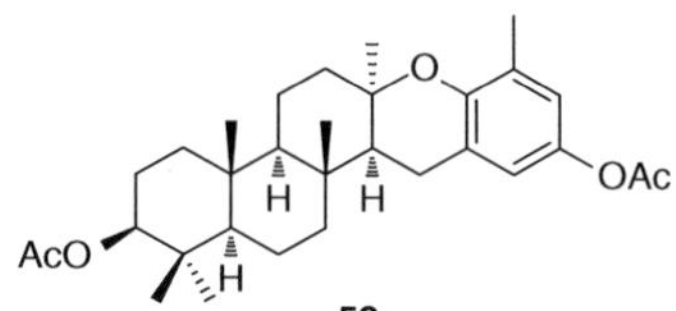 **58**
3	Rosmaridiphenol from *Rosmarinus officinalis* L. [26]. NMR, empirical ECD 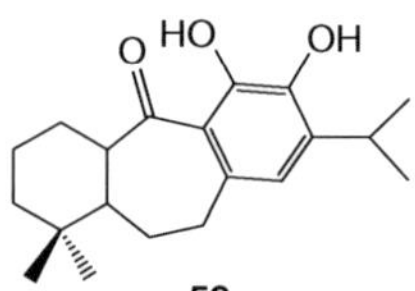 **59**	X-ray and VCD [27] 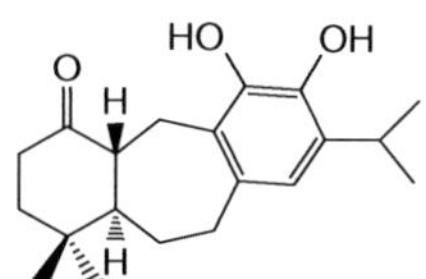 **60**
4	H3 receptor antagonist GT-2331 [28]. X-ray result 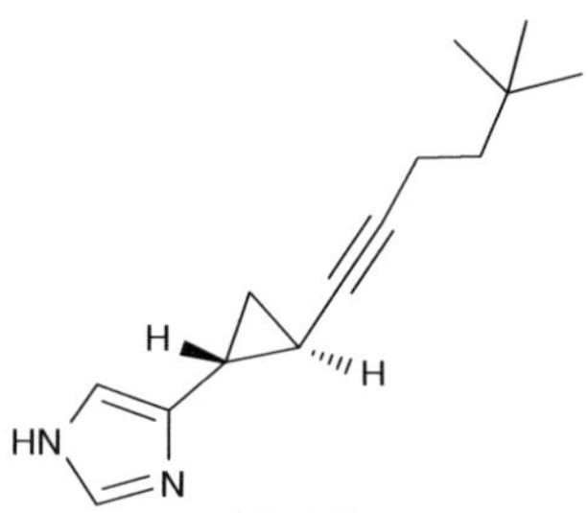 (1*R*, 2*R*)-**61**	Reassigned by VCD [29] 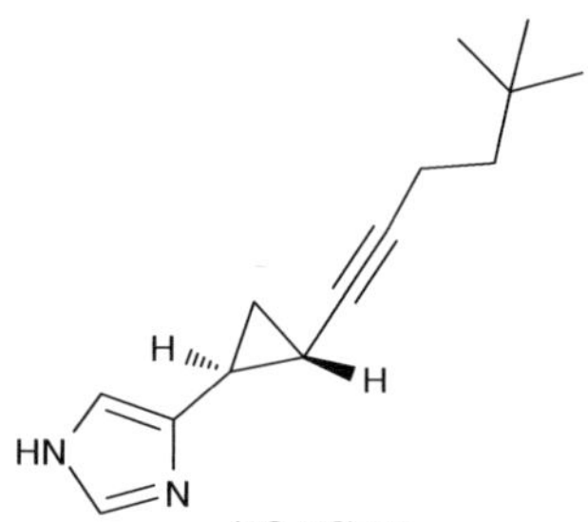 (1*S*, 2*S*)-**62**
5	Iridoids plumericin and isoplumericin [30], X-ray, semiempirical ECD 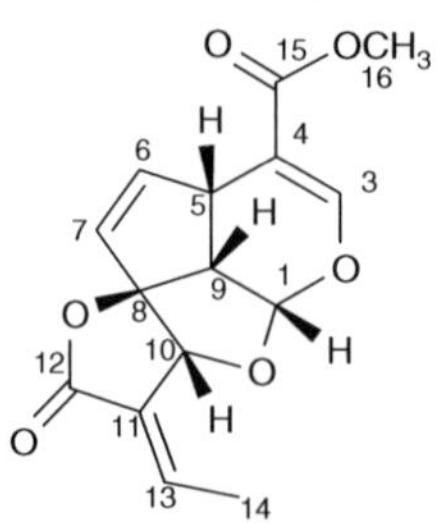 (1*S*, 5*R*, 8*R*, 9*R*, 10*R*)-**63**	VCD, ORD in the range 365–589 nm [31] 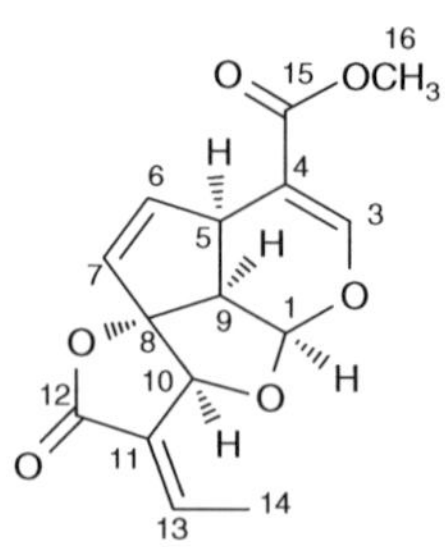 (1*R*, 5*S*, 8*S*, 9*S*, 10*S*)-**64**

Table 6.6 (*Continued*)

Entry	Original structure and methods used for AC identification	Corrected structure and methods for assignment of AC
	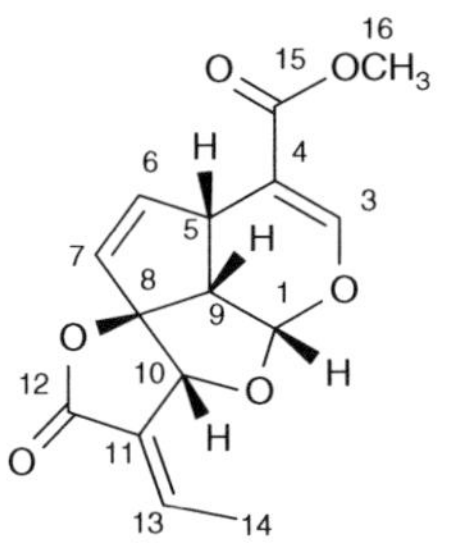 (1*S*, 5*R*, 8*R*, 9*R*, 10*R*)-**65**	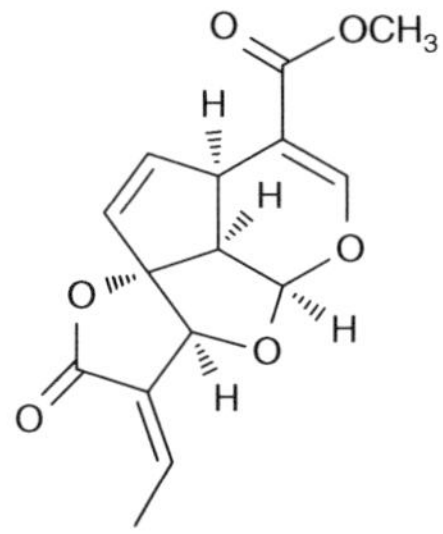 (1*R*, 5*S*, 8*S*, 9*S*,10*S*)-**66**
6	Oruwacin from *Morinda lucida* [32]. Comparison with a known compound plumericin	The naturally occurring oruwacin and plumericin have opposite ACs [33]
	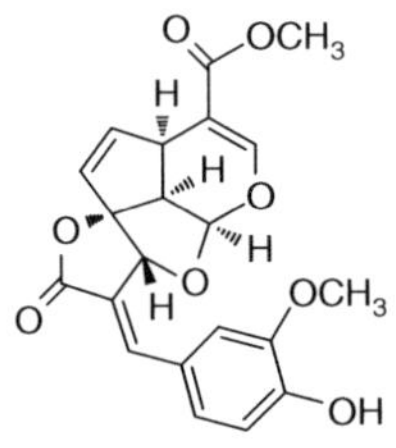 (1*S*, 5*S*, 8*S*, 9*S*, 10*S*)-**67**	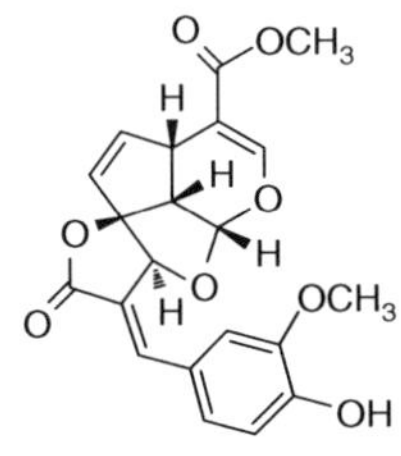 (1*S*, 5*S*, 8*S*, 9*S*, 10*S*)-**68**
7	Elatenyne isolated from the marine alga *Laurencia elata* [34]. NMR	NMR computations, total synthesis [35]
	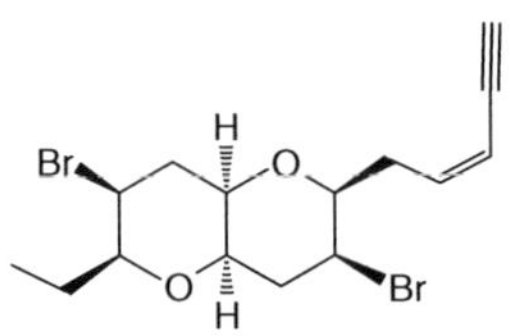 **69**	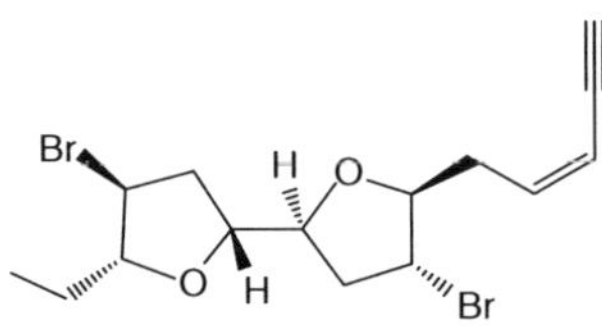 **70**
8	Aquatolide [36]	Computed ^{1}H and ^{13}C NMR, crystal [36]
	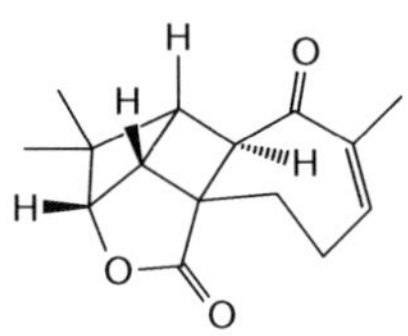 **71**	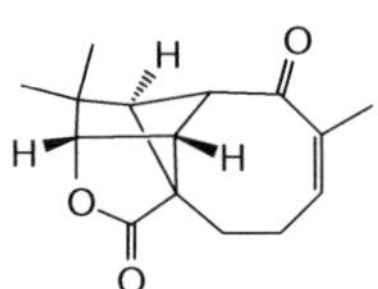 **72**

(*continued overleaf*)

Table 6.6 (*Continued*)

Entry	Original structure and methods used for AC identification	Corrected structure and methods for assignment of AC
9	Nobilisitine A, from *Clivia nobilis* [37] 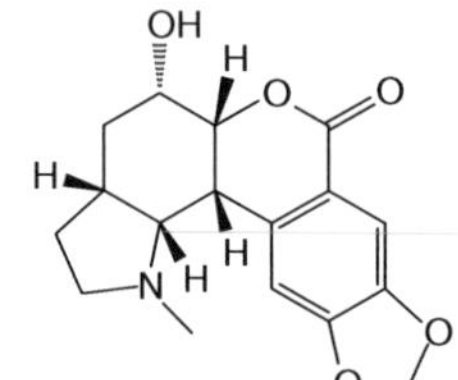 **73**	Computed ^{1}H and ^{13}C chemical shifts [38] 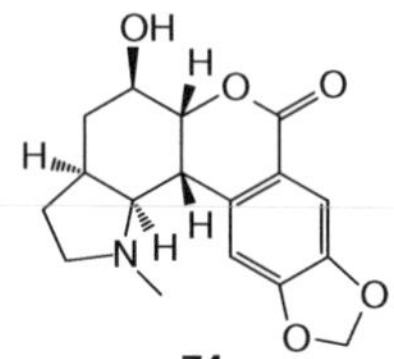 **74**
10	Plakotenin from a Plakortis marine sponge [39] 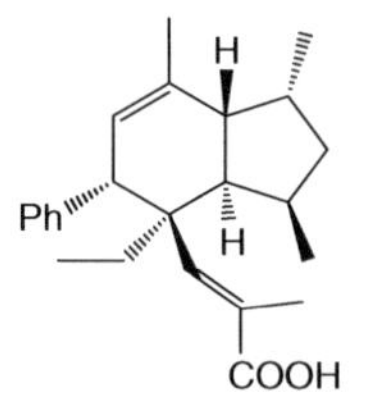 **75**	Calculated NMR [36] 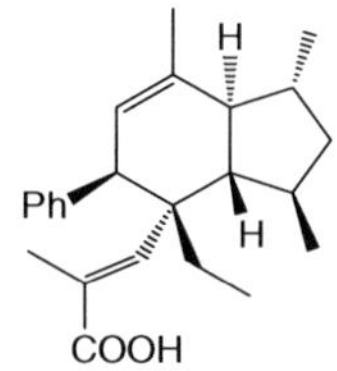 **76**
11	Hexacyclinol from the fungus *Panus rudis* [40] 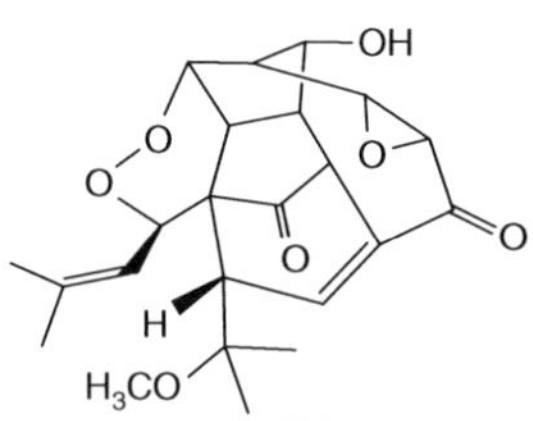 **77**	Calculated NMR, synthesis, and crystal [41] 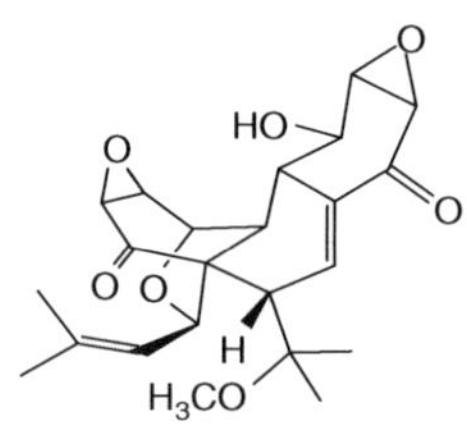 **78**
12	Vannusal B from the marine ciliate *Euplotes vannus* [42] 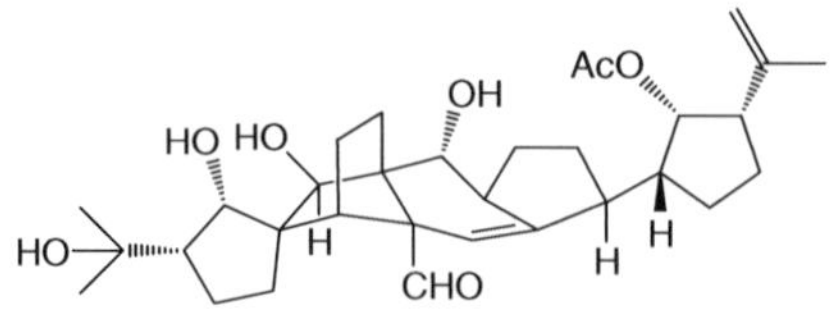 **79**	^{13}C chemical shifts and coupling constants [43, 44] 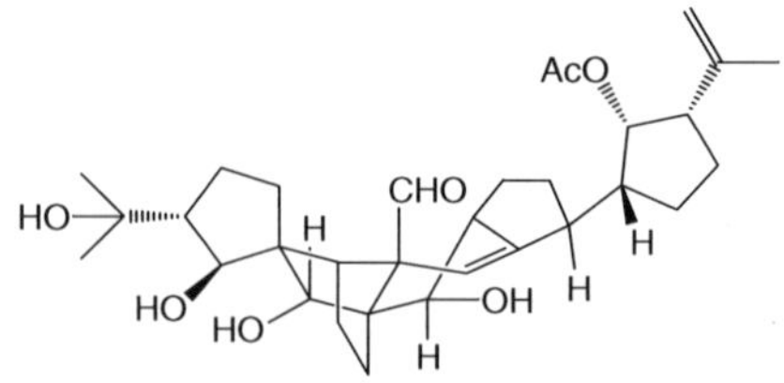 **80**

Table 6.6 (*Continued*)

Entry	Original structure and methods used for AC identification	Corrected structure and methods for assignment of AC
13	Griseusins ferment solution of *Nocardiopsis* sp. NMR empirical ECD [45a] 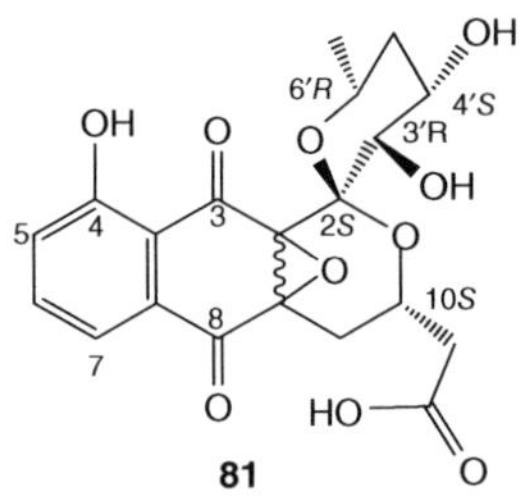 **81**	OR and ECD using QM [45b] 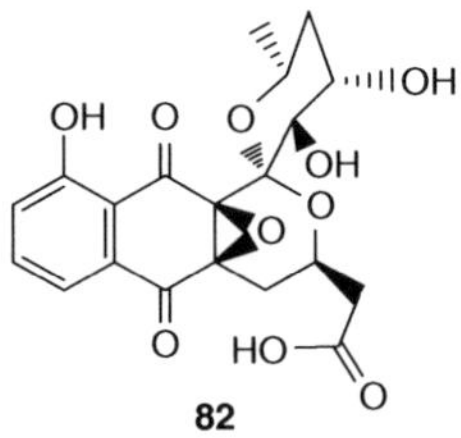 **82** (same reference for **85–90**)
14	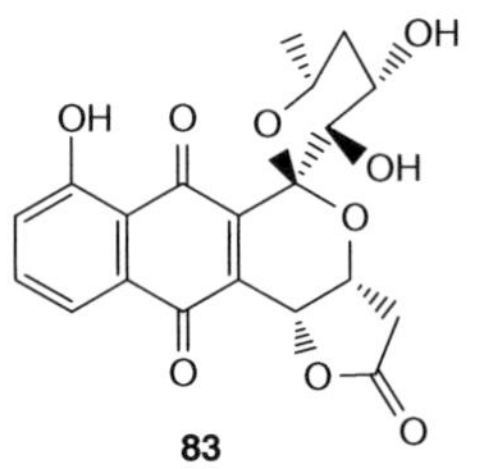 **83**	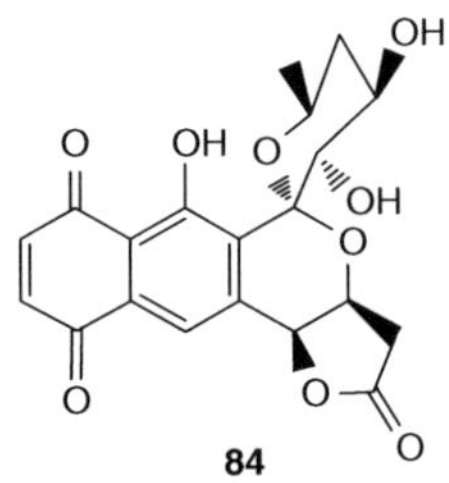 **84**
15	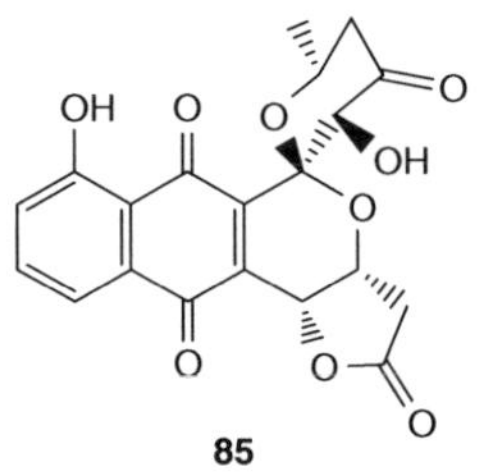 **85**	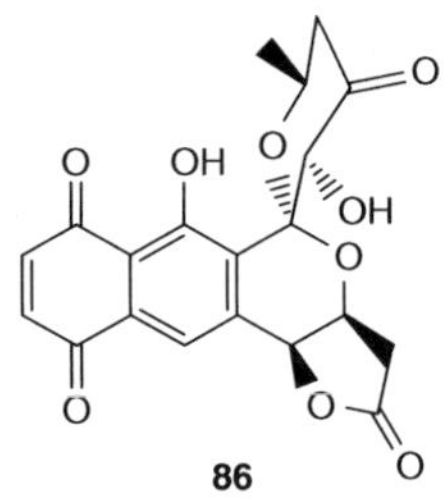 **86**
16	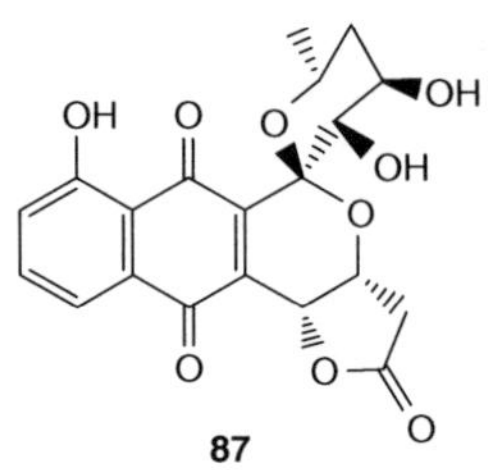 **87**	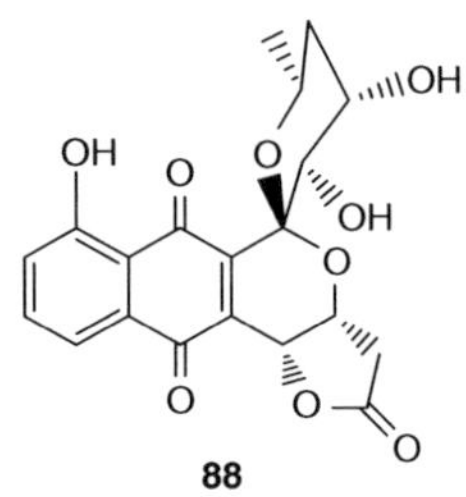 **88**

QM: quantum mechanics.

References

1. (a) Lee, T.J. and Scuseria, G.E. (1995) *Quantum Mechanical Electronic Structure Calculations with Chemical Accuracy* (ed. S.R. Langhoff), Kluwer Academic Publishers, Dordrecht; (b) Helgaker, T., Jørgensen, P., and Olsen, J. (2000) *Molecular Electronic Structure Theory*, John Wiley & Sons, Inc., New York; (c) Liu, D.-Z., Wang, F., Liao, T.-G., Tang, J.-G., Steglich, W., Zhu, H.J., and Liu, J.-K. (2006) *Org. Lett.*, **8**, 5749–5752; (d) Forsyth, D.A. and Sebag, A.B. (1997) *J. Am. Chem. Soc.*, **119**, 9483.
2. Stephens, P., Pan, J.J., Devlin, F.J., Urbanova, M., Julinek, O., and Hájíček, J. (2008) *Chirality*, **20**, 454–470.
3. (a) Brewster, J.H. (1959) *J. Am. Chem. Soc.*, **81**, 5475; (b) Zhu, H.J. (2009) *Modern Organic Stereochemistry*, Chapter 1, Science Press of China; (c) Zhang, L., Wang, S.Q., Li, X.J., Zhang, A.L., Zhang, Q., and Gao, J.M. (2012) *J. Mol. Struct.*, **1016**, 72.
4. Delvin, F.J., Stepehens, P.J., and Besse, P. (2005) *J. Org. Chem.*, **70**, 2980–2993.
5. Urena, F.P., Moreno, R.A., and Gonzalz, J.J.L. (2008) *J. Phys. Chem. A*, **112**, 7887–7893.
6. Moreno, J.R.A., Urena, F.P., and Gonzalz, J.J.L. (2009) *Vib. Spectrosc.*, **51**, 318–325.
7. Li, X.N., Zhang, Y., Cai, X.H., Feng, T., Liu, Y.P., Li, Y., Ren, J., Zhu, H.J., and Luo, X.D. (2011) *Org. Lett.*, **13**, 5896.
8. Atsufumi, N., Yoshihiro, Y., Nobuaki, M., Makoto, E., and Kenji, M. (2011) *J. Nat. Prod.*, **74**, 707–711.
9. Batista, J.M., Batista, A.N.L. Jr.,, Rinaldo, D., Vilegas, W., Cass, Q.B., Bolzani, V.S., Kato, M.J., Lopez, S.N., Furlan, M., and Nafie, L.A. (2010) *Tetrahedron: Asymmetry*, **21**, 2402–2407.
10. Nicolaou, K.C. and Snyder, S.A. (2005) *Angew. Chem. Int. Ed.*, **44**, 1012–1044.
11. Grajewska, A., Rozwadowska, M.D., and Gzella, A. (2007) *Pol. J. Chem.*, **81**, 1861–1868.
12. Giorgio, E., Roje, M., Tanaka, K., Hamersak, Z., Sunjic, V., Nakanishi, K., Rosini, C., and Berova, N. (2005) *J. Org. Chem.*, **70**, 6557–6563.
13. Huang, X., He, J., Niu, X., Menzel, K.D., Dahse, H.M., Grabley, S., Fiedler, H.P., Sattler, I., and Hertweck, C. (2008) *Angew. Chem. Int. Ed.*, **47**, 3995–3998.
14. Hosokawa, S., Mukaeda, Y., Kawahara, R., and Tatsuta, K. (2009) *Tetrahedron Lett.*, **50**, 6701–6704.
15. (a) Ohtani, I., Kusumi, T., Kashman, Y., and Kakisawa, H. (1991) *J. Am. Chem. Soc.*, **113**, 4092–4096; (b) Ohtani, I., Kusumi, T., Kashman, Y., and Kakisawa, H. (1991) *J. Org. Chem.*, **56**, 1296–1298; (c) Li, Q.-M., Ren, J., Shen, L., Bai, B., Li, Q.-M., Liu, X.-C., Wen, M.-L., and Zhu, H.-J. (2013) *Tetrahedron*, **69**, 3067–3074; (d) He, P., Wang, X., Guo, X., Zhou, C., Shen, S., Hu, D., Yang, X., Luo, D., Dukor, R., and Zhu, H.-J. (2014) *Tetrahedron Lett.*, **55**, 2965–2968.
16. (a) Li, G.Y., Yang, T., Luo, Y.G., Chen, X.Z., Fang, D.M., and Zhang, G.L. (2009) *Org. Lett.*, **11**, 3714–3717; (b) Ren, J., Li, G.-L., Shen, L., Zhang, G.-L., Nafie, L., and Zhu, H.-J. (2013) *Tetrahedron.*, **69**, 10351–10356.
17. Li, L.C., Ren, J., Jiang, J.X., and Zhu, H.J. (2007) *Eur. J. Org. Chem.*, **2007**, 1026–1030.
18. (a) Malkov, A.V., Stewart-Liddon, A.J.P., McGeoch, G.D., Ramirez-Lopez, P., and Kocovsky, P. (2012) *Org. Biomol. Chem.*, **10**, 4864–4877; (b) Carpentier, J.F. and Bette, V. (2002) *Curr. Org. Chem.*, **6**, 913–916; (c) Malkov, A.V., Stončius, S., and Kočovský, P. (2007) *Angew. Chem. Int. Ed.*, **46**, 3722–3724; (d) Zhou, L., Wang, Z., Wei, S., and Sun, J. (2007) *Chem. Commun.*, **28**, 2977–2979; (e) Jiang, Y., Chen, X., Zheng, Y., Xue, Z., Shu, C., Yuan, W., and Zhang, X. (2011) *Angew. Chem. Int. Ed.*, **50**, 7304–7307.
19. (a) Malkov, A.V. and Mariani, A. (2004) *Org. Lett.*, **6**, 2253–2256; (b) Malkov, A.V. and Liddon, A.J.P.S. (2006) *Angew. Chem. Int. Ed.*, **45**, 1432–1435; (c) Storer, R.I. and Carrera, D.E. (2005) *J. Am. Chem. Soc.*, **128**, 84–86; (d) Pan, W., Deng, Y., He, J.B., Bai, B., and Zhu, H.J. (2013) *Tetrahedron*, **69**, 7253.
20. Pan, W., Ma, W.G., Yang, X.D., Zhen, Y.H., Song, B.Q., Niu, Y.Z., Gu, G., Hu, D.B., Yang, Q., and Zhu, H.J. (2015)

Chem. J. Chin. Univ., **36**(2), 325–329. The absolute configuration determination of the products 52 and 53 was selected from the Ph. D Thesis of Dr. Pan W. It is not published until now.
21. (a) Landor, S.R., Sonola, O.O., and Tatchell, A.R. (1984) *Bull. Chem. Soc. Jpn.*, **57**, 1658–1661; (b) Witting, G. and Thiele, U. (1969) *Liebigs Ann. Chem.*, **726**, 1–12.
22. Mota, J.S., Leite, A.C., Batista, J.M., López, S.N. Jr.,, Ambrósio, D.L., Passerini, G.D., Kato, M.J., Bolzani, V.S., Cicarelli, R.M.B., and Furlan, M. (2009) *Planta Med.*, **75**, 620–623.
23. Batista, J.M. Jr.,, López, S.N., Mota, J.S., Vanzolini, K.L., Cass, Q.B., Rinaldo, D., Vilegas, W., Bolzani, V.S., Kato, M.J., and Furlan, M. (2009) *Chirality*, **21**, 799–801.
24. Rovirosa, J., Sepulveda, M., Quezada, E., and San-Martin, A. (1992) *Phytochemistry*, **31**, 2679–2681.
25. Areche, C., San-Martin, A., Rovirosa, J., Muñoz, M.A., Hernández-Barragán, A., Bucio, M.A., and Joseph-Nathan, P. (2010) *J. Nat. Prod.*, **73**, 79–82.
26. (a) Chang, S.S., Ho, C.T., and Houlihan, C.M. (1987) Isolation of a novel antioxidant rosmaridiphenol from Rosmarinus officinalis L. US Patent 4, 638, 095. (b) Pertino, M.W. and Schmeda-Hirschmann, G. (2010) *Planta Med.*, **76**, 629–632.
27. Muñoz, M.A., Perez-Hernandez, N., Pertino, M.W., Schmeda-Hirschmann, G., and Joseph-Nathan, P. (2012) *J. Nat. Prod.*, **75**, 779–783.
28. Khan, M.A., Yates, S.L., Tedford, C.E., Kirschbaum, K., and Phillips, J.G. (1997) *Bioorg. Med. Chem. Lett.*, **7**, 3017–3022.
29. Minick, D.J., Copley, R.C.B., Szewczyk, J.R., Rutkowske, R.D., and Miller, L.A. (2007) *Chirality*, **19**, 731–740.
30. (a) Albers-Schonberg, G. and Schmid, H. (1960) *Chimia*, **14**, 127–128; (b) Albers-Schonberg, G. and Schmid, H. (1961) *Helv. Chim. Acta*, **44**, 1447–1473.
31. Stephens, P.J., Pan, J.J., Devlin, F.J., Krohn, K., and Kurtan, T. (2007) *J. Org. Chem.*, **72**, 3521–3536.
32. Adesogan, E.K. (1979) *Phytochemistry*, **18**, 175–176.
33. Stephens, P.J., Pan, J.J., Devlin, F.J., and Cheeseman, J.R. (2008) *J. Nat. Prod.*, **71**, 285–288.
34. Hall, J.G. and Reiss, J.A. (1986) *Aust. J. Chem.*, **39**, 1401–1409.
35. Sheldrake, H.M., Jamieson, C., and Burton, J.W. (2006) *Angew. Chem. Int. Ed.*, **45**, 7199–7202.
36. Lodewyk, M.W., Soldi, C., Jones, P.B., Olmstead, M.M., Larrucea, J.R., Shaw, J.T., and Tantillo, D.J. (2012) *J. Am. Chem. Soc.*, **134**, 18550–18553.
37. Evidente, A., Abou-Donia, A.H., Darwish, F.A., Amer, M.E., Kassem, F.F., Hammoda, H.A.M., and Motta, A. (1999) *Phytochemistry*, **51**, 1151–1155.
38. Schwartz, B.D., Jones, M.T., Banwell, M.G., and Cade, I.A. (2010) *Org. Lett.*, **12**, 5210–5213.
39. Kobayashi, J., Takeuchi, S., Ishibashi, M., Shigemori, H., and Sasaki, T. (1992) *Tetrahedron Lett.*, **33**, 2579–2580.
40. Schlegel, B., Hartl, A., Dahse, H.M., Gollmick, F.A., and Grafe, U. (2002) *J. Antibiot.*, **55**, 814–817.
41. Rychnovsky, S.D. (2006) *Org. Lett.*, **8**, 2895–2898.
42. Guella, G., Dini, F., and Pietra, F. (1999) *Angew. Chem. Int. Ed.*, **38**, 1134–1136.
43. Saielli, G., Nicolaou, K.C., Ortiz, A., Zhang, H., and Bagno, A. (2011) *J. Am. Chem. Soc.*, **133**, 6072–6077.
44. Tantillo, D. (2013) *J. Nat. Prod. Rep.*, **30**, 1079.
45. (a) He, J., Roemer, E., Lange, C., Huang, X., Maier, A., Kelter, G., Jiang, Y., Xu, L.-H., Menzel, K.-D., Grabley, S., Fiebig, H.-H., Jiang, C.-L., and Sattler, I. (2007) *J. Med. Chem.*, **50**, 5168; (b) Zhu, H.J., Li, W.X., Hu, D.B., and Wen, M.L. (2014) *Tetrahedron*, **70**, 8236–8243.

Part III
Reactions

7
Enantioselective Reaction

An enantioselective reaction is the simplest stereoselective reaction. It just involves (*R*)- or (*S*)-product formation catalyzed by chiral catalysts or auxiliaries. Many outstanding results have been achieved in this area.

Enantioselective reaction includes three major reaction types: one is addition and another is reduction, or in some reactions, it is hydrosilylation; the third one is oxidation, or more specifically Sharpless epoxidation. All are important reactions.

To control one stereogenic center formation in reactions, it is a key step to design and synthesize suitable chiral catalysts or auxiliaries used in reactions. The catalytic procedure would form an asymmetric space during reaction, which leads to various transition state (TS) barriers and sequentially brings different quantities of (*R*) or (*S*) enantiomers. The ee value of the product can be measured by optical rotation (OR), high-performance liquid chromatography (HPLC), or gas chromatography (GC). The more the TS barrier difference, the larger will be the ee value.

7.1
Enantioselective Addition

7.1.1
Organic Zn- or Zn-Ti Reagent

The use of diethylzinc was reported very early by Frankland, in 1849 [1]. However, because of its low reactivity, diethylzinc has been of limited use until it was successfully used in addition to aldehydes in the presence of chiral catalysts [2]. The enantioselective diethylzinc addition to aldehydes has been well studied during the 1990s, and many excellent results have been achieved, including its application for the total synthesis of natural products [3]. The typical reaction equation is illustrated below (Eq. (7.1)).

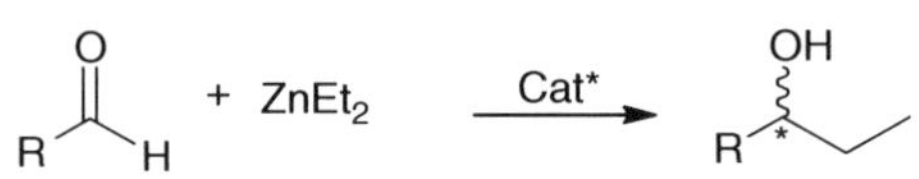

(7.1)

Organic Stereochemistry: Experimental and Computational Methods, First Edition. Hua-Jie Zhu.

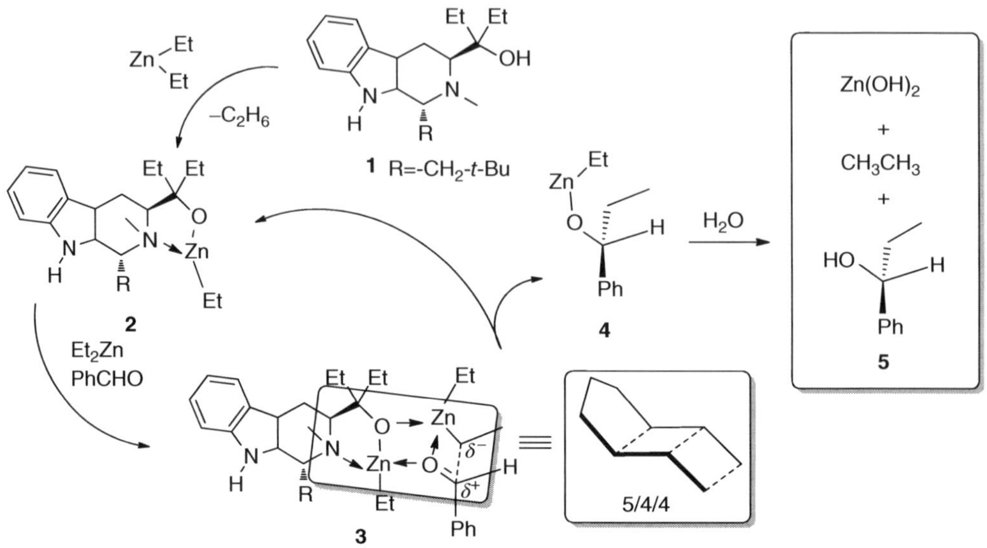

Figure 7.1 The plausible TS structure for addition of diethylzinc to benzaldehyde **4** [3b].

In this addition procedure, a traditional chiral catalyst is β-amino alcohol. It can chelate with the Zn^{++} to form a Zn complex first, based on early reports (Figure 7.1), and then this complex would catalyze asymmetric synthesis [4]. For example, the chiral β-amino alcohol catalyst **1**, which was derived from natural alkaloid abrine, can afford about 99% ee in the addition [5]. According to the mechanistic study results, this β-amino alcohol would react with one molecule of Et_2Zn to form complex **2**, and this would react with another molecule of $ZeEt_2$ and a benzaldehyde to form a TS state structure **3**. It is a 5/4/4 ring system. After the negative carbon of the ethyl group is transferred to the carbonyl group, it would decompose into an intermediate **4** and complex **2**. Intermediate **4** can be converted into the addition product by hydrolysis after the reaction is finished. A typical catalytic procedure is illustrated in Figure 7.1.

In this catalytic procedure, the quantity of $ZnEt_2$ should be at least twice that of the aldehydes. Experimental results have confirmed this assumption. Thus, in a traditional addition of diethylzinc to aldehyde, the mole quantity of Et_2Zn is double that of the aldehydes. Under basic condition, the catalyst can exhibit its enantioselective ability. For example, chiral ligand **6** can catalyze the addition with high ee values (Eq. (7.1) and Table 7.1) [6].

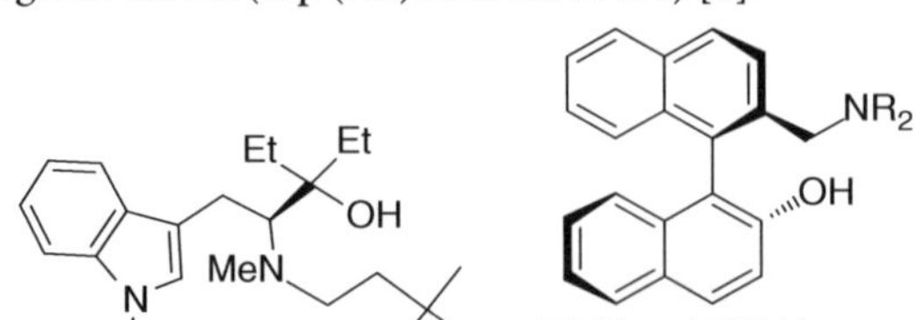

6

7 $NR_2 = N(CH_2)_4$
8 $NR_2 = N[(CH_2)_2O(CH_2)_2]$

Table 7.1 Addition of diethylzinc to aldehyde using ligand **6** [6].

Entry	Aldehyde	Yield (%)	op% ee	Configuration
1	4-MeOPh	90	92	*R*
2	4-ChPh	99	94	*R*
3	4-MePh	90	91	*R*
4	2-Naph	99	100	*R*
5	*c*-Hexyl	65	65	*R*

Table 7.2 Addition of diethylzinc to aldehydes using ligand **7** and **8** [7].

Entry	R	Catalyzed by 7		Catalyzed by 8	
		Yield (%)	ee (%)	Yield (%)	*ee* (%)[a]
1	Ph	89	94[a]	—	—
2	4-MeOPh	97	94[b]	97	98[c]
3	2-ClPh	97	93[d]	95	96[b]
	4-MePh	—	—	97	97[b]
4	3-ClPh	95	96[b]	95	92[d]
5	4-ClPh	97	99[a]	97	95[e]
6	1-Naph	92	96[b]	—	—
7	2-Naph	95	96[b]	98	97[b]
8	Hexyl	96	94[d]	94	66[b]
9	*c*-Hexyl	89	98[f]	97	68[g]
10	(*E*)–Ph–CH=CH	95	86[b]	93	75[b] (*S*)
11	Ph–C≡C	97	76[b]	—	—
12	TIPS–C≡C	84	92[h]	72	85[i] (*R*)

a) Determined by chiral GC (Chiraldex G-Ta Column).
b) Determined by chiral HPLC (Chiralcel OD).
c) Determined by HPLC (Chiralcel OJ).
d) Determined by chiral HPLC [(*R*,*R*)-Welk-O1].
e) Determined by HPLC (Chiralcel OB-H).
f) Determined by chiral GC (Chiraldex G-TA) of the corresponding acetate derivatives.
g) Determined by HPLC (Chiralcel AD-H).
h) Determined by ^{19}F nuclear magnetic resonance (NMR) of the corresponding (*R*)-MTPA ester derivative.
i) Determined by ^{1}H NMR of the corresponding (*R*)-MTRPA ester derivatives.

Many excellent reports are available on the design and synthesis of various chiral catalysts, such as the axial catalysts **7** and **8**; both promoted the addition of diethylzinc to aldehydes in high enantioselectivities in diethylzinc to various aldehydes (Table 7.2) [7].

The chiral axial catalyst **9** was also reported with a high 94% ee [8]. The active section (1,4-diol structure) near the center of the axis provides a big wall-like space exclusion for asymmetric addition. Indeed, smaller groups are still capable of catalyzing the addition with high ee values. For example, ligand **10** gives almost

90–99% ee for most aromatic aldehydes except for aldehydes with a triple bond, such as Ph–C≡C–CHO (76% ee). This could be improved if the readily available ligand 1,1′-bi-2,2′-naphthol (BINOL) (*S*)-**11** was used [9]. It could afford up to 99% ee for a series of aldehyde addition with terminal alkynes using diethylzinc (Eq. (7.2), Table 7.3).

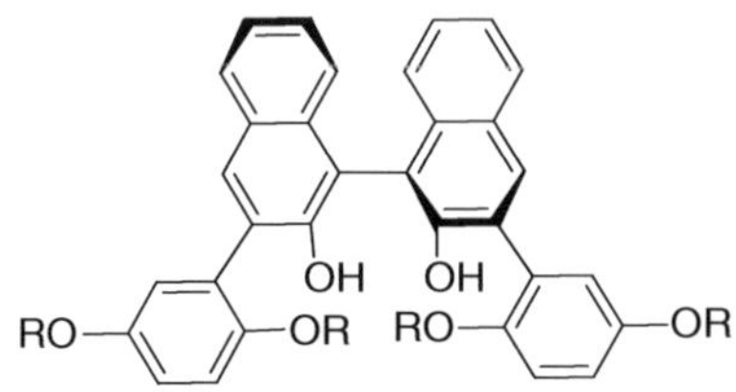

9 R=*n*-C_6H_{13}

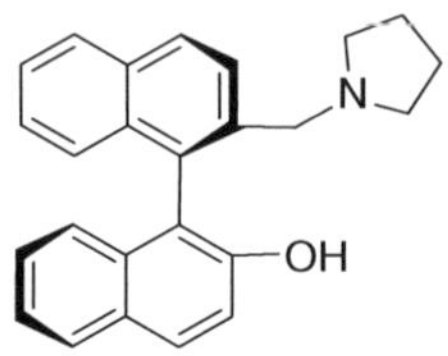

(*R*)-**10**

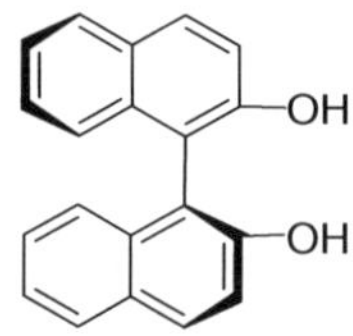

(*S*)-**11**

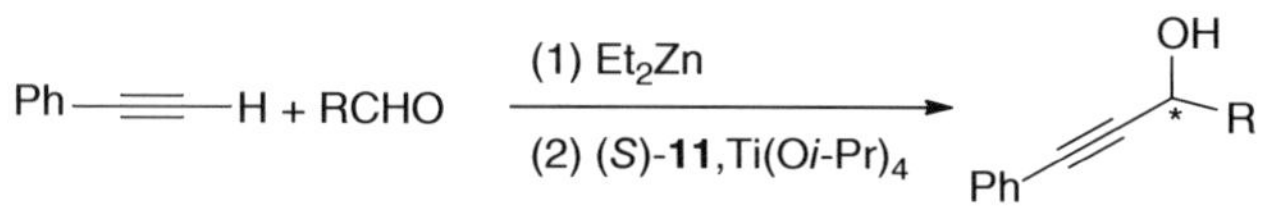

(7.2)

In the catalytic reaction (Eq. (7.2)), the ee values obtained are very high (up to 99% ee) in both aromatic and aliphatic aldehydes. $Ti(Oi\text{-}Pr)_4$ is needed for the

Table 7.3 Reactions of terminal alkynes with various aldehydes [9].

Entry	R	Yield (%)	*ee* (%)	Entry	R	Yield (%)	ee (%)
1	Ph	77	96	13[a)]	*n*-C_8H_{17}	96	91
2	4-FPh	74	96	14	*n*-C_7H_{15}	70	93
3	4-NO_2Ph	79[b)]	97	15	*n*-C_4H_7	91	93
4	4-ClPh	81	92	16	*i*-Pr	84	97
5	3-ClPh	79	92	17	Et	60	94
6	2-ClPh	95[c),d)]	92	18	*c*-Hexyl	58	95
7[c),d)]	4-MePh	93	97	19	PhCH_2CH_2	99	93
8	2-MeOPh	73	93	20[e)]	PhCH_2	93	91
9[d)]	2-Furyl	72	92	21	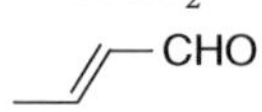	92	96
10	1-Naph	71[b)]	92	22	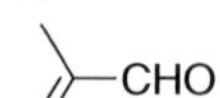	93	96
11	2-Naph	77	98	23	(*E*)–PhCH=CH	89	97
12[f)]	Ph	75	92	24	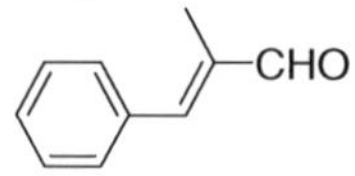	96	99

a) Redistilled aldehyde was used, 2 ml of toluene was used in the first step.
b) Determined by ^1H NMR.
c) Phenylacetylene: Et_2Zn: $Ti(Oi\text{-}Pr)_4$:BINOL:aldehyde = 4 : 4 : 1 : 0.4 : 1.
d) Previously unpublished result.
e) Aldehyde was added dropwise.
f) Triisopropylsilylacetylene was used in place of phenylacetylene.

catalysis. $ZnEt_2$ here is used as a moderate base instead of an addition reagent. It can remove the proton from the terminal alkyne to form an anion of Ph–C≡C–. This anion can attack the carbonyl carbon of aldehydes to afford the corresponding alcohols in the additions.

$ZnPh_2$ is also used in the additions [8]. However, it is not as widely used as Et_2Zn. When it is used in the addition to aromatic aldehydes catalyzed by **9**, the phenylation cannot be performed if $ZnEt_2$ is not used. Only in the presence of $ZnEt_2$, the phenylation can be achieved with up to 80–90% ee at room temperature or at −30 °C (Eq. (7.3)). In this case, $ZnEt_2$ acts as a promoter instead of an addition reagent.

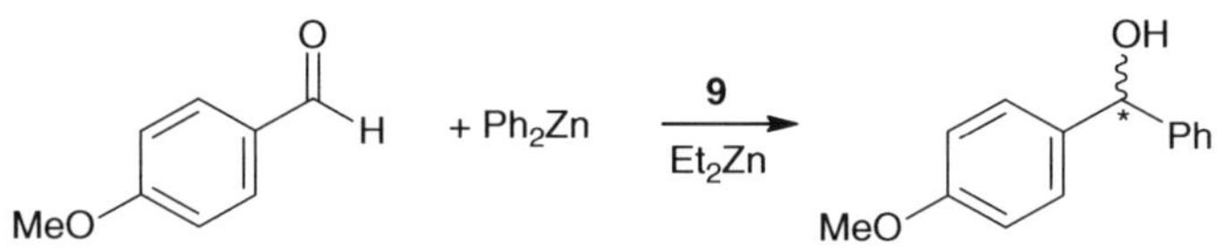

(7.3)

7.1.2
Organic Cu–Zn, Cu–Li Reagent

Organic Cu-promoted conjugate additions are well known; the recorded ee values can be up to 95% or higher [10]. The combined use of a Cu salt with an organic Zn reagent can promote1,4-additions efficiently. About 1 mol% of chiral ligand used in the addition could afford good enantioselectivities (up to 90% ee). A Cu salt, such as $Cu(OTf)_2$, is required in this catalytic procedure.

Conjugate additions of dialkyzinc reagents with Cu salts can be achieved directly. For example, the chiral ligand **12** induced excellent enantioselectivities in the additions of ZnR_2 to acyclic oxazolidinones (Eq. (7.4) and Table 7.4) [11]. A quantity of 1 mol% of the $(CuOTf)_2$–C_6H_6 is enough to promote the addition with 92% ee selectivity. A 1% mole is also the minimum quantity for this addition. Once it is reduced to 0.5%, the ee values decreased greatly to 76% ee (Table 7.4, entry 3).

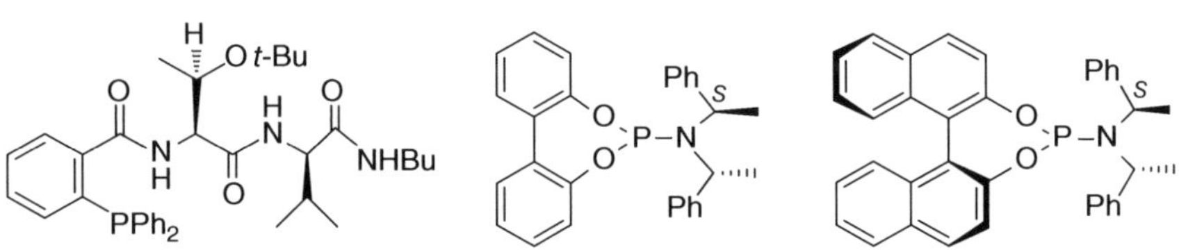

12 **13** **14**

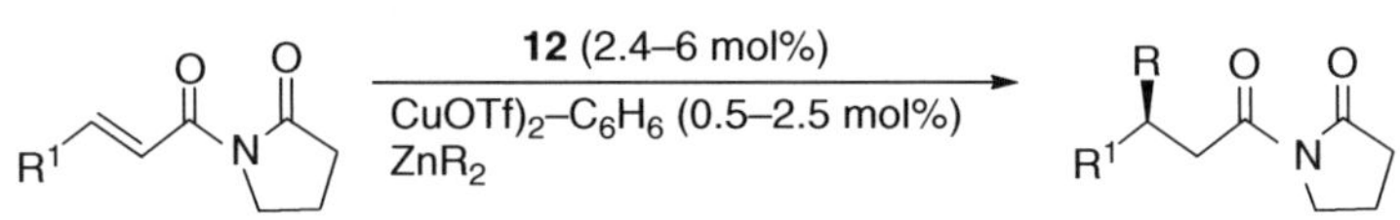

(7.4)

Derivatives of BINOL have been widely used in different catalytic processes [11b]. They showed good enantioselectivities in the enantioselective 1,4-additions

Table 7.4 Cu-catalyzed additions of alkyl zinc reagents to unsaturated oxazolidinones using catalyst **12** [11].

Entry	R^1	R	mol%	mol% Cu	Yield (%)[a)]	ee (%)[b)]
1	Me	Et	2.4	1.0	95	95
2	Me	i-Pr(CH_2)$_3$	5.0	1.0	61	93
3	Me	i-Pr	2.4	0.5	95	76
4	n-Pr	Et	2.4	1.0	86	94
5	n-Pr	i-Pr(CH_2)$_3$	2.4	1.0	89	95
6	–(CH_2)$_3$OTBS	Et	2.4	1.0	95	>98
7	i-Pr	Et	2.4	1.0	88	92

a) Isolated yields. All reactions proceeded to at least 95% conversion (GLC analysis).
b) Determined by GLC (β-DEX column for entries 1–5 and 7, CDGTA column for entry 6).

Table 7.5 Diethylzinc conjugate additions to nitropropene acetals [12].

Entry	Substrate	Ligand	Product	Yield (%)[a)]	ee (%)[b)]
1	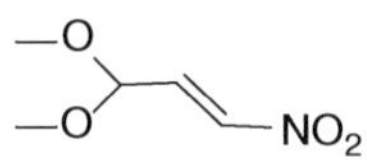	**13**	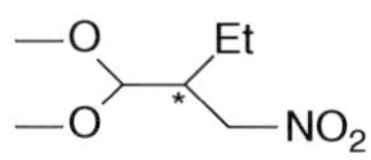	27[c)]	87
2		**14**		28[c)]	93
3	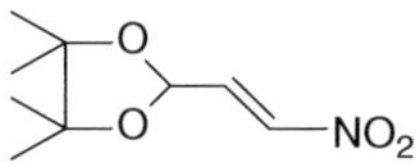	**13**	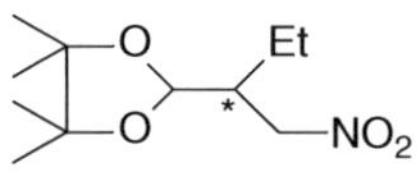	70	76
4		**14**		72	91
5	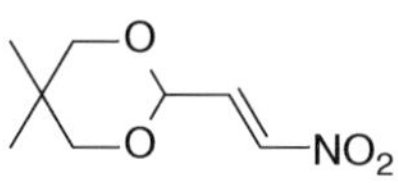	**13**	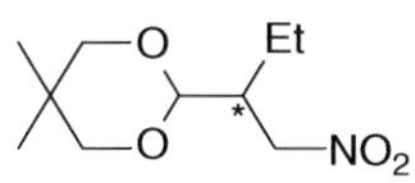	74	79
6		**14**		79	92
7	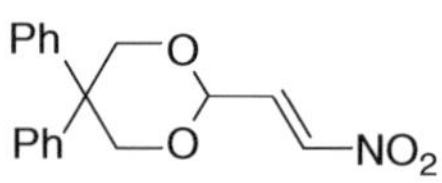	**13**	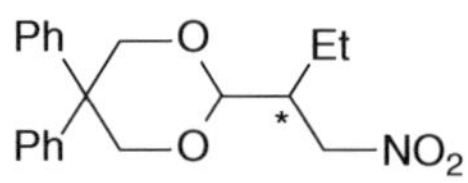	72	84[d)]

a) Isolated yield.
b) Determined by chiral GC.
c) Nonoptimized conditions.
d) Determined by chiral HPLC.

of diethylzinc to acyclic nitroalkenes (Eq. (7.5)) [12]. Ligand (aS,R,R)-**14** produced a very good enantioselectivity (up to 93% ee) in diethylzinc conjugate additions to nitroalkenes. Table 7.5 presents the details of the reactions using ligands **13** and **14**.

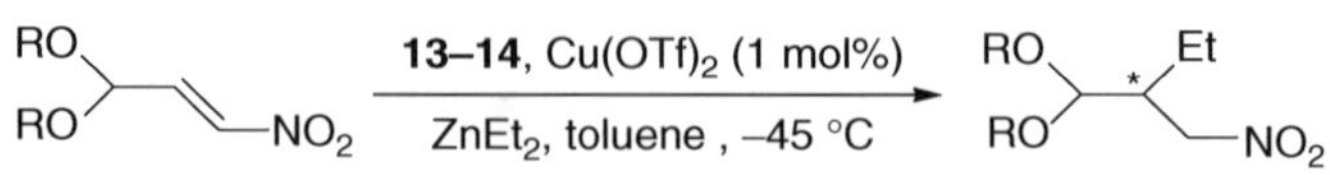

(7.5)

Table 7.6 Enantioselective 1,4-conjugate additions of $ZnEt_2$ to acyclic enones using **15** [13].

Entry	R[1]	R[2]	Yield (%)[a)]	ee (%)[b)]	Configuration[c)]
1	Ph	Ph	82	97	*S*
2	4-ClC_6H_4	Ph	76	97	+[d)]
3	4-MeC_6H_4	Ph	86	97	+[d)]
4	4-$MeOC_6H_4$	Ph	80	97	*S*
5	Ph	4-ClC_6H_4	75	95	–[d)]
6	Ph	4-MeC_6H_4	71	89	+[d)]
7	Ph	4-$MeOC_6H_4$	31	74	–[d)]
8	Ph	Me	48	58[e)]	*S*
9[f)]	Ph	Me	64	90[e)]	*S*

a) Isolated yield.
b) Determined by HPLC (Chiralpak-AD).
c) Absolute configuration was assigned by comparison of optical rotation with reported data.
d) Sign of the optical rotation of addition product.
e) Determined by GC (Gamma-DEX-225).
f) Reaction was carried out with **15** as the ligand.

In acyclic conjugate additions (Eq. (7.6)), a P,N,O-containing ligand **15**, which is also derived from binaphthanol [13], showed good catalytic effects. Table 7.6 lists the results of addition of ethyl functions to the substituted aromatic acyclic enones employing diethylzinc in the presence of $[Cu(CH_3CN)_4]BF_4$. Up to 97% ee was achieved. The catalysts contained two axial ligand sections, and exhibited big space exclusion in catalytic procedure and showed excellent enantioselectivities.

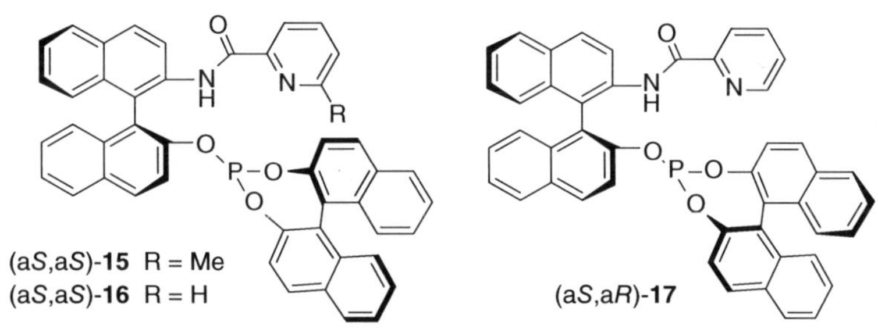

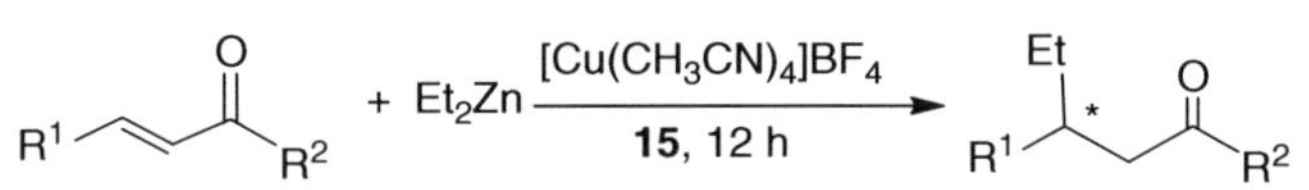

(7.6)

If the size of the substituent R of catalyst **15** is decreased, such as **16**, only a 54% ee was achieved in this addition; its another derivative (a*S*,a*R*)-**17** induced the lowest ee of 26% under the same conditions. This result is different from the reported results using catalyst **6–8**, in which the smaller group exhibited good enantioselectivities in 1,2-addition of diethylzinc to aldehydes. Obviously, big space exclusion is necessary in chiral catalyst design. However, it is not enough; other suitable conditions are also required.

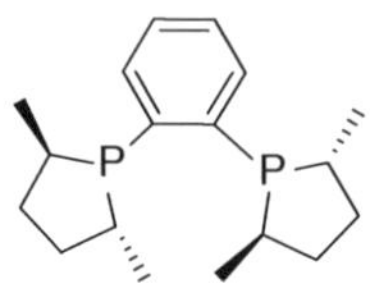

18

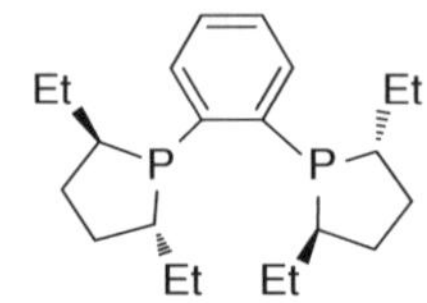

19

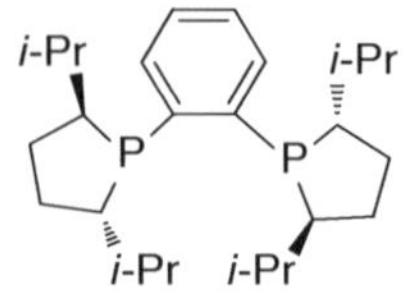

20

The chiral P-containing ligand **18** can efficiently promote the dialkylzinc addition to *N*-diphenylphosphinoylimines [14]. In experiments, (*R*,*R*)-**18** was determined to be a good chiral catalyst in the presence of $Cu(OTf)_2$ with 5 mol% (Table 7.7, Eq. (7.7)).

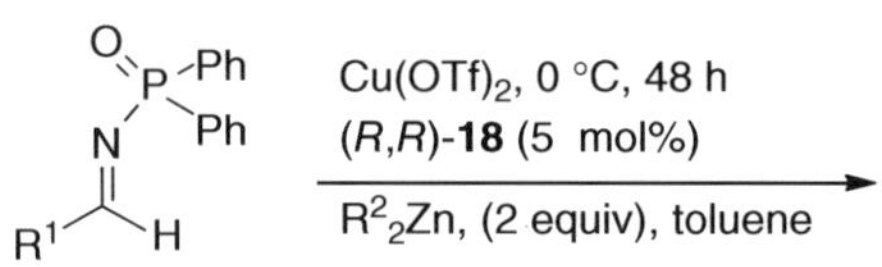

(7.7)

In this example, when the methyl group was changed to an ethyl group (**19**), the induced ee decreased sharply from 93 to 38%. When it was isopropyl (**20**), no enantioselectivity was observed and the yield also decreased to 10% only. The substituent effect is very great. Currently, it looks as if there is no regular rule to design an effective chiral catalyst. This is a fact. It is also the hot point that attracts researchers' attention in methodology study. Possibly, it may require good space matching between the catalyst, substrates, and/or other auxiliary.

Table 7.7 Addition of $ZnEt_2$ to *N*-diphenylphosphinoylimines using **18** [14].

Entry	R^1	R^2	Yield (%)	ee (%)[a]
1	Ph	Et	94	96
2	4-MeC_6H_4	Et	91	95
3	4-ClC_6H_4	Et	95	90
4	4-BrC_6H_4	Et	96	92
5	4-$MeOC_6H_4$	Et	74	95
6	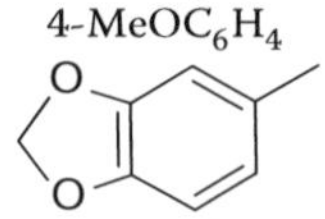	Et	81	95
7	1-Naph	Et	93	92
8	2-Furyl	Et	97	89
9	*c*-Pr	Et	82	85
10	Ph	Me[b]	51	90
11	Ph	Bu	71	91

a) Determined by HPLC on chiral stationary phase.
b) $(CuOTf)_2$–$C_6H_5CH_3$ was used.

Michael additions of organo-lithium or magnesium reagents to enones could be catalyzed by Cu and Co salts, respectively. Without the participation of Cu ion, it is hard to achieve good ee values. The reported Michael additions exhibited quite good ee magnitudes using alkyl Cu–Li reagents under the promotion of the ligand **14**. The ee values for the 1,4-additions of α,β-unsaturated cyclohexanone could high, up to 98% [15], except for isopropylation. The results are illustrated below.

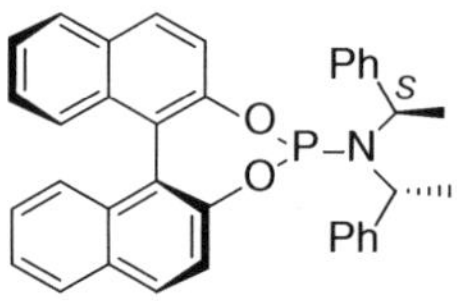

14

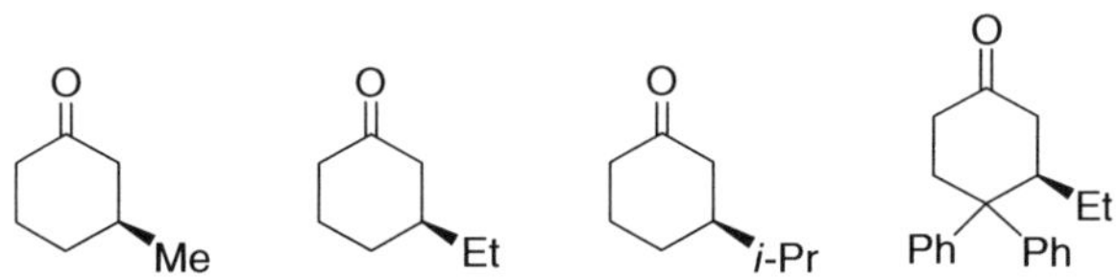

>98% ee   >98% ee   94% ee   >98% ee

7.1.3
Organo-Fe Complexes

Among the various chiral catalysts, ferrocene ligands have also been well studied in conjugate additions. For example, the chiral catalysts **21** and **22** [16] could afford up to 90% ee with yields over 89% in the addition of diethylzinc to α,β-unsaturated aromatic ketones. The copper salt $[CuOTf]_2–C_6H_6$ was used in this reaction.

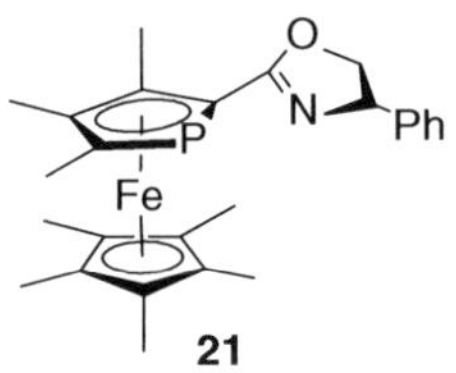

21

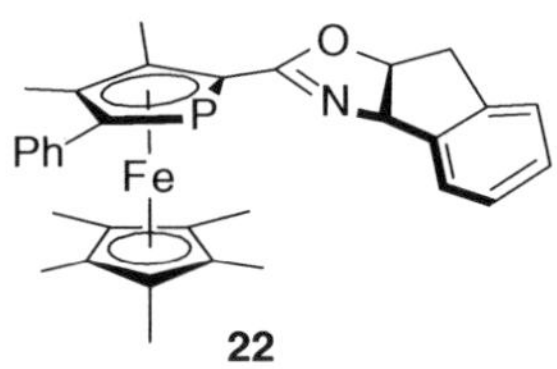

22

Indeed, ferrocene-containing ligands are large chiral catalysts that have been well used in different catalytic reactions. Their combinations with Pd, Cr, and other ions have been frequently used in various reactions [17]. In the addition of diethylzinc to aldehydes (Eq. (7.1)), the ligands **23–25** afforded up to 95% ee [18], while the catalysts **26** and **27** induced 97 and 99% ee, respectively, with almost 100% conversions.

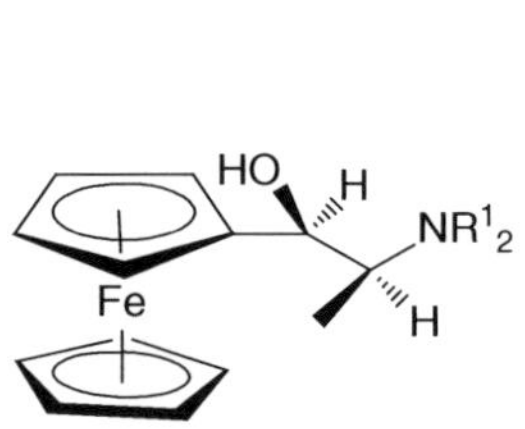

23 R = Me
24 R = Bu
25 R = $-(CH_2)_4-$

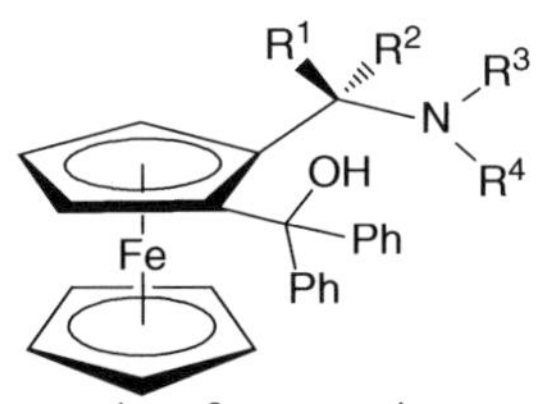

26 $R^1 = R^3 = Me$, $R^4 = H$
27 $R^1 = Me$, $R^2 = H$,
$R^3,R^4 = -(CH_2)_5-$

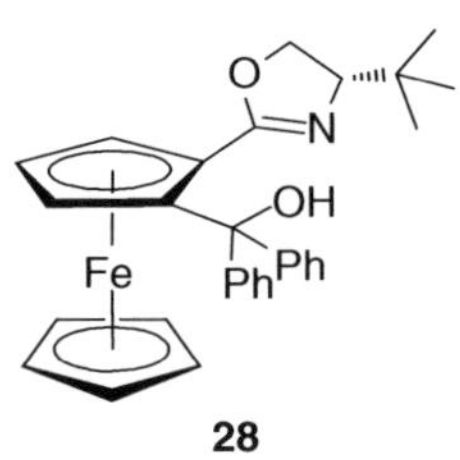

28

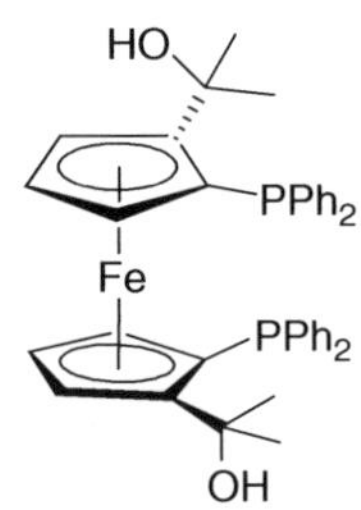

29

Table 7.8 Phenyl transfer from $PhB(OH)_2$ to various aldehydes using **28** [19].

Entry	Ar	Yield (%)[a]	ee (%)[b]
1	4-ClPh	95	92
2	4-PhPh	93	95
3	4-MePh	88	90
4	Ferrocenyl	33	94
5	2-MeOPh	79	90

a) Isolated yield.
b) Determined by HPLC using chiral stationary phase.

An enantioselective phenylation was achieved in the presence of an aryl aldehyde $PhB(OH)_3$ and a threefold excess of diethylzinc at 60 °C for 12 h prior to adding 10 mol% of **28** at 10 °C with optically active diaryl methanols, providing very high enantioselectivity (Eq. (7.8), Table 7.8) [19a].

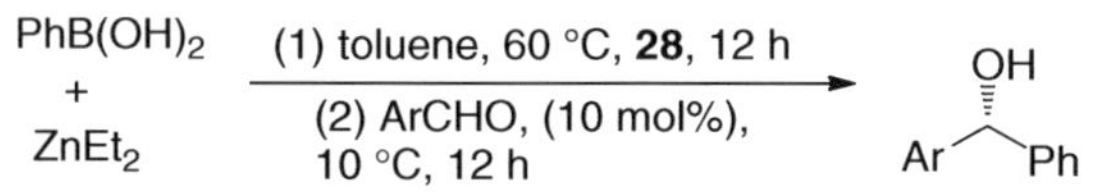

(7.8)

An asymmetric substitution of the acetate derivatives was derived from α,β-unsaturated enol (Eq. (7.9)) using Pd complex of ligand **29**, which could be generated by mixing $[Pd(\pi\text{-allyl})Cl]_2$ and **29**. The –OAc could be replaced in high enantioselectivities (92–96% ee) (Table 7.9) [19b].

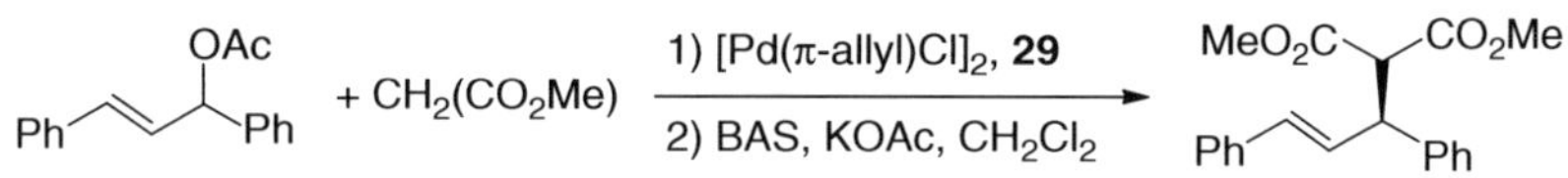

(7.9)

The chiral ferrocene derivatives **30–33**, used as nucleophilic ligands, could catalyze the addition of enolates to aldehydes (i.e., aldol reaction) in the absence of

Table 7.9 Asymmetric allylic alkylation catalyzed by chiral Pd complex **29** [19b].

Entry	Base	*T* (°C)	*t* (min)	Yield (%)[a]	ee (%)[b]	Configuration[c]
1	NaH	20	10	>99	92.3	*S*-(−)
2	BSA	20	5	>99	94.2	*S*-(−)
3	BSA	0	5	>99	95.5	*S*-(−)
4	BSA	−30	360	>98	96.3[d]	*S*-(−)
5[e]	BSA	0	150	>98	95.1	*S*-(−)

a) Isolated yield.
b) Determined by HPLC (Chiralpak AD).
c) Confirmed by comparing the specific rotation with the literature [20].
d) $[\alpha]_D$ −19.6 (*c*, 1.4 EtOH).
e) The reaction was carried out in the presence of 0.2 mol% of a Pd catalyst.

Table 7.10 Catalytic enantioselective intermolecular C-acylation of silyl ketene acetals [21].

Entry	R	R^1	ee (%)[a]	Yield (%)[a]
1	Me	Ph	90	80
2	Me	4-MeOPh	95	78
3	H	4-CF_3Ph	90	84
4	Me	*o*-Tolyl	95	89
5	Me	1-Naph	99	82
6	Me	2-Thienyl	76	84
7	H	3-Thienyl	80	73
8	Me	3-(*N*-Me-indolyl)	94	92

a) Average of two runs.

other metallic ions. Ligand **33** induced high (up to 90%) ee in this intramolecular C-acylation. This catalytic nucleophilic intermolecular C-acylation was used to give high (up 92%) ee for other substrates (Eq. (7.10), Table 7.10) [21].

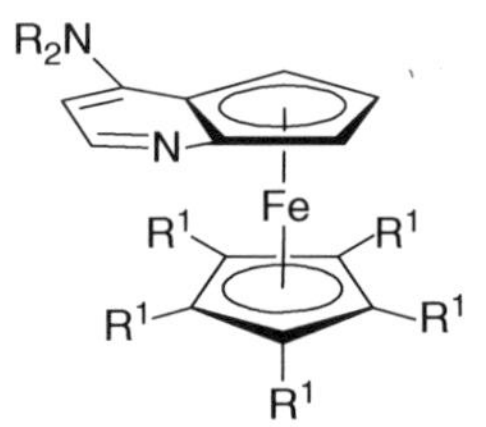

30 R^1 = Me NR_2 = dimethylamino
31 R^1 = Me pyrrolidino
32 R^1 = Me dimethylamino
33 R^1 = Ph pyrrolidino

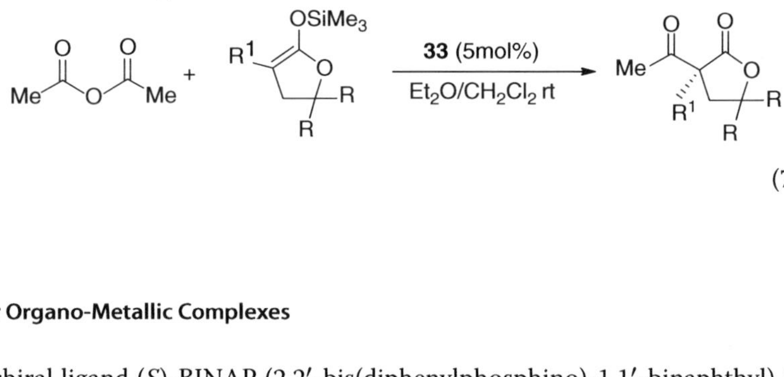

(7.10)

7.1.4 Other Organo-Metallic Complexes

The chiral ligand (*S*)-BINAP (2,2′-bis(diphenylphosphino)-1,1′-binaphthyl) (**34**), after being chelated with Ag, was an efficient catalyst for promoting the allylation of aromatic aldehydes with allyltributyltin [22]. High (up to 97%) ee was recorded (Table 7.11, entry 3) in THF at −20 °C. In the presence of AgOTf, ligand **34** could catalyze the reaction to produce 96% ee with yields up to 88%. In contrast, while $AgClO_4$ afforded only a 26% ee with only 1% yield in the addition of allyltributyltin to benzaldehyde (Eq. (7.11)) [22], the silver salt AgOTf led to high (up to 98%) ee. Unfortunately, other ligands, such as (*R*,*R*)-**35**, (*S*,*S*)-**36**, and (*S*,*S*)-**37**, afforded only 2, 48, and 3% ee, respectively, even when AgOTf was used.

Table 7.11 Allylations of aldehydes with allyltributyltin catalyzed by **34**-AgOTf complex [22].

Entry	R	Yield (%)[a)]	ee (%)[b)]
1	Ph	88	96
2[c)]	(*E*)–PhCH=CH	83	88
3[c)]	1-Naph	89	97
4[d)]	2-Furayl	94	93
5[e)]	(*E*)-*n*–C_3H_7=CH	72	93[f)]
6	2-Me–Ph	85	97
7	4-MeO–Ph	59	97
8	4-Br–Ph	95	96
9[e)]	$PhCH_2CH_2$	47	88

a) Isolated yield.
b) Determined by HPLC (Chiralcel OD-H, AD, or OJ).
c) 3 equiv of allyltributyltin and 0.15 equiv of (*S*)-BINAP-AgOTf were used.
d) 4 equiv of allyltributyltin and 0.2 equiv of (*S*)-BINAP-AgOTf were used.
e) The reaction was started using 2 equiv of allyltributyltin and 0.1 equiv of the catalyst for 4 h.
f) Determined by HPLC (Chiralcel AD) of the benzoate ester of the product.

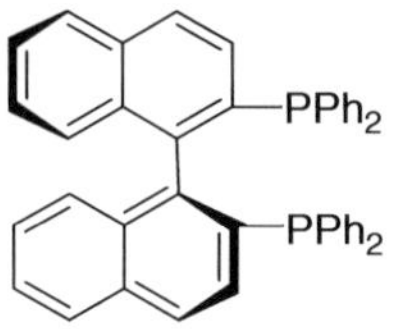

34

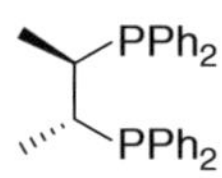

(*R*,*R*)-**35**

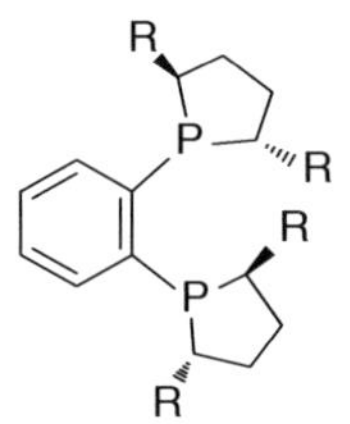

(*S*,*S*)-**36**, R = Me
(*S*,*S*)-**37**, R = Et

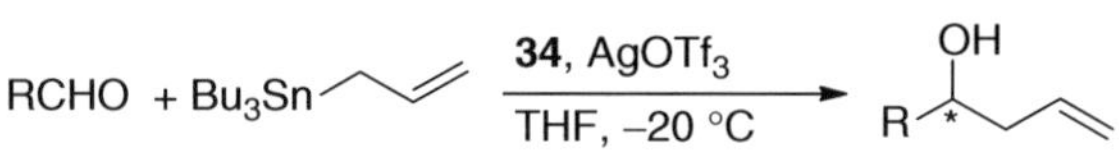

(7.11)

The element Pd has an enormous number of catalytic uses in organic syntheses [23]. For example, the novel chiral sulfinyl imine/Pd complexes were used in asymmetric allylic alkylations [24]. A 93% ee was observed in this aldol addition when the ligand (*R*)-**38** was used and the Pd source was $[Pd(allyl)Cl]_2$. This ligand afforded very high enantioselectivities (up to 96% ee) in the reaction in Eq. (7.11).

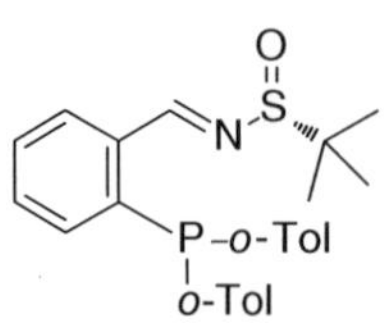

(*R*)-**38**

Enantioselective nucleophilic addition to imines was performed and catalyzed by 20 mol% Yb–binaphthanol complexes, as shown in Eq. (7.12) [25]. The highest ee was 91% in this addition reaction.

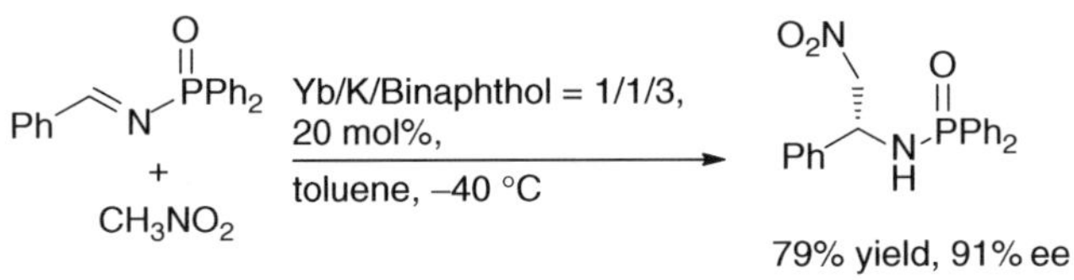

(7.12)

The addition of $MeNO_2$ to aldehydes (Henry reaction) has been well investigated [26]. Chiral products from Henry reactions can be transformed into many valuable chiral blocks by addition to aldehydes, such as nitro alkenes, amino alcohols, amino acids, and so on [27]. These intermediates can be further used for the synthesis of natural products such as amino alcohols [28]. Various types of chiral metallic and organo-catalytic catalysts have been developed [29]. Most chiral metallic catalysts developed contain copper [30] because of its excellent ability to chelate with various chiral ligands. Other metallic ions such as Zn(II) [31], Co(II) [32], and Pd(II) [33] have also been developed.

The addition of diethylzinc to aldehydes could be a good way to synthesize some chiral bioactive compounds. For example, it is reported that concentricolide (**39**) has some definite anti-HIV activity [34a]; the synthesis of its analogs **40** and **41** could be performed using chiral catalysts **1** and **6**, respectively. An ee value of 99% was recorded after reaction optimization [34b]. Bioactivity studies showed that the product (**40**) with (*R*)-AC had better activity than the corresponding (*S*)-AC compounds [34]. Formation of byproducts **A** and **B** was unavoidable. For example, at LiCl/Et_2Zn/substrate/ligand = 1/1.1/1/0.05 (mole ratio), no compound **40** was obtained, **A** was obtained in 87% yield and **B** in 13% at −10 to −15 °C (Eq. (7.13)).

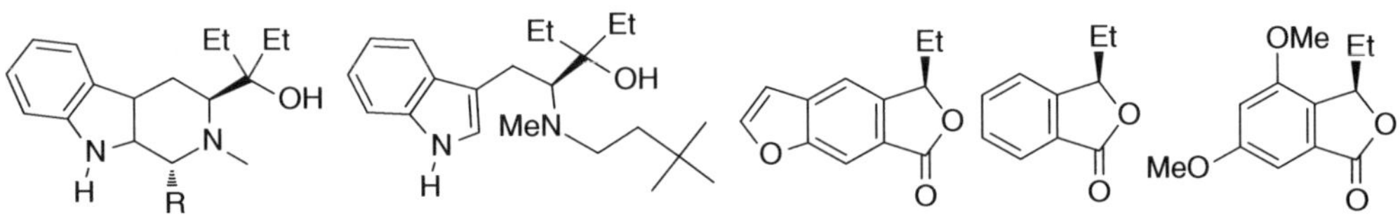

1 R=-CH2-*t*-Bu **6** **39** **40** **41**

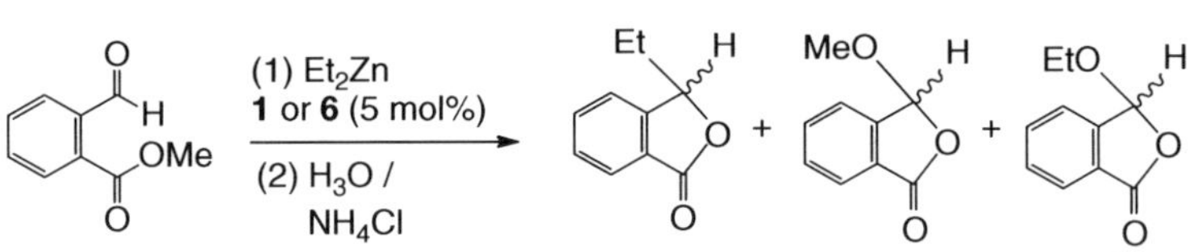

40 A B

(7.13)

7.1.5
Organo-Si Reagents

Organo-Si reagents, such as CH_2=CH–CH_2SiCl_3, are good addition reagents used in addition to aldehydes, which so far has been one of the most efficient methods to form homoallylic alcohols using allyltrialkylsilanes [35]. Chiral Lewis or Brønsted acids [36], especially some binaphthyl derivatives, were used as chiral catalysts. Chiral axial 2,2′-bipyridine *N,N′*-dioxides were effective as catalysts for the asymmetric allylation [37]. The typical reaction is illustrated in Eq. (7.14).

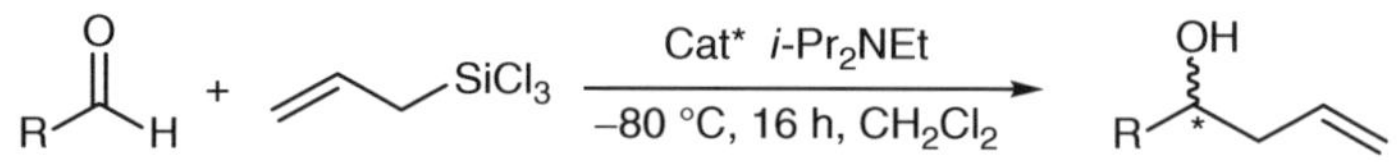

(7.14)

For example, a chiral formamide **42** could catalyze the addition reaction of allyltrialkylsilane to benzaldehyde with almost 98% ee value when 20–40 mol% of catalyst was used [38]. In contrast, chiral 2,2′-bipyridine *N,N′*-dioxides (**43**) can catalyze the additions with up to 98% ee in the presence of 0.1 mol% of catalyst [39]. A chiral ligand (**44**) derived from L-tryptophan exhibited high enantioselectivity in the additions, and the results are included in Table 7.12 [40]. Only 1 mol% of catalyst was required for the catalytic procedure. It showed a high selectivity for aliphatic aldehydes (Table 7.12).

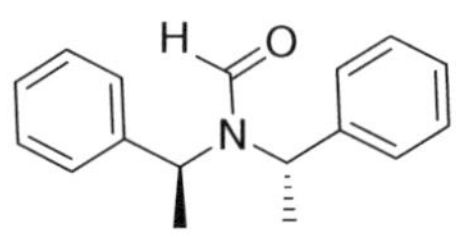

20–40 mol% used
98% ee recorded

42

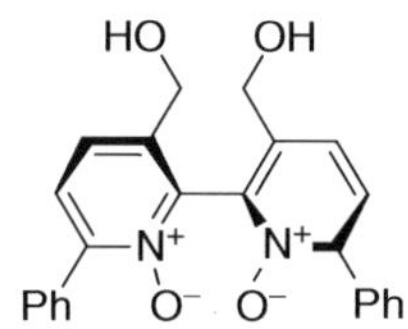

0.1 mol% used
up to 98% ee
recorded

43

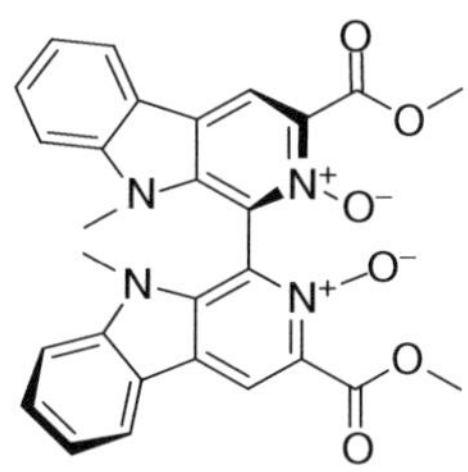

(*R*)-**44**

The possible mechanism for the addition can be illustrated as in Figure 7.2. The Si atom can form six coordination bonds in TS structure. Cleavage of Si–CH_2CH=CH_2 would happen when the C=C double bond approaches the carbonyl of aldehyde; at the same time, a pair of electrons of C=O would be transferred to the Si atom to form one Si–O bond. This intermediate would decompose into the (*S*)-AC product and the catalyst itself. The catalyst would take part in another catalytic cycle.

7.2
Enantioselective Reduction

Enantioselective reductions of compounds containing >C=C<, >C=O, >C=N– groups can provide a new chiral center in a molecule under catalysis. In this methodology study, acetophenones were used as a model molecule, just like the

Table 7.12 Asymmetric allylation to aldehydes catalyzed by (*R*)-(+)-**44** [40].[a]

Entry	R	Yield (%)[b]	ee (%)[c]	Configuration[d]
1	Ph	88	95	(*S*)
2	4-MeO–Ph	75	99	(*S*)
3	3-Cl–Ph	74	97	(*S*)
4	4-F–Ph	76	97	(*S*)
5	4-Me–Ph	74	95	(*S*)
6	2-Thiophenyl	83	97	(*S*)
7	1-Naph	82	96	(*S*)
8	PhCH=CH_2	90	91	(*S*)
9	$PhCH_2CH_2$	73	92	(*R*)
10	*c*-C_6H_{11}	53	97	(*S*)[e]

a) Reaction condition: aldehyde (0.5 mmol), allyltrichlorosilane (0.6 mmol), (+)-**44** (1 mol%).
b) Isolated yield. Some products are fairly volatile.
c) Determined by HPLC using chiral column.
d) Determined by comparing OR data and retention times using HPLC.
e) Determined after converting to 3,5-dinitrobenzoate ester.

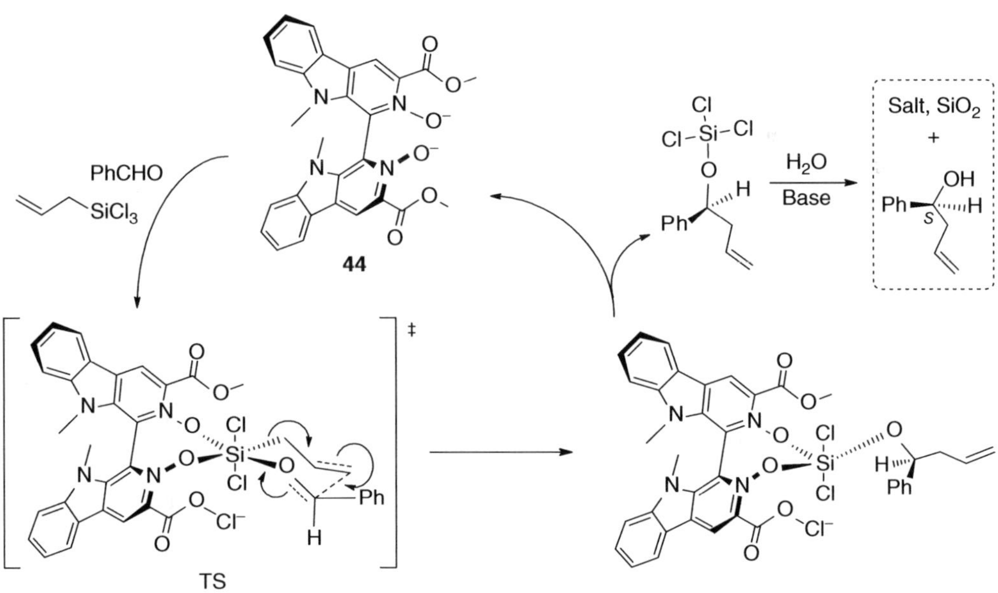

Figure 7.2 The proposed TS mechanism for allylation [40].

model molecule benzaldehyde used in enantioselective diethylzinc addition. The H_2/MeOH system, formic acids/triethylamine, silanes, or boranes could be used as hydrogen sources.

Chiral ruthenium catalysts were generally used for the hydrogenation of >C=C< bonds [19]. Ruthenium complexes could be prepared containing P-stereogenic ligands. Examples include chiral ligand **45–49** [41].

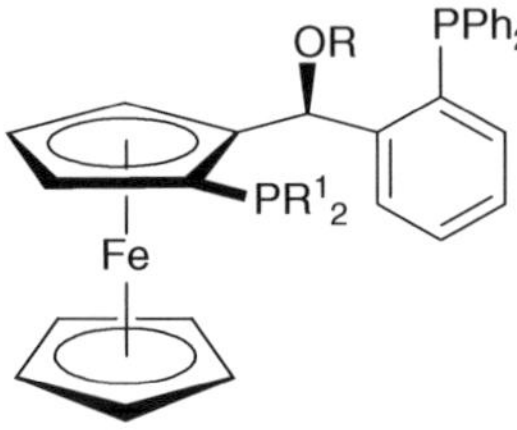

45 R = Me, R[1] = Ph
46 R = Me, R[1] = Xyl

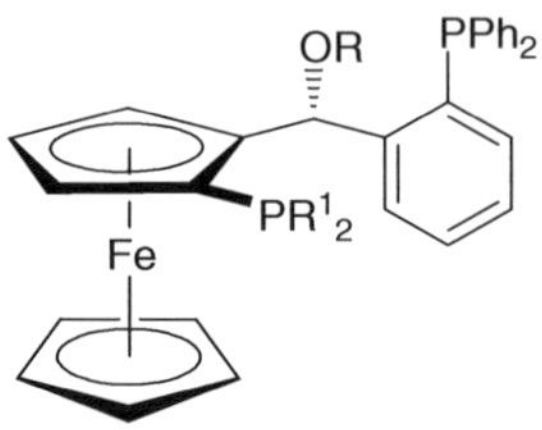

47 R = Me, R[1] = Ph
48 R = Me, R[1] = Xyl

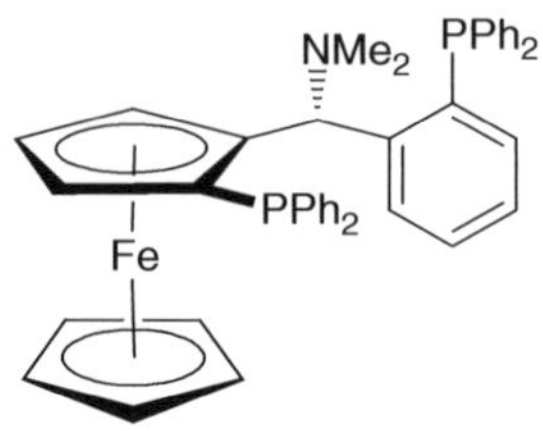

49

Three different substrates were tested using the Rh-ferrocene complex as catalysts. High ee values were recorded in the hydrogenations of α- and/or β-substituted unsaturated esters (Eq. (7.15)) [41]. Table 7.13 summarizes the effect of different ligands (**45–49**) on the ee values.

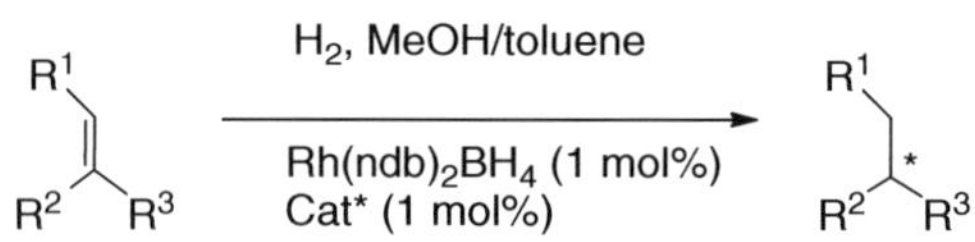

(7.15)

Another metal Ru has a wide range of oxidation states (from −2 to +8), which can form many coordination geometries; this provides it with a unique opportunity to participate in various catalyses. Ruthenium has now been involved in transfer hydrogenation and direct hydrogenations [42].

Table 7.13 Asymmetric hydrogenation of different substrates using chiral rhodium catalysts [41].

Entry	R^1	R^2	R^3	Catalyst	*t* (h)	ee (%)[a)]
1	Ph	NHAc	CO_2Me	**45**	1.5	99(*S*)
2	Ph	NHAc	CO_2Me	**46**	1.5	99(*S*)
3	Ph	NHAc	CO_2Me	**47**	2	94(*S*)
4	Ph	NHAc	CO_2Me	**48**	1.5	92(*S*)
5	Ph	NHAc	CO_2Me	**49**	0.5	95(*R*)
6	H	CH_2CO_2Me	CO_2Me	**45**	0.5	98(*R*)
7	H	CH_2CO_2Me	CO_2Me	**46**	1	90(*R*)
8	H	CH_2CO_2Me	CO_2Me	**47**	2.5	95(*R*)
9	H	CH_2CO_2Me	CO_2Me	**48**	14	91(*S*)

a) Entries 1–5, ee (%) values were determined by means of chiral GC (Chirasil-L-Val), entries 6–15, by HPLC (Daicel Chiralcel OD-H), entries 16–20, by chiral GC (Chirasil-L-Val), or HPLC (Daicel Chiralcel OD-H).

7.2.1
Green Chemistry

Green chemistry focuses on environment-friendly reagents and reactions. The water-soluble chiral ligands are one kind of important catalysts. For example, a vicinal diamine, **50**, was found to promote the hydrogenation of ketones to the corresponding chiral alcohols with very good enantioselectivity in the presence of $[RuCl_2(p\text{-cymene})]_2$ in aqueous media (Eq. (7.16)) [43]. A phase-transfer catalyst (PTC) needs to be used. Table 7.14 lists the reduction results using ligand **50**. Formic acid/triethylamine was used here as the hydrogen source.

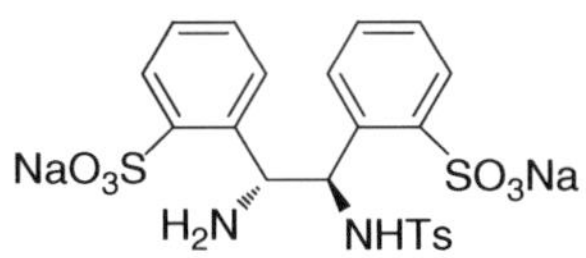

50

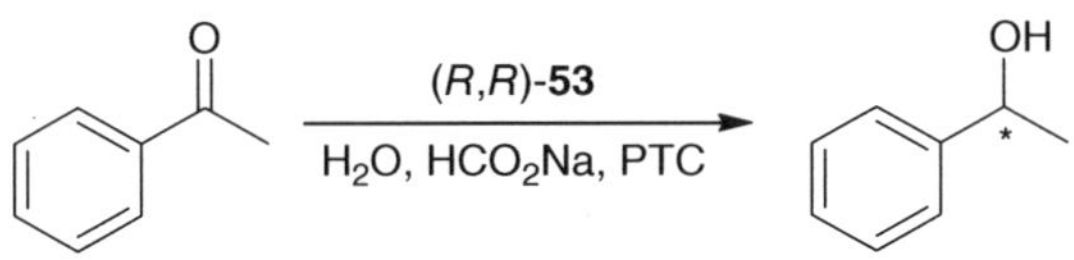

(7.16)

Table 7.14 Asymmetric hydrogenation of acetophenone in aqueous media [43].

Entry	Metal complexes[a)]	HCO_2Na (equiv)	Conv. (%)[b)]	ee (%)	Configuration[c)]
1	$[RuCl_2(p\text{-Cy})]_2$	5	>99	95	*R*
2[d)]	$[RuCl_2(p\text{-Cy})]_2$	5	>99	93	*R*
3[e)]	$[RuCl_2(p\text{-Cy})]_2$	5	34	89	*R*
4[f)]	$[RuCl_2(p\text{-Cy})]_2$	5	61	94	*R*
5[f),g)]	$[RuCl_2(p\text{-Cy})]_2$	5	95	93	*R*
6[h)]	$[RuCl_2(p\text{-Cy})]_2$	5	<1	ND	*R*
7	$[RuCl_2(p\text{-Cy})]_2$	2	47	90	*R*
8	$[RuCl_2(p\text{-Cy})]_2$	10	>99(75)[i)]	94(94)[i)]	*R*
9	$[RuCl_2(p\text{-Cy})]_2$	HCO_2NH_4 (5)	3	7	*R*
10	$[RuCl_2(p\text{-Cy})]_2$	HCO_2NH_4(10)	5	29	*R*
11[j)]	$[RuCl_2(p\text{-Cy})]_2$	5	17	94	*R*
12	$[RuCl_2(PhH)]_2$	5	66	71	*R*

a) Cy = cymene.
b) The conversion and ee (%) values were determined by GLC (CPcyclodex B-236 M).
c) Configuration was determined by the sign of rotation of the isolated product.
d) 15-Crown-5 was added as phase-transfer catalyst.
e) The reaction was conducted without PTC or surfactant.
f) S/C = 200.
g) The reaction time was 48 h.
h) S/C = 100.
i) The data in parentheses obtained from the second recycling.
j) The reaction was carried out at 28 °C.

The distance from hetero atom to the metallic center would affect the catalyst activity. For example, Ru-chelating diphosphine complexes, such as **51**, give poor enantioselectivities (16.6% ee) in hydrogenations of methyl benzoylformate. (*R*)-Mandelate is formed in the procedure [44]. In contrast, the chiral Ru complex **52** could afford over 99% ee. The shorter distance between the two P atoms in **51** resulted the strong repulsions in TS structure. This steric repulsion was avoided in the TS structure by using **52** as a ligand.

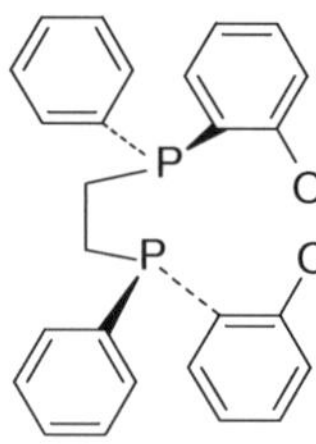

51

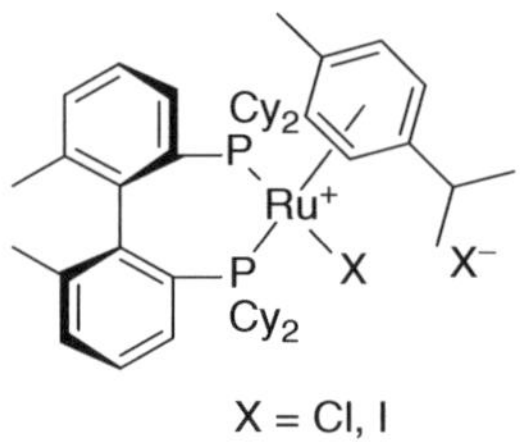

52

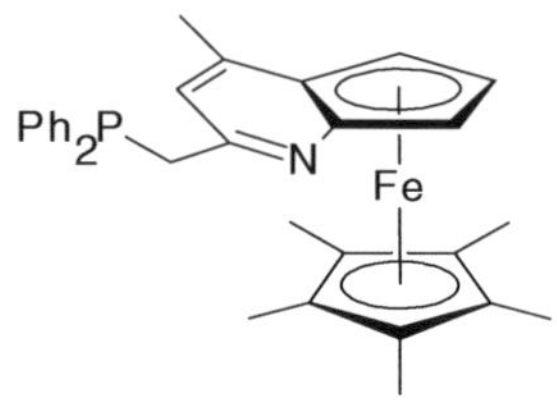

53

Many chiral ferrocenyl ligands have been synthesized and used in hydrogenation reactions. The use of 1–2.5 mol% of $[RhCl(cod)]_2$ and 2–6 mol% of catalyst **53** is made in the enantioselective reactions of acetophenone (Eq. (7.17)). High (up 98%) ee was achieved in the presence of Si derivatives used as the hydrogen source [45]. Table 7.15 gives the details of the reductions.

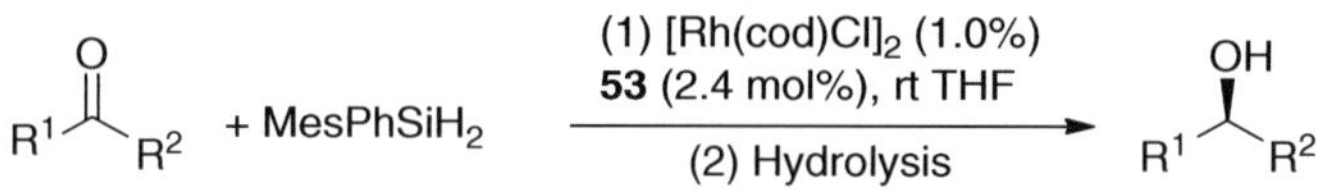

(7.17)

Enantioselective hydrosilylation of >C=N– with $HSiCl_3$ has proven to be a powerful method for producing chiral amines (Eq. (7.18)) [46]. Some effective chiral organocatalysts have been developed, such as *N*-formyl derivative (**54**) [47], picolinamide analog (**55**) [46d, 48], and pyridyl-oxazoline (**56**) [49].

Table 7.15 Catalytic asymmetric hydrosilylation of aryl ketones4 [45].

Entry	R^1	R^2	ee (%)$^{a)}$ yield (%)$^{a)}$	Entry	R^1	R^2	ee (%)$^{a)}$ yield (%)$^{a)}$
1	Ph	Me	98(94)	6	2,4,6-triMePh	Me	98(99)
2	4-MeOPh	Me	97(97)	7	Tetralone		98(95)
3	4-CF_3Ph	Me	96(88)	8	Ph	Et	98(96)
4	1-Naph	Me	99(97)	9	Ph	D	95$^{b)}$ (74)
5	2,4-diMePh	Me	95(97)	—	—	—	—

a) Average of two runs.
b) (*R*)-[*α*-D]-Benzyl alcohol is generated preferentially.

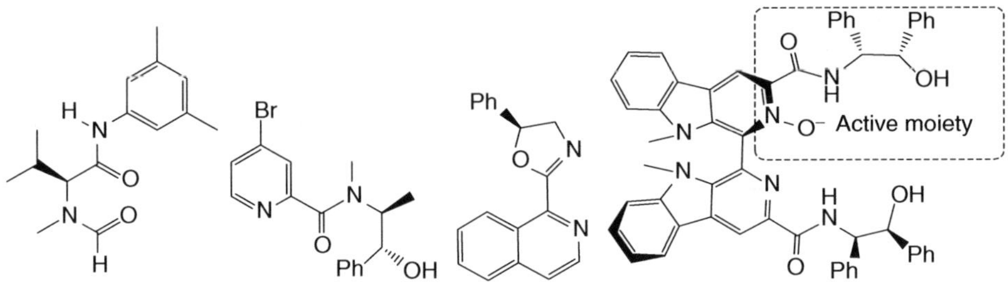

54 **55** **56** **57**

An asymmetric-axle-supported chiral N–O amide (**57**) was synthesized and used in the catalytic enantioselective hydrosilylation of *N*-aryl ketoimine with $HSiCl_3$ at room temperature (Eq. (7.18)) [50]. High conversion yield and high enantioselectivity up to 96% were achieved. The results are summarized in Table 7.16.

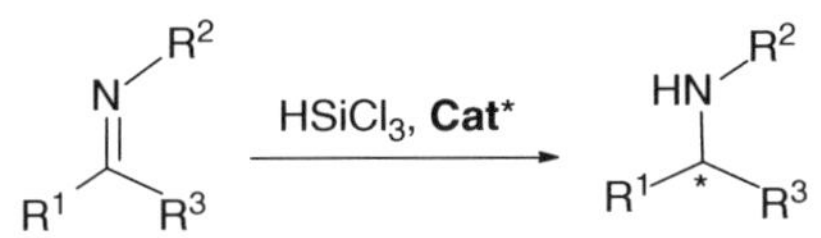

(7.18)

Table 7.16 Enantioselective reductions of ketoimine catalyzed by 20 mol% of **57** [50].

Entry	R^1, R^2, R^3	Yield (%)[a)]	ee (%)[b)]	OR
1	Ph, Ph, Me	95	93	−13.5
2	4-F–Ph, Ph, Me	98	95	−17.1
3	4-Cl–Ph, Ph, Me	97	94	−12.1
4	4-Br–Ph, Ph, Me	96	93	−17.1
5	4-F_3C–Ph, Ph, Me	95	84	−40.0
6	4-NO_2–Ph, Ph, Me	95	78	+16.5
7	4-Br–Ph,PMP ,Me	98	76	+11.7
8	4-MeO–Ph,Ph,Me	94	77	−13.9
9	Ph, PMP, Me	97	96	−2.0[c)]
10	Ph, 4-EtO–Ph,Me	95	94	−18.2
11	Ph, 4-Me–Ph,Me	95	93	+5.7
12	Ph, 4-Et–Ph,Me	95	89	−1.74
13	Ph, 4-Br–Ph,Me	98	75	+26.6
14	Ph, Ph, Et	95	94	+40

a) Isolated yield.
b) Determined using HPLC.
c) The sign of OR was not very stable during measurement due to resolution limitation of the polarimeter.
d) Unless specified, reactions were carried out with 2 equiv of $HSiCl_3$ and 20% of **57** in 2 ml of $CHCl_3$ at room temperature for 4 h.

7.3 Enantioselective Oxidation

Enantioselective oxidation generally happens in Sharpless oxidation. Normal alcohol oxidation cannot afford any enantiomer since this procedure converts an sp^3 carbon into an sp^2 carbon. However, oxidation of >C=C< should form at least one stereogenic center. Thus, a catalytic oxidation of >C=C< could generate enantioselective products under the catalytic conditions. The general reaction is illustrated in Eq. (7.19). Sharpless oxidations frequently are applied in total organic synthesis.

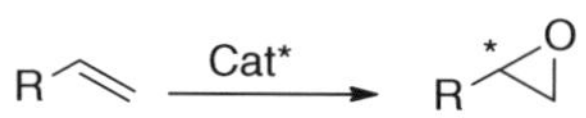

(7.19)

One of the widely used catalysts is chiral tartaric acid derivatives. For example, (−)-diethyl tartrate ((−)-DET, **58**) was used in the oxidation of (*E*)-3,7-dimethylocta-2,6-dien-1-ol (geraniol) to the corresponding epoxy alcohol with 91% ee (Eq. (7.20)) [51]. This was used in the total synthesis of *N*-Boc-(2*R*,3*R*)-3-methyl-3-hydroxypipecolic acid. Indeed, many examples reported employed epoxyl alcohols, which were used as intermediates for the total synthesis of various natural products [51].

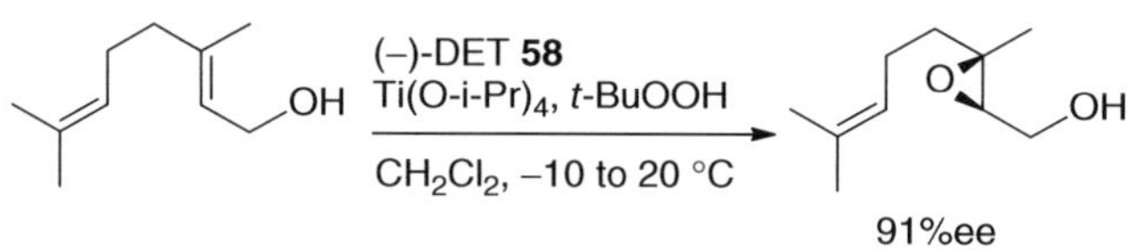

(7.20)

Another example is the use of (−)-DET (**58**) to promote the oxidation of an intermediate, as illustrated below during synthesis of the C31-C40/C43-C52 unit of amphidinol [52].

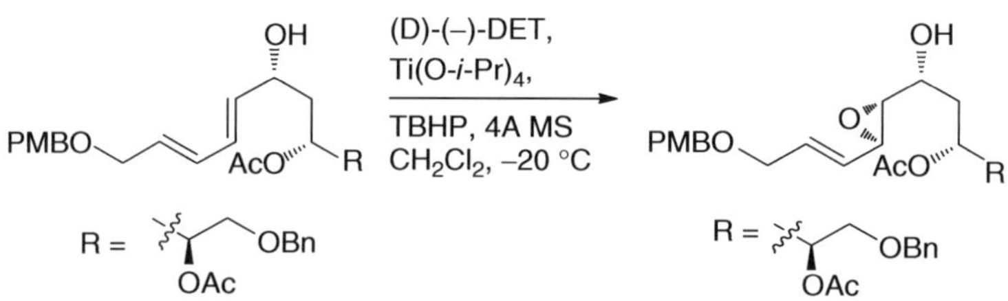

(7.21)

In the preparation of epoxyl from allylic alcohol, (+)-DET (**59**) was used as a catalyst; this procedure could afford the product with a diasteriomeric ratio (dr) of 97 : 3 [53]. It is almost the enantioselective Sharpless oxidation reaction, although this allylic alcohol had two chiral enters.

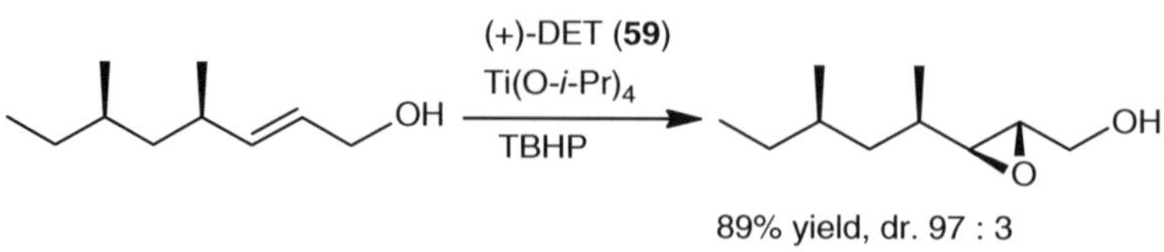

(7.22)

Zhou's modified Sharpless asymmetric oxidation condition is as follows: $Ti(Oi\text{-}Pr)_4$, L-(+)-diisopropyl tartaric acid (diisopropyl tartaric acid (DIPT), **60**), $t\text{-}BuO_2H$, CaH_2, SiO_2, CH_2Cl_2, −25 °C, 4 days [54]. This modification could increase the yields. For example, in the following epoxyl preparation, the rough product had 85% ee with 95% yield (Eq. (7.23)) [55]. The enantiomeric excess increased to 99% after recrystallization.

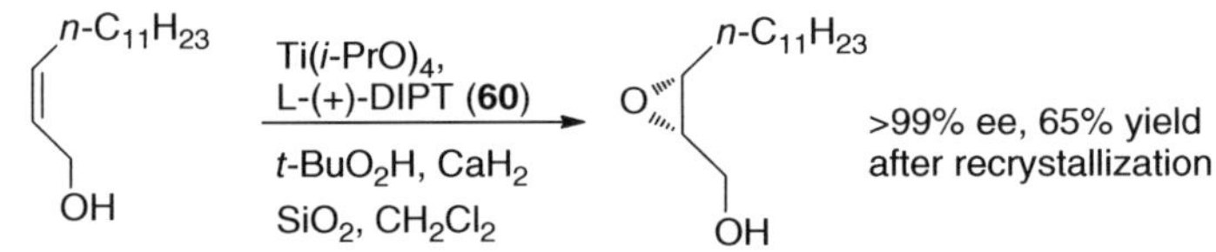

(7.23)

There are many valuable reports on sharpless epoxidations. Interested readers may refer to the corresponding literature.

7.4
Prediction of ee Using Calculations

Until now, predicting the selectivity for a specific asymmetric reaction has been a big challenge; successful examples are not many. Theoretically, computation of the corresponding TS barriers can predict the selectivity for a reaction. This needs a clear reaction mechanism, and the computation can be performed at a relatively high quantum level. A good example is mechanistic study of the addition of diethylzinc to aldehyde. The mechanism is shown in Figure 7.1 [4].

An early reported combinational method is applied for study of the enantioselective addition of diethylzinc to benzaldehyde by use of quantum mechanics (QM)/molecular mechanics (MM) [56]. The calculations were carried out using Gaussian92/density functional theory (DFT) and MM3(92). The TS activation energies of each of the three TS structures were analyzed. Correct predictions of enantioselectivity in these diethylzinc additions catalyzed by (*R*)-2-piperidine-1,1,2-triphenylethanol, **61**, were achieved. The calculated ee (after Boltzmann assumption) and the experimental ee were found to be 97.9 and 98%, respectively.

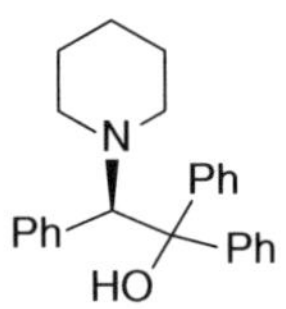

61

As a typical example, the procedure of catalytic reductions using catalyst **57** is analyzed carefully in this section. In solution, catalyst **57** exists in two conformations, namely **57** and **62** (epi-**57**) [57]. The difference between the two conformations is that the lower indole moiety is front in **57** and at the back in **62**. The moiety without the N–O bond acts as a big wall to block the other groups

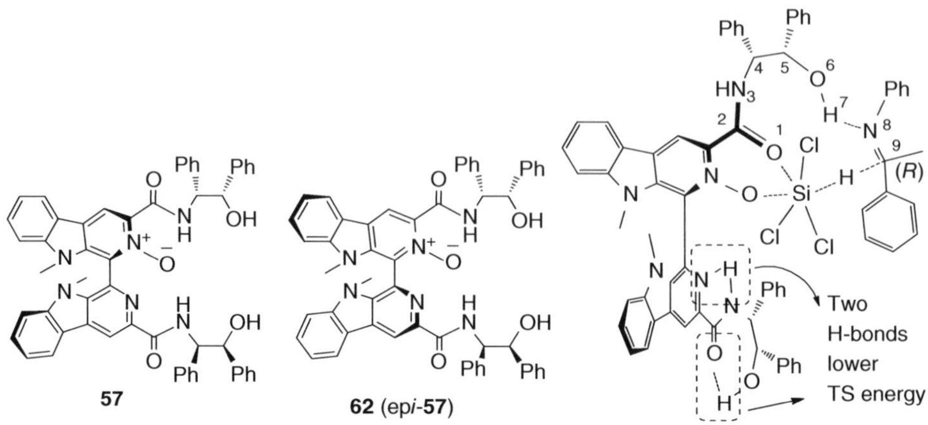

Figure 7.3 TS structure in reductions using **57** [57].

from approaching. Thus, the reduction of $HSiCl_3$ to the C=N bond can take place in a limited space. Both TS structures using **57** and **62** were considered in the TS calculation [57]. Since the ring system is quite large, it is important to construct the TS geometry for obtaining the right TS barriers. To this aim, the H-bonds in the TS must be fixed, since H-bond formation can decrease the molecular energy (Figure 7.3). Then, by changing the position of aldehyde in TS, reasonable TS structures using GaussView could be constructed. On the other hand, it is necessary to investigate the population of **57** and **62**, since both can form the corresponding TS structures and have contributions to the ee. The starting material (*E*)-*N*-(1-phenylethylidene)aniline was used in TS constructions; its reduction product is *N*-(1-phenylethyl)aniline. The ee value is 93% catalyzed by **57** (Table 7.16, entry 1).

Three basic TS types are plausible based on the analysis above and the position of H on Si: type A (inside), type B (up), and type C (front). They could be obtained by rotating the corresponding dihedral angles O1–C2–N3–C4, C2–N3–C4–C5, and C3–C4–C5–O6, changing the position of Ph of aldehydes, and orienting H on the Si atom (Figure 7.4) [57].

Notice

To compute such a big TS system at a very high level, such as at the B3LYP/6-311+G(d) level, the first time is not a good idea. It is suggested that the computations be performed at a low level first, such as at the HF/3-21G(d) level. Once the TS structure is obtained with a correct imaginary frequency, this TS structure could be used as a new structure for computation at a higher basis set, such as B3LYP/6-31G(d). All possible TS structures should be investigated for looking for the lowest TS geometry and its barrier. In this example, HF and B3LYP theories were used.

HF/6-31G(d) theory, which is useful in predicting barrier sequences in proton-transfer reactions, such as in reductions using sodium borohydride [58], was used

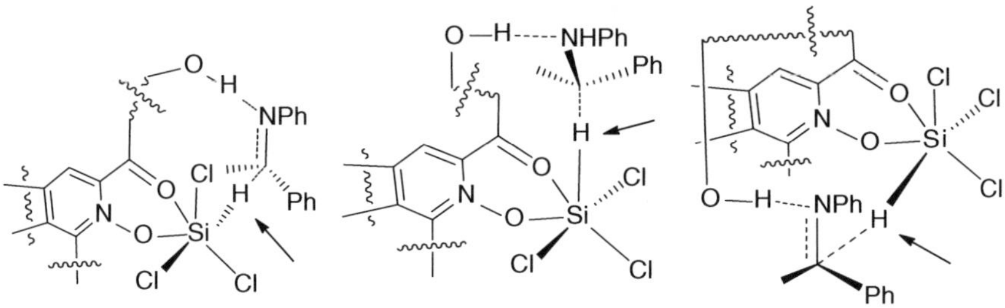

Figure 7.4 Three possible TS types involved in the hydrosilylations (only the key sections are illustrated here, others are omitted for clarity) [57].

Table 7.17 The relative TS barriers that are found in types A–C [57].

Type	ΔE	ΔE_0	ΔG
A (**TS-1**)	3.606	4.354	4.837
A (**TS-2**)	0.000	0.000	0.373
A (**TS-3**)	0.292	0.952	0.879
A (**TS-4**)	1.028	1.239	0.000
B	24.865	24.633	22.577
B	19.023	—a)	—a)
B	19.168	19.126	17.447
C	23.320	23.372	20.900
C	16.715	16.632	16.469
C	19.383	—b)	—b)

a) The Gibbs free energy and zero-point energy were not computed since the barriers using total electronic energy are very high.
b) Since the TS barrier energy were much higher, their frequency was not computed. The eigenvalues were used to test their TS structures.

for TS computations for all possible structures that can be found. All the barriers are summarized in Table 7.17. The energy unit is kilocalories per mole for all barriers.

The barriers in type A were much smaller than those in types B and C. Therefore, only type A was selected for further TS computations. Two TSs with low barriers from 0 to 4 kcal mol^{-1} using **57** or **62** were found via type A, respectively. In total four TS geometries were found. The total electronic energy (TEE), zero-point energy (ZPE), and Gibbs free energy (GFE) data were used in barrier computations.

DFT theory was also used for the barrier computations using the corresponding TS geometries obtained from HF theory. The barriers computed at the B3LYP/6-31G(d) level are listed in Table 7.18. At the same time, since the percentage of **57** and **62** will affect the ee values, it is necessary to consider them in ee calculations. After conformational searches, optimizations at the B3LYP/6-31G(d) level in the

Table 7.18 The relative barriers and predicted ee (%) values [57].

	ΔE		ΔE_0		ΔG	
	DFT[a)]	HF[b)]	DFT	HF	DFT	HF
TS-1	2.688	3.606	3.285	4.354	3.555	4.837
TS-2	0.000	0.000	0.000	0.000	0.000	0.373
TS-3	0.653	0.292	1.264	0.952	1.034	0.879
TS-4	0.559	1.028	0.762	1.239	0.201	0.000
AC	*S*	*S*	*S*	*S*	*S*	*S*
ee (%)[c)]	16	34	82	70	81	76
ee (%)[d)]	80	84	97	88	97	95

a) The full materials were used in TS computations at the B3LYP/6-31G(d) level in the gas phase.
b) The full materials were used in TS computations at the HF/6-31G(d) level in the gas phase.
c) The ee (%) was predicted just using the relative energy data.
d) The ee (%) was predicted using the relative barrier energy and the quantity of catalyst **57** and **62**.

gas phase for all geometries were carried out. The conformations with relative energy 0–1.5 kcal mol^{-1} were further computed at the B3LYP/6-311+G(d) level. It was found that conformation **57** had 87% in the gas phase and its epimer (**62**) had 13% based on B3LYP/6-311+G(d)-optimized geometry energy. This percentage was used for the ee computations in both HF and DFT methods. The TEE, ZPE, and GFE data are selected in ee calculations. Their ee values using the two theories are summarized in Table 7.18. The energy unit is kilocalories per mole for all barriers.

When the quantities of catalyst **57** and **62** were not used in ee calculations, the predicted ee values were smaller than the experimental data (93%, Table 7.16, entry 1). The closest result is 82% using ZPE. Using GFE could predict 81% ee. The bad prediction is to use TEE data; only 16% ee was predicted. However, if the distributions of **57** (87%) and **62** (13%) were used in the computations, all methods predicted quite well. The closest prediction (95% ee) was when HF/6-31G(d,p) method and GFE data were used.

In another example, when Ph_2Zn and Et_2Zn were both added into the addition reaction, the addition product is the phenylation product, as mentioned earlier (Eq. (7.3)) [59]. By theoretical investigation of the TS study (favorite structure is the anti-trans configuration), it was found that the barrier for phenylation was about 30–40 kJ/mol lower in energy than ethylation by using DFT theory (Eq. (7.24)) [59]. This is due to the overlap with the π-system of transferring phenyl, which significantly lowers the barrier to the bending of the C–Zn bond out of the plane of the phenyl group. On the other hand, in the presence of Et_2Zn in the system, its selectivity can be improved since the ethyl group takes part in the TS procedure.

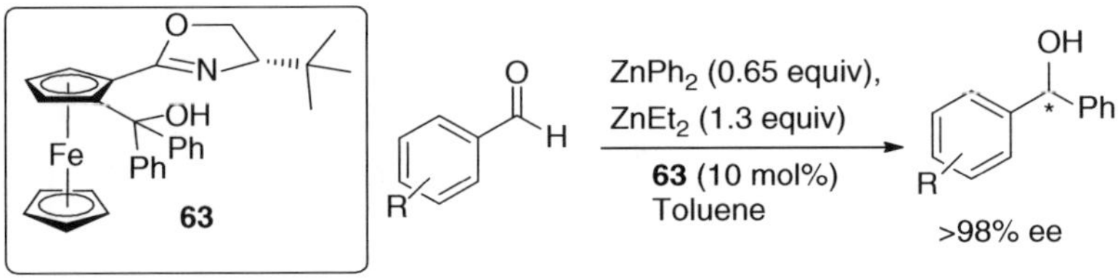

(7.24)

Another methods instead of TS barrier calculation, a grid-based QSAR (quantitative structure activity relationship) approach, quantitative structure selectivity relationship (QSSR) approach, was reported using diethylzinc to benzaldehyde (Eq. (7.1)) [60]. To arrive at a reasonable result, a total of 18 known β-amino alcohol catalysts with similar reactivities and encompassing a wide enantioselectivity range were used to construct the relationship between the obtained ΔG_{fit} (computed from the ee magnitudes in the experiments) and the calculated ΔG_{cal}. By using this relationship, prediction of the ee becomes possible.

7.5 Catalyst Types

Catalysts can be divided into several classes. Generally, it could be homogeneous or heterogeneous catalyst based on its solubility, such as a polymer-supported chiral catalyst. Homogeneous catalysts includes amino alcohol, N–O-containing catalysts, chiral ferrocene derivatives, and metallic complexes according to their functional group characteristics. It could be axial catalysts, helical catalysts, and asymmetric-axle-supported chiral catalysts on the basis of their structural characteristics. Each structurally different chiral catalyst may contain different functional groups.

7.5.1 Amino Alcohols

Amino alcohols became popular chiral catalysts after β-amino alcohols were used in the catalytic addition of diethylzinc to benzaldehyde. Many chiral catalysts have been reported. In addition to the β-amino alcohols mentioned above, such as **1** and **6**, γ-amino alcohol **10** was also used. Also, diol ligands **9** and **11** are important catalysts [5, 6, 8, 9]. Some chiral compounds containing N, P, S, or O were also reported with good enantioselectivity [61]. The chiral amino alcohols **64** and **65** could afford high (up to 99%) ee in diethylzinc addition to aldehydes [62].

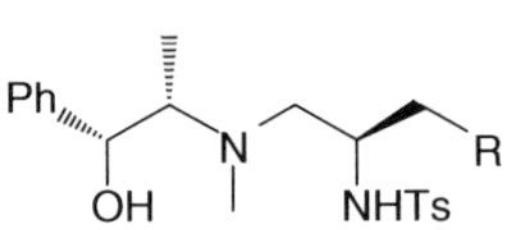

(1*R*,1′*R*,2′*S*)-
64 R = Ph

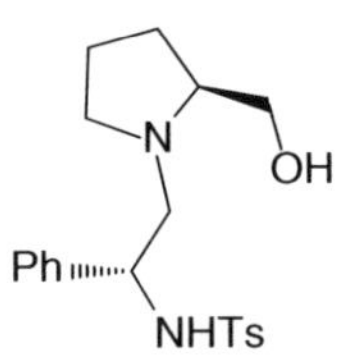

(*S*,*R*)-**65**

In the absence of Cu salts, ligands **66–68** can catalyze diethylzinc to *N*-diphenylphosphinoyl benzalimine to take part in 1,4-addition reactions with high ee values (>90% ee) [63].

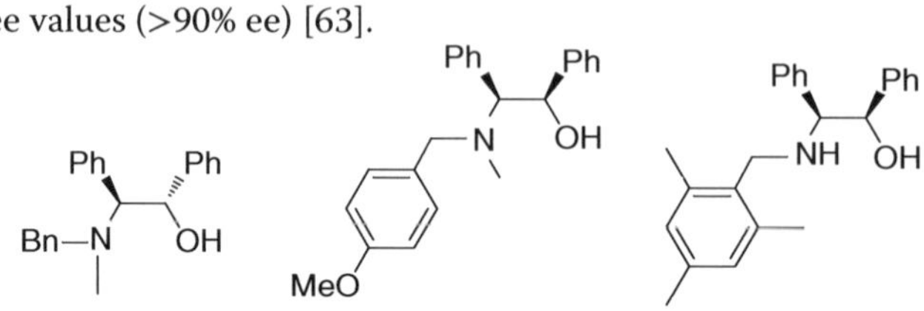

66 **67** **68**

7.5.2
Chiral Ligands Containing N–O Group

Chiral ligands containing the N–O group is an important chiral catalysts. After the first reports, *N,N′*-dioxide chiral ligands have been used widely in different reactions [37a]. Several kinds of *N*-oxide derivatives from pyridines and tertiary amines have been synthesized and used for asymmetric allylation [64]. These catalysts exhibit good enantioselectivity in many reactions, such as in additions or reductions. For example, (*R*)-**69** can promote the addition of allyltrichlorosilane to aldehyde in a relatively high enantioselectivity efficiently (mostly 85–93% ee) while (*R*)-**70** can catalyze the reduction of ketoimines to the corresponding amines in good ee values (70–85% ee) [65]. Another chiral catalyst, **71**, can promote the enantioselective Henry reactions with very high ee values (up to 98 and 99% yield) [26]. This catalyst can form metallic complexes of Cu(II) to promote ene/Prins cyclization reaction with a high ee values [66] including dr magnitudes.

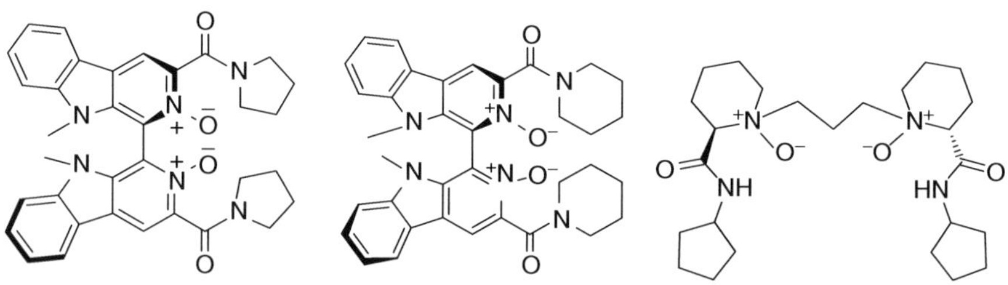

(*R*)-**69** (*R*)-**70** **71**

Other chiral N–O catalysts, such as **72–74**, also exhibited good enantioselectivity in different reactions [39, 67–69].

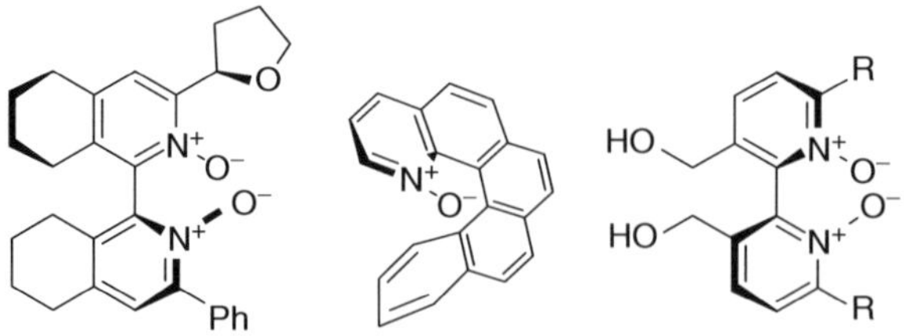

(*S,R*)-**72** **73** (*R*)-**74** R = PhOMe

7.5.3
Chiral Axial Catalysts

Chiral axial catalysts are other important ligands used in various reactions. Some catalysts containing an S atom, such as ligands **75–77**, exhibited very high enantioselectivity in the addition of diethylzinc to α,β-unsaturated ketones (Eq. (7.25)) [70]. Among the three ligands, catalyst **75** gave no enantioselectivity in the addition reaction (Table 7.19).

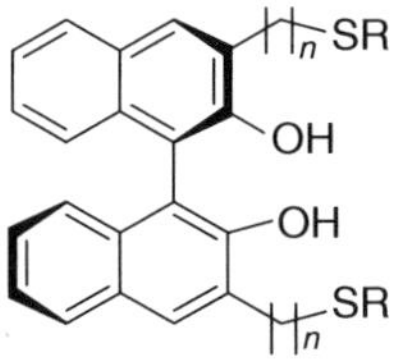

75 $n = 1$, R = Me

76 $n = 2$, R = Me

77 $n = 2$, R = *t*-Bu

$$\text{Substrate} \xrightarrow[\text{Et}_2\text{Zn (3.0 equiv), solvent (0.1 M), 0 °C, 2 h}]{\text{Cu salt (0.02 equiv), ligand (0.024 equiv)}} \text{Product} \tag{7.25}$$

Table 7.19 Cu-catalyzed conjugated addition of Et_2Zn to various enones [70].

Entry[a)]	Substrate	Ligand	Cu salt	Yield (%)[b)]	ee (%)	Configuration
1	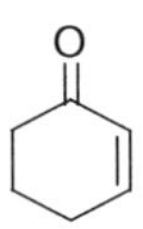	**76**	$Cu(OTf)_2$	95	85[c)]	*R*
2		**76**	$Cu(OAc)_2$	86	22[c)]	*R*
3		**76**	$Cu(acac)_2$	58	8[c)]	*R*
4		**77**	$Cu(OTf)_2$	93	77[c)]	*R*
5		**77**	$Cu(OAc)_2$	49	13[c)]	*R*
6	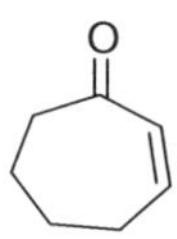	**76**	$Cu(OTf)_2$	96	81[d)]	*R*[e)]
7		**76**	$Cu(OAc)_2$	23	21[d)]	*R*[e)]
8		**77**	$Cu(OTf)_2$	86	76[d)]	*R*[e)]
9		**77**	$Cu(OAc)_2$	10	44[d)]	*R*[e)]
10	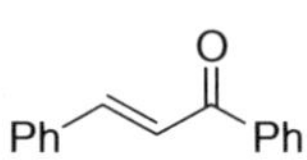	**76**	$Cu(OTf)_2$	22	17[f)]	*R*
11		**76**	$Cu(OAc)_2$	53	44[f)]	*S*
12		**76**	$Cu(acac)_2$	24	40[f)]	*S*
13		**77**	$Cu(OTf)_2$	72	92[f)]	*S*
14		**77**	$Cu(OAc)_2$	91	96[f),g)]	*S*
15[h)]		**77**	$Cu(OAc)_2$	85	96[f),g)]	*S*

a) Solvent: Methyl *t*-butyl ether (MTBE), Et_2Zn: 3.0 equiv.
b) Solvent: toluene, Et_2Zn: 3.0 equiv.
c) Isolated yield.
d) Determined by GC analysis of the chiral acetal derived from (2*R*,4*R*)-(−)-pentanediol (HP-FFAP).
e) Determined by GC analysis (CP-Cyclodextrin-B-2,3,6-M-19).
f) Absolute configuration was determined by comparison of the specific rotation with the literature value [70b,c].
g) Determined by HPLC (Chiralcel OD-H).
h) The reaction was carried out at −10 °C for 2 h.

Under the catalysis of **78–80**, the imines can be reduced to amines with high ee values (Eq. (7.26)) [71]. The reduction reagent used here is not the frequently used $HSiCl_3$, but was diethyl 2,6-dimethyl-1,4-dihydropyridine-3,5-dicarboxylate. Among the three catalysts, ligand **80** afforded the best efficiency; in Eq. (7.26), when R = Et, the highest ee can be reached (up to 96% ee) with a conversion of 99%.

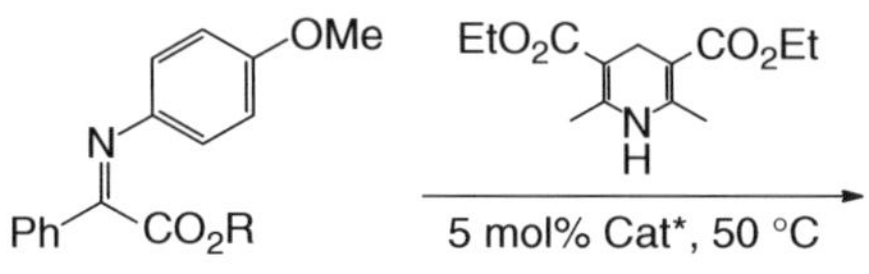

(7.26)

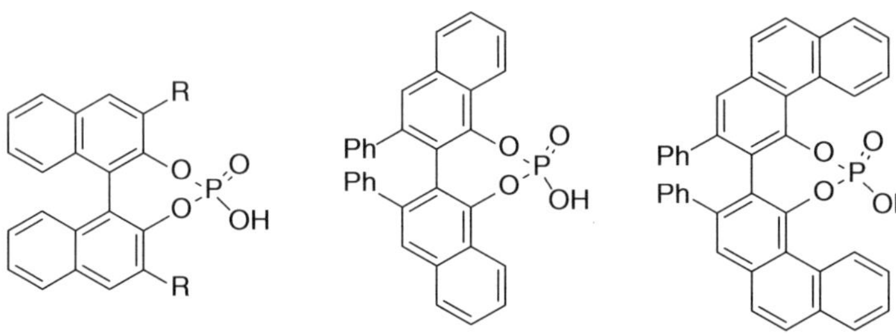

78 R = 2-Naph **79** **80**

7.5.4
Solid-Supported Chiral Compounds

Solid-supported chiral compounds can provide economical and convenient catalysts. Separation of the products from the catalyst is very easy, just by filtration. This convenience has attracted many researchers' attention in their study. For example, chiral ligands **81–83**, which connect three kinds of small chiral catalysts to the polymer support, have been reported. Ligand **82** exhibited quite high enantioselectivity (up to 97% ee) in diethylzinc addition to aromatic aldehydes [72]. Catalyst **83** afforded about 80–93% ee in the same addition, mostly ~85% ee [73].

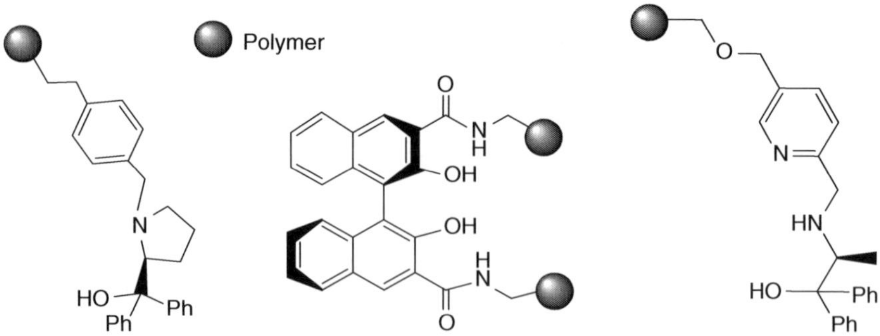

81 **82** **83**

Fixing a chiral catalyst on surface of SiO_2 (MCM-41, or SBA-15) is another way to produce the corresponding solid-supported catalysts. E xamples include **84** and **85**. Ligand **85** provided 81% ee in diethylzinc to benzaldehyde addition in the presence of Ti (O*i*-Pr)$_4$. The ee decreased by about 3% in the additions if the recovered catalyst **85** was used [74].

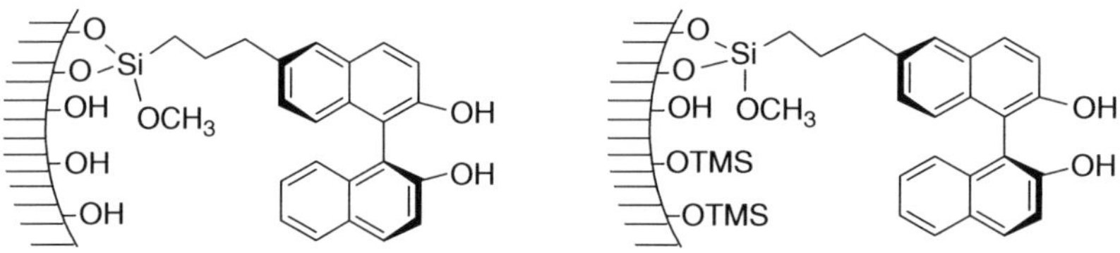

84 **85**

85 connected to ^{TM}MCM-41, **86** connected to ^{TM}SBA-15

It is possible to synthesize a polymer chiral catalyst that has good solubility in definite solvents, such as BINO-polymerized chiral ligand **86**; its enantioselectivity can reach up to 95% in the addition of diethylzinc to benzaldehyde in the presence of Ti(O*i*-Pr)$_4$ [75].

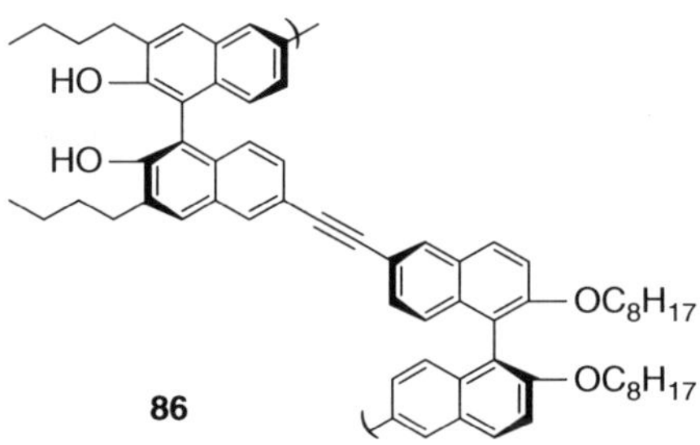

86

7.5.5 Spiral Chiral Compounds

Spiral chiral compounds were developed for use as chiral catalysts in various reactions. Its structure contains the spiral atom and forms an asymmetric space. For example, in reductive coupling of diene with aldehyde, the spiral chiral catalyst **87** could provide high (up to 90–96%) ee. The major product was the cis product (up to 98–99%) with high enantioselectivity [76]. The specific reaction is listed in Eq. (7.27) and partial experimental results are summarized in Table 7.20. Other similar spiral chiral catalysts such as **88** were developed in enantioselective copper-catalyzed B–H bond insertion reaction with high ee values (Eq. (7.28)) [77].

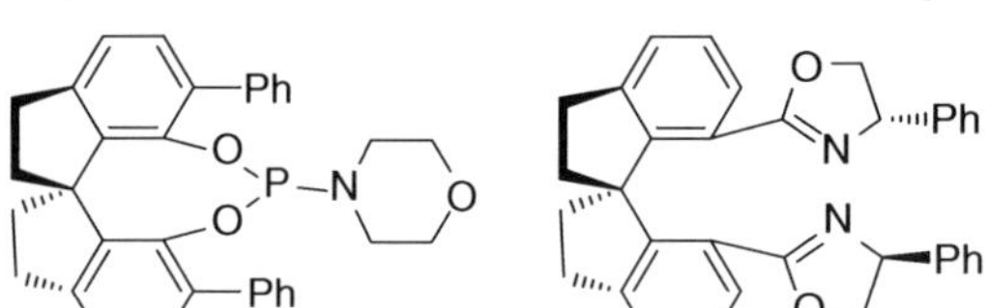

(*R*)-**87** **88**

Table 7.20 Enantioselective reductive coupling additions of diene with aldehydes catalyzed by (*R*)-**87** [76].

Entry	Ar	Yield (%)	ee (%)[a]
1	Ph	99	96
2	2-MePh	95	93
3	2-MeOPh	98	91
4	4-Me_2NPh	85	96
5	4-ClPh	92	90
6	4-CF_3Ph	98	85
7	2-Naph	96	86
8	2-Furyl	96	92
9	2-Thiophyl	98	91

a) The recoded ee (%) are for trans isomers. The cis isomer just occupied about 1–2%.

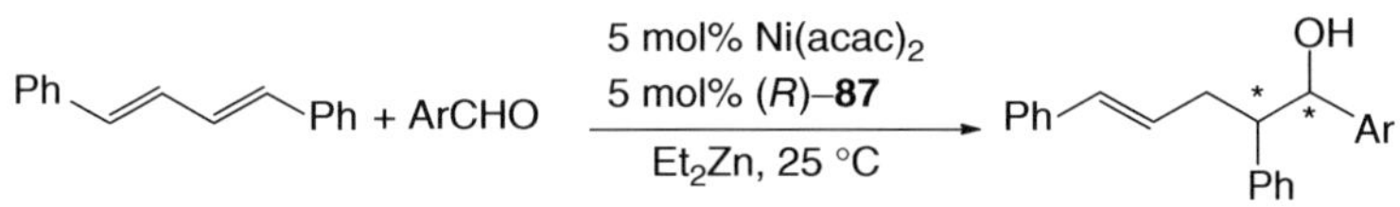

(7.27)

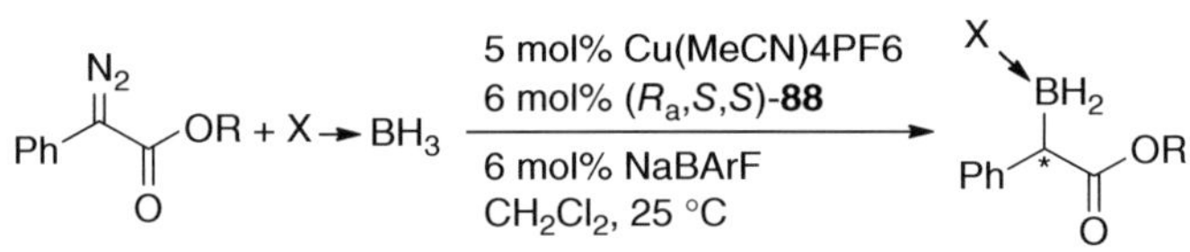

(7.28)

In this insertion reaction, when R = 2,6–Cl_2–Ph, X = Me_2PH, the ee could reach up to 92%. In most cases, the ee greatly depended upon the size of R. When X was Me_2PH and R was a methyl, the ee was 61%, and when R was a *tert*-Bu, its ee was 77%.

7.5.6
Asymmetric-Axle-Supported Chiral Catalyst

Asymmetric-axle-supported chiral catalysts, such as **57**, can be used efficiently [57]. Their design principle is illustrated below. Two planar molecules form a perpendicular base via a single bond, such as a C–C bond. The active moiety connects to one side or both sides (only one side is illustrated in Figure 7.5). The big planar structures may allow big substrates to take reactions near the active centers in high enantioselectivity.

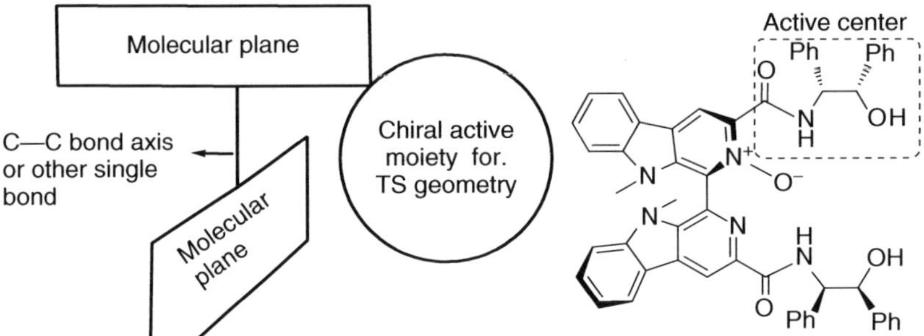

57

Figure 7.5 The diagrammatic sketch for design of asymmetric-axle-supported chiral catalysts [57].

7.5.7 Chiral Shiff-Base Ligands

Chiral Shiff-base ligands are important chiral ligands that have been used efficiently in reactions. Two special Shiff-base structures, **89** and **90**, which have been identified by X-ray, exhibited high ee values in diethylzinc to aldehydes addition (up to 90%) [78]. Shiff bases **91** and **92** showed a high enantioselectivity in the reduction of F-substituted ketones (Eq. (7.29)) [79] in the presence of the reduction reagent $LiBH_4$ and auxiliary EtOH or (tetrahydrofuran-2-yl) methanol. The highest ee can reach up to 97% under 1 mol% of catalyst **91**.

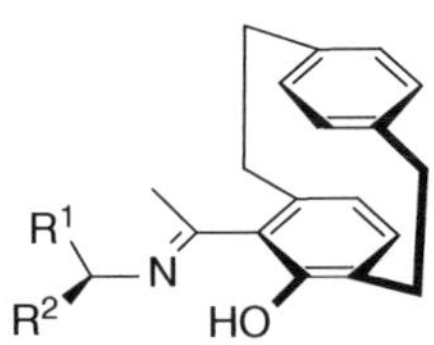

(S_P,S)-**89** R^1 = Cy, R^2 = Me

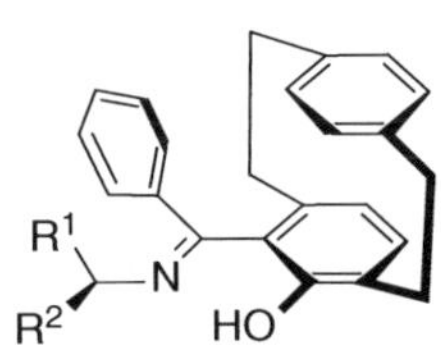

(S_P,S)-**90** R^1 = Cy, R^2 = Me

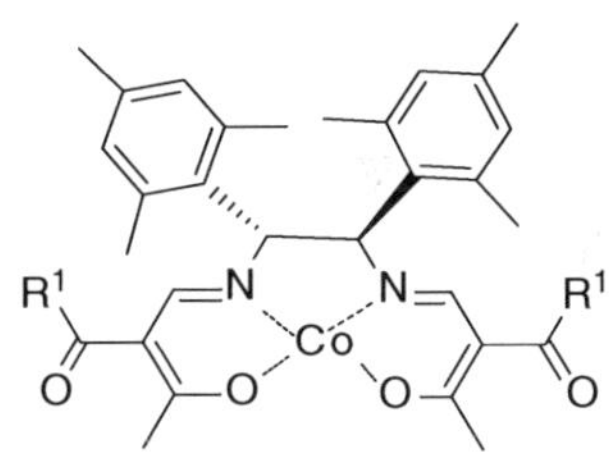

91 R^1 = Me

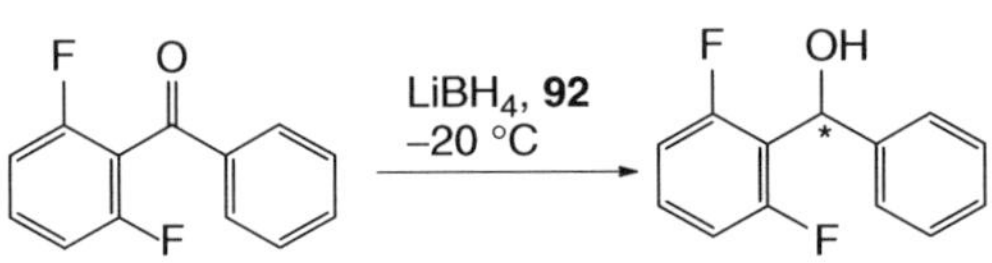

96% ee (7.29)

Asymmetric oxyamination of azlactone and oxaziridines could afford high ee values values under the catalysis with the Shiff-base ligand **92** (Eq. (7.30)) [80]. This reaction was accompanied by cis and trans product selection. In most cases, the cis product was the major one (Table 7.21).

Table 7.21 Substrate scope of **A** for asymmetric oxyamination of azlactone with oxaziridines [80].

Entry	R^1	R^2	cis:trans	ee (%) of A	ee (%) of B
1	4-MeC_6H_4	C_6H_5	95 : 5	84	95
2	C_6H_5	C_6H_5	94 : 6	80	93
3	$4\text{-Cl } C_6H_4$	C_6H_5	94 : 6	80	90
4	$4\text{-NO}_2\ C_6H_4$	C_6H_5	85 : 15	43	98
5	4-MeC_6H_4	2-MeC_6H_4	>95 : 5	83	87
6	4-MeC_6H_4	3-FC_6H_4	95 : 5	80	98
7	4-MeC_6H_4	3-Furyl	95 : 5	85	85
8	4-MeC_6H_4	1-Naph	95 : 5	82	85
9	4-MeC_6H_4	Bn	90 : 10	71	82
10	4-MeC_6H_4	Et	95 : 5	88	82

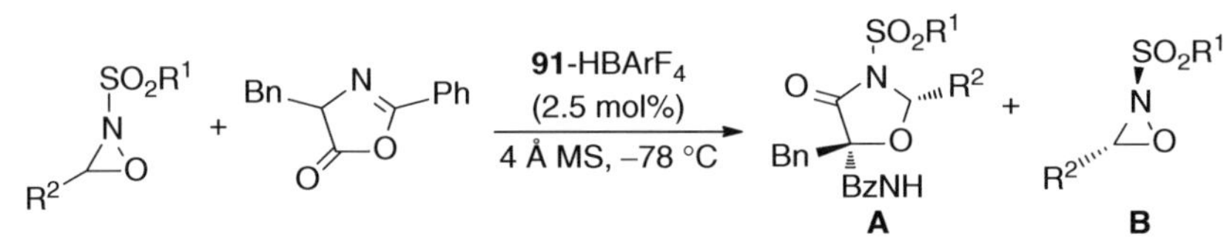

(7.30)

7.5.8
Some Asymmetric Lewis Acids

Some asymmetric Lewis acids could catalyze Diels–Alder reaction, which is an important tool for constructing large, new cyclic compounds. Cu-containing complexes play an important role. For example, use of the Cu-bisoxazoline catalyst **93**, which is easily derived from **94** in the Diels–Alder reaction of α-sulfenylacrylates with the cyclopentadiene, gave a high product yield with high enantioselectivity (Eq. (7.31)) [81]. Table 7.22 provides the details using the ligand **93**.

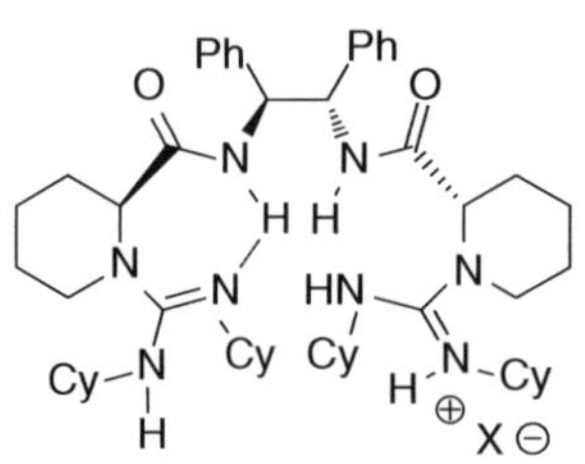

92 HX Cy = cyclohexyl

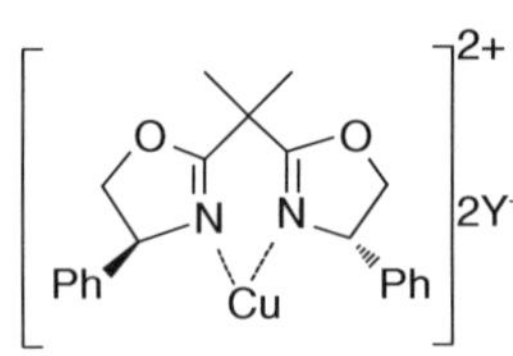

93

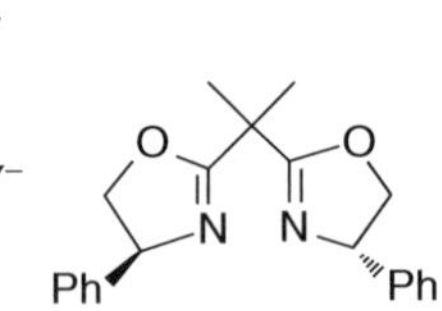

94

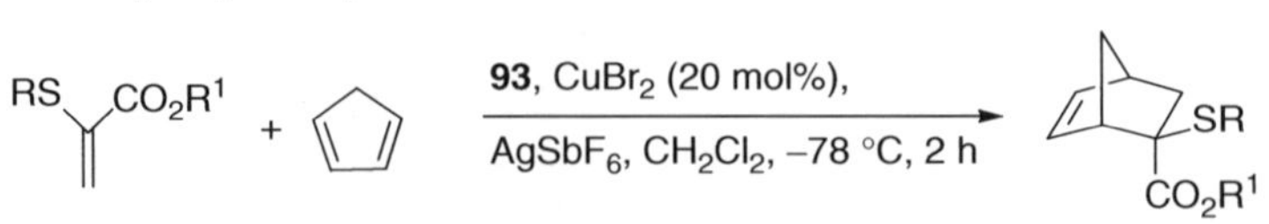

(7.31)

Table 7.22 Diels–Alder reactions of α-sulfenylacrylates with the cyclopentadiene catalyzed by **93** [81].

Entry	R	R^1	Cu-salt (Y)[a)]	Time (h)	T (°C)	exo : endo[b)]	ee (%)[c)]
1	Et	Me	$Cu(OTf)_2$	6	−40	1 : 2.4	40
2	Me	Ph	$Cu(OTf)_2$	6	−40	1 : 3.7	84
3	Et	Ph	$Cu(OTf)_2$	6	−40	1 : 4	80
4	Et	Ph	$Cu(OTf)_2$	9	−78	1 : 7	>95
5	Et	Ph	$CuBr_2/AgSbF_6$[d)]	1	−78	1 : 15	>95
6	*i*-Pr	Ph	$Cu(OTf)_2$	4	−40	1 : 2.3	85
7	*i*-Pr	Ph	$CuBr_2/AgSbF_6$[d)]	2.5	−78	1 : 5	81
8	*t*-Bu	Ph	$Cu(OTf)_2$	5.5	0	1 : 2.5	26
9	CH_2CF_3	Ph	$CuBr_2/AgSbF_6$[d)]	2	−78	1 : 13	>95

a) 20 mol% of $Cu(OTf)_2$, 30 mol% ligand **95**, 1 equiv of dienophole and 4 equiv of cyclopentadiene.
b) Determined by NMR integration of crude reaction mixtures.
c) Determined by NMR integration in presence of Pirkle' reagent, (*R*)-(−)-2,2,2-trifluoro-1-(9-anthryl)ethanol.
d) 10 mol% of $CuBr_2/AgSbF_6$. 10 mol% bisoxazoline **95**, 1 equiv dienophile and 4 equiv cyclopentadiene.

7.5.9
Organic P-Containing Ligands

Organic P-containing ligands are important type of catalysts in methodology study. The typical ligands listed above included **13–20**, **34–38**, **51–52**, **78–80**, and **87**. Some of them are monodentate phosphorus ligands, such as **87**. Monodentate hosphoramidite (DpenPhos) catalysts have been developed [82], such as catalyst **95**. Its synthesis could be from the corresponding diphenol reaction with $P(NMe_2)_3$ or $P(NEt_2)_3$ in high yields (75–95%) as illustrated below (Eq. (7.32)) [82c]. Its Rh(I) complex could catalyze the enantioselective hydrogenation of dehydro-α-amino acid methyl esters with very high ee values (up to 99.9%, Eq. (7.33)). Excellent ee values up to 99.7% were recorded.

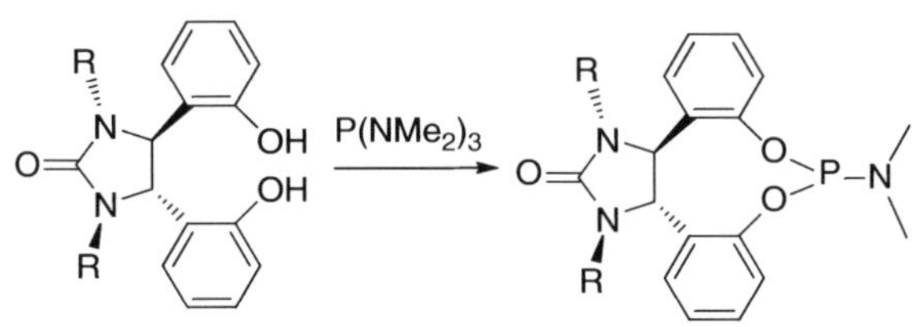

(7.32)

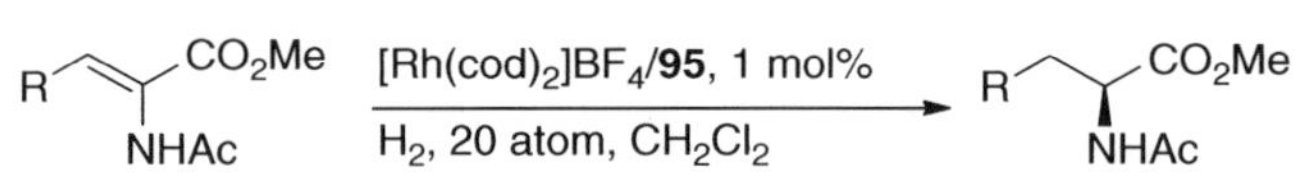

(7.33)

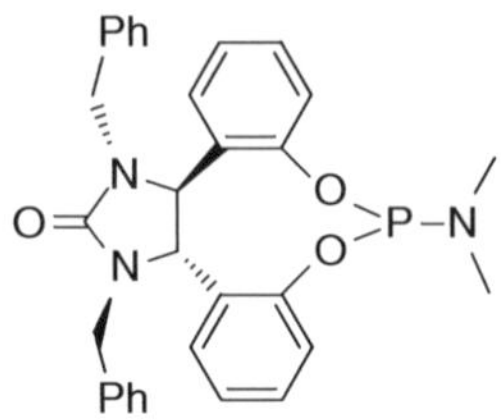

95 DpenPhos

7.6
Three Phenomena

During study of diethylzinc addition to aldehydes, three phenomena were noted with interest. They are chirality amplification (nonlinear effect), auto-self catalysis, and odd–even carbon effect.

7.6.1
Chirality Amplification (Nonlinear Effect)

When camphor derivatives catalyze diethylzinc to benzaldehyde addition, it was found that the ee values of the addition product could change with the ee values of the catalysts. When the ee value of camphor catalyst was small, the addition product's ee value was larger than the catalyst's ee [83]. This looks like the phenomenon in electronics where a triode can amplify electronic signals. The results are summarized in Figure 7.6.

However, this amplification does not happen frequently. In some cases, the ee values of the addition product are smaller than the catalysts' ee values. This is not amplification. Therefore, some researchers call both cases as nonlinear effects (Figure 7.7). For example, the ee values on line *b* (under line *a*) should be smaller

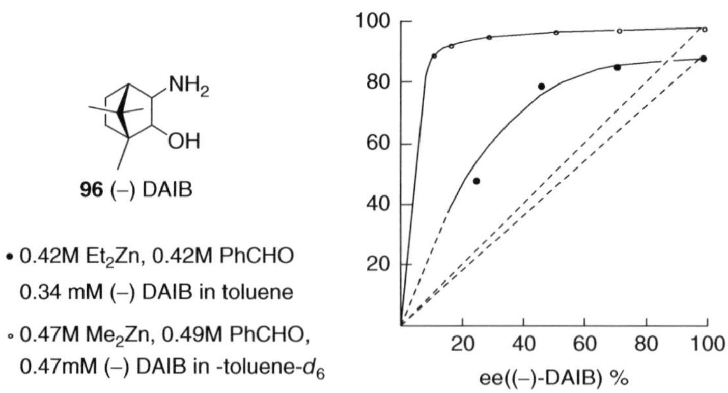

Figure 7.6 Chirality amplification (nonlinear effect) [3b, 83a].

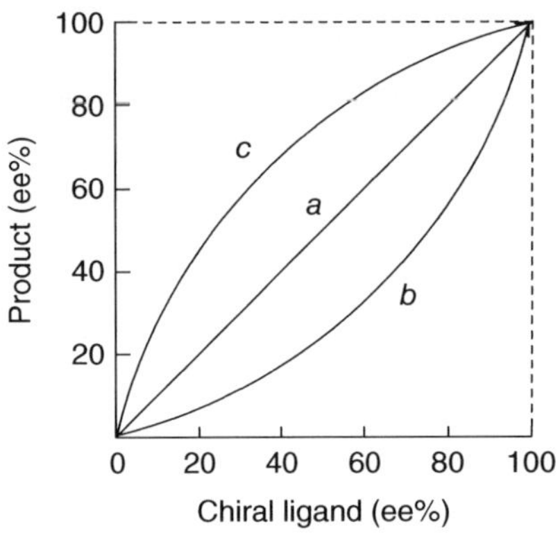

Figure 7.7 The chirality amplification area in enantioselective addition [3b].

than the catalyst's ee. Only the ee values on line *c* (above the line *a*) are larger (amplification).

7.6.2
Auto-Self Catalysis

In the addition of diisopropylzinc to some aldehydes, such as the pyrimidine-5-carbaldehyde derivative, the product formed by the catalysis could catalyze the addition reaction again. Thus, after the reaction happens for a while, more and more of the product is formed, which means the catalyst becomes more and more active and the catalytic process gets speeded up. This is auto-self catalysis [84]. A typical example is listed below (Eq. (7.34)).

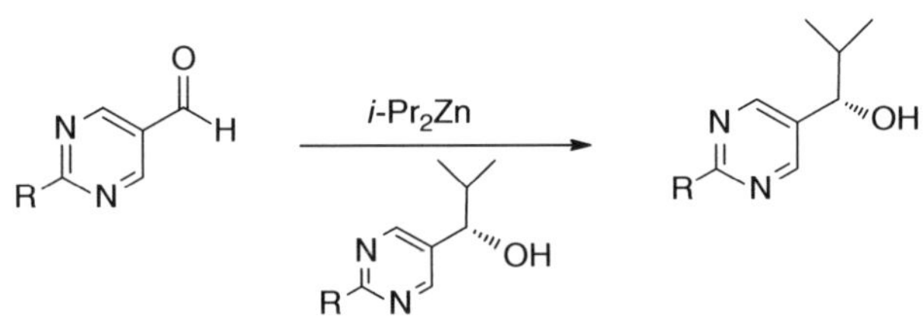

(7.34)

The catalytic process is illustrated in Figure 7.8.

7.6.3
Odd–Even Carbon Effect

During the study of the addition of diethylzinc to benzaldehyde, it was found that the ee values of the addition product showed a regular change; for example, when catalyzed by **98–101** and **102–105**, the addition product's ee values showed regular changes if the substituent R was changed from –Me,–Et, *n*-Pr to *n*-Bu (Figure 7.9) [85].

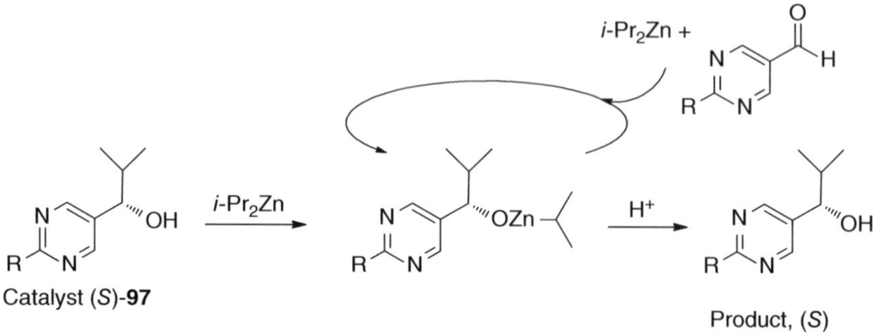

Figure 7.8 Catalytic procedure of auto-self catalysis [84].

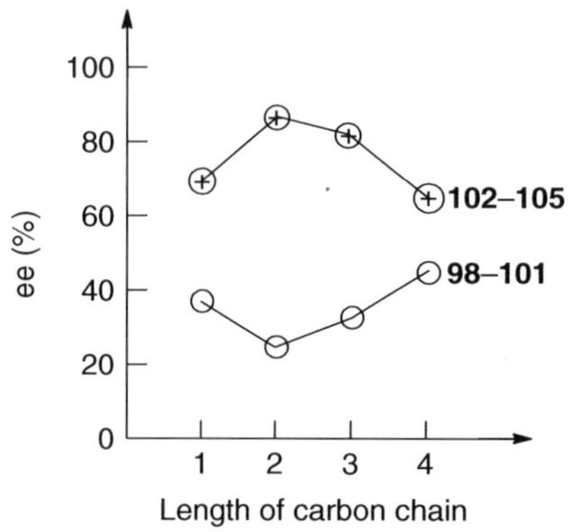

Figure 7.9 Value of ee changing with the length of the side chain on C3 [85].

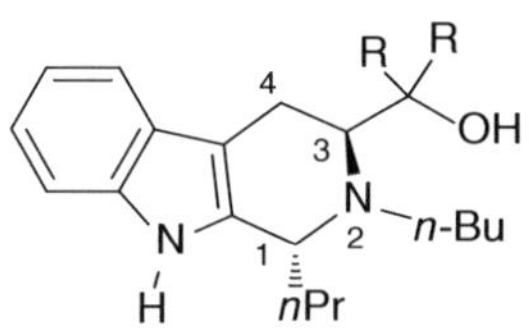

98 R = Me
99 R = Et
100 R = *n*-Pr
101 R = *n*-Bu

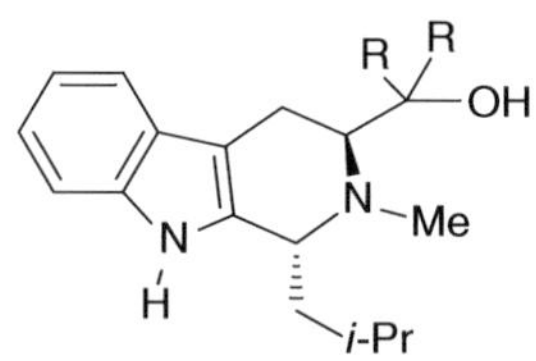

102 R = Me
103 R = Et
104 R = *n*-Pr
105 R = *n*-Bu

Indeed, other types of chiral catalyst have been tested. Different chiral ligands lead to similar results as mentioned above. This phenomenon may provide some clues in designing chiral catalysts.

This odd–even carbon effect was discovered from the addition of diethylzinc to aldehydes. However, a similar effect has been observed in other reactions. For example, in the reduction of sodium borohydride to aliphatic esters, with the same length of the carbon chain of the starting material, the initial reaction temperature (onset temperature) is very sensitive to the chain length (R^1) when changed from Me to *n*-Bu. In another reduction using phenyl esters, the same results were seen when R^2 changed from Me to *n*-Bu. The relationship between the onset temperature and the length of chain is illustrated in Figure 7.10a [86]. An interesting discovery was that the length of linear alcohols, such as from MeOH to *n*-BuOH,

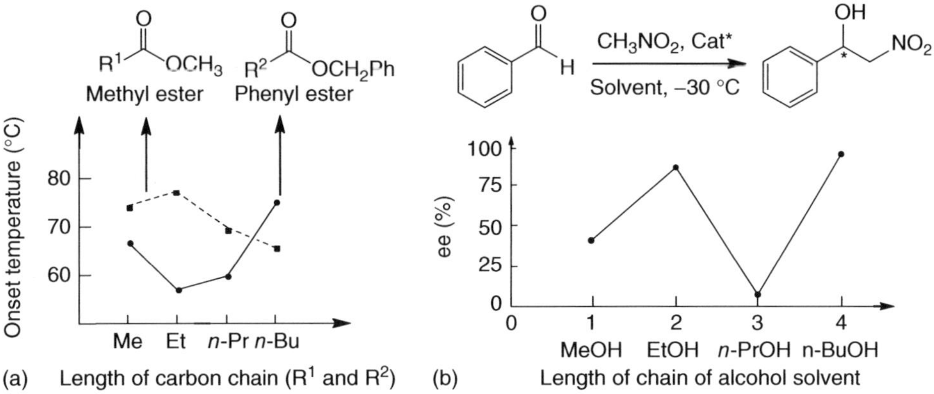

Figure 7.10 The relationship between the onset temperature and length of carbon chain (a) and that between ee and the length of linear alcohol solvent (b) [87].

had a strong effect on the enantioselectivity in Henry additions (Figure 7.10b) [87]. The results cannot be explained using traditional theory.

Whether in the design of chiral catalyst or in other reactions, the length of the carbon chain plays an important role. Therefore, paying much more attention to this area may bring us some valid benefits that are different from the traditional results. This may have some relevance when studying an animal's structure.

References

1. Frankland, E. (1849) *Liebigs Ann. Chem.*, **71**, 171–213.
2. Oguni, N. and Omi, T. (1984) *Tetrahedron Lett.*, **25**, 2823–2824.
3. (a) Kim, H.J., Pongdee, R., Wu, Q.Q., Hong, L., and Liu, H.W. (2007) *J. Am. Chem. Soc.*, **129**, 14582–14584; (b) Kitamura, M., Suga, S., Niwa, M., and Noyori, R. (1995) *J. Am. Chem. Soc.*, **117**, 4832.
4. Yamakawa, M. and Noyori, R. (1995) *J. Am. Chem. Soc.*, **117**, 6327.
5. Dai, W.M., Zhu, H.J., and Hao, X.J. (1996) *Tetrahedron Lett.*, **37**, 5971–5974.
6. Dai, W.M., Zhu, H.J., and Hao, X.J. (1995) *Tetrahedron: Asymmetry*, **6**, 1857.
7. Ko, D.H., Kim, K.H., and Ha, D.C. (2002) *Org. Lett.*, **4**, 3759–3762.
8. Huang, W.S. and Pu, L. (1999) *J. Org. Chem.*, **64**, 4222–4223.
9. Gao, G., Moore, D., Xie, R.G., and Pu, L. (2002) *Org. Lett.*, **4**, 4143–4146.
10. Boudier, A., Bromm, L.O., Lotz, M., and Knochel, P. (2000) *Angew. Chem. Int. Ed.*, **39**, 4414–4435.
11. (a) Hird, A.W. and Hoveyda, A.H. (2003) *Angew. Chem. Int. Ed.*, **42**, 1276–1279; (b) Zhu, H.J., Jiang, J.X., Ren, J., Yan, Y.M., and Pittman, C.U. (2005) *Curr. Org. Synth.*, **2**, 547–587.
12. Duursma, A., Minnaard, A.J., and Feringa, B.L. (2003) *J. Am. Chem. Soc.*, **125**, 3700–3701.
13. Hu, Y.C., Liang, X.M., Wang, J.W., Zheng, Z., and Hu, X.Q. (2003) *J. Org. Chem.*, **68**, 4542–4545.
14. Boezio, A.A. and Charette, A.B. (2003) *J. Am. Chem. Soc.*, **125**, 1692–1693.
15. Krause, N. (1998) *Angew. Chem. Int. Ed.*, **37**, 283–285.
16. Shintani, R. and Fu, G.C. (2002) *Org. Lett.*, **4**, 3699–3702.
17. (a) Pioda, G. and Togni, A. (1998) *Tetrahedron: Asymmetry*, **9**, 3903–3910; (b) Liu, D., Li, W., and Zhang, X.M. (2002) *Org. Lett.*, **4**, 4471–4474; (c) Alexakis,

A. and Benhaim, C. (2001) *Tetrahedron: Asymmetry*, **12**, 1151–1157.

18. Bastin, S., Ginj, M., Brocard, J., Pelinski, L., and Novogrocki, G. (2003) *Tetrahedron: Asymmetry*, **14**, 1701–1708.
19. (a) Bolm, C. and Rudolph, J. (2002) *J. Am. Chem. Soc.*, **124**, 14850–14851; (b) Kang, J., Lee, J.H., and Choi, J.S. (2001) *Tetrahedron: Asymmetry*, **12**, 33–35.
20. Hayashi, T., Yamamoto, A., Hagihara, T., and Ito, Y. (1986) *Tetrahedron Lett.*, **27**, 191–194.
21. Mermerian, A.H. and Fu, G.C. (2003) *J. Am. Chem. Soc.*, **125**, 4050–4051.
22. Yanagisawa, A., Nakashima, H., Ishiba, A., and Yamamoto, H. (1996) *J. Am. Chem. Soc.*, **118**, 4723–4724.
23. Reviews: (a) Poli, G., Giambastian, G., and Heumann, A. (2000) *Tetrahedron*, **56**, 5959–5989; For Ni-Pd mixture use, see microreview: (b) Dunach, E., Franco, D., and Olivero, S. (2003) *Eur. J. Org. Chem.*, **2003**, 1605.
24. Schenkel, L.B. and Ellman, J.A. (2003) *Org. Lett.*, **5**, 545–548.
25. Yamada, K.I., Harwood, S.J., Groger, H., and Shibasaki, M. (1999) *Angew. Chem. Int. Ed.*, **38**, 3504–3506.
26. Qin, B., Xiao, X., Liu, X., Huang, J., Wen, Y., and Feng, X.M. (2007) *J. Org. Chem.*, **72**, 9323–9328.
27. Ballini, R. and Petrini, M. (2004) *Tetrahedron*, **60**, 1017–1047.
28. (a) Xu, K., La, C.Y., Zha, Z.G., Pan, S.S., Chen, H.W., and Wang, Z.Y. (2012) *Chem. Eur. J.*, **18**, 12357–12362; (b) Zhou, Y.R., Dong, J.F., Zhang, F.G., and Gong, Y.G. (2011) *J. Org. Chem.*, **76**, 588–600.
29. (a) Marcelli, T., van der Haas, R.N.S., van Maarseveen, J.H., and Hiemstra, H. (2006) *Angew. Chem. Int. Ed.*, **45**, 929–931; (b) Wang, C.G., Zhou, Z.H., and Tang, C.C. (2008) *Org. Lett.*, **10**, 1707–1710.
30. (a) Chougnet, A., Zhang, G.Q., Liu, K.G., Haussinger, D., Kagi, A., Allmendinger, T., and Woggona, W. (2011) *Adv. Synth. Catal.*, **353**, 1797–1806; (b) Arai, T., Takashita, R., Endo, Y., Watanabe, M., and Yanagisawa, A. (2008) *J. Org. Chem.*, **73**, 4903–4906.
31. Palomo, C., Oiarbide, M., and Laso, A. (2005) *Angew. Chem. Int. Ed.*, **44**, 3881–3884.
32. Lang, K., Park, J., and Hong, S. (2012) *Angew. Chem. Int. Ed.*, **51**, 1620–1624.
33. Handa, S., Nagawa, K., Sohtome, Y., Matsunaga, S., and Shibasaki, M. (2008) *Angew. Chem. Int. Ed.*, **47**, 3230–3233.
34. (a) Qin, X.D., Dong, Z.J., and Liu, J.K. (2006) *Helv. Chim. Acta*, **89**, 127–133; (b) Ren, J., Jiang, J.X., Li, L.B., Liao, T.G., Tian, R.R., Chen, X.L., Jiang, S.P., Pittman, C.U. Jr., and Zhu, H.J. (2009) *Eur. J. Org. Chem.*, **2009**, 3987–3991.
35. (a) Hosomi, A. and Sakurai, H. (1976) *Tetrahedron Lett.*, **16**, 1295–1298; (b) Denmark, S.E. and Fu, J.P. (2002) *Org. Lett.*, **4**, 1951–1953.
36. (a) Cesarotti, E., Araneo, S., Rimoldi, I., and Tassi, S. (2003) *J. Mol. Catal. A: Chem.*, **204-205**, 221–226; (b) Jain, P. and Antilla, J.C. (2010) *J. Am. Chem. Soc.*, **132**, 11884–11886.
37. (a) Nakajima, M., Saito, M., Shiro, M., and Hashimoto, S. (1998) *J. Am. Chem. Soc.*, **120**, 6419–6420; (b) Nakajima, M., Yokota, T., Saito, M., and Hashimoto, S. (2004) *Tetrahedron Lett.*, **45**, 61–64; (c) Jiao, Z.G., Feng, X.M., Liu, B., Chen, F.X., Zhang, G.L., and Jiang, Y.Z. (2003) *Eur. J. Org. Chem.*, **2003**, 3818–3826; (d) Chelucci, G., Belmonte, N., Benagliab, M., and Pignataro, L. (2007) *Tetrahedron Lett.*, **48**, 4037–4041.
38. Iseki, K., Mizuno, S., Kuroki, Y., and Kobayashi, Y. (1998) *Tetrahedron Lett.*, **39**, 2767–2770.
39. Shimada, T., Kina, A., Ikeda, S., and Hayashi, T. (2002) *Org. Lett.*, **4**, 2799–2801.
40. B. Bai, L. Shen, J. Ren, H. J. Zhu (2012) *Adv. Synth. Catal.*, **354**, 354-358.
41. Lotz, M., Polborn, K., and Knochel, P. (2002) *Angew. Chem. Int. Ed.*, **41**, 4708–4711.
42. Patti, A. and Pedotti, S. (2003) *Tetrahedron: Asymmetry*, **14**, 597–602.
43. Ma, Y.P., Liu, H., Chen, L., Cui, X., Zhu, J., and Deng, J.G. (2003) *Org. Lett.*, **5**, 2103–2106.
44. (a) Maienza, F., Santoro, F., Spindler, F., Malan, C., and Mezzetti, A. (2002) *Tetrahedron: Asymmetry*, **13**, 1817–1824; (b) Chiba, T., Miyashita, A., Nohira, H.,

and Takaya, H. (1993) *Tetrahedron Lett.*, **34**, 2351–2354.

45. Tao, B. and Fu, G.C. (2002) *Angew. Chem. Int. Ed.*, **41**, 3892–3894.
46. (a) Malkov, A.V., Stewart-Liddon, A.J.P., McGeoch, G.D., Ramirez-Lopez, P., and Kocovsky, P. (2012) *Org. Biomol. Chem.*, **10**, 4864–4877; (b) Carpentier, J.F. and Bette, V. (2002) *Curr. Org. Chem.*, **6**, 913–916; (c) Zhou, L., Wang, Z., Wei, S., and Sun, J. (2007) *Chem. Commun.*, **28**, 2977–2979; (d) Jiang, Y., Chen, X., Zheng, Y., Xue, Z., Shu, C., Yuan, W., and Zhang, X. (2011) *Angew. Chem. Int. Ed.*, **50**, 7304–7307.
47. Iwasaki, F., Onomura, O., Mishima, K., Maki, T., and Matsumura, Y. (1999) *Tetrahedron Lett.*, **40**, 7507–7511.
48. Guizzetti, S., Benaglia, M., Cozzi, F., and Annunziata, R. (2009) *Tetrahedron*, **65**, 6354–6363.
49. Malkov, A.V., Liddon, A.J.P.S., Ramírez-López, P., Bendová, L., Haigh, D., and Kočovský, P. (2006) *Angew. Chem. Int. Ed.*, **45**, 1432–1435.
50. Pan, W., Deng, Y., He, J.B., Bai, B., and Zhu, H.J. (2013) *Tetrahedron*, **69**, 7253.
51. Noe, M.C., Hawkins, J.M., Snow, S.L., and Wolf-Gouveia, L. (2008) *J. Org. Chem.*, **73**, 3295–3298.
52. Kanemoto, M., Murata, M., and Oishi, T. (2009) *J. Org. Chem.*, **74**, 8810–8813.
53. Reiss, T. and Breit, B. (2009) *Org. Lett.*, **11**, 3286–3289.
54. (a) Wang, Z.-M., Zhou, W.-S., and Lin, G.-Q. (1985) *Tetrahedron Lett.*, **26**, 6221–6224; (b) Wang, Z.-M. and Zhou, W.-S. (1989) *Synth. Commun.*, **19**, 2627–2632.
55. Du, Y., Zheng, J.-F., Wang, Z.-G., Jiang, L.-J., Ruan, Y.-P., and Huang, P.-Q. (2010) *J. Org. Chem.*, **75**, 4619–4622.
56. Vazquez, J., Pericas, M.A., Maseras, F., and Lledos, A. (2000) *J. Org. Chem.*, **65**, 7303–7309.
57. Pan, W., Ma, W.G., Yang, X.D., Zhen, Y.H., Song, B.Q., Niu, Y.Z., Gu, G., Hu, D.B., Yang, Q., and Zhu, H.J. (2015) *Chem. J. Chin. Univ.*, **36**(2), 325–329.
58. (a) Li, L.C., Jiang, J.X., Ren, J., Ren, Y., Pittman, C.U., and Zhu, H.J. (2006) *Eur. J. Org. Chem.*, **2006**, 1981; (b) Ren, J., Li, L.C., Liu, J.K., Zhu, H.J., and Pittman, C.U. (2006) *Eur. J. Org. Chem.*, 1991.
59. Rudolph, J., Bolm, C., and Norrby, P.O. (2005) *J. Am. Chem. Soc.*, **127**, 1548–1552.
60. Kozlowski, M.C., Dixon, S.L., Panda, M., and Lauri, G. (2003) *J. Am. Chem. Soc.*, **125**, 6615.
61. Yang, W.G., Wu, S., and Shi, M. (2007) *Org. Chem. (Chin.)*, **27**, 197–208.
62. Mao, J.C., Wan, B.S., Wang, R.L., Wu, F., and Lu, S.W. (2004) *J. Org. Chem.*, **69**, 9123–9127.
63. Zhang, H.L., Jiang, F., Zhang, X.M., Cui, X., Gong, L.Z., Mi, A.Q., Jiang, Y.Z., and Wu, Y.D. (2004) *Chem. Eur. J.*, **10**, 1481–1492.
64. (a) Malkov, A.V., Barlog, M., Jewkes, Y., Mikusek, J., and Kocovsky, P. (2011) *J. Org. Chem.*, **76**, 4800–4804; (b) Hrdina, R., Dracinsky, M., Valterova, I., Hodacova, J., Cisarova, I., and Kotora, M. (2008) *Adv. Synth. Catal.*, **350**, 1449–1456.
65. For (*R*)-**69**, see: (a) Deng, Y., Pan, W., Pei, Y.N., Li, J.L., Liu, X.C., and Zhu, H.J. (2013) *Tetrahedron*, **69**, 10431–10437; (b) For (*R*)-**70**, see: Pei, Y.N., Deng, Y., Li, J.L., and Zhu, H.J. (2014) *Tetrahedron Lett.*, **55**, 2948–2952.
66. Zheng, K.; Liu, X.; Qin, S.; Xie, M.; Lin, L.; Hu, C.; Feng, X. *J. Am. Chem. Soc.* 2012, **134**, 17564–17573.
67. (a) Hrdina, R., Valterová, I., Hodačová, J., Císařová, I., and Kotora, M. (2007) *Adv. Synth. Catal.*, **349**, 822; (b) Kadlcikova, A., Valterová, I., Duchácková, L., Roithová, J., and Kotora, M. (2010) *Chem. Eur. J.*, **16**, 9442.
68. Takenaka, N., Sarangthem, R.S., and Captain, B. (2008) *Angew. Chem.*, **120**, 9854.
69. Kina, A., Shimada, T., and Hayashi, T. (2004) *Adv. Synth. Catal.*, **346**, 1169.
70. (a) Kang, J., Lee, J.H., and Lim, D.S. (2003) *Tetrahedron: Asymmetry*, **14**, 305–315; (b) Degrado, S.J., Mizutani, H., and Hoveyda, A.H. (2001) *J. Am. Chem. Soc.*, **123**, 755; (c) Soai, K., Hayasaka, T., and Ugajin, S. (1989) *J. Chem. Soc., Chem. Commun.*, 516.
71. Li, G.L., Liang, Y.X., and Anyilla, J. (2007) *J. Am. Chem. Soc.*, **129**, 5830–5831.
72. Yang, X.W., Sheng, J.H., Da, C.S., Wang, H.S., Su, W., Wang, R., and

Chan, A.S.C. (2000) *J. Org. Chem.*, **65**, 295–296.

73. Kelsen, V., Pierrat, P., and Gros, P.P. (2007) *Tetrahedron*, **63**, 10693–10697.
74. Pathak, K., Bhatt, A.P., Abdi, S.H.R., Kureshy, R.I., Khan, N.H., Ahmad, I., and Jasra, R.V. (2006) *Tetrahedron: Asymmetry*, **17**, 1506–1513.
75. Zou, X.W., Zhang, S.E., Cheng, Y.X., Liu, Y., Huang, H., and Wang, C.Y. (2007) *J. Appl. Polym. Sci.*, **106**, 821–827.
76. Yang, Y., Zhu, S.F., Duan, H.F., Zhou, C.Y., Wang, L.X., and Zhou, Q.L. (2007) *J. Am. Chem. Soc.*, **129**, 2248–2249.
77. Cheng, Q. Q.; Zhu, S. F.; Zhang, Y. Z.; Xie, X. L.; Zhou, Q. L. *J. Am. Chem. Soc.* 2013, **135**, 14094–14097.
78. Lauterwasser, F., Nieger, M., Mansikkamäki, H., Nättinen, K., and Bräse, S. (2005) *Chem. Eur. J.*, **11**, 4509–4525.
79. (a) Kokura, A., Tanaka, S., Ikeno, T., and Yamada, T. (2006) *Org. Lett.*, **8**, 3025–3027; Theoretical study, see (b) Sun, L., Tang, M., Wang, H., Wei, D., and Liu, L. (2008) *Tetrahedron: Asymmetry*, **19**, 779–787; Experimental study, see: (c) Zagozda, M. and Plenkiewicz, J. (2006) *Tetrahedron: Asymmetry*, **17**, 1958–1962; (d) Barkakaty, B., Takaguchi, Y., and Tsuboi, S. (2007) *Tetrahedron*, **63**, 970–976.
80. Dong, S.; Liu, X.; Zhu, Y.; He, P.; Lin, L.; Feng, X. *J. Am. Chem. Soc.* 2013, **135**, 10026–10029
81. Aggarwal, V.K., Jones, D.E., and Martin-Castro, A.M. (2000) *Eur. J. Org. Chem.*, **2000**, 2945–2939.
82. (a) Reetz, M.T. and Mehler, G. (2000) *Angew. Chem. Int. Ed.*, **39**, 3889; (b) Komarov, I.V. and Borner, A. (2001) *Angew. Chem. Int. Ed.*, **40**, 1197; (c) Liu, Y. and Ding, K. (2005) *J. Am. Chem. Soc.*, **127**, 10488–10489.
83. (a) Kitamura, M., Suga, S., Kawai, K., and Noyori, R. (1986) *J. Am. Chem. Soc.*, **108**, 6071–6072; (b) Puchot, C., Samuel, O., Dunach, E., Zhao, S., Agami, C., and Kagan, H.B. (1986) *J. Am. Chem. Soc.*, **108**, 2353–2357; (c) Oguni, N., Matsuda, Y., and Kaneko, T. (1988) *J. Am. Chem. Soc.*, **110**, 7877–7878.
84. (a) Soai, K., Shibata, T., Morioka, H., and Choji, K. (1995) *Nature*, **378**, 767–768; (b) Shibata, T., Morioka, H., Hayase, T., Choji, K., and Soai, K. (1996) *J. Am. Chem. Soc.*, **118**, 471–472; (c) Shibata, T. *et al.* (1996) *Tetrahedron Lett.*, **37**, 8783.
85. Zhu, H.J., Jiang, J.X., Saobo, S., and Pittman, C.U. (2005) *J. Org. Chem.*, **70**, 261–267.
86. Lv, X.J., Jiang, J.X., Ren, J., and Zhu, H.J. (2008) *Chem. J. Chin. Univ.*, **29**, 537–541.
87. Li, J.L., Liu, L., Pei, Y.N., and Zhu, H.J. (2014) *Tetrahedron*, **70**, 9077–9083.

8
Chemoselective Reaction

It would appear that chemoselective reaction has little relationship with organic stereochemistry. Indeed, when put on a platform of total organic synthesis, it is very much concerned with the stereochemistry accompanying asymmetric technology. It is necessary to combine a chemoselective reaction with asymmetric synthesis to present a full and valid method in front of the modern readers.

Generally, chemoselective reactions include chemoselective addition, reduction, and oxidation, including some named reactions, which are useful in organic stereochemistry study.

8.1
Chemoselective Additions

If there are two or more functional groups such as >C=O in a molecule, an added reagent just reacts with one of them and affords a new product; this is chemoselective addition action. Compared to chemoselective reductions, chemoselective addition is not very common. Compounds with the functional groups aldehyde or ketone (–CHO, >C=O), ester (–CO_2R), amide (–CONHR), acid (–CO_2H), and acid anhydride (–C=O(O)C=O–), containing >C=C< bonds, could be potential substrates to be used in a chemoselective reaction. A ketone or an aldehyde could easily take part in chemoselective additions. Compared to this, the addition to an acid or acid anhydride is relatively difficult.

8.1.1
Addition to C=O or C=C Groups

Recently, the addition of dialkylzinc to acid hydride has been developed [1]. When there is a stereogenic center, this reaction does not change the existing AC of the chiral center. This is suitable for application for further reactions. It needs the presence of elements such as Ni or Rh to react with dialkylzinc or zinc salt (Eq. (8.1)). Experimental results have shown that this reaction gives high yields and chemoselectivity. Some results are listed in Table 8.1 [2].

Organic Stereochemistry: Experimental and Computational Methods, First Edition. Hua-Jie Zhu.

Table 8.1 Addition of diethylzinc to acid anhydrides with five-membered ring under the presence of Ni salt [2].

Entry[a)]	Anhydride	Product	Yield (%)
1	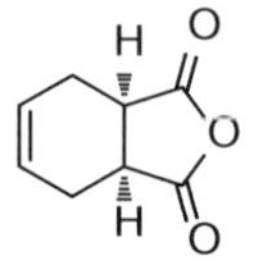	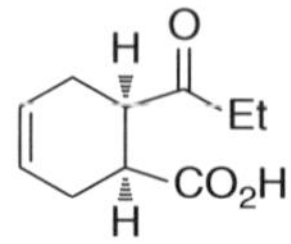	95[b),c)]
2	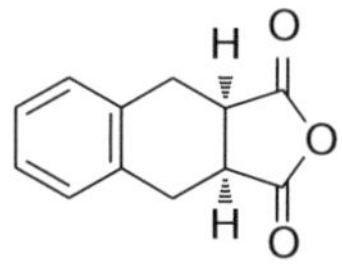	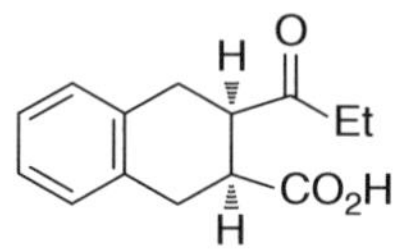	90[d)]
3	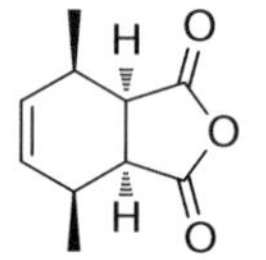	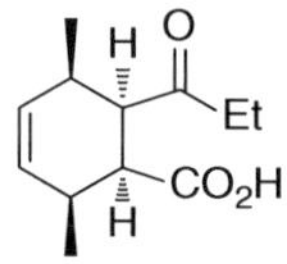	61[d)]
4	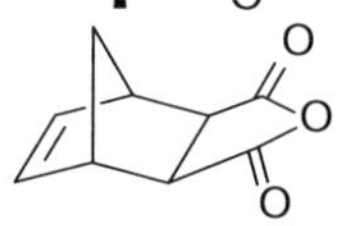	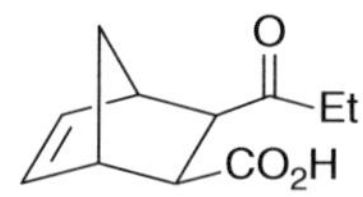	91[b),c)]
5	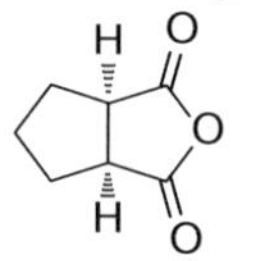	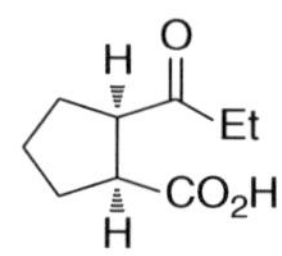	71[d)]
6	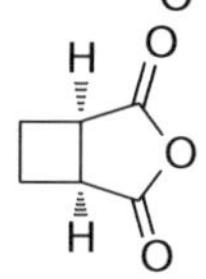	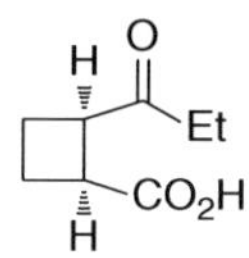	61[b),c)]

a) Reaction was performed using $Ni(COD)_2$ (5 mol%), bpy (6 mol%), 4-F-sty (10 mol%), and Et_2Zn (1.2 equiv) at 0 °C unless noted otherwise.
b) Reaction was performed using $Ni(COD)_2$ (10 mol%), bpy (12 mol%), and 4-F-sty (20 mol%).
c) 4-CF_3-sty was used as promoter.
d) Isolated methyl ester.

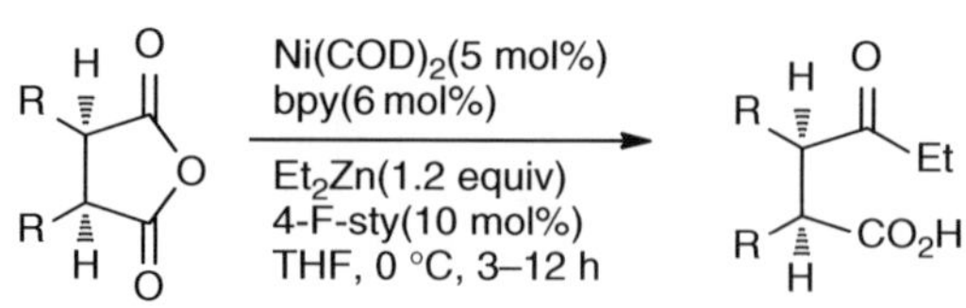

(8.1)

When faced with an anhydride with a six-membered ring, this addition could lead to the same chemoselectivity. This is illustrated in Eq. (8.2). Only one major product was obtained from the reaction [3]. In all reactions, the stereogenic centers do not change.

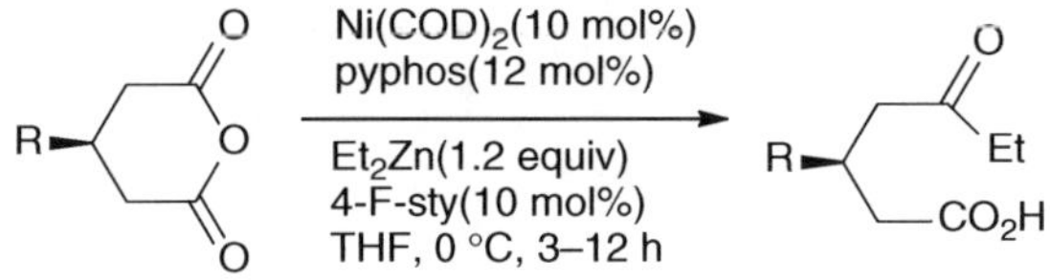

(8.2)

An addition–cyclization reaction could take place and be catalyzed by a base (Eq. (8.3)). In this example, the α-H of –CN could be removed by a base, and this carbon anion could attack the carbonyl carbon to afford a new ring structure. Formation of –OH was achieved first as an intermediate, and then loss of –OH forms a new double bond [3].

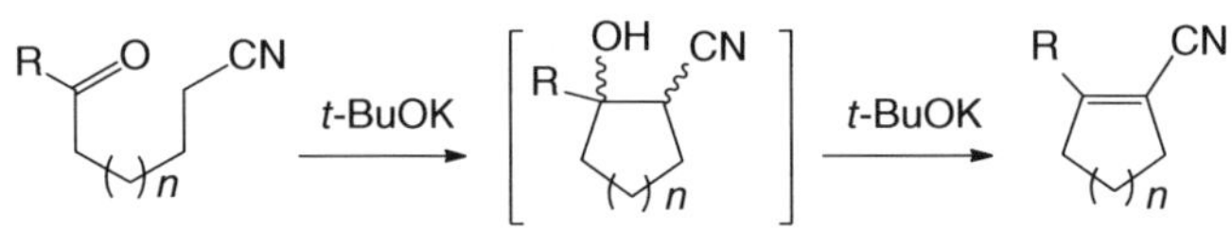

(8.3)

This is a way to synthesize α,β-unsaturated nitrile, which could be further converted into the corresponding amides or esters by hydrolysis. Some typical results are listed in Table 8.2.

Table 8.2 Addition product of cyano-ketones under *t*-BuOK catalysis [3].

Entry	Substrate	Product	Yield (%)
1	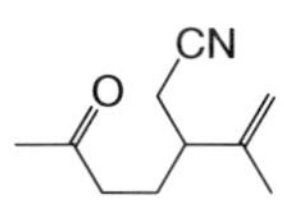	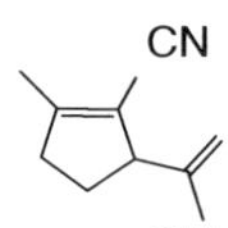	60
2	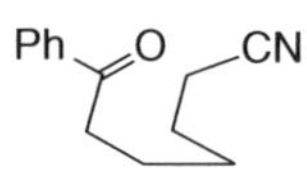	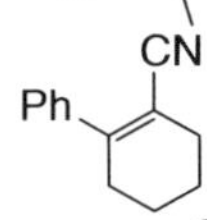	66
3	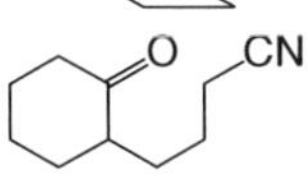	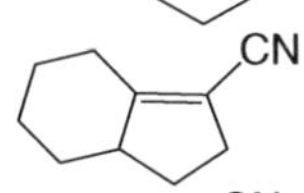	79
4	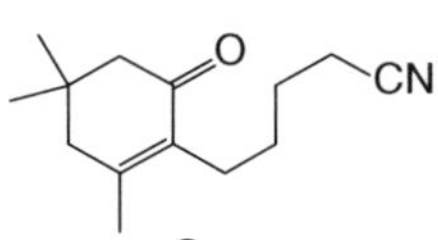	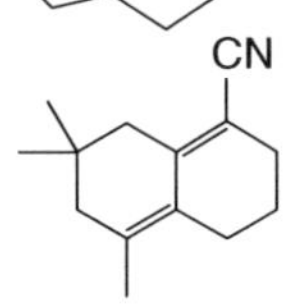	61
5	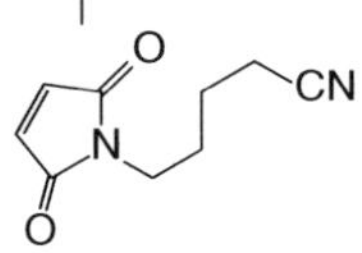	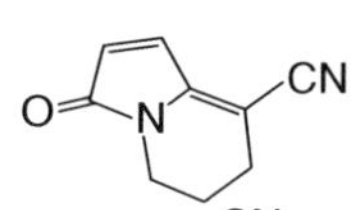	62
6	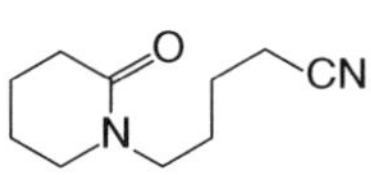	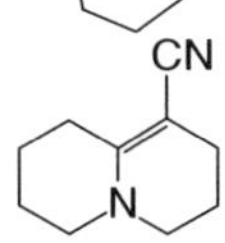	60

A chemoselective addition was carried out using Me_2Zn and Et_3B after the condensation reaction of the aldehyde with primary amine is completed (Scheme 8.1) [4]. When Me_2Zn was used in the mixed solution, the major product was a condensation–addition product. However, once reagent Et_3B was used, the major compound was the addition product of THF to benzaldehyde.

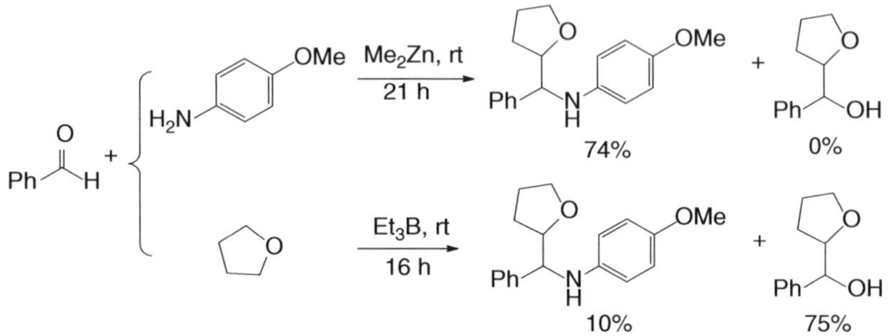

Scheme 8.1 Chemoselective additions using either Me_2Zn or Et_3B [4].

In a nucleophilic addition/Birch reaction, some widely used amides containing the N–O group could be smoothly converted into the corresponding ketones [5]. The reaction is shown in Eq. (8.4). This procedure involved the carbon anion attacking carbonyl carbon first to form the addition intermediate **A** under the interaction of ammonia electronic solution (Li–NH_3(liquid)) and loss of Li(–N(Me)–OMe) group to give the corresponding ketone. The experimental results are summarized in Table 8.3.

Table 8.3 Nucleophilic addition/Birch reduction of >CON–OMe to ketone [5].

Entry	Substrate	Product	Yield (%)
1	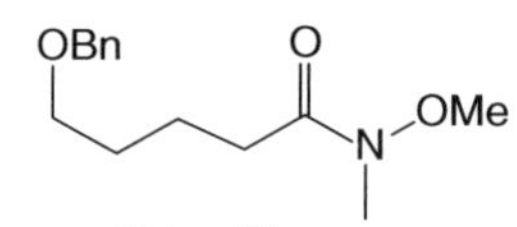	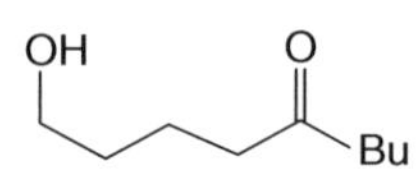	72
3	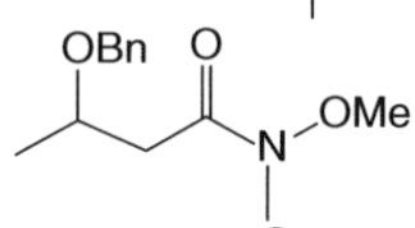	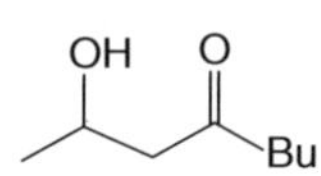	58
5[a)]	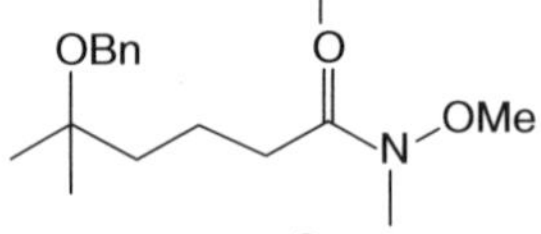	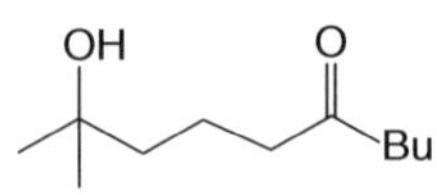	92
7[b)]	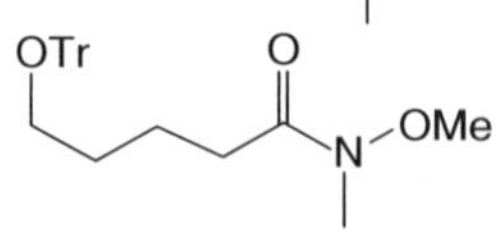	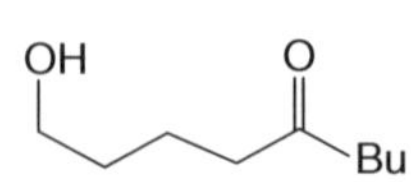	70

a) *t*-BuOH was not used.
b) Tr = Ph_3C–.

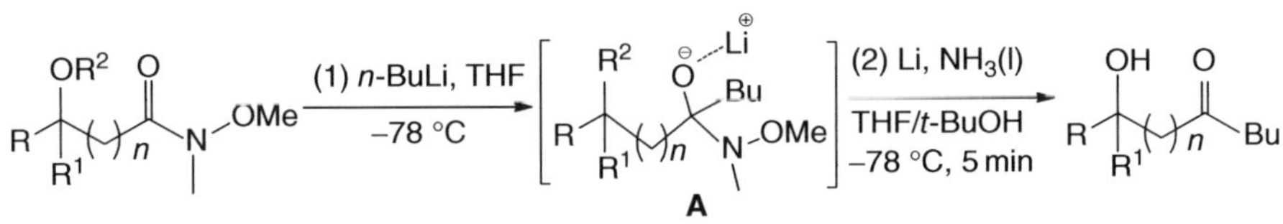

(8.4)

In the different chemoselective reactions, the key factor is the magnitude of their transition state (TS) barrier. The smaller the barrier, the stronger the competition ability, and the more the major product that is formed. For example, in addition reactions of an epoxy compound with amine, the aliphatic amine would quickly react with the epoxy derivative to afford the major product (Eq. (8.5)) [6]. The major product has the anti structure.

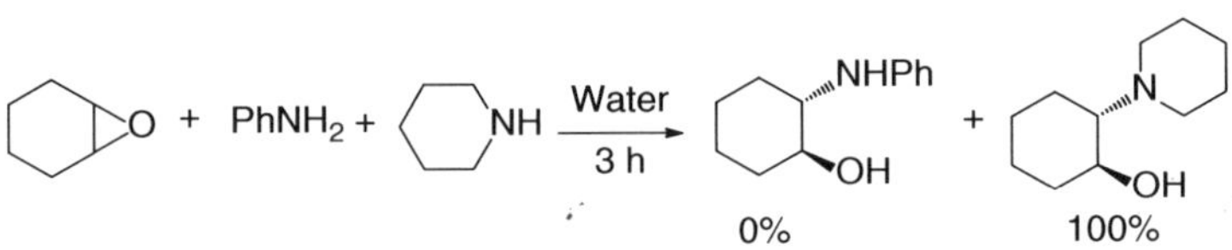

(8.5)

In addition to aldehyde using chloride, a new stereogenic center is formed. Because of the reaction of Li with O atom to afford enol-Li structure (Scheme 8.2), and then Li chelating with the S atom, the *anti*/*syn* ratio could reach 13 : 1 with a yield of 88% [7].

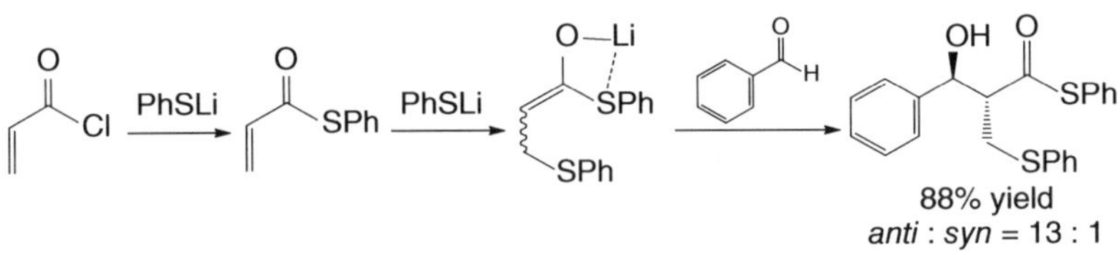

Scheme 8.2 Addition of Li–S complex to benzaldehyde.

Apart from the addition to >C=O, there are many other chemoselective additions to different C=C bonds. Asymmetric [3 + 2] annulation of Morita–Baylis–Hillman carbonates of Isatins with propargyl sulfones could afford spirocyclic 2-oxindoles with an unusual cyclopentadiene moiety (Eq. (8.6)) [8]. This addition could be catalyzed by a chiral catalyst **1**, resulting in outstanding ee values (up to >99%).

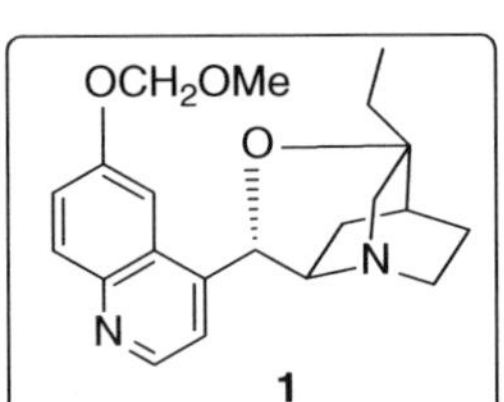

1

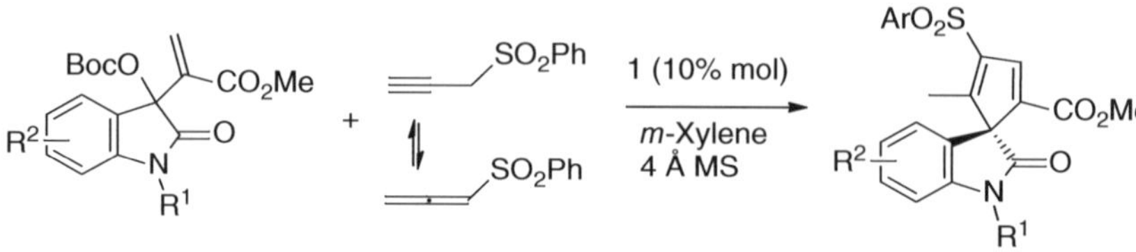

(8.6)

The effect of different substituents on the reactivity was investigated. When R^2 was 5-OMe, the ee value was over 99%. A strong electron-withdrawing group such as F on C5 and C7 made the ee values to decrease to 95 and 91%, respectively.

The reaction proceeds via formal dipolar cycloaddition of *in situ* generated allylic *N*-ylide and allenyl sulfone followed by a C=C bond isomerization sequence, giving an efficient protocol to construct spirocyclic 2-oxindoles incorporating an unusual cyclopentadiene moiety. The plausible mechanism is illustrated in Scheme 8.3 [8]. Two ways were possible in the procedure, but only approach ***b*** via [3 + 2] cycloaddition could afford the correct product.

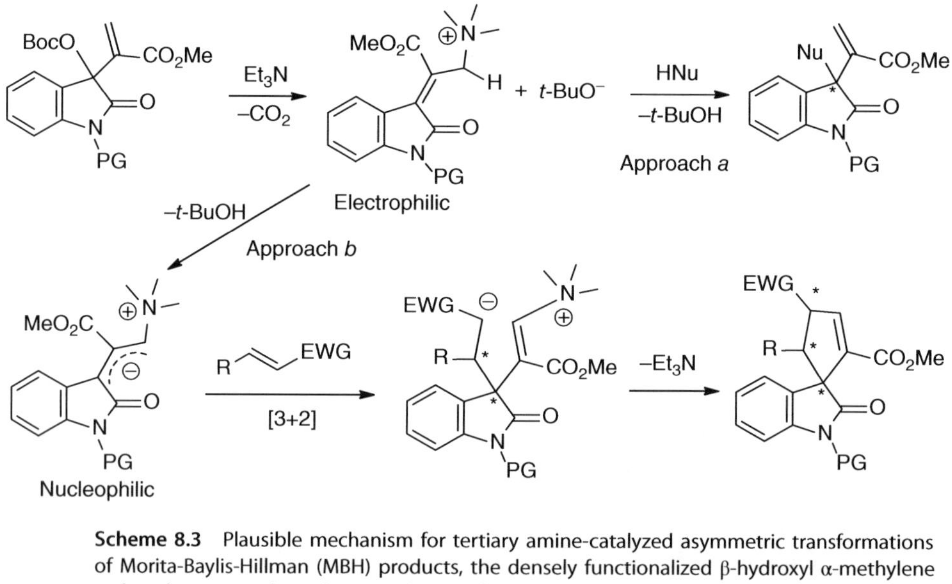

Scheme 8.3 Plausible mechanism for tertiary amine-catalyzed asymmetric transformations of Morita-Baylis-Hillman (MBH) products, the densely functionalized β-hydroxyl α-methylene carbonyl compounds. carbonates derived from Isatins [8].

8.2
Chemoselective Reduction

Any unsaturated functional groups could be a potential starting material used in reductions, such as ketone, imine, >C=C<, $-NO_2$, and others. The reduction reagents include H_2, hydride, and other proton resources such as $HSiCl_3$. In many cases, various metals or metallic salts are required as catalysts. When there are two or more functional groups in a compound, one functional group could be reduced chemoselectively using suitable reagents.

Specifically, hydrogenation using H_2 is very general; it requires the participation of transition metals such as Pt, Pd, Ni, and Rh. The metals are adsorbed on the surface of a support such as active C, or they are prepared in a very small particle size to increase the surface area. H_2 could reduce many unsaturated compounds. In some cases, H_2 is replaced by hydrazine, ammonium formate, or cyclohexene in reductions [9].

Hydrides can exist in many forms. The frequently used reagents include $LiAlH_4$ and $NaBH_4$. The latter has many advantages that $LiAlH_4$ does not have. For example, $NaBH_4$ can be used in a water system or in a basic solution. Alcohol or ether system has also been investigated as hydrogen sources [10]. In many cases, both reductants are treated with various alcohols to replace one or more hydrides to decrease their activity [11]. In the case of $NaBH_4$, Na can be also replaced by K, Ca, or Li; one to three hydrides can be replaced by an alkoxyl group [12].

Chemoselective reductions can be achieved by using catalysts or other reaction conditions [13, 14]. For example, hydrides from Si-containing compounds can be activated by chelating with transition-metal coordinates, F^-, or Lewis acids; they can afford a proton anion (hydride) in the reductions. Hydrides from silicane can reduce the aldehyde or ketone group to form the corresponding alcohol after it chelates with Li salt in many cases [15].

When two functional groups have very different activities, it is easy to carry out chemoselective reductions. For example, when a molecule contains an aldehyde and a ketone group, the reduction of aldehyde can be easily carried out, but the ketone carbonyl group cannot be reduced even by selecting a suitable condition. There are many reports on chemoselective reactions [16]. A typical reduction used $Zn(BH_4)_2$ to reduce the aldehyde group, but the ketone was not reduced [17] (Eq. (8.7)). This reduction gave a high yield of up to ~90%.

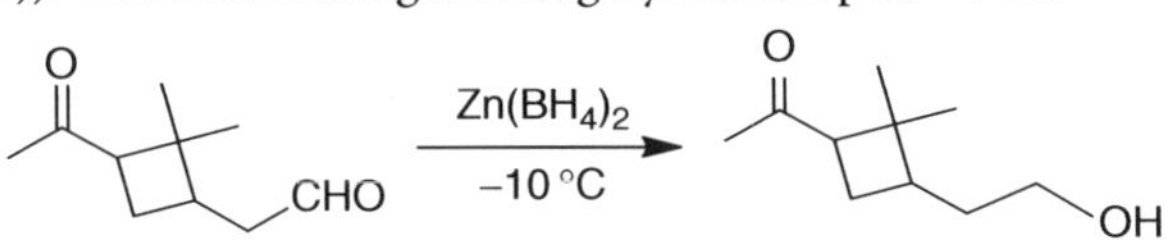

(8.7)

Among the various >C=O-containing compounds, aldehydes can be easily reduced using many reductants. For α,β-unsaturated aldehydes, their reductions generally afford the corresponding alcohols by using $B_{10}H_{14}/CeCl_3{-}7H_2O$/pyrrolidine. Only the aldehyde group could be reduced. Other groups, such as $-NO_2$ and >C=C<, could not be reduced [18a]. Table 8.4 summarizes some typical results.

Different orientation of the same functional group in a molecule, such as >C=O, would lead to different activity. Chemoselectivity can be used to reduce one of them. For example, a >C=O that is not on the ring is readily reduced using Al powder in an aqueous NaOH solution (Eq. (8.8)) to afford the corresponding alcohol with a yield of 76% [18b]. It is to be noted that it could not happen if no water was used.

Table 8.4 Chemoselective reduction of aldehydes using $B_{10}H_{14}/CeCl_3{-}7H_2O$/pyrrolidine [18a].

Entry	Substrate	Product	*t* (h)	Yield (%)
1	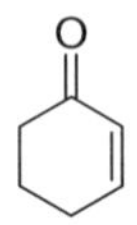	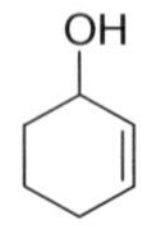	4	92
2	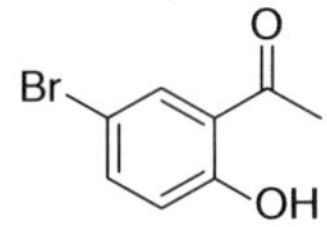	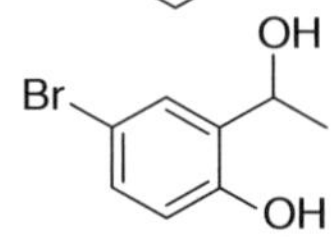	>4	96
3	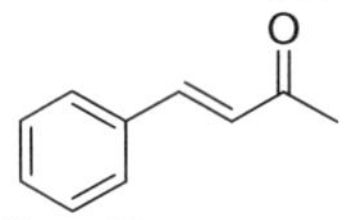	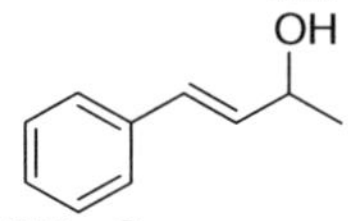	5	98
4	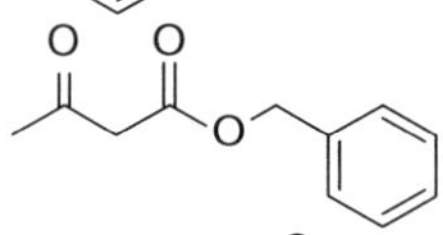	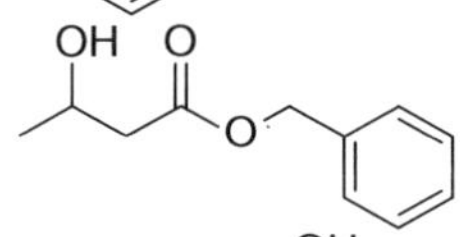	1.5	98
5	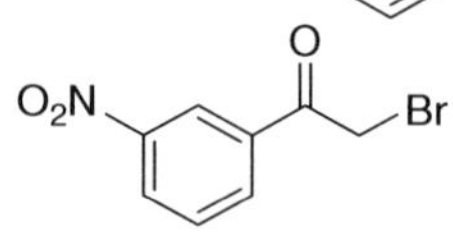	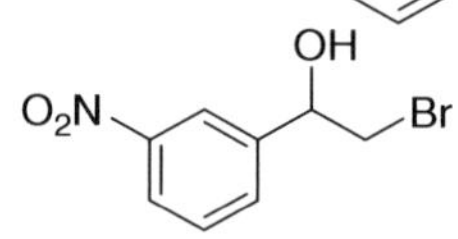	2	92

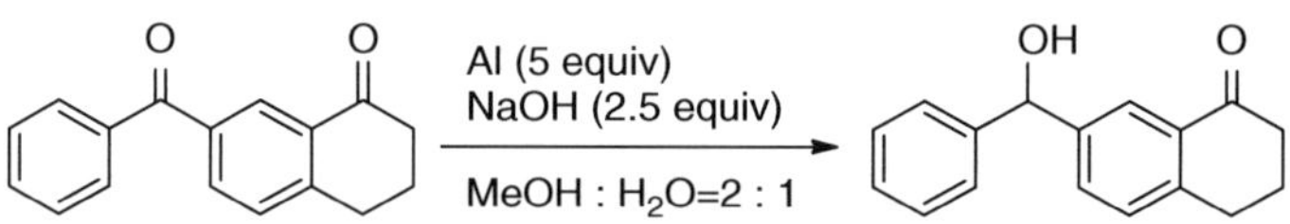

(8.8)

Under promotion of a chiral catalyst, >C=O could be reduced with a definite enantioselectivity. In the reduction-coupling reaction of 1-(4-acetylphenyl) propan-1-one catalyzed by the chiral ligand **2**, it afforded **A** and **B** in the ratio 18 : 1 (Eq. (8.9)) [18c]. The chemoselective reduction coupling took place on the acetyl carbon.

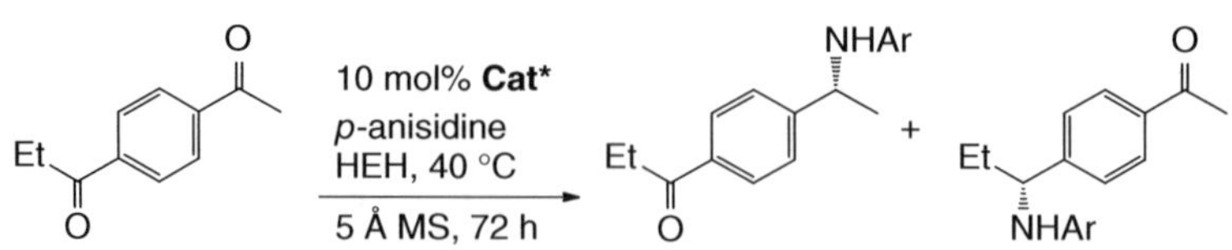

A **B**
(8.9)

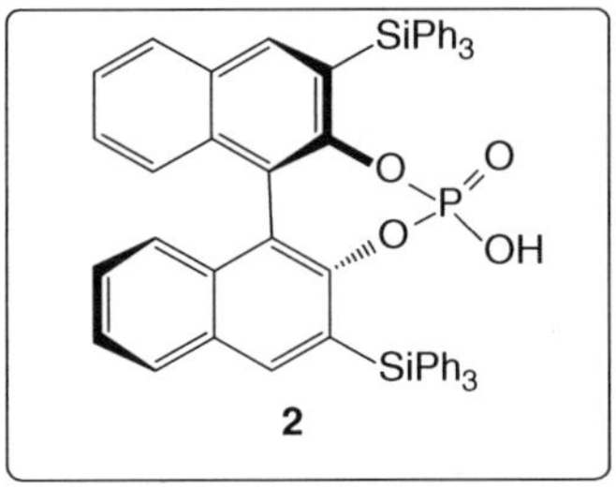

Iodotrichlorosilane (ITCS) reagents can be obtained from the reaction of $SiCl_4$ with NaI, which can reduce the C=C bond of α,β-unsaturated ketones to afford the corresponding carbonyl compounds at room temperature in acetonitrile in 1 h (Eq. (8.10)). No byproduct was formed [19]. Other reagents such as $Co_2(CO)_8$–H_2O can also chemoselectively reduce the C=C bonds of α,β-unsaturated aldehydes or ketones with high yield (Eq. (8.11)) [20].

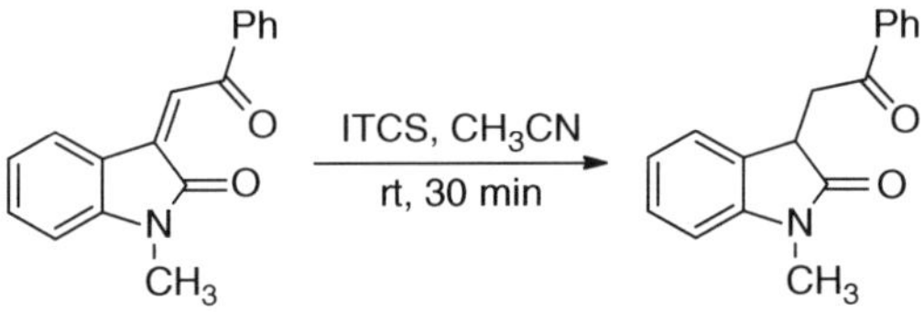

(8.10)

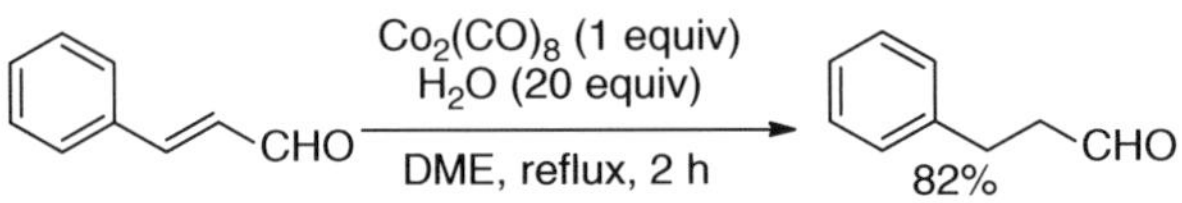

(8.11)

The difficulty to reduce a >C=O to the corresponding $-CH_2-$ remained for a long time. A valid method used for this reduction is Clemmensen reduction or Huang Min-lon reduction. Use of a hydride to reduce >C=O is difficult unless the –OH formed in the first step is easily removed. If there is a strong electron-donating group on the aromatic ring, this reduction can happen. Under trifluoroacetic acid (TFA) interaction, >C=O could be reduced into $-CH_2-$ by $NaBH_4$ (Eq. (8.12)) with a yield of 90% [21a].

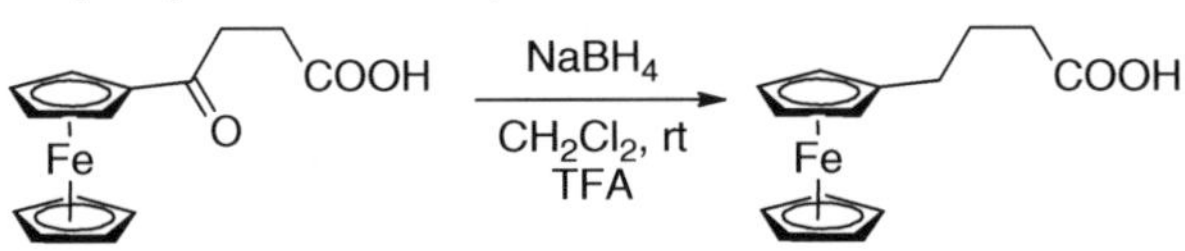

(8.12)

The carbonyl carbon of esters could not be easily reduced because of its inertness. In many cases, if the ester could be reduced, other functional groups may be reduced as well. To chemoselectively reduce $>CO_2R$ is possible in organic synthesis. The reduction system $LiBH_4$–MeOH–EtOH could selectively reduce $>CO_2R$ in the presence of $-NO_2$, –Cl, $-CO_2H$, and –CONHR with high yields (90–100%, Table 8.5) [21b].

The reagent BOP-$NaBH_4$ could reduce acids to the corresponding alcohols; groups such as $-NO_2$, $-CO_2R$, –CN, and $-N_3$ would not be affected in these

Table 8.5 Chemoselective reduction of esters [21b].

Entry	Substrates (A + B)		Products C (yield in %)[a]	Recovered B (%)[b]
	A	B		
1	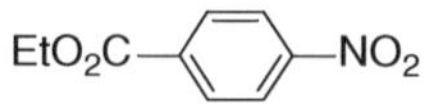	—	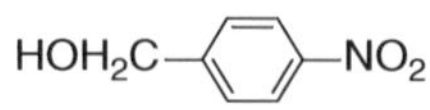 (90)	—
2	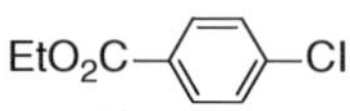	—	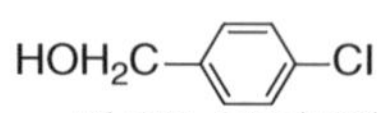 (90)	—
3	$PhCO_2Et$	$PhCO_2NH_2$	$PhCH_2OH$ (100)	89
4	$PhCO_2Me$	n-$C_9H_{19}CO_2H$	$PhCH_2OH$ (92)	98

a) Yield in percent.
b) Recovered material B in percent.

reductions (Eqs. (8.13) and (8.14)) [22]. Use of the reagent $NaBH_4$–I_2 could reduce an acid to the corresponding alcohols (Eq. (8.15)) [23].

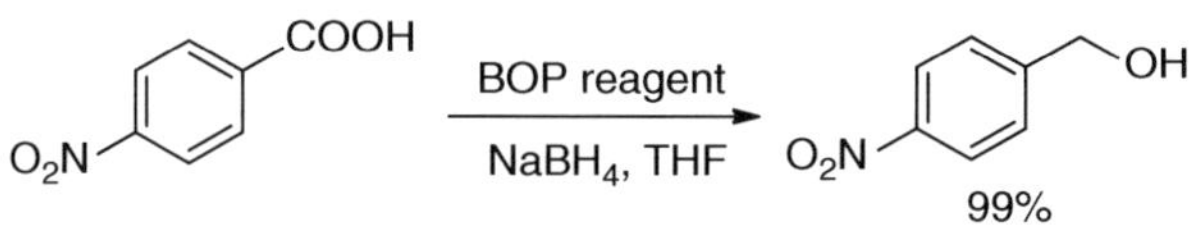

(8.13)

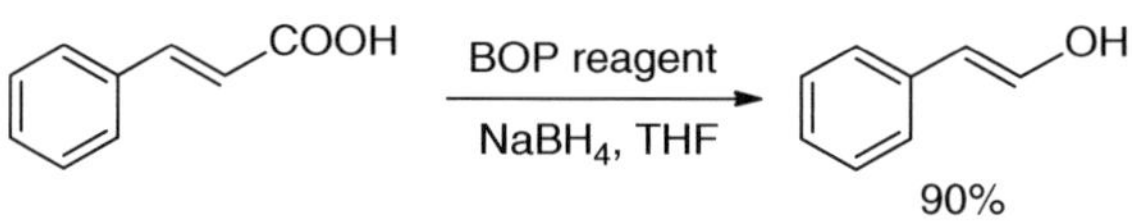

(8.14)

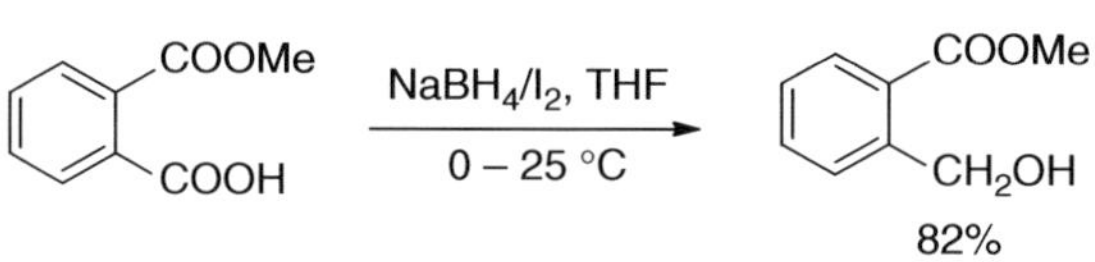

(8.15)

The reagent $NaBH_4$–I_2 could reduce $-CO_2H$ of amino acids without any effect of the t-butoxycarbonyl (Boc) group on the N atom [24]. It is a safe and valid reductant, especially for the synthesis of chiral amino alcohols on a large scale.

Generally, there are two or more carboxy groups in a natural compound. To chemoselectively reduce one of them is a valid trial. However, it is not readily available to directly study the chemoselective reduction of natural products. The use of models is relatively reliable and simple. For example, the following models were used for the study [25].

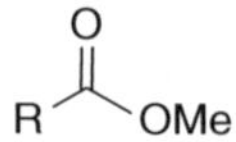

R = $-CH_2OH$	R = $-CH_2NHMe$	R = $-CH_2CH=CH_2$	R = $-CH_2Cl$
R = $-CH(Me)OH$	R = $-CHMe_2$	R = $-Ph$	R = $-CH_2Br$
R = $-CH_2NH_2$	R = $-H$	R = $-CH_2Ph$	
R = $-CH(Me)NH_2$	R = $-CH=CH_2$	R = $-CH_2F$	

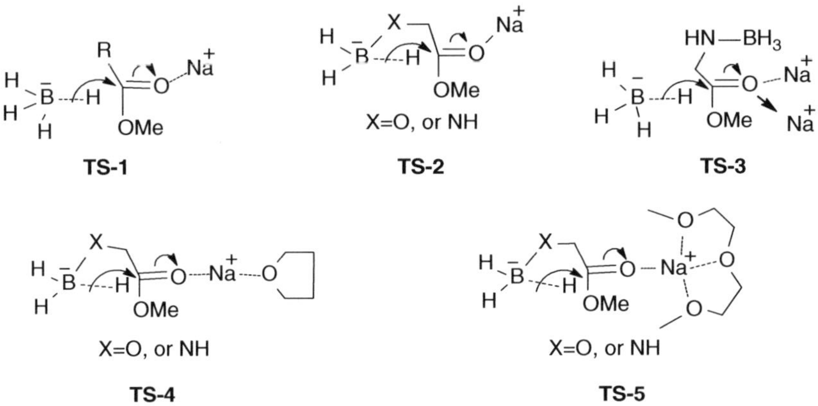

Figure 8.1 Possible TS structures used in TS calculations [25].

Considering the complexity of natural products and their large molecular weight, TS barrier computation using density function theory (DFT)/6-311 + G(d,p) is very difficult. The Hatree–Fock (HF) theory at the 6-31G(d,p) level is a reasonable choice. The TS structures **TS-1–TS-5** were used for all possible TS barrier computations (Figure 8.1). The intrinsic reaction coordinate (IRC) was calculated for checking whether the TS structure computed was correct or not.

All TS structures were investigated, and the computed TS barriers are listed in Table 8.6. During TS barrier calculations for halogenated esters (R = CH_2F, CH_2Cl, CH_2Br, respectively), it was necessary to consider the interaction of the halide with the cation Na^+ in TS structure. There are three possible TS structures (I, II, and III, Figure 8.2). The calculated barriers were 21.8, 20.0, and 19.2 kcal mol^{-1}, respectively, for the three esters (Table 8.7, entries 12–14). The activation energy clearly shows that strong interaction takes place between the halide and the cation Na^+. The barriers via TS general structures II and III have larger activation energies than that via the TS structure I.

In traditional organic theory, the F atom has the largest electronegativity among F, Cl, and Br atoms. This leads to the carbon connected to the F atom having most positive charge among those connected to Br, Cl or I atom. Therefore, that a hydride attacking the carbon connected to F would have the smallest activation energy in sodium borohydride reductions, generally.

However, an interesting discovery is that the barrier predicted for the α-F ester has the highest activation energy in the gas phase. The second highest barrier is for the α-Cl ester, and the smallest barrier happens in reduction of α-Br ester with sodium borohydride. When solvent effect was considered in barrier calculations, it was found that the highest activation energy was still for the reduction of the α-F ester with sodium borohydride. The results are listed in Table 8.6. This conclusion is in conflict with the traditional organic theory analyzed above using electronegativity data.

Table 8.6 TS barriers for all TS models computed at the HF/6-31G(d) level [25].

Entry	R=	χ of R[a]	$\Delta E^{\neq}$[b] (in gas)	$\Delta E^{\neq}$ (in THF)	$\Delta(\Delta E^{\neq})$ (in gas)[c]	$\Delta(\Delta E^{\neq})$ (in THF)	>C(=O)---H$^-$ in TS (Å)	>C=O---Na$^+$ (Å) in TS
1	$-CH_2OH$	2.59	24.9[d]	23.2[d]	6.5	2.4	1.314	2.134
2	$-CH(Me)OH$	2.59	23.4[d]	21.4[d]	5.0	0.6	1.308	2.138
3	$-CH_2NH_2$	2.54	31.7, 32.3[d]	33.0, 32.1[d]	13.3, 13.9[d]	12.2, 11.3[d]	1.155, 1.404[d]	2.076, 2.131[d]
4	$-CH(Me)NH_2$	2.54	–[e], 31.2[d]	–, 30.8[d]	–, 12.8[d]	–, 10.0[d]	–, 1.416[d]	–, 2.132[d]
5	$-CH_2NHMe$	2.54	31.6, 32.3[d]	32.7, 33.4[d]	13.2, 13.9[d],	11.9, 12.6[d]	1.159, 1.444[d]	2.075, 2.126[d]
6	$-CH_2NMe_2$	2.54	34.3	36.7	15.9	15.9	1.164	2.082
7	$-H$	2.1	18.4	20.8	0.0	0.0	1.275	2.141
8	$-CH=CH_2$	2.78	33.6	34.3	15.2	13.5	1.219	2.145
9	$-CH_2CH=CH_2$	2.51	30.2	27.8	11.8	7.0	1.235	2.149
10	$-Ph$	2.78	44.8	42.1	26.4	21.3	1.129	2.012
11	$-CH_2-Ph$	2.51	31.8	36.4	13.4	15.6	1.217	2.117
12	$-CH_2F$	2.64	21.8 (**I**)	24.5 (**I**)	3.4 (**I**)	3.7 (**I**)	1.184 (**I**)	2.105 (**I**)
13	$-CH_2Cl$	2.54	20.0 (**I**)	21.8 (**I**)	1.6 (**I**)	1.0 (**I**)	1.230 (**I**)	2.103 (**I**)
14	$-CH_2Br$	2.50	19.2 (**I**)	22.1 (**I**)	0.8 (**I**)	1.3 (**I**)	1.260 (**I**)	2.118 (**I**)

a) Regarding electronegativity, see Ref. [26].
b) Unit: kcal mol^{-1}. **TS-1** was used until noticed.
c) Unit: kcal mol^{-1}. From entry 1–12, the lowest barrier of 9.3 kcal mol^{-1} was used to compute $\Delta(\Delta E)^{\neq}$ in the gas phase. Minimum of 22.0 kcal mol^{-1} was used to compute $\Delta(\Delta E)^{\neq}$ in THF.
d) **TS-2** was used.
e) **TS-1** model was not obtained in computations.

Table 8.7 Chemoselective reduction of nitrines by Fe/NiCl$_2$6H$_2$O–THF reductant [27].

Entry	Substrates	Products	*t* (min)	Yield (%)
1	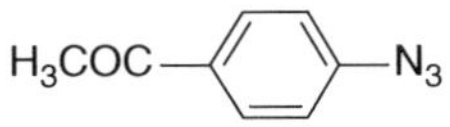	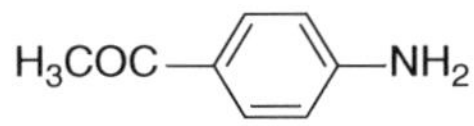	50	80
2		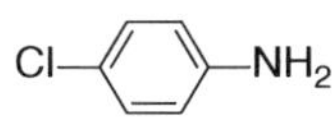	50	85
3	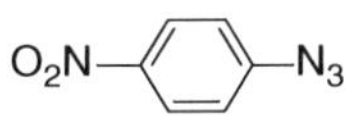	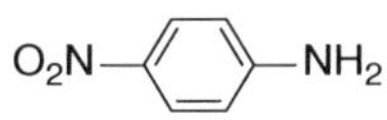	30	85
4	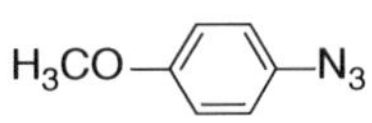	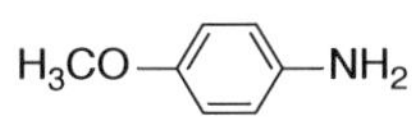	30	90
5	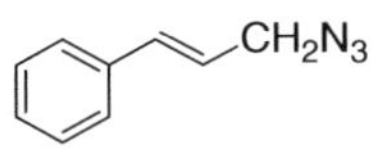	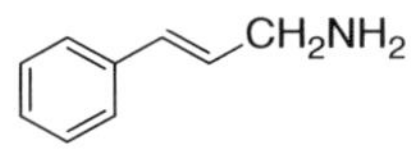	50	78

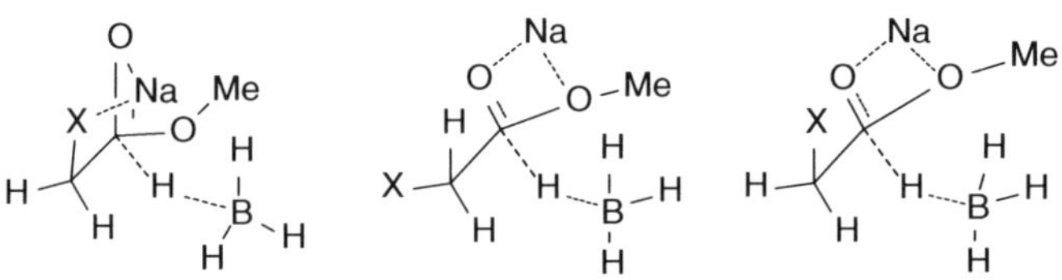

I II III

Figure 8.2 Three possible TS structures considering interaction of halide with Na$^+$ anion [25].

An experimental study of the onset temperature for the same compounds was performed in diglyme using sodium borohydride. The measured onset temperatures were −15 °C (α-F ester),−23 °C (α-Cl ester), and −26 °C (α-Br ester). This sequence is the same as that of barriers predicted in the gas phase.

After analyzing the relationship between the onset reaction temperatures and the barriers predicted at the HF/6-31G(d) level for the nine reductions, a linear relationship could be concluded as in Eq. (8.16) and illustrated in Figure 8.3.

$$\Delta E^{\neq}_{\mathrm{THF}} = -\frac{11\,383.0}{T} + 66.6\ (R^2 = 0.975) \tag{8.16}$$

Therefore, it is possible to use this formula to predict the onset temperature for the selected carboxyl group after its barrier is obtained using the same methods (HF/6-31G(d,p)). If the predicted onset temperatures are large enough for two carboxyl groups, such as above 15 °C, the chemoselective reduction for the carboxyl at the lower onset temperature should be possible by carefully raising experimental temperature and checking with TLC tests.

Historically, the different chemoselectivity observed in the reductions of primary amides, secondary amides, and tertiary amides with sodium borohydride has been a phenomenon. Namely, a primary amide reacted with $NaBH_4$ to afford nitrile instead of amine; the secondary amide did not react with $NaBH_4$; and only

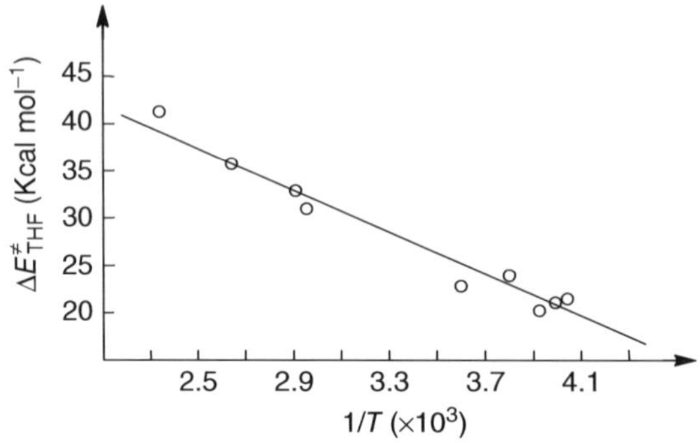

Figure 8.3 The relationship between the onset temperature and barriers predicted by HF/6-31G(d,p) method [25].

the tertiary amide could be reduced to the corresponding amines (Scheme 8.4) [28]. Although this is an old question in chemistry, it is still important and basic to most experimental chemists. $NaBH_4$ reductions are so widely used that an understanding of the specific mechanisms that operate during each type of functional group reaction clearly represents an important goal.

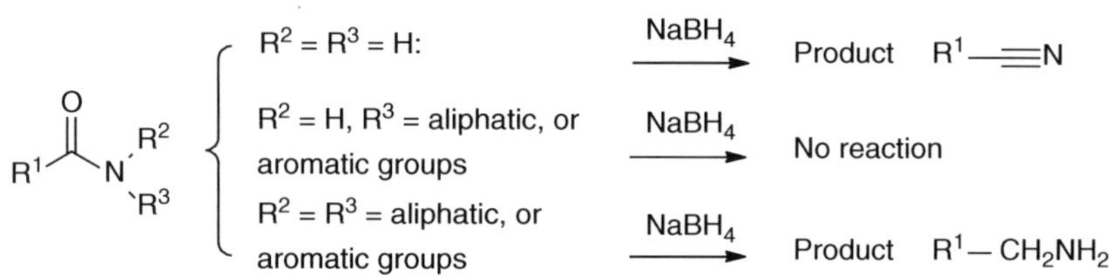

Scheme 8.4 Different chemoselectivities recorded in amide reduction with $NaBH_4$ [28].

This study of different chemoselectivities in amide reductions was performed using models via quantum theory. Calculations were performed both at the B3LYP/6-31G(d,p) and HF/6-31G(d,p) levels of theory, and single-point energy was computed at the B3LYP/6-31++G(d,p) level for energy computations. The calculation results clearly revealed the secrets [28].

In the reduction of $NaBH_4$ with acetamide, it would form a complex **a.** To reduce **a** to the corresponding amine, it needs to overcome a barrier of 38.8 kcal mol^{-1}. However, 1 mol of H_2 formed just needed to overcome a barrier of 31.3 kcal mol^{-1} via **TS-1.** This is different from Newman's study, in which 2 mol of hydrogen was proposed to be formed per 1 mol of primary amide [29]. Because the formation of H_2 is irreversible, the followed procedures must have two cases: intermediate **b** should be reduced into the corresponding compound, or it should stay at this stage. This will depend on the corresponding TS activation energy. This is illustrated in Scheme 8.5.

Intermediate **b** could undergo tautomerization to **c** via **TS-2**, or could be reduced to an amine. The barrier of isomerization is high, up to 46.0 kcal mol^{-1}, but to form an amine needs to overcome an energy barrier of 48.5 kcal mol^{-1}.

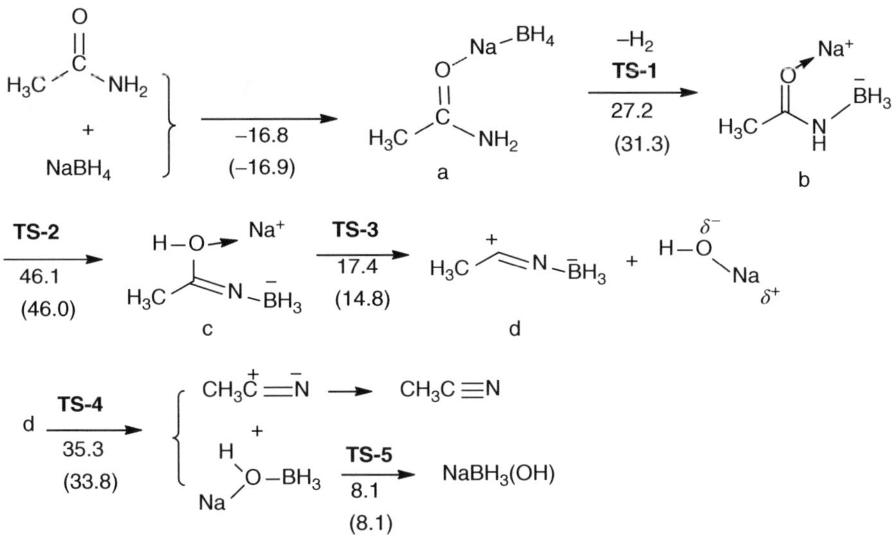

Scheme 8.5 Procedure to form nitrile from acetamide [28].

Thus, it would form the intermediate **c**, and then it quickly forms the nitrile via **TS-3** and **TS-4**, respectively, since the barriers after the isomerization are very small. The tautomerization should have the lowest rate. In the experiments, to finish the reaction it took ~2 h at ~154 °C (boiling point of diglyme).

Similarly, *N*-methylacetamide would undergo the same reaction to afford a molecule of H_2 and the intermediate **b′**. However, the formed **b′** has a methyl group on the N atom. It is almost impossible to convert it into another intermediate **c′**, as illustrated below. Thus, it has to wait for reduction to afford an amine. Since there is a methyl on the N atom, it would require that the TS barrier must be higher than 48.5 kcal mol^{-1}, which is the barrier to reduce **b** into the corresponding amine. It is very difficult to convert **b′** to **c′** using a solvent such as diglyme (Eq. (8.17)). A solvent with a much higher boiling point, such as triglyme, may have to be chosen for the reductions.

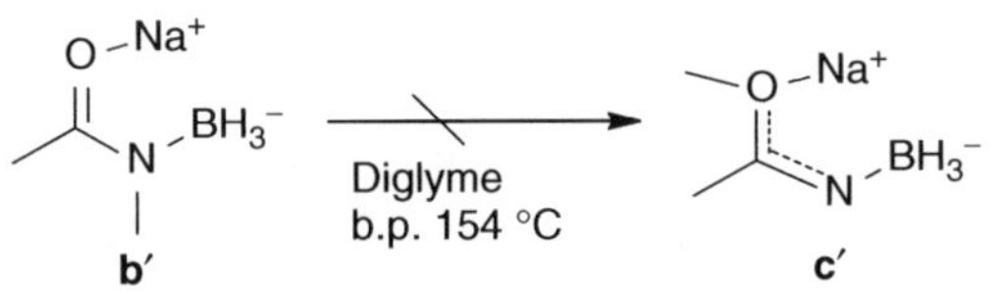

(8.17)

A tertiary amide structure has no proton on the N atom, so it cannot react with $NaBH_4$ to afford H_2. Therefore, the hydride would directly attack the carboxyl carbon. If the energy barrier is not high enough, the amide should be reduced into amine; however, if the barrier is too high to be overcome in the reduction, the amide will not be reduced. Thus, *N*,*N*-dimethylformamide was used as a model in TS calculations, and the predicted reduction activation energy was 36.3 kcal mol^{-1}. When *N*,*N*-dimethylacetamide was used in calculation, it had a barrier of 42.4 kcal mol^{-1}. Since the barrier of 46.0 kcal mol^{-1} for isomerization of **b** to **c** could be overcome in diglyme, the two barriers are not high enough for their reduction in diglyme to afford the corresponding amines. The results are as expected and agree with the experiments.

Because of the difficulty in the reduction of amides under mild conditions, more reductants were searched. $NaBH_4$–TFA reagents could effectively reduce amides or lactams to the corresponding amines [30]. Like the synthesis of the side chain of delphinium and aconitum, $NaBH_4$ could be used to cleave the N–C bond of CF_3CONH– (Eq. (8.18)) [31].

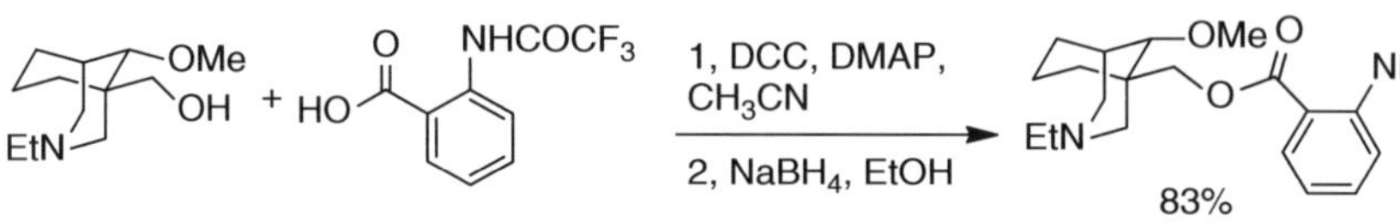

(8.18)

The reagent Ph_2SiH_2 with catalytic amount of $RhH(CO)(PPh_3)_3$ could selectively reduce lactams without any effect on the ester groups. This is an effective way for the reduction of a new amino acid, and the yield could reach up to 94% [32] (Scheme 8.6).

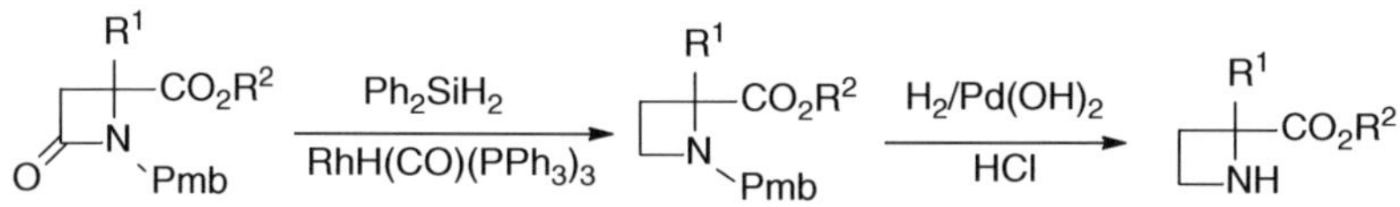

Scheme 8.6 Two reaction steps to the new amino acids [32].

Reduction of pyrimidine could be well used in the synthesis of β-amino acids. However, the use of reductants $LiAlH_4$ or DIBAL-H afforded the corresponding products with only a yield of 40% [33]. The use of BH_3–THF could increase the reaction yield up to 87% (Eq. (8.19)) [34].

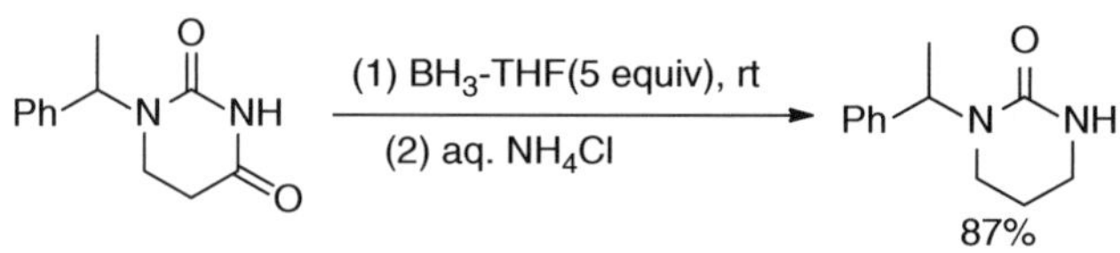

(8.19)

For a lactam with *N*-Boc, the reductant $LiEt_3BH/Et_3SiH/Et_2O–BF_3$ is a good choice; the lactam carbonyl carbon could be chemoselectively reduced. However, the groups like ester, >C=C<, –CN would not be affected (Eq. (8.20)) [35]. Another reductant $NaBH_4/I_2$ could also chemoselectively reduce the lactam but did not reduce the ester group under reflux in THF [36].

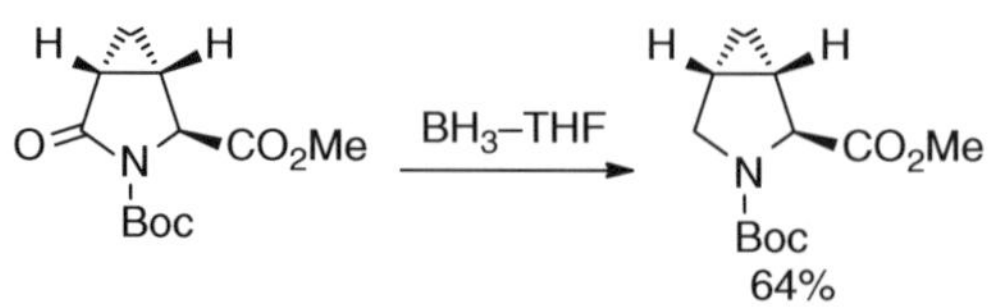

(8.20)

Indole-containing derivatives could be reduced into the corresponding indoline analogs under an acidic reaction condition (TFA or AcOH) [37]. However, an exceptional case is the chemoselective reduction of the >C=C< bond of tetrahydropyridine instead of the C=C of indole. $NaBH_4$/AcOH could not reduce the halide or $-NO_2$ on the ring of indole (Eq. (8.21)) [38].

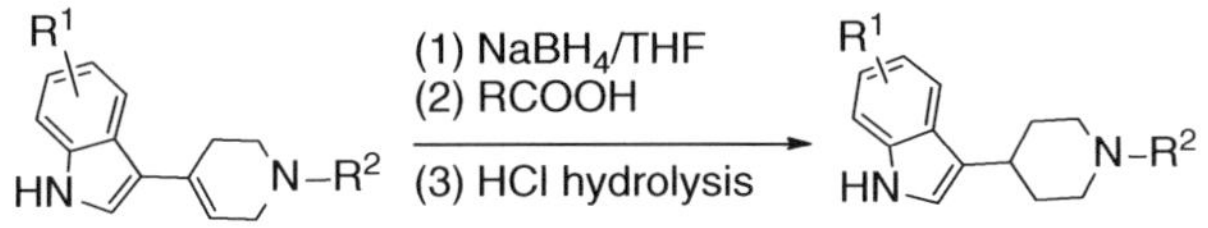

(8.21)

Reduction of nitrines could afford the corresponding amines [39]. There are many efficient methods for asymmetric or regioselective synthesis [40]. The chemoselective reduction of nitrines is focused on B-containing reagents because $LiAlH_4$ has low chemoselectivity in the reductions [41]. Reductants $BHCl_2-SMe_2$, BH_3-SMe_2, and BF_3-OEt_2 were well investigated [42]. Reductant $BHCl_2-SMe_2$ could selectively reduce the nitrine group without any effect on the ketone, ester, or $-NO_2$ structures. On the other hand, Fe/$NiCl_2-6H_2O$–THF could reduce aromatic nitrine with a yield of 80–90% (Table 8.7) [27].

Nitrine could be converted into *N*-formamide derivatives by using Pd/C and $HCOONH_4$ via phase-transfer catalysis refluxed in acetonitrile. The reaction could be performed under mild conditions with high yield (Eq. (8.22)) [43].

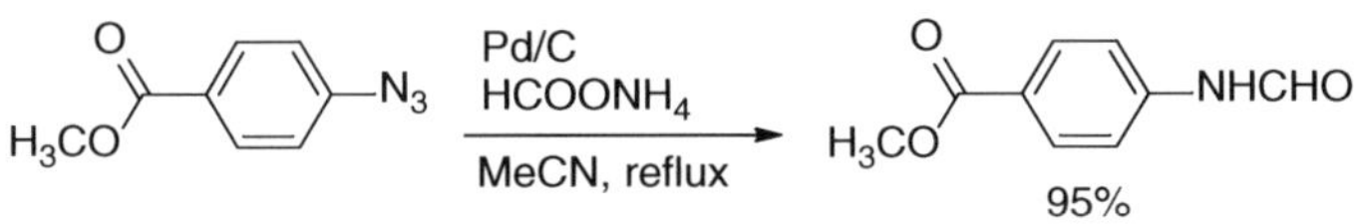

(8.22)

The reductant $InCl_3$/$NaBH_4$ could selectively reduce >C=C< bonds; but the carbonyl groups of ester or ketone or –CN could not be reduced (Eq. (8.23)) [44]. Reagents $NaBH_4$/$BiCl_3$ could reduce >C=C< of α,β-unsaturated ester [45] or amides [46], while other >C=C< bonds were not be affected. The reaction condition is quite mild (Eq. (8.24)).

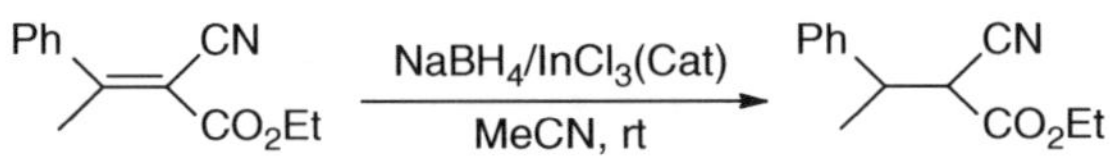

(8.23)

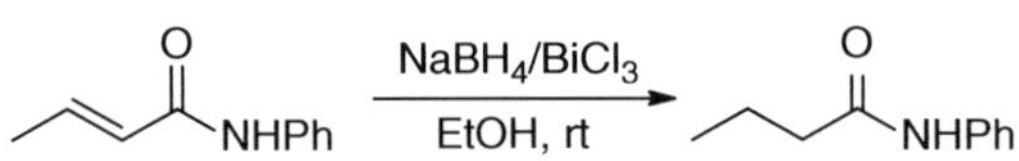

(8.24)

It is possible to chemoselectively synthesize the (*R*)- or (*S*)-enantiomer in the presence of a chiral catalyst or chiral auxiliary in some reductions. For example, Et_2Zn could be used as a reducing reagent in the reduction of sulfones [47]. The C=C bonds could be chemoselectively reduced to form a new stereogenic center with 98% ee when catalyst **3** was used (Eq. (8.25)) [47].

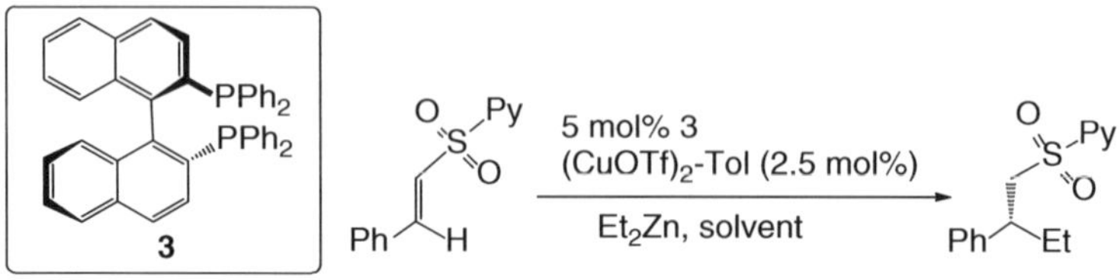

(8.25)

A frequently used model reduction for C=C bonds in an α,β-unsaturated compound involves α,β-unsaturated ketones (Eq. (8.26)). In the reduction, it was found that $NaCO_2H$ could serve as a proton donor [48].

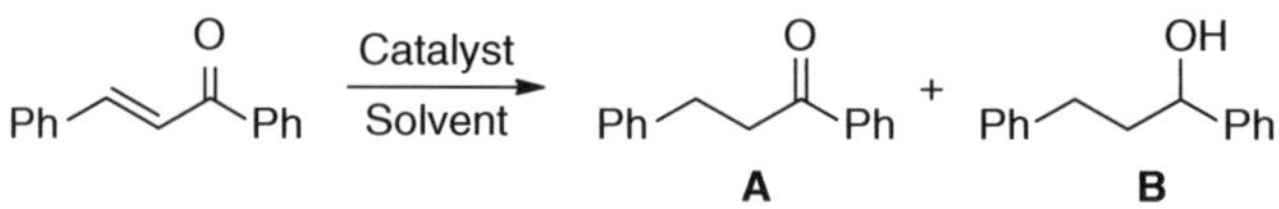

(8.26)

A catalyst could be fixed on a polymer (**4**). This conversion is dependent on the reaction temperature and time. Water is a good hydrogen donor. Under optimized reaction conditions of 12 h at 100 °C, the ratio of **A** : **B** could reach from 97 : 3 to 99 : 1 (Eq. (8.26)) [48]. It could be extended to other substrates, and high conversion (up to 99%) and chemoselectivity (83 : 17 – 100 : 0) were recorded.

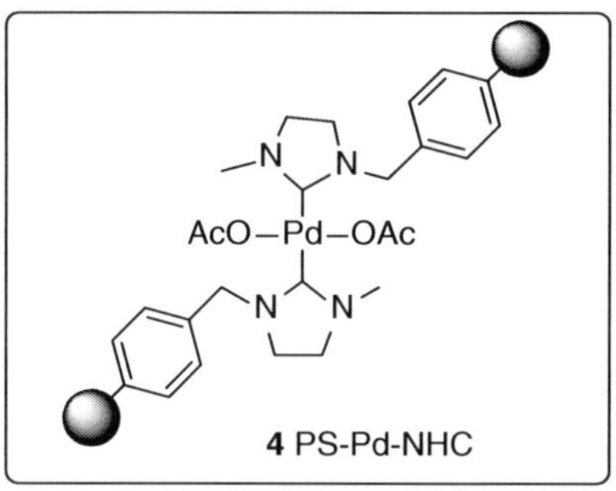

Reduction of nitro-containing compounds could form the corresponding aromatic amines. This is an important organic conversion. After investigation of porphyrin/$NaBH_4$ and metallic porphyrin/$NaBH_4$, it was found that phthalocyanatoiron(II)/$NaBH_4$ (PcFe(II)/$NaBH_4$) had the ability to chemoselectively reduce $-NO_2$ of 4-substituted nitrobenzene. By addition of 2-Br ethanol, the yield could be increased. Other functional groups such as –CN, >CON–, or ether were not affected (Eq. (8.27)) [14].

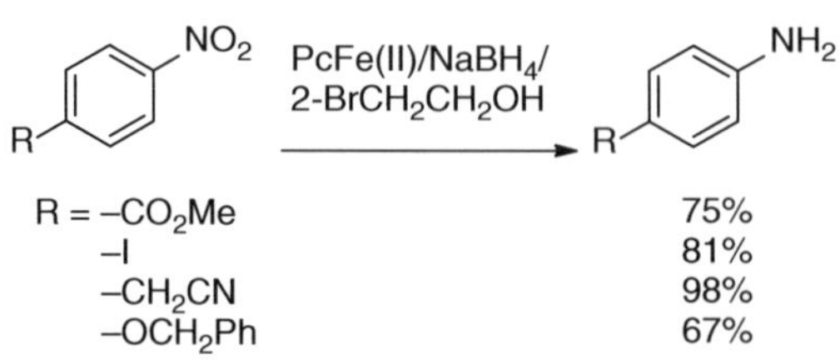

(8.27)

In Rh/C, $Ni(NO_3)_2{\cdot}6H_2O$, or $Fe(OAc)_2$ catalytic systems, aromatic nitro compounds could be reduced into the corresponding amine without the reduction of ether or halide [49]. In a methanol solution, $Me_3N{-}BH_3$, $Pd(OH)_2/C$ could chemoselectively catalyze the reduction of aromatic nitro compounds with high yield (99%) (Eq. (8.28)) [50].

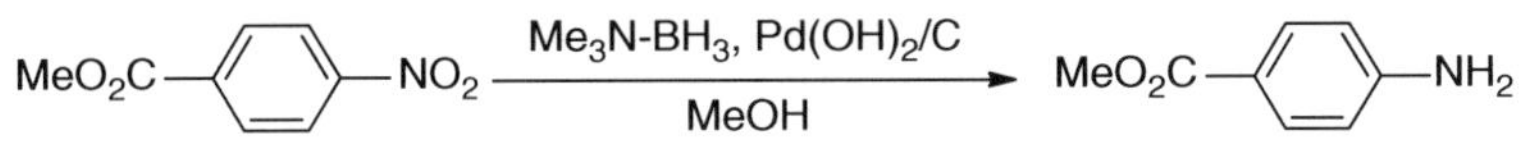

(8.28)

A novel core–shell AgNPs–CeO_2 nanocomposite was designed that consisted of core AgNPs (10 nm in diameter) and a shell assembled with spherical CeO_2 NPs (3–5 nm in diameter). The shell had nano-sized spaces between the CeO_2 NPs, which permitted the access of the reactants to the active center of the AgNPs.

Table 8.8 Chemoselective reduction of nitro compounds bearing >C=C< bonds catalyzed by AgNPs–CeO_2 [51].[a)]

Entry	Substrate	Product	*t* (h)	Yield [b)] (%)
1	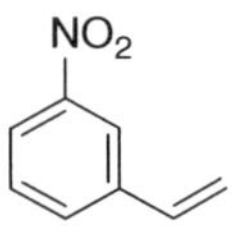	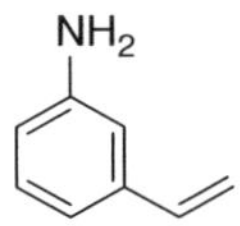	6	98
2	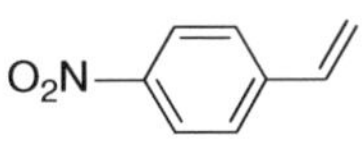	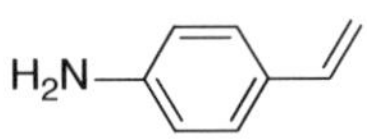	6	98
3[c)]	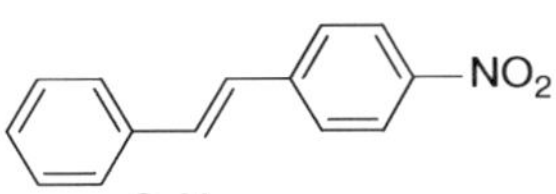	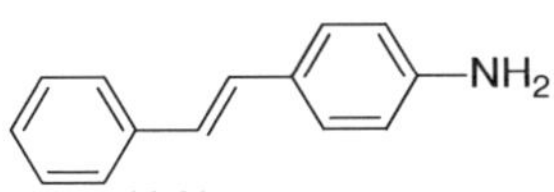	24	97
4[c)]	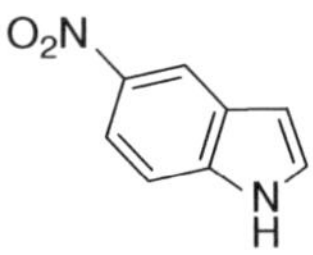	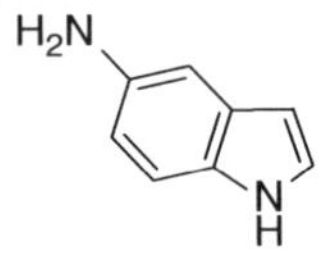	24	95
5	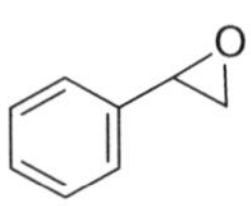	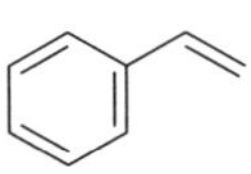	6	97
6	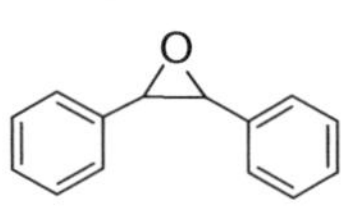	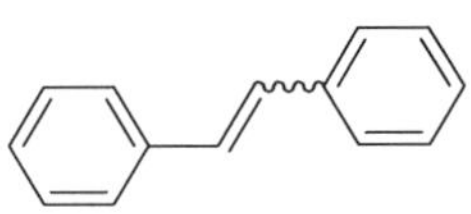	12	96 (*E*:*Z* = 3 : 1)
7	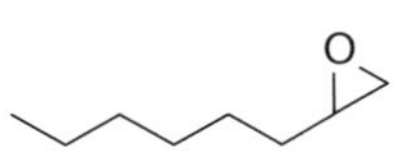	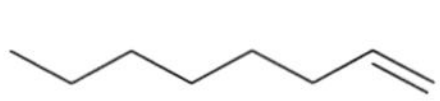	24	95

a) Reaction conditions for reduction of $-NO_2$: AgNPs–CeO_2 (25 mg), substrate (0.5 mmol), dodecane (5 ml), H_2 (6 atm), 110 °C.
b) Determined by GC and LC using an internal standard.
c) Determined by GC and LC using an internal standard.

The catalyst could completely promote the conversion of both nitrostyrenes to aminostyrenes (Table 8.8, entries 1–4) and epoxides to alkenes (entries 5–7) using H_2. However, the >C=C< bond was not reduced [51]. Table 8.8 summarizes the results.

There are several works involving the reduction of triple bonds to (*E*) or (*Z*) alkene structures. The reagent $[Cp^*Ru(MeCN)_3]PF_6$ could catalyze compounds with triple bonds to afford (*E*)-alkenes in the presence of CuI (Scheme 8.7). It does not affect the functional groups R^1 or R^2,which could be an ester, ketone, halide, or other groups [52].

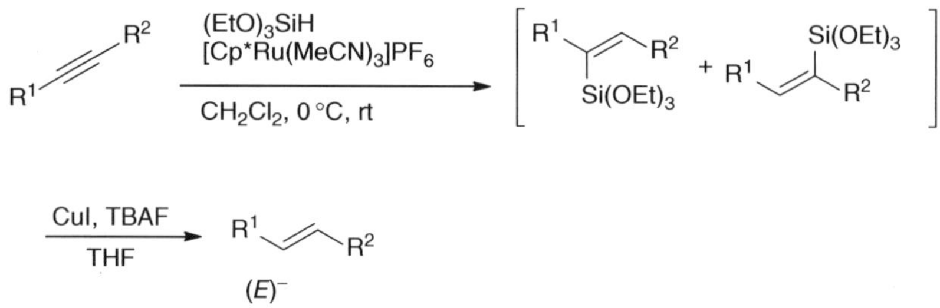

Scheme 8.7 Conversion of alkynes to the corresponding (*E*)-alkenes [52].

Reduction of imines (C=N) that were derived from the corresponding amine and aldehyde (or ketone) has been well investigated. This C=N could be easily reduced. Therefore, methods to chemoselectively reduce this group are also readily available. The use of $Zn(BH_4)_2$ (Eq. (8.29)) [53] or $NaBH_4/BiCl_3$ (Eq. (8.30)) [54] could well afford the corresponding amines with high yields (80–90%). Substituents R, R^1, and R^2 could be aliphatic or aromatic groups.

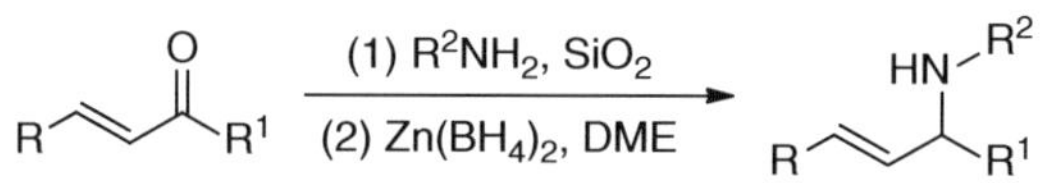

(8.29)

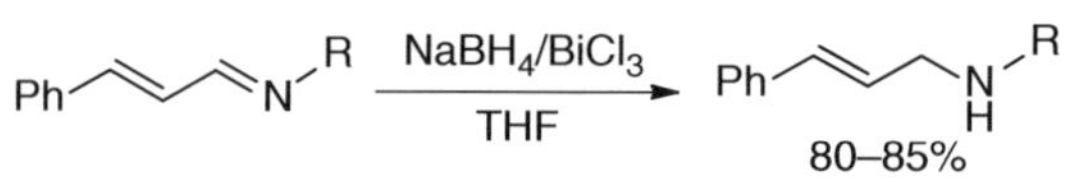

(8.30)

The reduction of sulfoxides to the corresponding sulfoethers is a standard reaction. Many reductions have been tested for the sulfoxides. However, because of serious side reactions, low yields, or low chemoselectivity, the methods are limited in their applications [55]. A series of studies showed that the complex $Mo(VI)O_2$ might catalyze sulfoxides to the corresponding sulfides [56]. For example, MoO_2Cl_2 was found to be effective in reductions [57]. Now, $PhSiH_3/MoO_2Cl_2$ could effect chemoselective reductions of sulfoxides, but the functional groups >C=C<, ester, and halide in a molecule could not be reduced (Table 8.9) [57].

Table 8.9 Reduction of sulfoxides to sulfoethers by the complex $PhSiH_3/MoO_2Cl_2$ [57].

Entry	Substrates	Products	*t* (h)	Yield (%)
1	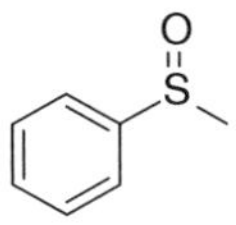	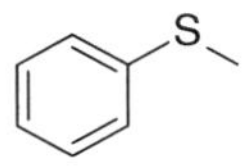	2	97
2	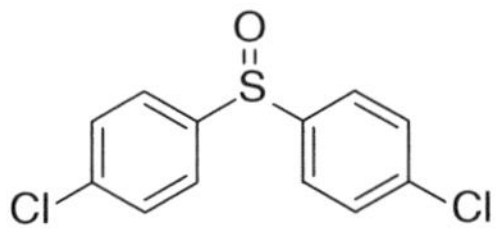	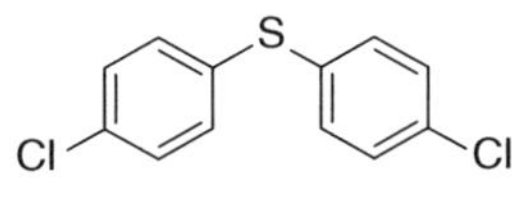	2	96
3	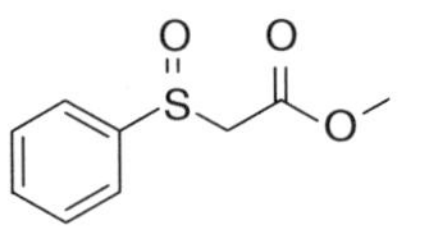	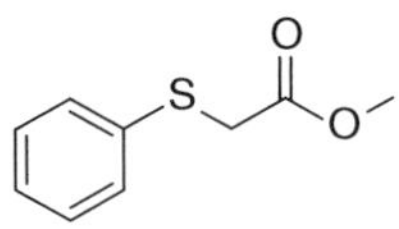	2.5	96
4	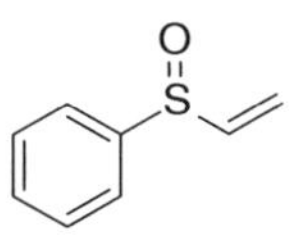	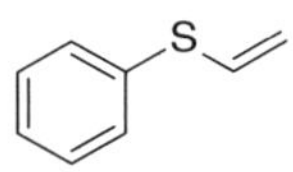	2	92

In the synthesis of the alkaloid peripentadenia, Ni/Al alloy could reduce the –CN group in a basic ethanol solution [58]. This reaction afforded an unstable amine with a yield of 94% (Scheme 8.8). When other reductants such as BH_3–SMe_2 or $LiAlH_4$ were used, only unseparated mixtures were obtained.

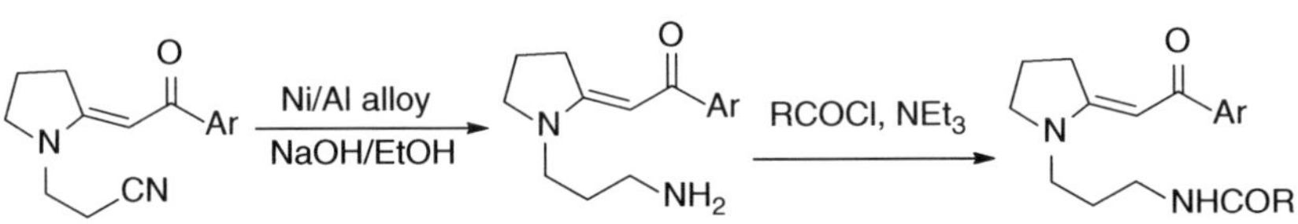

Scheme 8.8 The chemoselective reduction of nitrile to amine in basic ethanol solution.

The reductant *n*-Bu_2SnIH could chemoselectively reduce >C=C< in a molecule, but the aldehyde group could be preserved. This procedure is illustrated below (Scheme 8.9) [59].

8.3 Chemoselective Oxidation

Apart from inorganic reagents such as Cr_2O_3, SeO_2, and MnO_2, organic oxidants are well investigated in organic reactions. For example, the Swern reaction is a well-used reaction [60]. However, its low chemoselectivity has limited its wide application. Another factor is that it forms the byproduct sulfoether, so it is not easy in a workup. Another one is the Dess–Martin reaction [61], which has

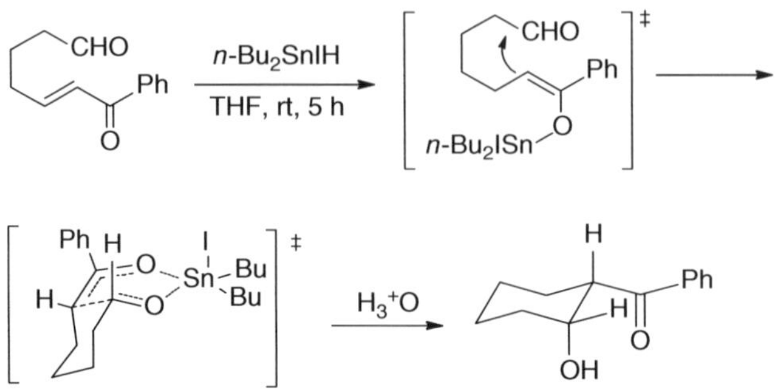

Scheme 8.9 Procedure of chemoselective reduction of α,β-unsaturated ketone [59].

been widely used. Some cheap oxidants include H_2O_2, *t*-BuOOH (TBHP), and *m*-chloroperoxybenzoic acid (*m*-CPBA).

Complexes of Cr(III) with an organic molecule, such as Salen, could be applied for different alcohol oxidations; the oxidant is $PhI(OAc)_2$ (BAIB) [62]. The reaction can be performed at 20 °C in CH_2Cl_2. The quantity of $PhI(OAc)_2$ should be about 1.2 equiv of the alcohol, and 10% of the complex is used. Partial results are summarized in Table 8.10.

The use of 1,3,5-trichloro-1,3,5-triazinane-2,4,6-trione (TCTATO, **5**) and 2,2,6,6-tetramethyl-piperidine 1-oxyl (TEMPO, **6**) could afford quite good chemoselectivity in oxidation reactions [63]. The reaction could be performed under mild conditions. The conversion may reach 90% or higher. The quantity of TCTATO should be the same as the alcohol (in moles), but only 0.01 mol% of TEMPO was used. Partial results are listed in Table 8.11.

Table 8.10 Chemoselective oxidation of secondary alcohol to ketone using $PhI(OAc)_2$ [62].

Entry	Substrates	Products	*t* (h)	Conversion (%)	Yield (%)
1	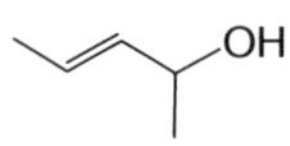	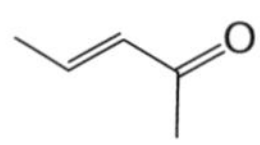	6	92	77
2	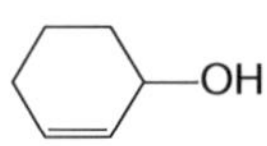	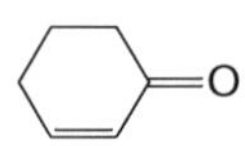	5	95	96
3	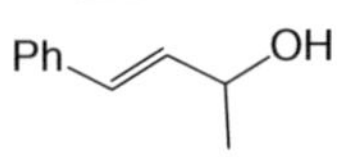	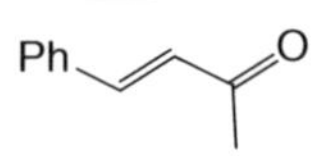	4	86	90
4	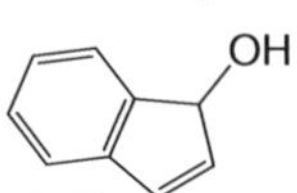	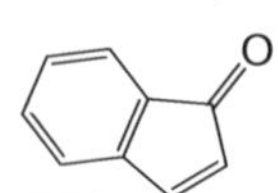	5	95	93
5	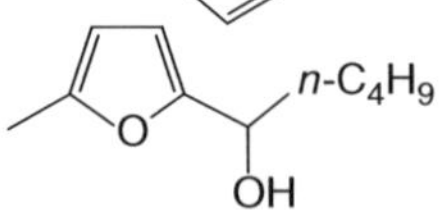	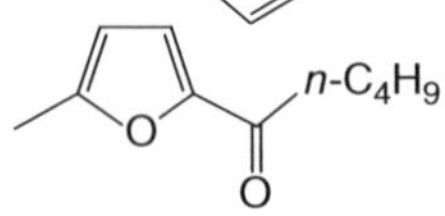	6	94	69
6	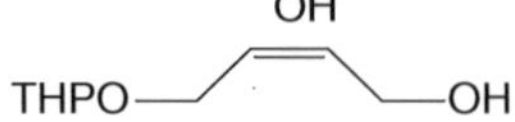	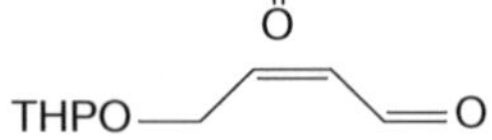	5	83	93

Table 8.11 Chemoselective oxidation of alcohols to the corresponding aldehydes or ketones [63].

Entry	Substrates	Products	Conversion (%)
1	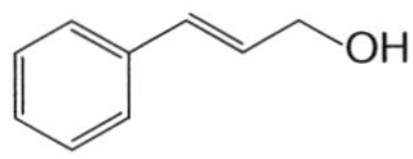	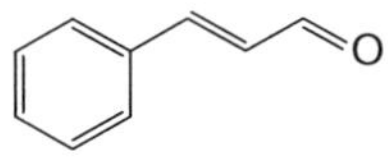	90
2	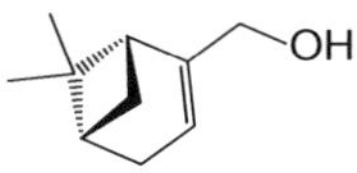	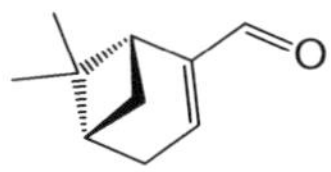	99
3	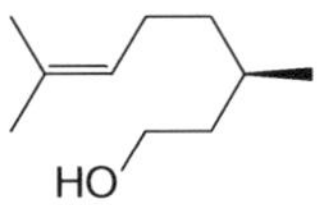	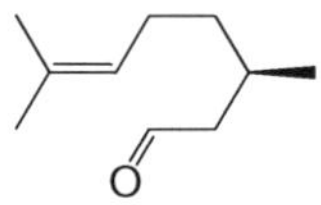	90
4	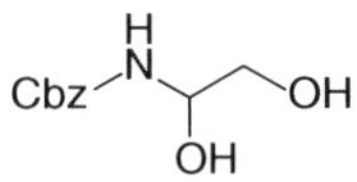	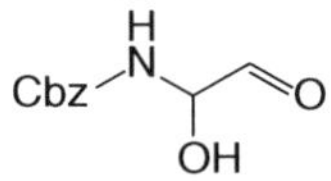	95
5	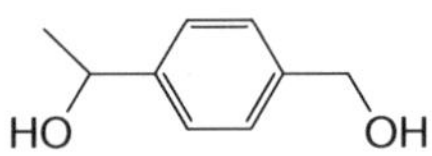	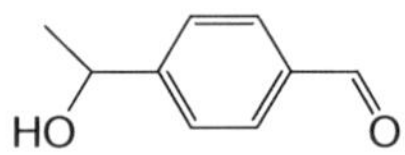	98

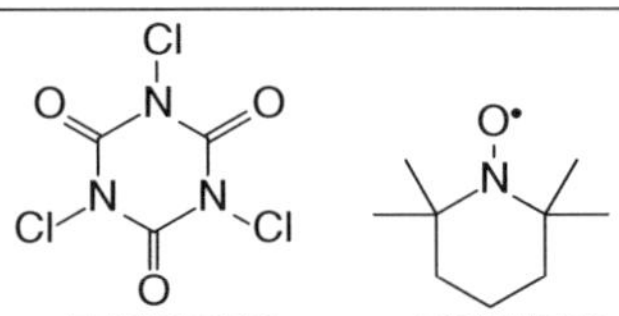

5 TATCTO **6** TEMPO

The oxidants used here could chemoselectively oxidize the primary alcohol to an aldehyde. However, the secondary alcohol could not be oxidized (Table 8.12, entries 4 and 5). Other groups such as >C=C< could not also be oxidized.

The combined use of BAIB and TEMPO could oxidize the primary alcohol to the corresponding acid, but the secondary alcohol could not be converted into ketone. This reaction can be used for the synthesis of some lactones (Eq. (8.31)) [64].

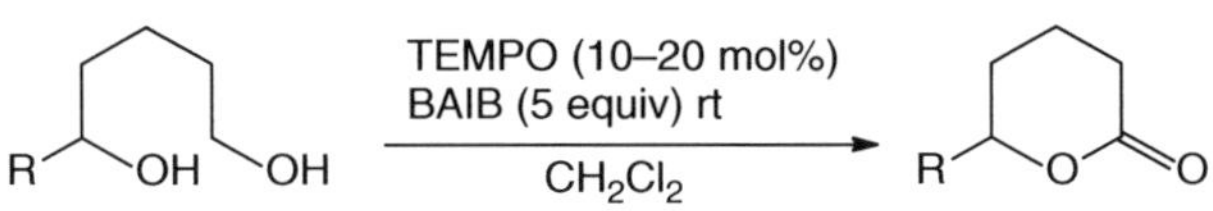

(8.31)

In this oxidation, TEMPO is used as the catalyst. The reaction could be completed in 20 h with a high yield (Table 8.12).

The Dess–Martin periodinane (DMP) reagent is 1,1,1-triacetoxy-1,1-dihydro-1,2-benziodoxol-3(8.1*H*)-one (**7**). It is also a derivative of iodine. Its usage is very wide in organic synthesis. Generally, the reactions mentioned above could be performed using DMP. The extent of its reactions shows its wide usage range. For

Table 8.12 Combinational use of BAIB-TEMPO for the oxidation of 1,5-diols to lactones [64].

Entry	Substrate	Product	*t* (h)	Yield (%)
1	R=TBDPSO–CH_2–	R=TBDPSO–CH_2–	20	94
2	R=PMBO–CH_2–	R=PMBO–CH_2–	16	77
3	R=BnO–CH_2–	R=BnO–CH_2–	1.5	96
4	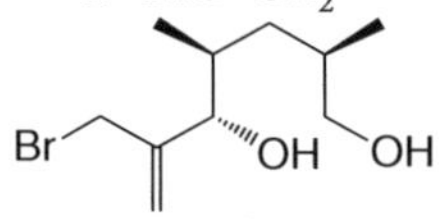	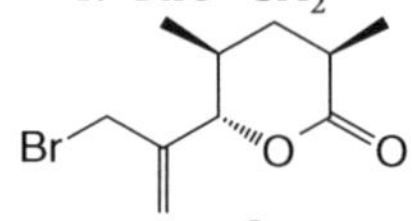	3.5	87
5	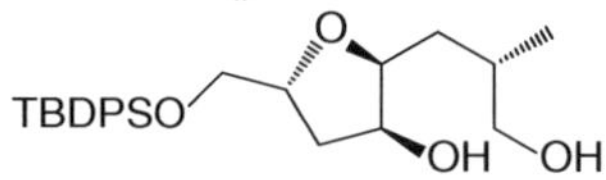	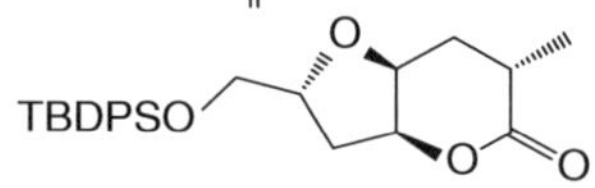	4	78
6	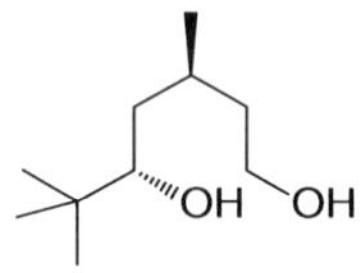	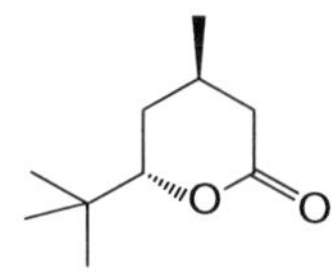	3.5	85

example, oxidative cyclization of aldehydes to pyrrolo[1,2-α]quinoxalines could be easily performed in the presence of DMP, although it is structurally specific in synthesis (Eq. (8.32)) [65]. In this reaction, if R = Ph, the reaction cannot be performed. The general yields are 36–81%.

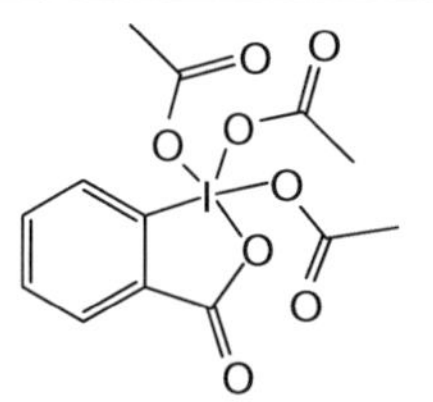

7

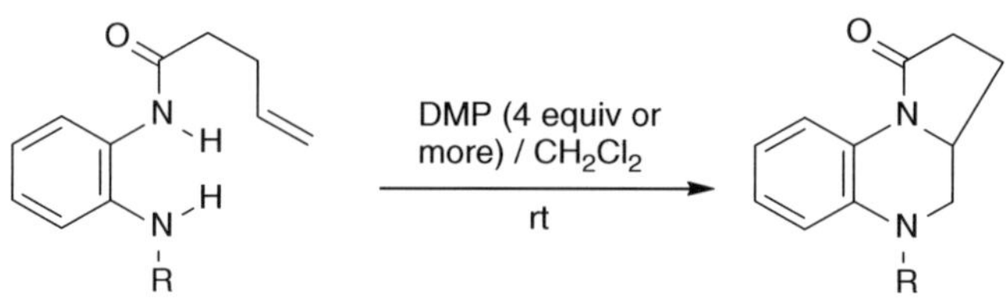

R = CH_2=CH–$(CH_2)_2$–CO–; (8.32)

Attempts to synthesize acyl azides from aldehydes were made using DMP and sodium azide. The reaction condition is mild, efficient, and general [66]. This method is readily applicable for the synthesis of aromatic or aliphatic acyl azides (Eq. (8.33), Table 8.13). The functional groups –NO_2, –OMe, or >C=C< were

Table 8.13 Synthesis of acyl azides from aldehydes and sodium azide [66].

Entry	R	*t* (h)	Yield (%) a)
1	Ph	1.5	89
2	4-OMe	1	95 b)
3	2-Cl	2.5	91
4	2-Naph	1	90
5	3-Py	3	86
6	$CH_3(CH_2)_4CHO$	4	88b)
7	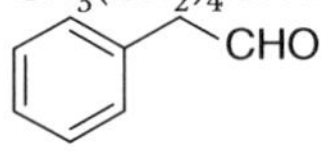	1	95b)

a) Isolated yield.
b) The acyl azides (neat) decompose upon standing in air.

preserved.

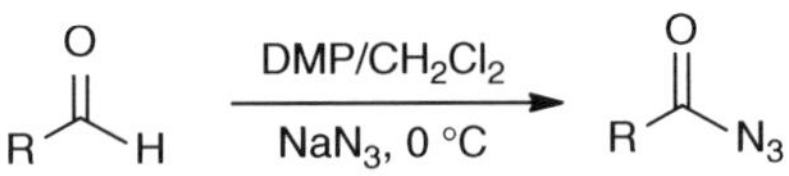

(8.33)

Notice
Azido compounds are dangerous (explode) when being concentrated under vacuum or stored neat. Operations in hood and safety shield are required.

A chemoselective oxidative cyclization to 1,3,4-oxadiazoles could be performed using DMP at room temperature [67]. One example is illustrated in Eq. (8.34).

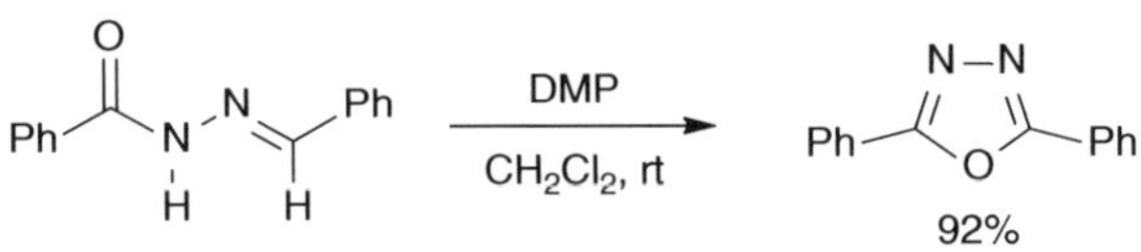

(8.34)

Sulfoether is an important intermediate in the pharmaceutical industry. In the oxidation of sulfoether, the cheap reagent H_2O_2 and inorganic salts could be used. Inorganic salts are used as catalysts to promote the oxidations. The major product is sulfoxide (**A**), and the side product is sulfone (**B**) (Eq. (8.35)) [68]. Partial experimental results are listed in Table 8.14.

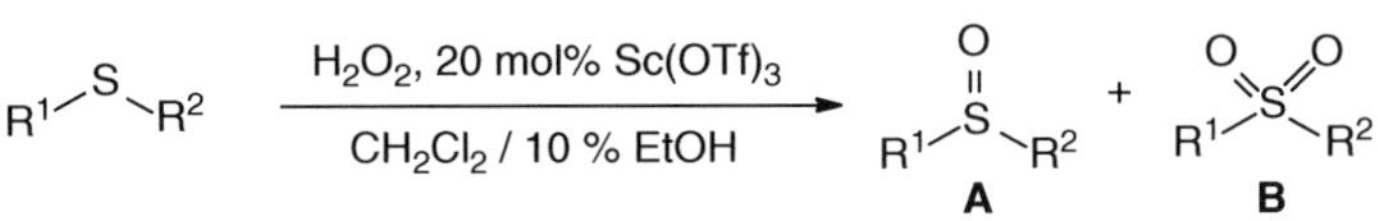

(8.35)

N-t-Bu *N*-chlorocyanamide could be used to oxidize sulfoethers to the corresponding sulfoxides in a mixture of acetonitrile and water. Good chemoselectivity

Table 8.14 Oxidation of four substrates [68].

Entry	Substrates	H_2O_2 [a)] (equiv)	*t* (h)	Yield (%)[b)] A	B
1	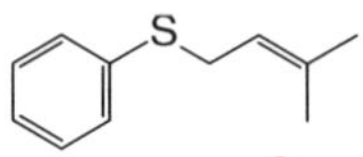	1.2	5	95	3
2	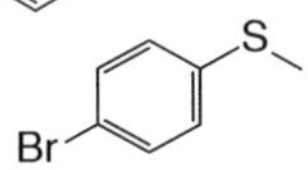	1.2	5.5	94	2
3	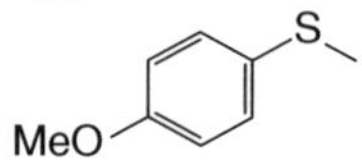	1.2	2.3	98	2
4	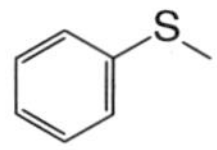	1.2	6	97	2

a) Aqueous H_2O_2 (60%) solution was used.
b) Determined by ^1H NMR.

Table 8.15 Oxidation of sulfoethers to sulfoxides using *N*-(*tert*-butyl)-*N*-chlorocyanamide [69].

Entry	Substrate	Product	Yield (%)
1	Ph–S–Me	Ph–S(=O)–Me	94
2	4Cl-Ph–S–Me	4Cl-Ph–S(=O)–Me	95
3	*i*-Pr–S–*i*-Pr	*i*-Pr–S(=O)–*i*-Pr	94
4	Et–S–CO_2Me	Et–S(=O)–CO_2Me	95
5	HOH_2CH_2C–S–Et	HOH_2CH_2C–S(=O)–Et	94

was achieved (Eq. (8.36) and Table 8.15) [69].

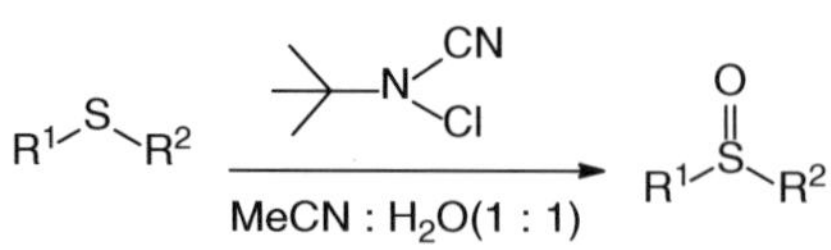

(8.36)

8.4 Other Chemoselective Reactions

A way to change the reaction condition is to carry out different chemoselective additions. This is convenient in lab experiments. Basic or acidic solution may change the experimental results. For example, the following additions could form different addition products under acidic or basic conditions (Scheme 8.10) [70].

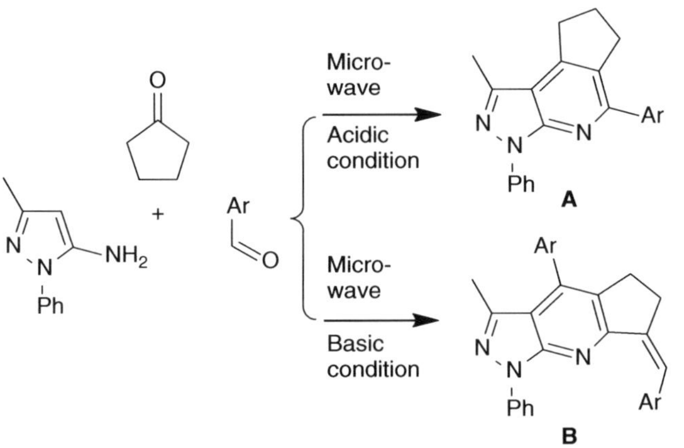

Scheme 8.10 Different reaction pathways for the condensation of 5-aminopyrazoles, aldehydes, and cyclic ketones [70].

This chemoselective condensation reaction could take place at about 80 °C. Under acidic condition, when all starting materials were mixed in a 1 : 1 : 1 ratio in acetic acid, the major product obtained was **A** with a yield range of 80–90%. When the ratio of cyclohexanone to aldehyde is 2 : 1, and 3-methyl-1-phenyl-1*H*-pyrazol-5-amine is in the same mole ratio with the aldehyde, the reaction under acidic condition would afford product **B**, and product **A** was isolated as a byproduct. If the condition changed to basic, only **B** was obtained. For example, when Ar was 4-Cl–Ph, the product **A** was obtained with 79% yield in acetic acid in 15 min. Once a base was used, product **B** was isolated with a yield of 86% using 0.2 equiv at 120 °C in *N*,*N*-dimethyl formamide (DMF).

The formation of >C=C< double bonds in reactions often leads to the (*E*) structure due to its lower molecular energy than (*Z*)-structure. However, it is possible to obtain (*Z*)-structure by use of catalysts, for example, Ru-catalyzed *Z*-selective olefin metathesis could afford the *cis* product using catalyst **8.** For example, the Ru-containing ligand catalyzes the formation of the (*Z*) product (Eq. (8.37)) [71]; the results are summarized in Table 8.16.

Table 8.16 Scope of cross-metathesis with *trans*-1,4-hexadiene [71].

Entry	R	Yield (%)[a)]	*Z* (%)[b)]
1	$-CH_2Ph$	63[c)]	>95
2	$-(CH_2)_8CHO$	70	>95
3	$-(CH_2)_2CO_2Me$	49	>95
4	$-(CH_2)_7CO_2Me$	82	>95
5	$-CH_2NHPh$	68	>95
6	$-CH_2NHBoc$	54	>95
7	$-CH_2OCO_2Me$	79	>95
8	$-CH_2BPin$	65	>95

a) Yield of isolated product.
b) Determined by ^{1}H NMR analysis.
c) Yield determined by ^{1}H NMR analysis with 1,3,5-trimethoxybenzene as an external standard.

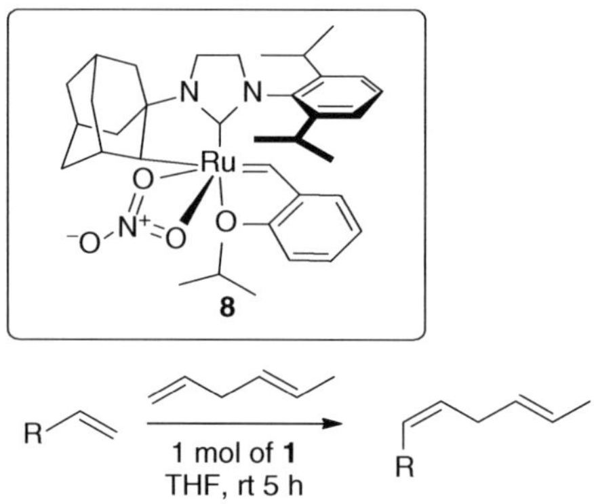

(8.37)

Wittig reaction is an important reaction in the synthesis of various compounds with a new C=C bond [72]. They generally happen between two molecules. An intramolecular Wittig reaction could provide another way to afford various complex molecules. For example, the formation of indole alkaloids could be accomplished in high yields ranging from 70 to 89% (Scheme 8.11) [72] Partial results are summarized in Table 8.17.

When the N atom of an amide is replaced by an O atom, the reaction could afford benzofuran derivatives with high yield; the products have the same molecular skeletons as their indole analogs.

Chemoselective exchange reaction is a useful reaction, such as the exchange reaction of glyceride with methanol to give the corresponding methyl ester and glycerol. Generally, it could be easily accomplished when the relative energy of the final product is lower than that of the starting material, or the final product could be removed from the reaction system such as by distillation. However, it is not easy to perform this procedure even if the final products' energy is lower than that of the starting material. For example, in the exchange reaction of acetamide

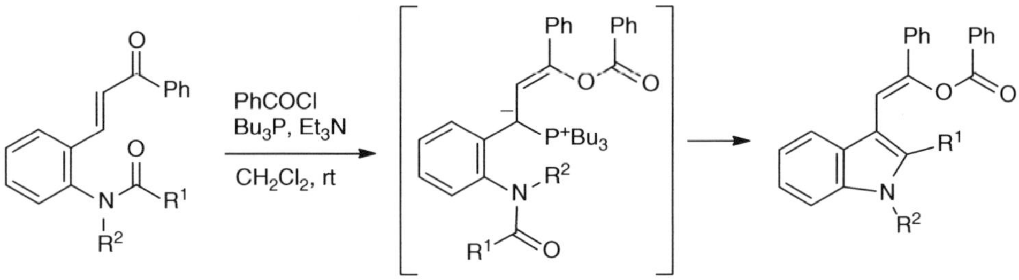

Scheme 8.11 Formation of indole derivatives via intramolecular Wittig reactions [72].

Table 8.17 Indole derivatives formed via intramolecular Wittig reactions [72].[a)]

Entry	R^1	R^2	*t* (h)	Yield (%)[b)]
1	Ph	Boc	1.5	80
2	Ph	Cbz	2	82
3	Ph	TS	8	81
4	4-MeOPh	Boc	1.5	77
5	2-BrPh	Boc	3	67
6	Me	Boc	3	75
7	*i*-Pr	Boc	5	80
8	CO_2Et	Boc	2	70

a) Reactions were performed with starting material (0.5 mmol), chloride (1.1 equiv), Bu_3P (1.2 equiv), and Et_3N (1.3 equiv) in dried THF (1 ml) under nitrogen.
b) Isolated yield.

(CH_3CONH_2) with methanol to form methyl acetate and ammonia, the relative energy of final products ($CH_3CO_2Me+NH_3$) is lower by 6.28 kcal mol^{-1} than the starting materials' ($CH_3CONH_2 + MeOH$) at the B3LYP/6-31G(d) level in the gas phase. However, the exchange reaction is difficult under the catalysis of HCl–methanol.

The situation changed when the $SOCl_2$–MeOH system was used for the conversions. The primary amide could be converted into the corresponding methyl ester (Eq. (8.38)) [73].

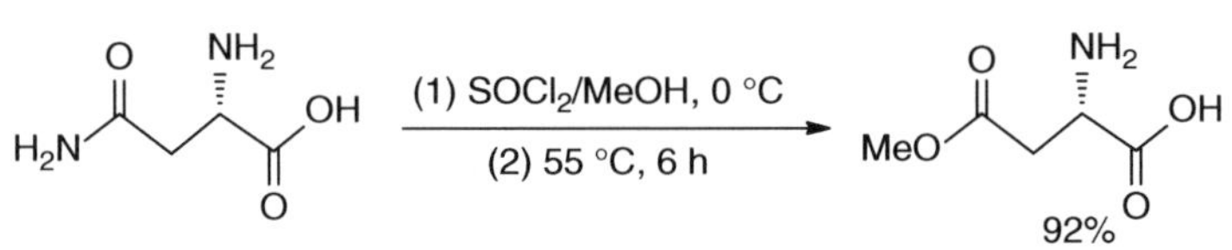

(8.38)

As mentioned above, experiments have shown that it is not easy to catalyze by HCl produced from $SOCl_2$ in methanol since the addition of HCl in the reaction cannot be carried out. Obviously, this is an exchange reaction involving a series of intermediates of $SOCl_2$ in methanol. Scheme 8.12 illustrates the whole procedure.

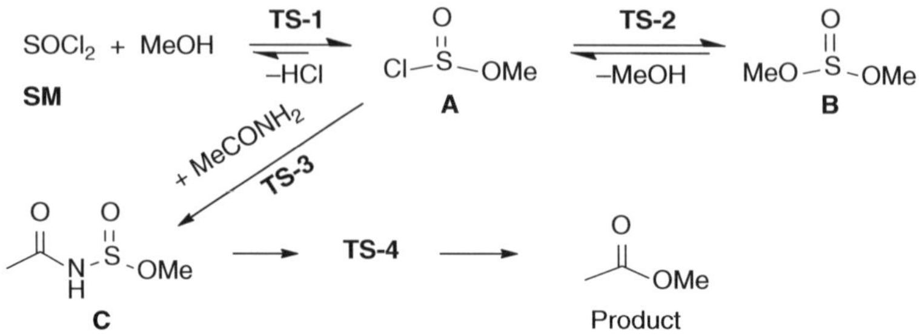

Scheme 8.12 Plausible mechanism for exchange reaction of primary amide to the corresponding methyl ester [73].

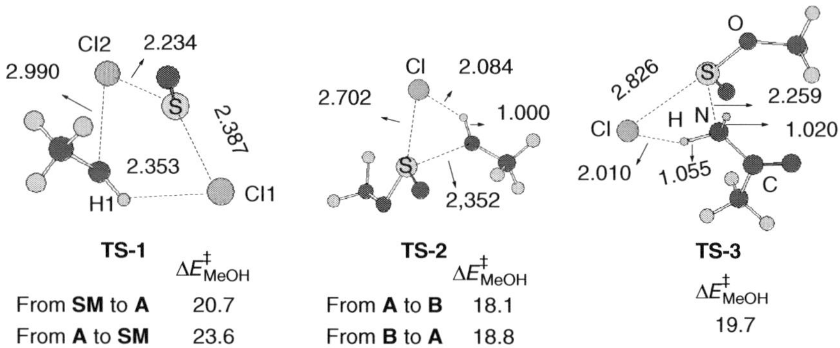

Figure 8.4 The predicted three TS structures and their barriers in methanol [73].

The study of the mechanism was conducted using $MeCONH_2$ as a model. Calculations were performed at the B3LYP/6-31++G(d,p) level. The TS barriers for the first three procedures were computed. The predicted TS structures for **TS-1** to **TS-3** and their barriers are summarized below their 3D structure in Figure 8.4.

Computational results showed that the intermediate **A** would form quickly in methanol. **A** and **B** coexisted in methanol. At the same time, intermediate **A** would react with amide to form the intermediate **C**, which would have four possible TS structures (**I–IV**) to form the final product.

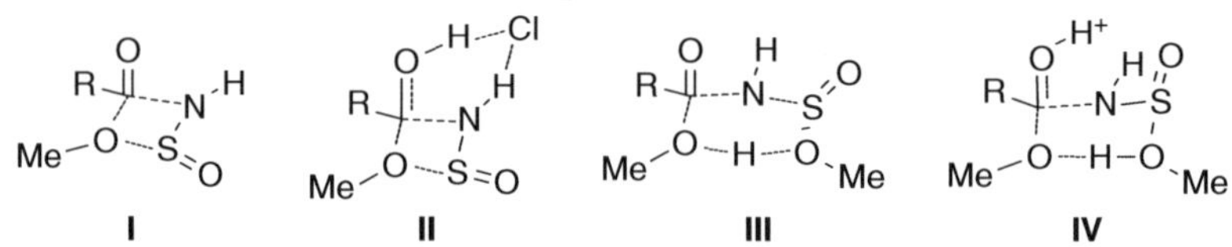

The predicted four TS structures (**TS-4-I** to **TS-4-IV**) are illustrated below. The calculation results showed that the barrier via **TS-4-II** was the lowest with 21.6 kcal mol^{-1}. The second lowest was via **TS-4-IV** with a barrier of 24.7 kcal mol^{-1} (Figure 8.5). It looks like the reaction would go through **TS-4-II**.

To further confirm which TS structure is preferred, more calculations need to be investigated. Logically, if one TS is close to the true one, when the substrate is

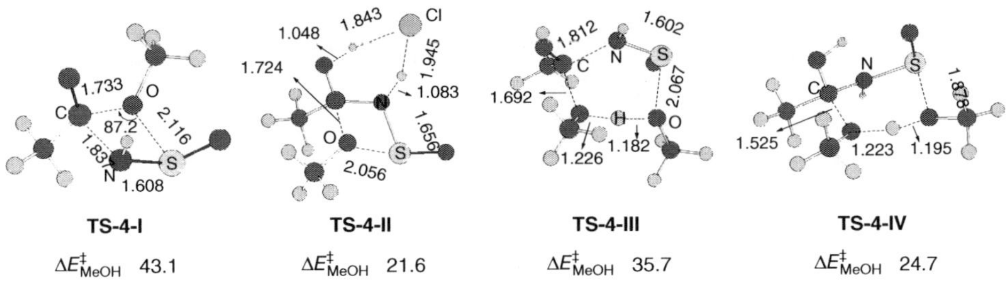

Figure 8.5 Four possible TS models and their TS barriers in methanol [73].

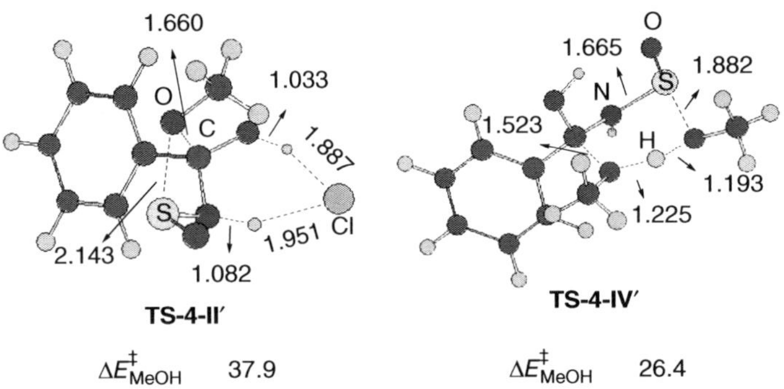

Figure 8.6 The 3D TS structures for models II and IV using benzamide and formamide in TS calculation and their TS barriers [73].

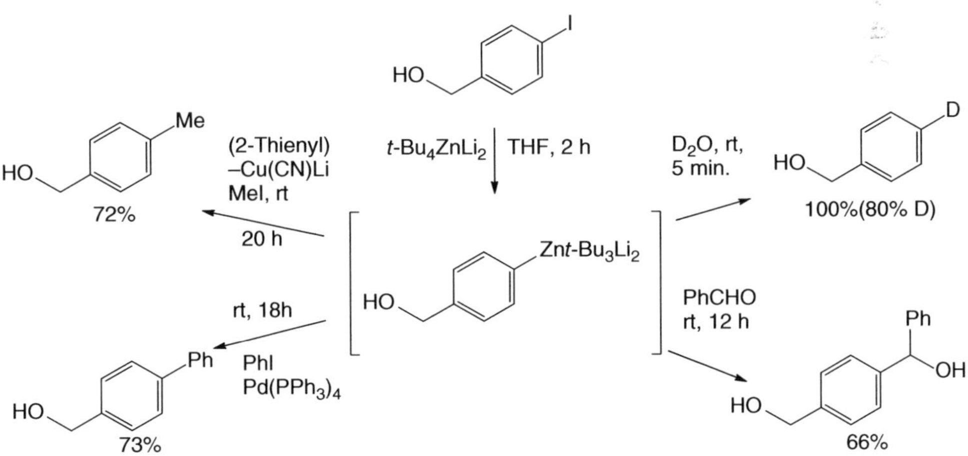

Scheme 8.13 Exchange reactions of (4-iodophenyl)methanol in presence of Bu_4ZnLi_2 [74].

changed to other, the TS barrier will change a little; however, this change should not be large. A benzamide was used in the calculations. The predicted barrier via model II is high up to 37.9 kcal mol^{-1} (**TS-4-II′**), which is higher in energy by 16.3 kcal mol^{-1} than that in acetamide reduction via using model II. However, when model IV was used, the predicted TS barrier was only 26.4 kcal mol^{-1} (**TS-4-IV′**), which is only higher in energy by 1.7 kcal mol^{-1} than in acetamide using model IV. Obviously, the six-membered ring TS structure is reliable. The predicted TS structures for both amides are illustrated in Figure 8.6.

In this exchange reaction, because of the formation of the ester, the equilibrium must be broken and, finally, the new equilibrium will form when the reaction finished, namely, most of the starting material would be converted to the methyl ester.

Unfortunately, the secondary and tertiary amides could not be converted into the esters in the presence of $SOCl_2$–MeOH solution.

Use of a halide in reaction is useful. For example, elemental iodine has strong reactivity, so it could be used in exchange reactions. Organic zinc reagents could be used to afford various derivatives [74]. Scheme 8.13 summarizes the results.

Traditional Schmidt reaction generally forms a mixture of two products using HN_3 catalyzed by H_2SO_4 (Eq. (8.39)). An improved and effective conversion from aldehydes to the corresponding nitriles is to use NaN_3 (Eq. (8.40)). This chemoselective Schmidt reaction could be performed under TFA catalysis in near-quantitative yields and tolerates a variety of electron-withdrawing and

Table 8.18 Substrate scope for Schmidt reaction [75].

Entry	R	Yield (%)	Entry	R	Yield (%)
1	Ph	82	13	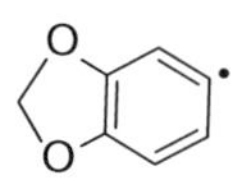	99
2	4-MePh	71			
3	4-MeOPh	97			
4	3,4-diMeOPh	97	14	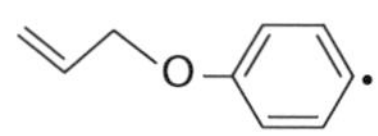	99
5	3,4,5-triMeOPh	97			
6	4-OHPh	99	15	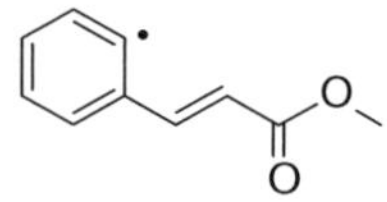	55
7	4-BnOPh	88			
8	3-PhOPh	71			
9	4-(*N*,*N*-diMe)Ph	98	16	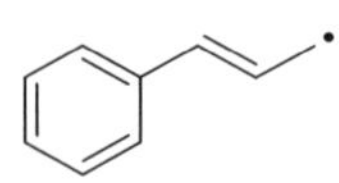	98
10	4-PhPh	85			
11	4-ClPh	73	17	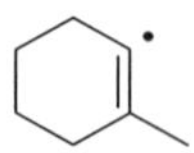	60
12	2-BrPh	93			

electron-donating substituents on the substrates [75]. Only nitrile was obtained in this reaction.

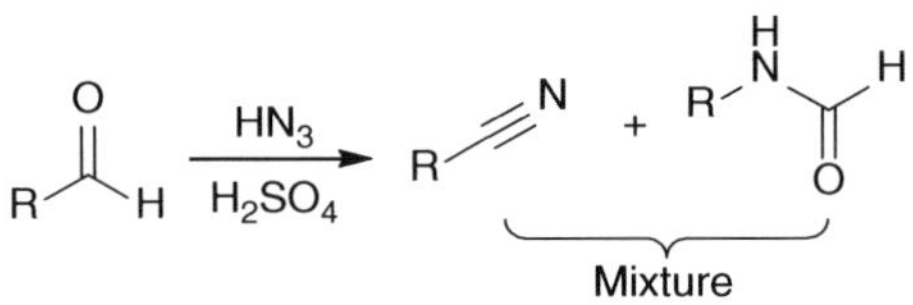

(8.39)

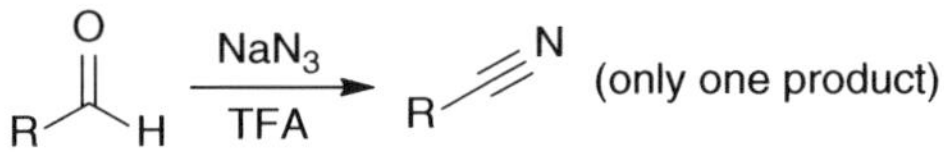

(8.40)

Only the aldehyde group could take part in the reaction in a molecule; others like –OH or $-CO_2H$ will not be affected. Other groups are also not affected. Its application scope is summarized in Table 8.18.

References

1. Atodiresei, I., Schiffers, I., and Bolm, C. (2007) *Chem. Rev.*, **107**, 5683–5712.
2. (a) Bercot, E.A. and Rovis, T. (2005) *J. Am. Chem. Soc.*, **127**, 247–254; (b) Cook, M.J. and Rovis, T. (2007) *J. Am. Chem. Soc.*, **129**, 9302–9303; (c) Johnson, J.B., Bercot, E.A., Rowley, J.M., Coates, G.W., and Rovis, T. (2007) *J. Am. Chem. Soc.*, **129**, 2718–2725.
3. Fleming, F.F., Funk, L.A., Altundas, R., and Sharief, V. (2002) *J. Org. Chem.*, **67**, 9414–9416.
4. Yamada, K.I., Yamamoto, Y., and Tomioka, K. (2003) *Org. Lett.*, **5**, 1797–1799.
5. Taillier, C., Bellosta, V., Meyer, C., and Cossy, J. (2004) *Org. Lett.*, **6**, 2145–2147.
6. Azizi, N. and Saidi, M.R. (2005) *Org. Lett.*, **7**, 3649–3651.
7. Zhou, G., Yost, J.M., Sauer, S.J., and Coltart, D.M. (2007) *Org. Lett.*, **9**, 4663–4665.
8. Peng, J., Huang, X., Jiang, L., Cui, H.L., and Chen, Y.C. (2011) *Org. Lett.*, **13**, 4584–4587.
9. Su, B.Z. and Li, R.T. (2000) *Youji Hecheng Jichu (The Basic of Organic Synthesis)*, Peking Medical University Presses, Beijing, pp. 107–124.
10. Brown, H.C. and Ichikawa, K. (1961) *J. Am. Chem. Soc.*, **83**, 4372.
11. Chary, K.P., Mohan, G.H., and Iyengar, D.S. (1999) *Chem. Lett.*, **28**, 1339.
12. Brown, H.C., Choi, Y.M., and Narasimhan, S. (1981) *Inorg. Chem.*, **20**, 4454.
13. (a) Liu, W.Y., Xu, Q.H., and Ma, Y.X. (2000) *Org. Prep. Proced. Int.*, **32**, 596; (b) Akisanya, J., Danks, T.N., and Oarman, R.N. (2000) *J. Organomet. Chem.*, **603**, 240.
14. Wilkinson, H.S., Tanoury, G.J., Wald, S.A., and Senanayake, C.H. (2001) *Tetrahedron Lett.*, **42**, 167–170.
15. Huang, X., Wang, Y., and Chen, Z.C. (2003) *Xinbian Youji Hecheng Huaxue (New Edition of Organic Synthesis)*, Chemical Industry Press, Beijing, p. 177.
16. (a) Chaikin, S.W. and Brown, W.G. (1949) *J. Am. Chem. Soc.*, **71**, 122–125; (b) Schlesinger, H.I., Brown, H.C., Hoekstra, H.R., and Rapp, L.R. (1953) *J. Am. Chem. Soc.*, **75**, 199–204.
17. Ranu, B.C. and Chakraborty, R. (1990) *Tetrahedron Lett.*, **31**, 7663.
18. (a) Bae, J.W., Lee, S.H., Jung, Y.J., Yoon, C.M., and Yoon, C.M. (2001) *Tetrahedron Lett.*, **42**, 2137–2139; (b) Bhar, S. and Guha, S. (2004) *Tetrahedron Lett.*, **45**, 3775–3777; (c) Storer, R.I., Carrera, D.E., Ni, Y., and MacMillan, D.W.C. (2006) *J. Am. Chem. Soc.*, **128**, 84–86.
19. Elmorsy, S.S., El-Ahl, A.A.S., Soliman, H., and Amer, F.A. (1996) *Tetrahedron Lett.*, **37**, 2297–2298.

20. Lee, H.Y. and An, M. (2003) *Tetrahedron Lett.*, **44**, 2775–2778.
21. (a) Bhattacharyya, S. (1996) *J. Chem. Soc., Perkin Trans. 1*, 1381; (b) Soai, K. and Ookawa, A. (1986) *J. Org. Chem.*, **51**, 4000.
22. McGeary, R.P. (1998) *Tetrahedron Lett.*, **39**, 3319.
23. Kanth, J.V.B. and Periasamy, M. (1991) *J. Org. Chem.*, **56**, 5964.
24. Naqvi, T., Bhattacharya, M., and Haq, W. (1999) *J. Chem. Res.*, **1**, 424–425.
25. Li, L.C., Jiang, J.X., Ren, J., Ren, Y., Pittman, C.U., and Zhu, H.J. (2006) *Eur. J. Org. Chem.*, **2006**, 1981.
26. Inamoto, N. and Masuda, S. (1982) *Chem. Lett.*, **11**(7), 1003–1006.
27. Baruah, M., Boruah, A., Prajapati, D., Sandhu, J.S., and Ghosh, A.C. (1996) *Tetrahedron Lett.*, **37**, 4559–4560.
28. Ren, J., Li, L.C., Liu, J.K., Zhu, H.J., and Pittman, C.U. Jr., (2006) *Eur. J. Org. Chem.*, **2006**, 1991.
29. Newman, M.S. and Fukunaga, T. (1960) *J. Am. Chem. Soc.*, **82**, 693.
30. (a) Umino, N., Iwakuma, T., and Itoh, M. (1976) *Tetrahedron Lett.*, **17**, 763–766; (b) Padwa, A., Harring, S.R., and Semones, M.A. (1998) *J. Org. Chem.*, **63**, 44–54.
31. Barker, D., Mcleod, M.D., and Brimble, M.A. (2001) *Tetrahedron Lett.*, **42**, 1785–1788.
32. Gerona-Navarro, G., Bonache, M.A., Alias, M., Vega, M.J.P.D., Garca-Lopez, M.T., Lopez, P., Cativielab, C., and Gonzalez-Muniz, R. (2004) *Tetrahedron Lett.*, **45**, 2193–2196.
33. (a) Kurtev, B.J., Lyapova, M.J., Mishev, S.M., Nakova, O.G., Orahovatz, A.S., and Pojarlieff, I.G. (1983) *Org. Magn. Reson.*, **21**, 334–338; (b) Kascheres, A., Kascheres, C., and Augusto, J. (1984) *Synth. Commun.*, **14**, 905–913.
34. Agami, C., Dechoux, L., and Melaimi, M. (2001) *Tetrahedron Lett.*, **42**, 8629–8631.
35. Meng, W.H., Wu, T.J., Zhang, H.K., and Huang, P.Q. (2004) *Tetrahedron: Asymmetry*, **15**, 3899–3910.
36. Haldar, P. and Ray, J.K. (2003) *Tetrahedron Lett.*, **44**, 8229–8231.
37. (a) Gribble, G.W. and Hoffman, J.H. (1977) *Synthesis*, **12**, 859–860; (b) Maryanoff, B.E., McComsey, D.F., and Nortey, S.O. (1981) *J. Org. Chem.*, **46**, 355–360; (c) Gribble, G.W., Lord, P.D., Skotnicki, J., Stephen, F., Eaton, J.T., and Johnson, J.L. (1974) *J. Am. Chem. Soc.*, **96**, 7812–7814; (d) Berger, J.G., Teller, S.R., Adams, C.D., and Guggenberger, L.J. (1975) *Tetrahedron Lett.*, **16**, 1807–1810; (e) Elliot, A.J. and Guzik, H. (1982) *Tetrahedron Lett.*, **23**, 1983–1984.
38. Borghese, A., Antoinea, L., and Stephenson, G. (2002) *Tetrahedron Lett.*, **43**, 8087–8090.
39. (a) Scriven, E.F.V. and Turnbull, K. (1988) *Chem. Rev.*, **88**, 297–368; (b) Kamal, A., Laxman, E., and Arifuddin, M. (2000) *Tetrahedron Lett.*, **41**, 7743–7746; (c) Lee, J.W. and Fuchs, P.L. (1999) *Org. Lett.*, **1**, 179–182.
40. (a) Blandy, C., Choukroun, R., and Gervais, D. (1983) *Tetrahedron Lett.*, **24**, 4189–4192; (b) Caron, M. and Sharpless, K.B. (1985) *J. Org. Chem.*, **50**, 1557–1560; (c) Chong, J.M. and Sharpless, K.B. (1985) *J. Org. Chem.*, **50**, 1560–1565; (d) Onak, M., Sugita, K., and Izumi, Y. (1986) *Chem. Lett.*, **15**, 1327–1328; (e) Sinou, D. and Emziane, M. (1986) *Tetrahedron Lett.*, **27**, 4423–4426; (f) Thompson, A.S., Hymphrey, G.R., DeMarco, A.M., Mathre, D.J., and Grabowski, E.J.J. (1993) *J. Org. Chem.*, **58**, 5886–5888.
41. (a) Boyer, J.H. (1951) *J. Am. Chem. Soc.*, **73**, 5865; (b) Boyer, J.H. and Canter, F.C. (1954) *Chem. Rev.*, **54**, 1; (c) Hojo, H., Kobayashi, S., Soai, J., Ikeda, S., and Mukaiyama, T. (1977) *Chem. Lett.*, 635–636; (d) Kyba, E.P. and John, A.M. (1977) *Tetrahedron Lett.*, **18**, 2737–2740; (e) Corey, E.J., Nicolaou, K.C., Balanson, R.D., and Machida, Y. (1975) *Synthesis*, **9**, 590.
42. Salunkhe, A.M., Ramachandran, P.V., and Brown, H.C. (2002) *Tetrahedron*, **58**, 10059–10064.
43. Reddy, P.G. and Baskaran, S. (2002) *Tetrahedron Lett.*, **43**, 1919–1922.
44. Ranu, B.C. and Samanta, S. (2002) *Tetrahedron Lett.*, **43**, 7405.
45. Ren, P.D., Pan, S.F., and Dong, T.W. (1995) *Synth. Commun.*, **25**, 3395.

46. Ren, P.D., Pan, S.F., and Dong, T.W. (1996) *Chin. Chem. Lett.*, **7**, 788.
47. Desrosiers, J.N., Bechara, W.S., and Charette, A.B. (2008) *Org. Lett.*, **10**, 2315–2318.
48. Bagal, D.B., Qureshi, Z.S., Dhake, K.P., Khan, S.R., and Bhanage, B.M. (2011) *Green Chem.*, **13**, 1490–1494.
49. Akao, A., Sato, K., Nonoyama, N., Masea, T., and Yasudab, N. (2006) *Tetrahedron Lett.*, **47**, 969–972.
50. Couturier, M., Tucker, J.L., Andresen, B.M., Dube, P., Brenek, S.J., and Negri, J.T. (2001) *Tetrahedron Lett.*, **42**, 2285–2288.
51. Mitsudome, T., Mikami, Y., Matoba, M., Mizugaki, T., Jitsukawa, K., and Kaneda, K. (2012) *Angew. Chem. Int. Ed.*, **51**, 136–139.
52. Trost, B.M., Ball, Z.T., and Ge, T.J. (2002) *J. Am. Chem. Soc.*, **124**, 7922–7923.
53. Ranu, B.C., Majee, A., and Sarkar, A. (1998) *J. Org. Chem.*, **63**, 370–373.
54. Borah, H.N., Prajapati, D., and Sandhu, J.S. (1994) *J. Chem. Res.*, 228.
55. (a) Madesclaire, M. (1988) *Tetrahedron*, **44**, 6537–6580; (b) Kukushkin, V.Y. (1995) *Coord. Chem. Rev.*, **139**, 375; (c) Espenson, J.H. (2005) *Coord. Chem. Rev.*, **249**, 329–341; (d) Raju, B.R., Devi, G., Nongpluh, Y.S., and Saikia, A.K. (2005) *Synlett*, **2**, 358; (e) Harrison, D.J., Tam, N.C., Vogels, C.M., Langler, R.F., Baker, R.T., Decken, A., and Westcott, S.A. (2004) *Tetrahedron Lett.*, **45**, 8493; (f) Yoo, B.W., Choi, K.H., Kim, D.Y., Choi, K.I., and Kim, J.H. (2003) *Synth. Commun.*, **33**, 53; (g) Nicolaou, K.C., Koumbis, A.E., Snyder, S.A., and Simonsen, K.B. (2000) *Angew. Chem. Int. Ed.*, **39**, 2529; (h) Yoo, B.W., Choi, J.W., and Yoon, C.M. (2006) *Tetrahedron Lett.*, **47**, 125.
56. (a) Most, K., Hoßbach, J., Vidović, D., Magull, J., and Mősch-Zanetti, N.C. (2005) *Adv. Synth. Catal.*, **347**, 463–472; (b) Arnáiz, F.J., Agudo, R., Pedrosa, M.R., and De Cian, A. (2003) *Inorg. Chim. Acta*, **347**, 33; (c) Arnáiz, F.J., Agudo, R., and Ilarduya, J.M.M. (1994) *Polyhedron*, **13**, 3257; (d) Enemark, J.H., Cooney, J.J.A., Wang, J.J., and Holm, R.H. (2004) *Chem. Rev.*, **104**, 1175.
57. Fernandes, A.C. and Romao, C.C. (2006) *Tetrahedron*, **62**, 9650–9654.
58. Michael, J.P. and Parsons, A.S. (1996) *Tetrahedron*, **52**, 2199–2216.
59. Suwa, T., Nishino, K., Miyatake, M., Shibata, I., and Baba, A. (2000) *Tetrahedron Lett.*, **41**, 3403–3406.
60. Lee, T.V. (1991) in *Comprehensive Organic Synthesis*, vol. 7 (eds B.M. Trost and I. Fleming), Pergamon, Oxford, p. 291.
61. Dess, D.B. and Martin, J.C. (1983) *J. Org. Chem.*, **48**, 4155.
62. Adam, W., Hajra, S., Herderich, M., and Saha-Mollller, C.R. (2000) *Org. Lett.*, **2**, 2773–2776.
63. Luca, L.D., Giacomelli, G., and Porcheddu, A. (2001) *Org. Lett.*, **3**, 3041–3043.
64. Hansen, T.M., Florence, G.J., Lugo-Mas, P., Chen, J., Abrams, J.N., and Forsyth, C.J. (2003) *Tetrahedron Lett.*, **44**, 57–59.
65. Dobrota, C., Graeupner, J., Dumitru, I., Matache, M., and Paraschivescu, C.C. (2010) *Tetrahedron Lett.*, **51**, 1262–1264.
66. Bose, D.S. and Reddy, A.V.N. (2003) *Tetrahedron Lett.*, **44**, 3543–3545.
67. Dobrotă, C., Paraschivescu, C.C., Dumitru, I., Matache, M., Baciu, I., and Rută, L.L. (2009) *Tetrahedron Lett.*, **50**, 1886–1888.
68. Matteucci, M., Bhalay, G., and Bradley, M. (2003) *Org. Lett.*, **5**, 235–237.
69. Kumar, V. and Kaushik, M.P. (2005) *Chem. Lett.*, **34**, 1230–1231.
70. Wang, S.L., Liu, Y.P., Xu, B.H., Wang, X.H., Jiang, B., and Tu, S.J. (2011) *Tetrahedron*, **67**, 9417–9425.
71. Cannon, J.S. and Grubbs, R.H. (2013) *Angew. Chem. Int. Ed.*, **52**, 9001–9004.
72. Lee, Y.T., Jang, Y.J., Syu, S., Chou, S.C., Lee, C.J., and Lin, W. (2012) *Chem. Commun.*, **48**, 8135–8137.
73. Li, L.C., Ren, J., Jiang, J.X., and Zhu, H.J. (2007) *Eur. J. Org. Chem.*, **2007**, 1026–1030.
74. Uchiyama, M., Furuyama, T., Kobayashi, M., Matsumoto, Y., and Tanaka, K. (2006) *J. Am. Chem. Soc.*, **128**, 8404–8405.
75. Rokade, B.V. and Prabhu, K.R. (2012) *J. Org. Chem.*, **77**, 5364–5370.

9
Stereoselective Reaction

Stereoselective reaction is a key research area. To perform a highly controlled stereoselective reaction, a chiral catalyst is necessary in the procedure. Similar to enantioselective reactions, the diastreomeric excess (de in percent) is used to express the excess quantity between two stereoisomers. When de is 100%, it is the so-called stereospecific reaction. This case mostly happens under enzyme catalysis. In organic synthesis, the values of de may be as high as 98%. Generally, if the de is >95%, this can be considered excellent diastereoselectivity.

Another term, the diastereomeric ratio (dr), is used to show the excessive quantity of one stereoisomer; it means the ratio between the two stereoisomers. Its calculation is similar to the ee calculations. For a mixture of de value 95%, the computed ratio is 97.5 : 2.5. To simplify the description, "stereoselective reaction" is used to represent stereospecific reaction, diastereoselective reaction, and others in this chapter.

Regioselective reactions may happen without any change in the stereogenic center. It depends on the position of the regioselective reaction in a molecule. In many cases, regioselective reactions accompany diastereoselective reactions.

9.1
Conformational Study

Originally, all reactions must have a correct "molecular posture" to "collide" and form a formal bond. A correct "posture" means a suitable geometry or conformation in solution. In other words, many collisions among different molecules are not effective. Transition state (TS) study is to investigate the collision with the correct gesture between two molecules or among different molecules. On the other hand, a free conformation distribution has a big effect on optical rotation (OR), optical rotatory dispersion (ORD), electronic circular dichroism (ECD), and vibrational circular dichroism (VCD), including nuclear magnetic resonance (NMR) spectra. Thus, conformational study is extremely important not only in asymmetric synthesis but also in chiroptical spectroscopy study.

The simplest example, ethane (**1**), has three stable conformations due to rotation about the C–C single bond (Figure 9.1). For example, if H_a has to rotate to

Organic Stereochemistry: Experimental and Computational Methods, First Edition. Hua-Jie Zhu.

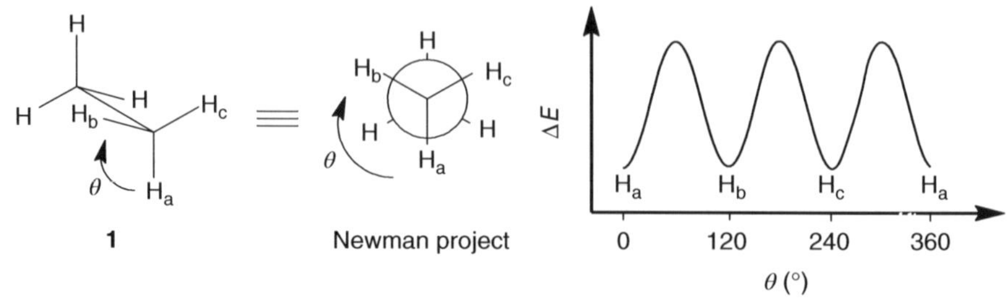

Figure 9.1 Energy changes of three conformations of ethane during C–C bond rotation.

the position H_b and then to H_c, H_a should overcome three energy barriers. If the rotation energy barrier is very low, it is impossible to separate the three conformations at room temperature. Every single C–C bond has three degrees of freedom; therefore, if a linear aliphatic alkane has n single bonds (no double bond inside), theoretically it would have $(n-1)^3$ geometries. Indeed, the real number of conformations is smaller than the the theoretical number. If there are double bonds or heteroatoms in a molecule, the conformations should be much smaller than theoretical estimates. Anyway, the more the number of single bonds, the more the number of stable conformations, and the more difficult it is to completely search out all stable conformations.

Historically, the well-known conformational study involves cyclohexane (**2**) and its derivatives. For example, there are two conformations possible for a cyclohexane molecule: one is the so-called chair conformation, and another is the boat conformation. The boat conformation can convert to another chair conformation. The boat conformation has a higher energy than the chair conformation and therefore it has a small distribution in solution. At room temperature, the two chair conformations convert so quickly that it is impossible to record their difference in ^{1}H NMR. The chemical shifts of protons (H_a and H_e) on bonds a and e are different. The recorded ^{1}H NMR at room temperature is the averaged values of the two chair conformations (1.44 ppm). However, when the measurement temperature is decreased to −110 °C, the difference between the two ^{1}H NMR would be recorded. Figure 9.2 illustrates the difference [1].

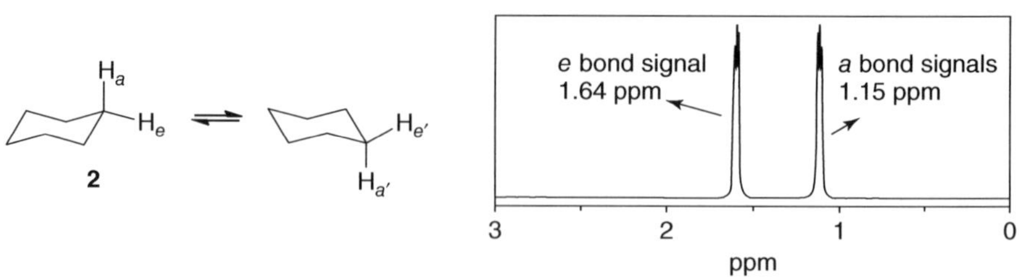

Figure 9.2 Exchange of two chair conformations and two proton ^{1}H NMR signals at −110 °C [1].

Table 9.1 Experimental and theoretical distribution for methyl hexane in solution.

Item	Me at e bond	Me at a bond
Experimental distribution (%)	95	5
Relative energy by experiments (kcal mol^{-1})	0.00	1.77
Predicted distribution (%)	98	2
Relative energy by calculation (kcal mol^{-1})	0.00	2.30

In derivative of hexane, such as 1-methylhexane (**3**), the methyl mainly takes up the *e* orientation and occupies about 95%; only about 5% is located on *a* orientation [2]. This case can be simulated using quantum theory. For example, when the B3LYP/6-31+G(d) basis set was used, the predicted distribution for e orientation goes up to 98% (Table 9.1).

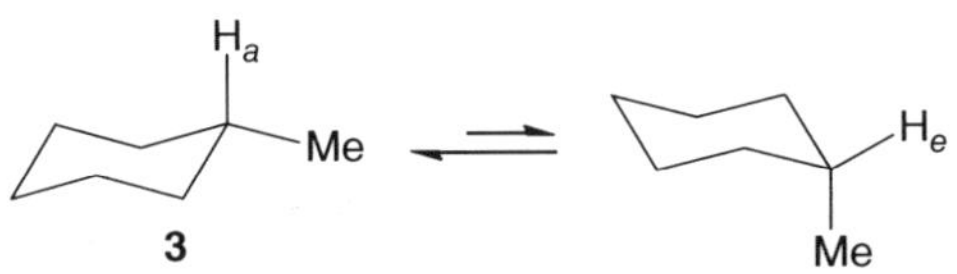

3

The two examples are very typical. It is difficult to find all stable conformations for a real couple compounds, especially a nonrigid compound. Conformational study is the most basic but one of the important tasks in modern stereochemistry.

The use of NMR and HPLC (high-performance liquid chromatography) techniques, especially NMR measured at different temperatures, have provided much progress in conformational studies, for example, the dynamics of the phenyl ring rotation in compound **4** [3]. The most stable conformation is the apical cofacial geometry, that is, a face-to-face conformation. Since the rotation of the (3′-methyl) phenyl ring (in broken circle) is very fast at room temperature, the atropisomers interconvert rapidly at ambient temperature, and no difference can be recorded between the chemical shifts of C1 and C5. The fast interconversion shows that its TS barrier is very low. The predicted TS barrier for rotation in the atropisomer conversion is only about 10.8 kcal mol^{-1}, which is very low.

However, if temperature is decreased to −60 °C, the sharp signals would become broad (Figure 9.3). When the temperature is decreased to −63 °C, the difference between the signals of the two carbons can be recorded. When temperature is decreased to −75 °C, the two signals get completely separated from each other. Once the temperature is decreased to −96 °C, the two signals become very sharp since the rotation speed of the 4-methoxy-phenyl becomes very low. This leads to observation of ^{13}C NMR signals for the corresponding pairs of ortho and meta carbons. It shows that the rotation of the (3′-methyl) phenyl ring is very slow. Indeed, with the decrease of temperature in measurement of ^{13}C NMR, the chemical shift difference between the two carbons (ortho and meta) on 4-MeOPh could be recorded as that while 3-MePh rotated.

Other than the C–C single bond restriction leading to axial chirality formation, any restriction of bond rotation would produce atropoisomers. For example,

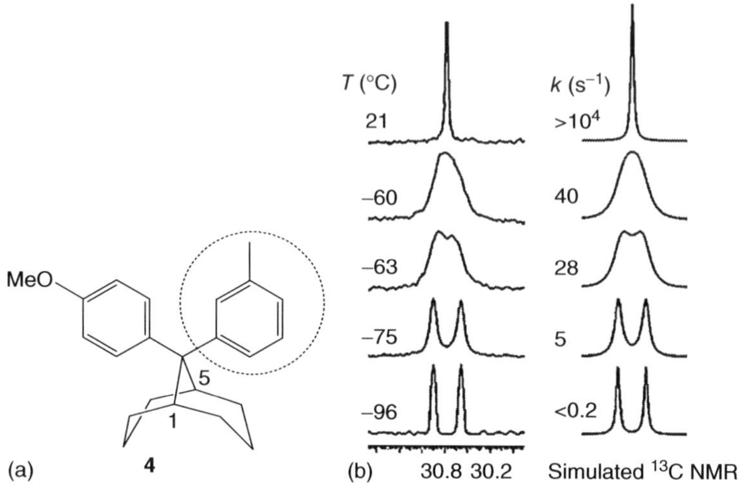

Figure 9.3 Experimental (a) and computed (b) variable NMR spectra (75.45 MHz in CD_2Cl_2) of the carbons in 1,5-positions of the nonane moiety in derivative **4** under different temperatures [3].

the C–P bond restriction gave rise to the chirality (**5**) that could be separated by HPLC under a low temperature of −65 °C [4]. Obviously, as the decrease in temperature can directly “freeze” a molecular behavior, it is easy to determine NMR signals under various temperatures [5]; the low temperature could lower the molecular internal energy and that could slow the rate of the single-bond free rotation. In other words, if the energy barrier is large enough, the difference in the physical characteristics of the molecule can be observed at a high temperature. This gives rise to a large family of chiral compounds, including axial chiral compounds, such as those mentioned in Chapters 2 and 7. Some of them could be used as chiral catalysts in asymmetric synthesis. Similar studies have been widely reported [6].

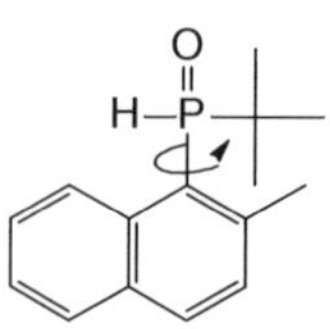

5

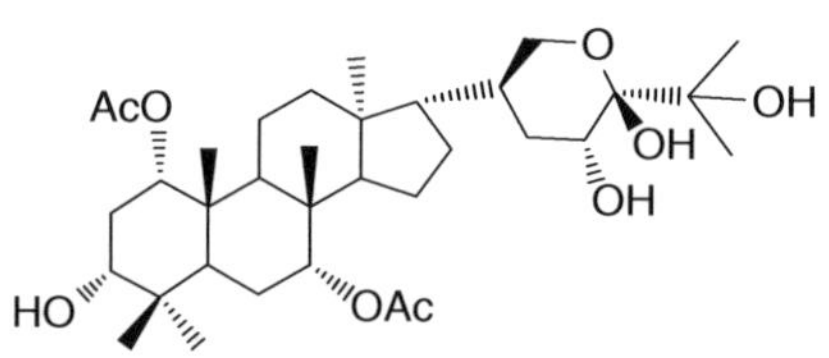

6

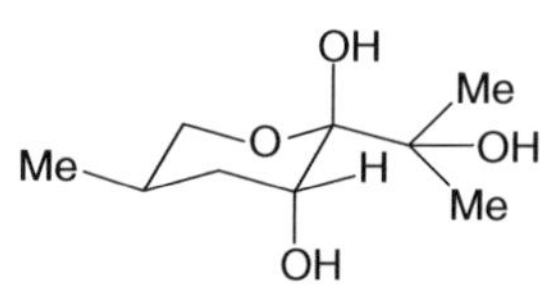

7

There is another methods to be used in conformational and configurational study involving ring-opening and ring-closing changes. This is a frequently used method to study the conformational distribution for natural products. The NMR data of compound **6** exhibited different values in different solvents; it looked as if two compounds existed with different ratios [7]. For example, the ratio of compound **6** in methanol, chloroform, and acetone is 7.8 : 1, 1.7 : 1, and 4.5 : 1, respectively. The higher the polarity, the larger the ratio. Since the compound can crystallize from water rather than from organic solvents, its structure was well established by X-ray.

It is difficult to compute the conformational distributions using the whole molecule. To use a model is a possible way. This molecule is simplified as **7**. It is possible to use the methyl group to represent the diterpene moiety; the isopropyl group could be also another choice.

The model compound can form various isomers such as **8–11** via ring-opening and ring-closing reactions (Eq. (9.1)); each of them can have several conformations. They (**9–11**) can exist in equilibrium in solution (Eqs. (9.2) and (9.3)) [7].

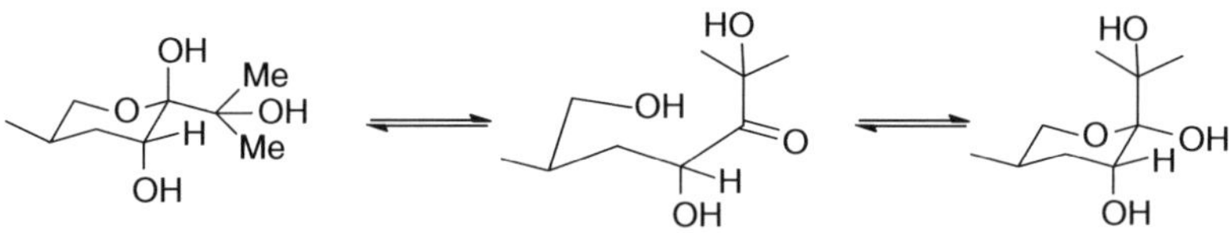

7 **8** **9** (9.1)

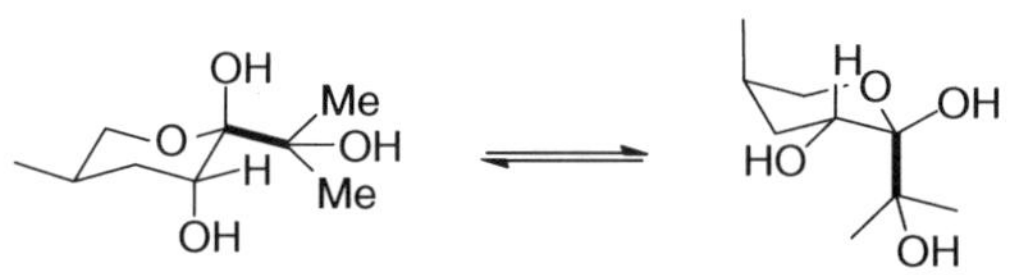

7 **10** (9.2)

9 **11** (9.3)

Because of the single-bond rotation (bold bond in **7, 9–11**), each one would form three conformations. For **7** and **10,** they form conformations **a, b,** and **c,** and for **9** and **11** they form **d, e,** and **f** (Figure 9.4).

First, optimizations were performed at the B3LYP/6-31G (d) level for the relative energy predictions. It was found that **9e, 10a, 10b,** and **10c** had energy magnitudes of 14.2, 10.2, 12.3, and 18.2 kcal mol^{-1} (Table 9.2, entries 2 and 3), respectively. The energy data are so large that their contribution in solution can be ignored.

Only the conformations with relative energy below 10 kcal mol^{-1} were used in further optimization; they are **7a–7c**, **9d, 9f,** and **11d–11f**. Different methods were used for further computations. One economical method could give reasonably good predictions. For example, all B3LYP/6-31G(d)-optimized geometries were used for single-point energy (SPE) computations using B3LPY and MP2 theory using the 6-31++G(d,p) basis of sets in methanol, acetone, and chloroform, respectively. The ratios of the two configuration isomers (**7** versus **9** and **11**) were

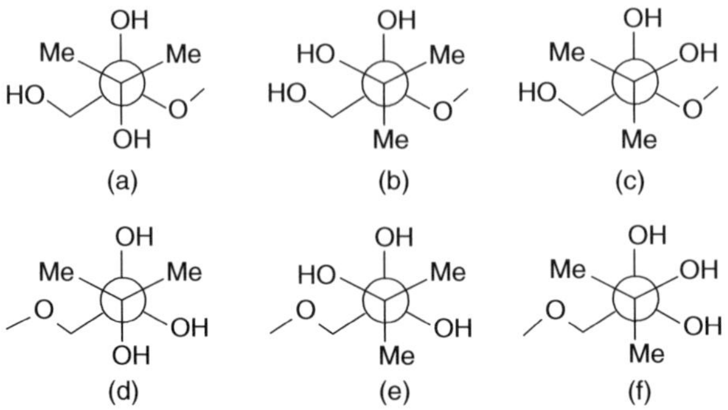

Figure 9.4 Six Newman projects for compounds **7** and **10** (**a**–**c**) and **9** and **11** (**d**–**f**) [7].

Table 9.2 Relative energy for all conformations at the B3LYP/6-31G(d) level (kcal mol^{-1}) [7].

Entry	Compound	Conformations and relative energy					
		a	b	c	d	e	f
1	**7**	5.186	4.637	1.992	—	—	—
2	**9**	—	—	—	4.839	14.227	4.158
3	**10**	10.227	12.336	18.223	—	—	—
4	**11**	—	—	—	1.206	3.547	0.000

Table 9.3 Relative SPEs using B3LPY and MP2 theories at the basis sets of 6-31++G(d,p) using B3LYP/6-31G(d)-optimized geometries in three solvents (with energy in kcal mol^{-1}) [7].

Entry	Conform	Δ*E* (MeOH)		Δ*E* ($CHCl_3$)		Δ*E* (acetone)	
		B3LYP	MP2	B3LYP	MP2	B3LYP	MP2
1	**7a**	2.348	3.012	2.698	3.338	2.127	2.791
2	**7b**	3.009	3.428	3.022	3.452	2.700	3.117
3	**7c**	0.000	0.000	0.000	0.000	0.000	0.000
4	**9d**	2.554	3.362	2.723	3.533	2.474	3.289
5	**9f**	3.657	4.458	2.936	3.720	3.223	4.022
6	**11d**	2.415	2.156	1.878	1.604	2.474	2.219
7	**11e**	2.623	2.716	2.606	2.693	2.527	0.618
8	**11f**	1.205	1.085	0.313	0.153	1.106	0.984
9	**7**:(**9**+**11**)	5.2:1.0	5.0:1.0	1.5:1.0	1.2:1.0	5.1:1.0	4.4:1

1.5 : 1, 5.1 : 1, and 5.2 : 1 using the B3LYP/6-311++G(d,p) method when the solvent was changed from chloroform to acetone to methanol, respectively (Table 9.3, entry 9). The ratios were 1.2 : 1, 4.4 : 1, and 5.0 : 1, respectively, using the MP2/6-311++G (d,p) method in the above case. It shows that the higher the polarity of solvent, the higher the fraction of **7** in the solvent.

MP2/6-31+G(d)-optimized geometries that were used for SPE computation at the B3LYP/6-311++G(d,p) level did not give good predictions. Even when all conformations were optimized in the three solutions respectively, they could not give good predictions. The use isopropyl in place of methyl in the model might improve the results.

9.2 Effect of Conformation on Reactions

The relationship between conformation and reaction activity was reported as early as in the 1940s [8]. However, this did not become a key point in asymmetric synthesis. The reason may be very complex. The major one may be that it is not easy to find out the correct conformation population, including the most and second most stable conformations (it also appears in the absolute configuration (AC) assignment study).

The quantity of a chiral catalyst used in reactions is generally <30 mol%. However, the axially chiral auxiliary should react with at least the same mole as those of substrate and form a new compound. The stereogenic center in the new molecule could control the chirality formation in the next reaction. For example, in a remote chirality control reaction [9], the aldehyde could react with the auxiliary ester of crotonic acid ((*E*)-but-2-enoic acid) of **12**, which was derived from **13** and crotonic acid (Eq. (9.4)). In this reaction, the chirality from the restricted rotation of the C–C bond controlled the new stereogenic center formation in the new lactone.

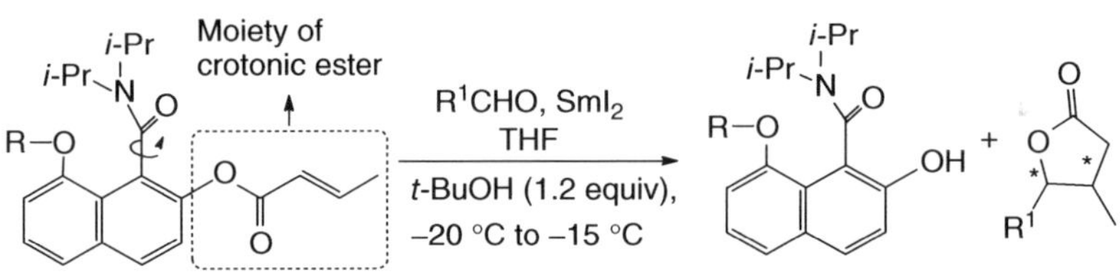

12 **13** **14** (9.4)

In the reactions, the substrate **12** played two roles: one as a substrate, and the other as a catalyst. Partial experimental results are given in Table 9.4.

In asymmetric synthesis, use of chiral catalysts provides researchers with a good tool to control different chirality formations. It is used to compute TS barriers to understand the experimental results. In many cases, because of the complexity of chiral catalyst, it is possible to use a simplified model in TS barrier calculations. The method has been widely used in various research areas. For example, some simplified models were used to explain the experimental results, and the calculations agreed well with experiments. However, this is not always possible. When used to explain experimental results using a series of chiral catalysts, one cannot use the simplified model. An example was given in Chapter 7: the odd–even carbon effect. The relationship between the free ligand conformations and final

Table 9.4 Effect of different axial chiralities on enantioselectivity [9].

Entry	Catalyst	R^1	ee (%; *cis,trans*) a)	*cis* $[\alpha]_D$, b) AC
1	(−)-(a*R*)-	*n*-Bu	>99, 96	+80.8, (3*R*,4*R*)
2	(+)-(a*S*)-	*n*-Bu	>99, 95	—
3	(+)-(a*S*)-	*i*-Pr	>99, 75	—
4	(+)-(a*S*)-	*t*-Bu	96, 61	—
5	(−)-(a*R*)-	Cy	80	+46.5

a) Determined by GC; Cyclosil B (30 m × 0.25 mm i.d. and 0.25 μm).
b) In MeOH (about 0.20–0.65). Reported [10]: (3*R*,4*R*)-, +73.8 (94% ee); (3*S*,4*S*)-, −74.3 (96% ee); (+)-, +55.7 (96% ee); (−)-, −56.8 (97% ee).

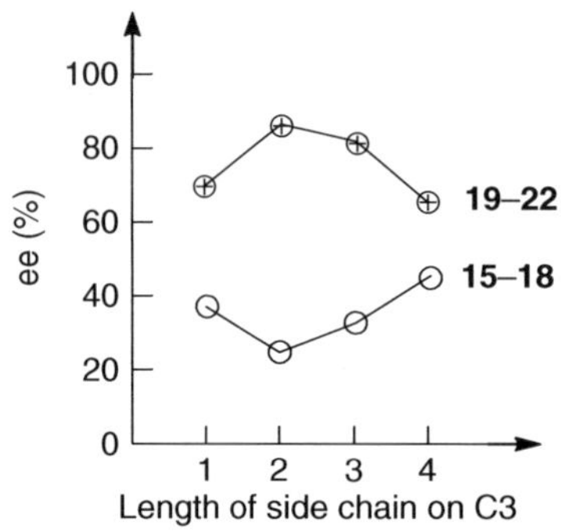

Figure 9.5 Two series of chiral catalyst structures and their ee values in the addition of diethylzinc to benzaldehyde [8].

ee values was constructed. It was found that when substituent of chiral ligands changed from Me and Et to *n*-Pr and *n*-Bu, the recorded ee values of the addition product **24** exhibited regular changes (Figure 9.5, Eq. (9.5)) [11]. Obviously, the use of a simple model cannot explain the relationship between the catalyst structure changes and their ee values [11].

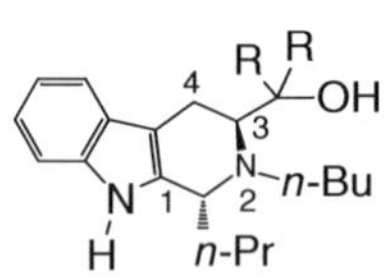

15 R = Me
16 R = Et
17 R = *n*-Pr
18 R = *n*-Bu

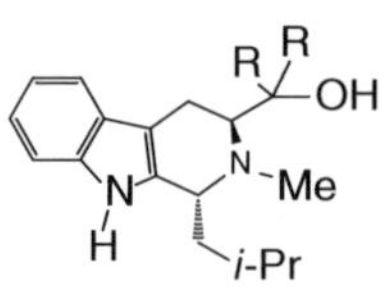

19 R = Me
20 R = Et
21 R = *n*-Pr
22 R = *n*-Bu

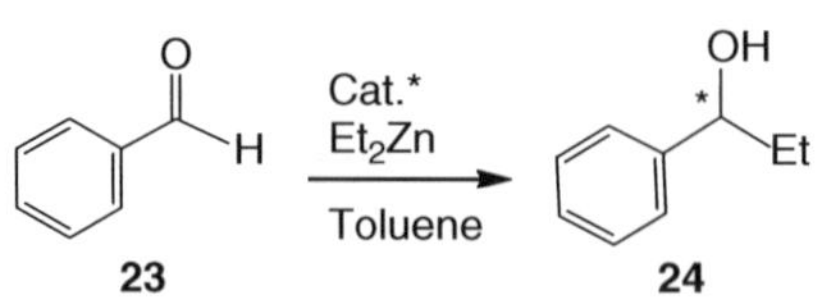

(9.5)

No doubt, the changes must be related to their TS barrier magnitudes. In this catalytic procedure (**TS-1**), there are many freely rotating single bonds, and this leads to the formation of a large number of TS structures.

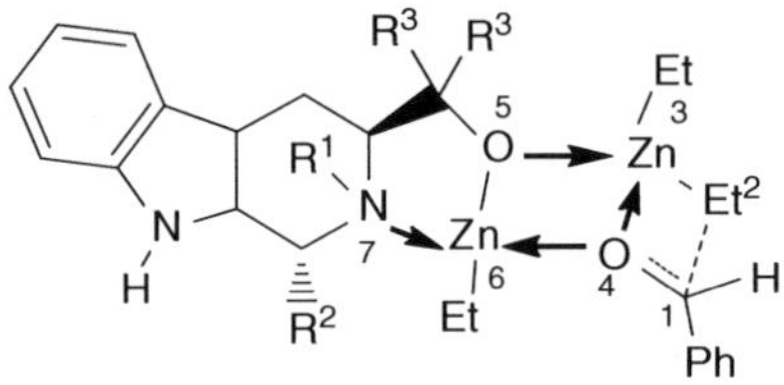

TS-1

In this case, even if the most unreasonable structures were removed, the number of TSs needed to be investigated should be in hundreds. If the TS calculations are performed at the (Hatree–Fock) HF/6-31G(d,p) level, the computation will take an enormous time. Such capacity is not readily available at the present time.

It was assumed that the distribution of free ligands in solution would affect the percentage of TS structures, and this effect finally turned into selectivity in reactions. Therefore, it may be possible to find the relationship between the distribution of free ligands and their ee values. The catalysts contain a N atom, which can overturn in the molecule. Because of the overturn, the catalyst contains cis and trans conformations. Possibly, the energy difference ($\Delta E_{(\text{trans-cis})}$, in kcal mol^{-1}) between the cis and trans conformations corresponds to the different TS geometry distributions, which is related to the recorded ee values. Because of the large molecular structure, the HF/6-31G(d,p) method was selected for the computation of energy. Other series of chiral catalysts (**25–28**, **29–32**) had similar catalytic effect. The computed energy differences predicted at the HF/6-31G(d,p) level are listed in Table 9.5 using the catalysts.

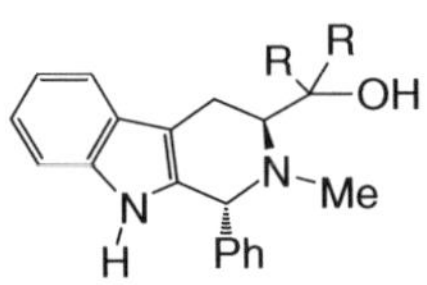

25 R = Me, 100%
26 R = Et, 69.5%
27 R = *n*-Pr, 87.6%
28 R = *n*-Bu, 67.8%

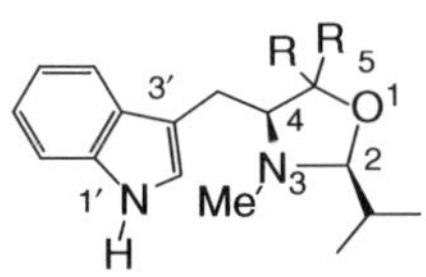

29 R = Me, 69.1%
30 R = Et, 89.4%
31 R = *n*-Pr, 72.0%
32 R = *n*-Bu, 71.3%

The relationship between the energy difference and the *R*% content is constructed on the basis of the experimental and theoretical analysis. The predicted ΔE and the *R*% are illustrated in Figure 9.6.

It was found that the conformational distribution has a large effect on various reactions. This study is extremely valid and important. However, to do the conformational search study is not easy even today. Many types of software have been developed. Some of them were used to test their function in the lab. Some of them may be easily used, but their function is limited. The major problem is that they cannot search find the stable conformation with low energy. Sometimes, it is the lowest conformation.

Table 9.5 The predicted energy differences between the different conformations [11].

Entry	R	$\Delta E^{a)}$	*R* (%)	Entry	R	$\Delta E^{a)}$	*R* (%)
1	**15** (Me)	−1.02	67.5	9	**25** (Me)	−4.27	79.2
2	**16** (Et)	−0.37	61.5	10	**26** (Et)	−1.44	77.9
3	**17** (*n*-Pr)	0.43	65.8	11	**27** (*n*-Pr)	−1.89	72.6
4	**18** (*n*-Bu)	−1.72	71.2	12	**28** (*n*-Bu)	+3.98	72.0
5	**19** (Me)	0.57	85.8	13	**29** (Me)$^{b)}$	0.08	51.7
6	**20** (Et)	2.46	92.6	14	**30** (Et)	2.57	80.9
7	**21** (*n*-Pr)	−0.35	90.9	15	**31** (*n*-Pr)	0.55	50.6
8	**22** (*n*-Bu)	−0.51	82.9	16	**32** (*n*-Bu)	1.33	64.4

a) Positive ΔE means the trans isomer is more stable than the cis isomer. It is the difference of trans and cis isomer from **15** to **27** ($\Delta E_{(cis\text{-}trans)}$).
b) Weighted average of ΔE values of two trans geometries from **29** to **32**. See ref. [11], for details.

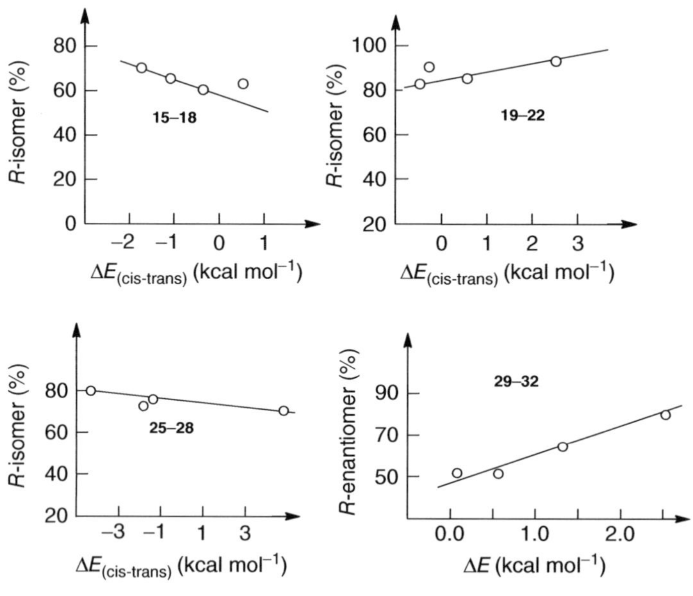

Figure 9.6 The relationship between the energy differences and *R* percentages [11].

Notice

It is suggested to use two or three different conformational search software for a same work. After the low-energy conformations are obtained, it is valuable to further optimize them using an economical quantum method, such as HF/6-31G(d). Then, by comparing all HF/6-31G(d)-optimized geometries using the different software, duplicate geometries can be removed, and all the conformations can be used for other purposes, such as optimization at the B3LYP/6-311+G(d) level for looking for the lowest energy conformations and the second lowest conformation for Mosher ester tests. Or they may be used for OR, ECD, VCD, and/or ROA (Raman optical activity) calculations.

9.3 Regioselective Reactions

When a molecule has different functional groups, the more active group may take part in the reaction while the other groups are preserved under the same reaction conditions. The regioselective reaction may accompany other reactions such as enantioselective reactions. This will greatly help organic synthetic chemists in different methodology study. For example, under the catalysis of ligand **33,** the ketone ester could react with dialkyne (reduction-coupling reaction). In this procedure, different R^1 groups can lead to form different products (Eq. (9.6)) [12].

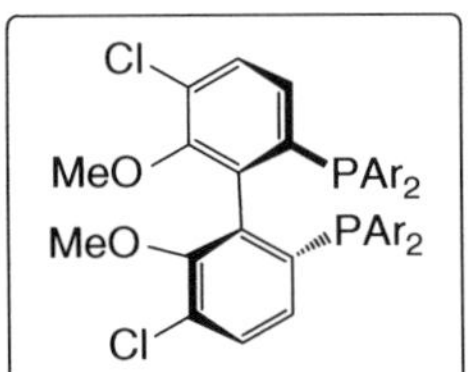

(*R*)-**33**

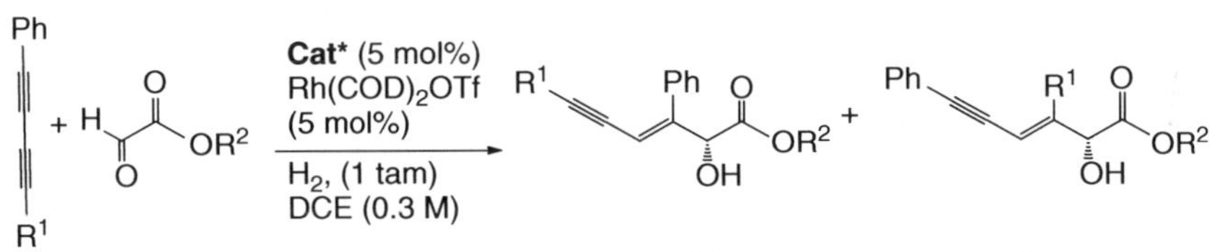

34 **35** **36** **37**

(9.6)

When the ligand (*R*)-**33** was used in the reduction-coupling reactions, different chemoselectivity of formation of **36** and **37** was obtained. The results are summarized in Table 9.6. With changes in the reaction conditions, the ratio of **36** : **37** showed different selectivity.

Regioselective reaction does not need assistance of a chiral ligand since the reaction does not involve the formation of new chirality centers. Because of the spatial interactions of various groups, different substrates have different advanced conformations. In reactions, the advanced (major) conformations have a large effect on the formation of products.

Table 9.6 Regioselective reaction results of reduction-coupling reaction [12].

Entry	R^1	R^2	T (°C)	36 : 37	Yield (%)
1	Ph	Et	25 (45)	1 : 1	83
2	Me	Et	25	3.3 : 1	78
3	Me	*t*-Bu	25	3 : 1	52
4	*t*-Bu	Et	40	>99 : 1	78
5	TMS	Et	25	1 : >99	74
6	TMS	Et	40	1 : >99	84

Not only addition reaction could produce regioselective products; reductive hydration of alkynes could also form regioselective compounds [13]. The different procedures led to two different products using catalysts **38, 39,** and **40,** respectively. Catalysts **38** and **39** were used for hydration catalysis, and **40** was used for transferring the hydrogenation procedure.

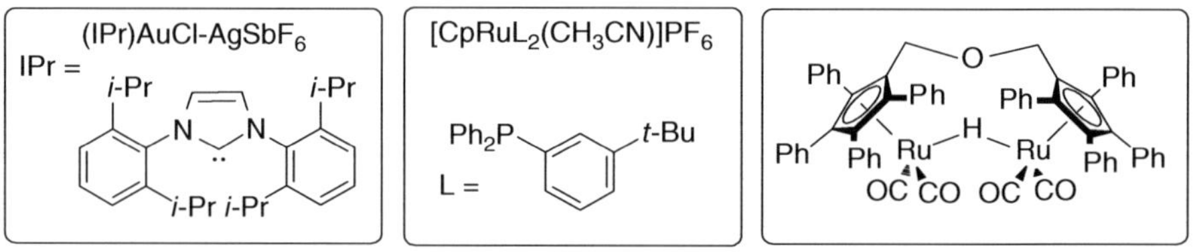

38 **39** **40**

The two reaction routes are illustrated in Scheme 9.1.

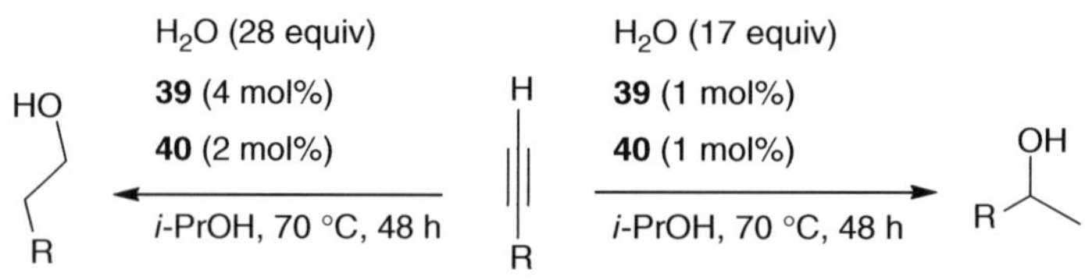

42 **41** **43**

Scheme 9.1 Two reaction routes to form the regioselective products [13].

When the reaction condition in Scheme 9.1 is relatively mild, its use may bring big convenience in conversion of alkyne to alcohols. Its scope of application involves different substrates and exhibits very good regioselectivities (Table 9.7). In this procedure, if a molecule has a C=C bond, this double bond cannot be reduced while the alkyne is converted into the corresponding functional groups (Table 9.7, entries 7 and 8) with high yields and regioselectivities.

Table 9.7 The selected substrates and their two regioselective products [13][a)] .

Entry	substrate	Product 42 (yield, %)	43 (yield,43 : 42)[b]
1	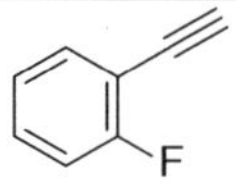	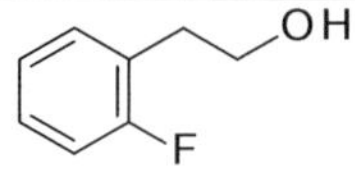 96%	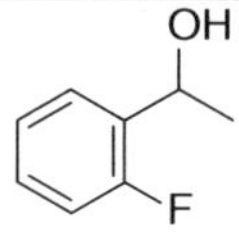 90%, 58 : 1
2	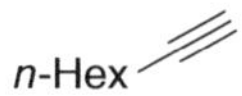	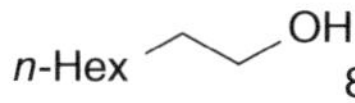 84%	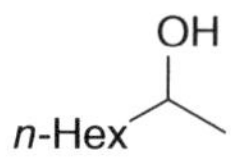 81%, 53 : 1
3	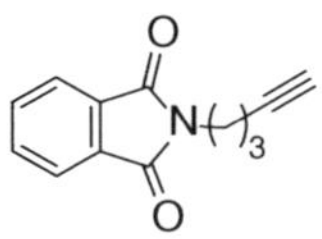	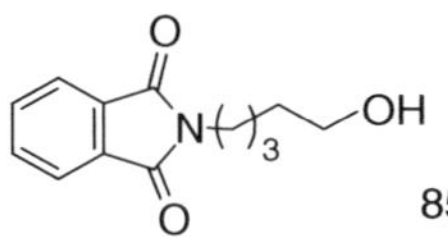 85%	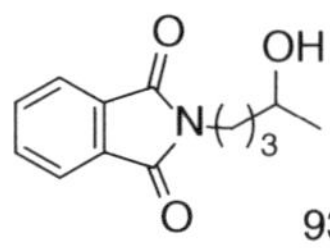 93%, 45 : 1
4		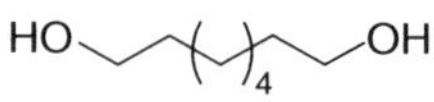 80%	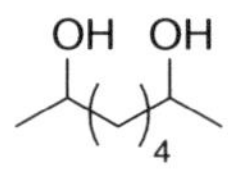 81%, 100 : 1[c]
5	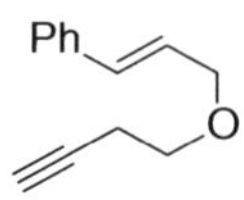	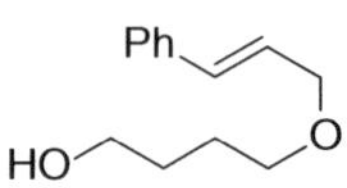 86%	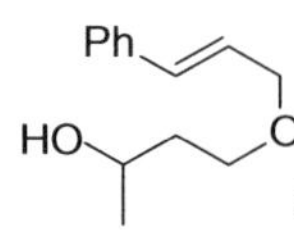 90%, 30 : 1

a) 0.5–1.0 mmol scale, 0.5 M of substrates in isopropanol solution.
b) Determined by ^{1}H NMR of unpurified product mixtures.
c) 0.2 mmol substrate, 85 equiv of H_2O, 5 mol% Au(IPr)Cl, 5 mol% AgO_2CCF_3, and 5 mol% **40** in 2 ml 2-propanol, 1 : 1 dr (^{13}C NMR).

Characteristics of a chiral molecule itself would affect the formation of the reaction products, such as the Cram rule found in reductions or additions. Another example is the use of a molecule with two (*E*)- or (*Z*)-double bonds to control the product's stereochemistry in cycloadditions [14]. The catalytic [5 + 2 + 1] cycloaddition has more advantages than the [5 + 2] reaction (Scheme 9.2) based on the predicted TS barriers listed in the scheme [14].

The experimental results confirmed the theoretical predictions. For example, the final products were all products formed via [5 + 2 + 1] cycloaddition, and the stereochemistry of the products obtained had good agreement with the contents of double bond of (*E*)- or (*Z*)-structure in starting materials. Partial results are listed in Scheme 9.3.

Ring-enlargement reaction is a good protocol to synthesize different substituted ring-containing derivatives. Cycloisomerization of cyclopropenyl ketones could afford two furan derivatives under the catalysis of metallic salts (Eq. (9.7)) [15a]. The effect of the salt on the derivative reactions is conclusive in many cases. The results are summarized in Table 9.8.

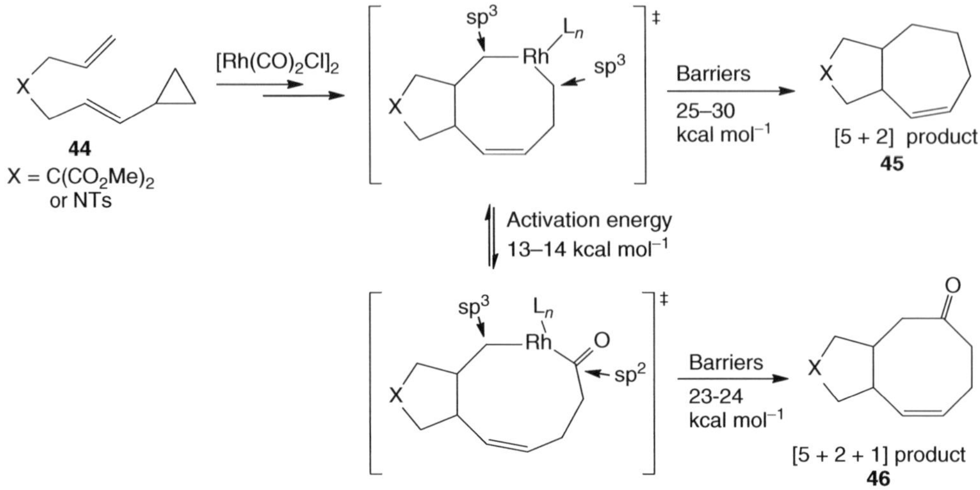

Scheme 9.2 The competition reaction between [5 + 2 + 1] and [5 + 2] catalyzed by Rh(I) [14].

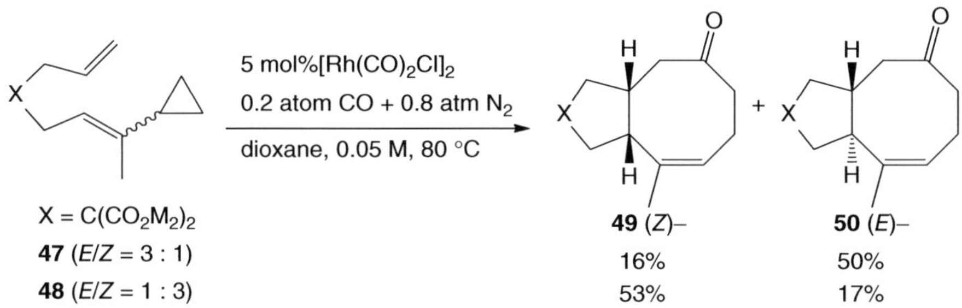

Scheme 9.3 Stereochemistry of products in [5 + 2 + 1] reactions [14].

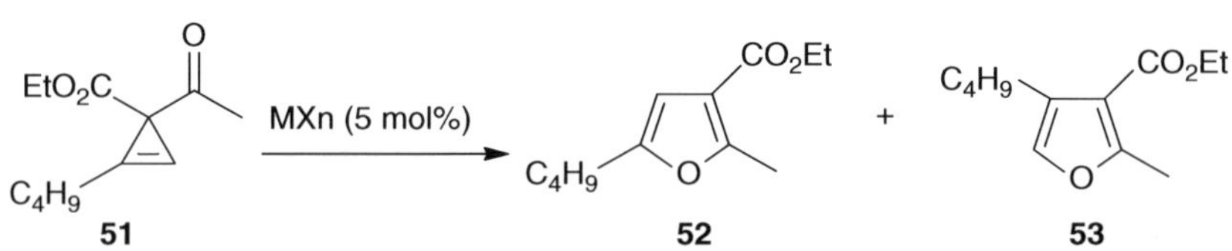

(9.7)

It was found that $FeCl_3$ could promote the formation of **52** and **53** with a ratio of 99 : 1 (entry 4), while CuI led to the ratio of 1 : 99. The chemoselectivity is highly obvious. Other salts such as $PdCl_2(CMe_3CN)_2$ could also afford high yields of 98%. A plausible mechanism was proposed on the basis of experimental results (Scheme 9.4) [15a].

Cyclopropenyl ketone **51** would form the intermediate int-**2** via path *a* (cycle A), and then undergo *β*-decarbopalladation to produce int-**3**. Subsequently, it

Table 9.8 Regioselectivity in cycloisomerization of cyclopropenyl ketones [15a][a)].

Entry	MXn	Solvent/T (°C)	$T^{b)}$	52 : 53[c)]	Yield (%)[d)]
1	Non	Acetone/reflux	19	—	0
2	$CoCl_2$	Acetone/reflux	24	65 : 35	7
3	$NiBr_2$	Acetone/reflux	24	20 : 80	20
4	$FeCl_3$	Acetone/reflux	12	99 : 1	25
5	$RuCl_3$-$3H_2O$	Acetone/reflux	14	98 : 2	61
6	$PdCl_2$	Acetone/reflux	11	94 : 6	60
7	$PdBr_2(PhCN)_2$	Acetone/reflux	17	94 : 6	53
8	PdI_2	Acetone/reflux	12	Trace	—
9	$PdCl_2(CH_3CN)_2$	CH_3CN/80	17	88 : 12	25
10	$CuCl_2$	Acetone/reflux	12	12 : 88	75
11	CuI	CH_3CN/80	9	<1 : 99	91[e)]

a) Reaction was carried out using **51** (0.5 mmol), catalyst (5 mol%) in solvent (2.0 ml).
b) Unit in hours.
c) The ratio was determined by ^1H NMR using reaction mixture.
d) Unless otherwise specified, isolated yields of two isomers.
e) Isolated yield of the major isomer.

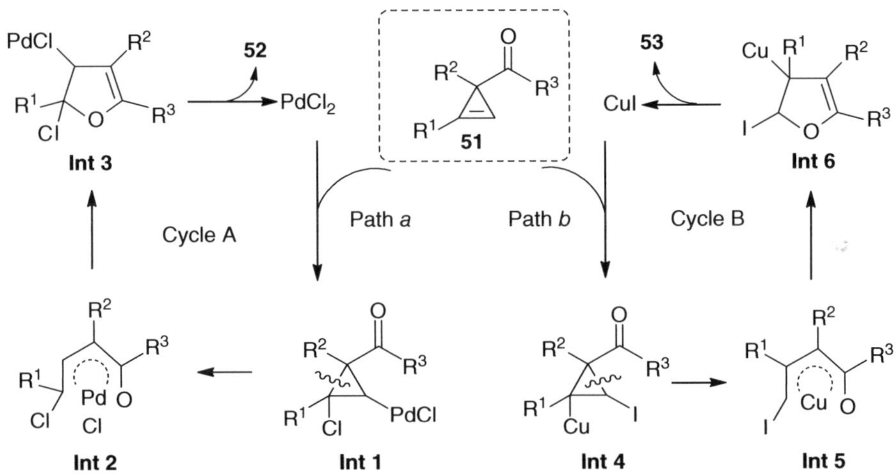

Scheme 9.4 The plausible catalytic routes leading to two products **52** and **53** in experiments [15a].

could form the final int-**4**, and form **52** via a β-halide elimination. In path *b* (cycle B), a β-decarbocupration would form delocalized int-**4**, which would lead to an intromular endo-mode insertion of the C=C bond into the oxygen–copper bond of int-**5**, subsequently to int-**6**, and finally to afford **53** via β-halide elimination.

Cyclization reaction can be used to construct novel poly-ring-containing compounds, such as regioselective aerobic oxidative cyclization of aromatic, heteroaromatic, and alkenyl acids with alkynes [15b]. This cyclization can be performed in the presence of catalytic amounts of [{$RuCl_2$(p-cymene)}$_2$], $AgSbF_6$, and $Cu(OAc)_2{\cdot}H_2O$. It affords the corresponding isocoumarin derivatives (Eq. (9.8)). The reaction exhibited good regioselectivity in the formation of the corresponding lactones.

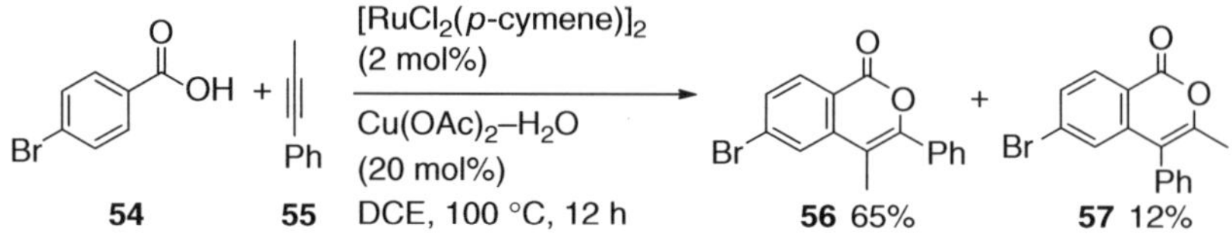

(9.8)

The enantioselective Friedel–Crafts alkylation of indoles with α,β-unsaturated aldehydes and ketones is an important approach for the synthesis of biologically active indole alkaloids. A regioselective and enantioselective 1,2- and 1,4-addition reaction was reported (Scheme 9.5) [16]. This reaction leads to formation of two chiral products with high ee values but different AC when chiral catalysts **58** and

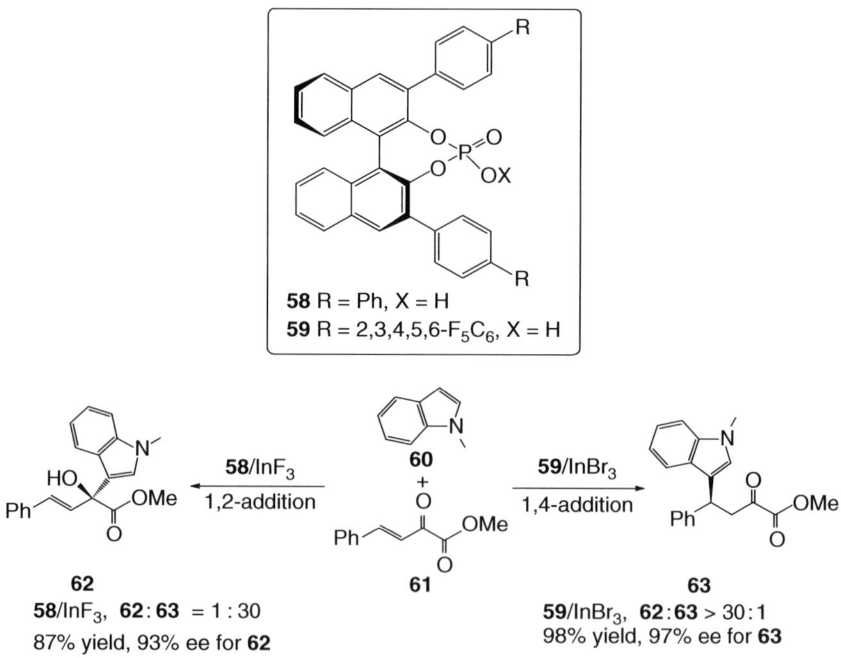

Scheme 9.5 Regioselective and enantioselective 1,2- and 1,4-addition reactions using catalysts **58** and **59** [16].

59 were used, respectively. The difference of the two catalysts is in the use of salts. InF_3 was used to convert the reaction to **62**, while $InBr_3$ was used to transfer it to **63.**

The reaction condition is relatively mild; for example, **60** (0.10 mmol), **61** (0.12 mmol), and **62** (10 mol%) with Lewis acid (10 mol%) were mixed together and heated to 70 °C after molecular sieves were added (4 Å, 20 mg). The yields could be high, up to 99% but mostly 85–90%.

A infrequently used starting material, allene (propa-1,2-diene) derivative, was selected in additions under catalysis of a Rh-containing reagent. This reaction exhibited good regioselectivity when the substituent at the α-position of ketone was either an alkyl or a carbonyl group (–COR) (Scheme 9.6). When the substituent is alkyl group, the addition product is **66** [17]. Once it is a group such as $-COR^1$, the product should be **67** [18].

In the decarboxylative addition of β-ketoacids to allenes, the choice of the right organic auxiliary is a key step. The favorite one is the so-called DPPF (**68**). The catalyst $[Rh(cod)Cl]_2$ was used in the addition (Eq. (9.9)) [18]. Several samples are illustrated below. Moderate yields were achieved in the additions. In addition to the singly substituted **65,** disubstituted allene **69** could be used in the additions; similar results were achieved, and the products are summarized below.

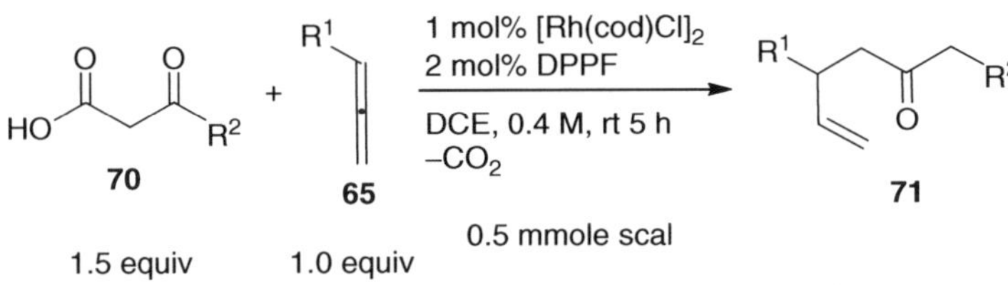

(9.9)

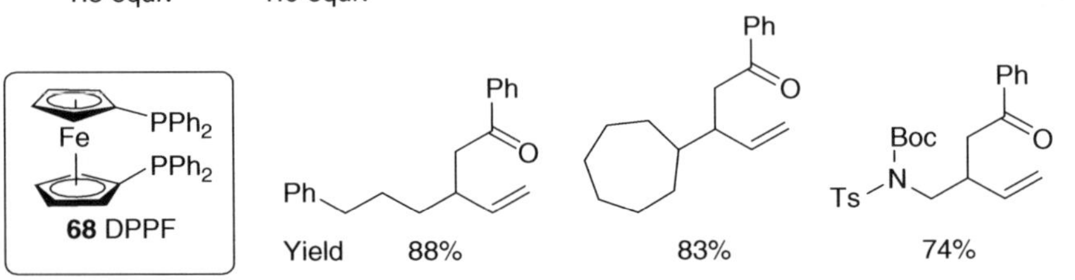

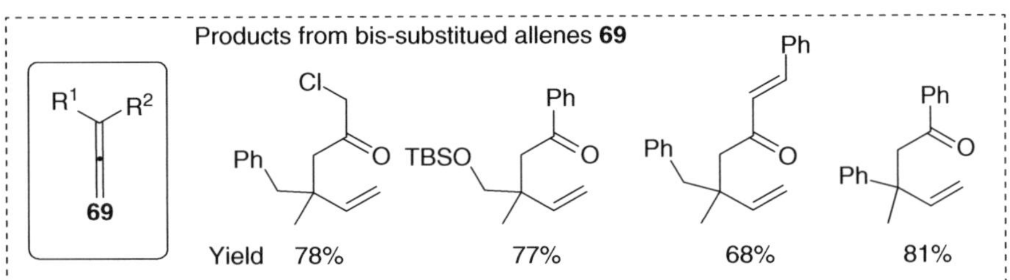

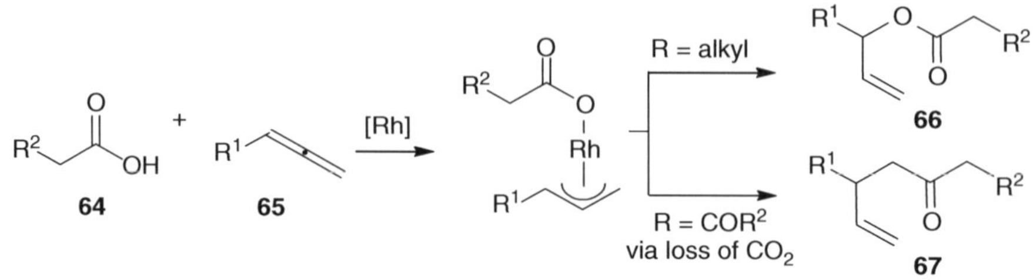

Scheme 9.6 Regioselective decarboxylative addition of β-ketoacids to terminal allenes [18].

Friedel–Crafts alkylation of benzene derivatives is an important protocol to afford the corresponding poly-ring-containing compounds. Metallic salts, such as $AlCl_3$, are necessary. $FeCl_3$ is not a frequently used salt in these reactions. However, it could play an unexpected role in some cases [19]. When sulfonamide **72** and disubstituted alkynes **73** were mixed together, $FeCl_3$ was found to be an effective catalyst to promote the Friedel–Crafts cyclization, while $ZnCl_2$, $CuCl_2$, $Pd(OAc)_2$, $AlCl_3$, or $Bi_2(SO_4)_3$ did not exhibit valid catalysis (Eq. (9.10)).

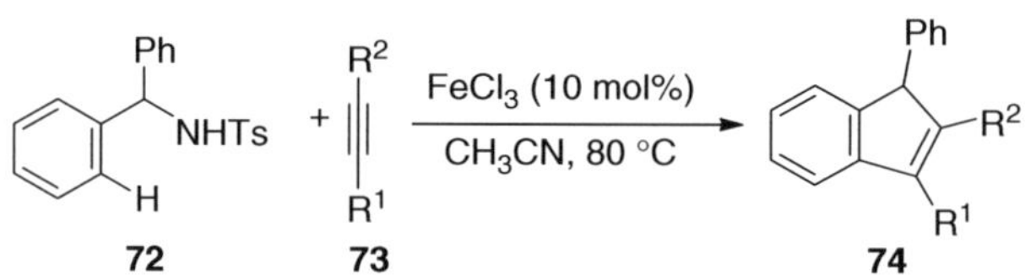

(9.10)

Generally, if R^2 is electron-withdrawing groups its position in product should be oriented as the illustrated in Eq. 9.10. Partial experimental results are listed in Table 9.9. This is a traditional Friedel–Crafts alkylation via cleavage of the sp^3 carbon–nitrogen bond to generate benzyl cation intermediates.

Zinc carbenoids from diethylzinc reaction with MeI can be used for the synthesis of complex natural products. In this procedure, different reaction conditions lead to similar ring enlargement reactions but different products (Simmons–Smith reaction). For example, the following reaction gave regioselective products (Eq. (9.11)) [20].

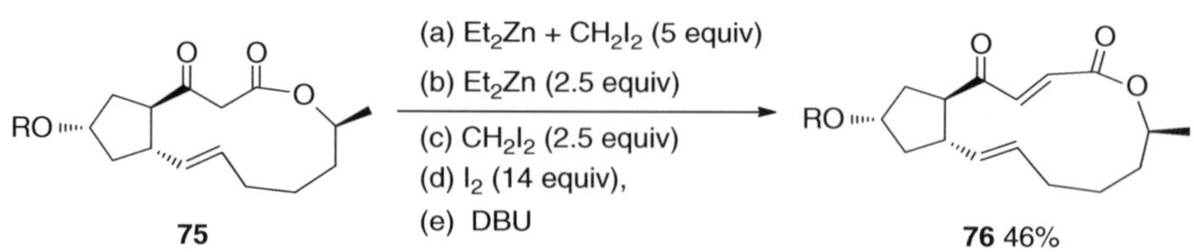

(9.11)

Table 9.9 $FeCl_3$-catalyzed regioselective synthesis of indene derivatives from sulfonamide and disubstituted alkynes [19][a].

Entry	R^1	R^2	Time (h)	Yield[b] (%)
1	Ph	Ph	12	66
2	4-$MeOC_6H_4$	Ph	5	75
3	Ph	4-$NO_2C_6H_4$	24	53
4	Ph	*n*-Pr	10	60
5	Ph	COOEt	24	74
6	Ph	COPh	24	66
7	SPh	Ph	3	43
8	SePh	Ph	6	58
9	Br	*n*-Bu	24	61

a) Reaction conditions: sulfonamide **70** (0.20 mmol), alkyne **71** (0.24 mmol), $FeCl_3$ (10 mol%), nitromethane (2.0 ml), 80 °C.
b) Isolated yields.

An interesting regioselective deprotonation to afford head-to-tail-type oligothiophenes **79–81** from 3-substituted thiophene **77** is introduced using TMPMgCl-LiCl (**82**) and 2,2,6,6-tetramethylpiperidine (TMPH) as a base [21]. Scheme 9.7 summarizes the relevant results. No other isomer such as **83** or **84** formed during the procedure.

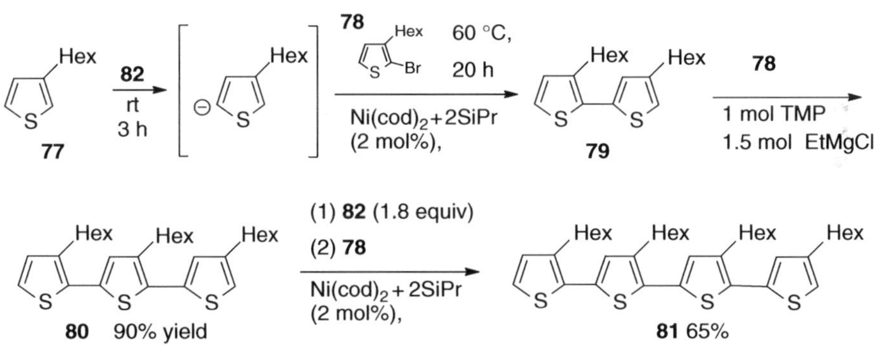

Scheme 9.7 Formation of head-to-tail-type oligothiophenes [21].

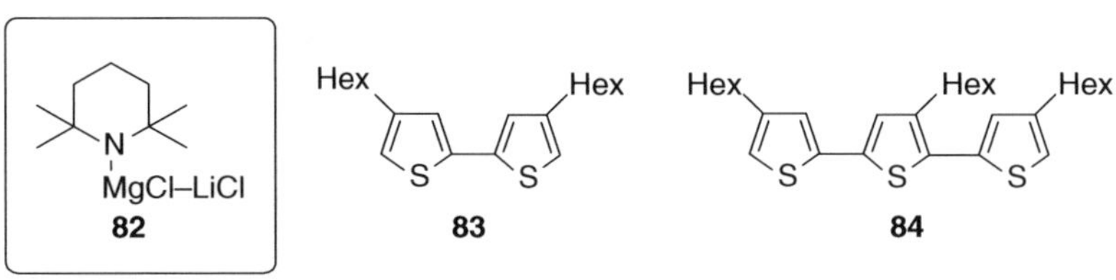

9.4
Diastereoselective Reactions

As an important reaction in organic stereochemistry, diastereoselective reactions occupy a unique position. By use of various chiral catalysts, or by the molecular characteristics themseves, this reaction could afford different chiral compounds used widely in chemistry, pharmaceutical studies, and similar areas.

9.4.1
Diastereoselective Additions

The ring-enlargement reaction is a useful method to afford various chiral cyclic compounds promoted by chiral catalysts. Imides with the α-cyclopropane structure could react with an aldehyde to form the corresponding lactams using $Sc(OTf)_3$ (Eq. (9.12)) [22a]. The *dr* values can reach up to 94 : 6; the results are listed in Table 9.10. In this example, no catalyst was used. Therefore, no *ee* values are reported. The obtained diastereomer is the mixture of two isomers.

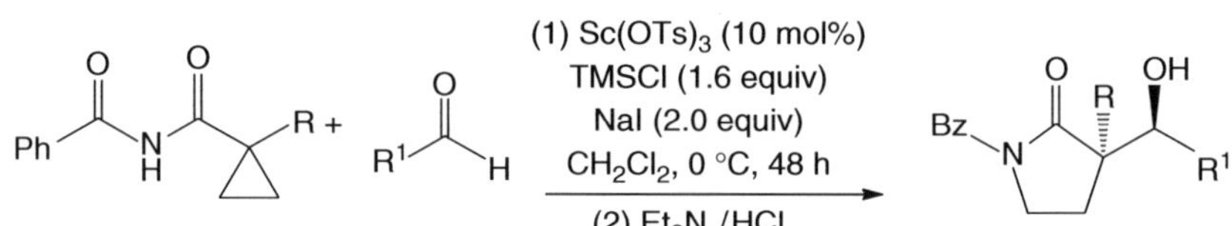

85 86 87

(9.12)

Chiral ligand **88** is a kind of B-containing catalyst that is widely used in asymmetric synthesis [22b]. The lactone **89** was converted into the iridoid **90** by the reaction with vinyl lithium. Addition of **90** to α,β-unsaturated ketone afforded a new aldehyde **91** under the catalysis of **88.** It could be further transformed to ketone **92** (Scheme 9.8) [22b].

Table 9.10 Diastereoselective synthesis of chiral lactams [22a].

Entry	R	R¹	Yield (%)	dr
1	Me	Ph	97	89 : 11
2	Me	4-Cl-Ph	87	88 : 12
3	Me	2-Furyl	70	83 : 17
4	Me	(*E*)-PhCH=CH	78	81 : 19
5	Me	*c*-C_6H_{12}	87	87 : 13
6	Me	(*Z*)-1-hex-3-enyl	90	90 : 10
7	Allyl	Ph	82	83 : 17
8	H	Ph	84	94 : 6

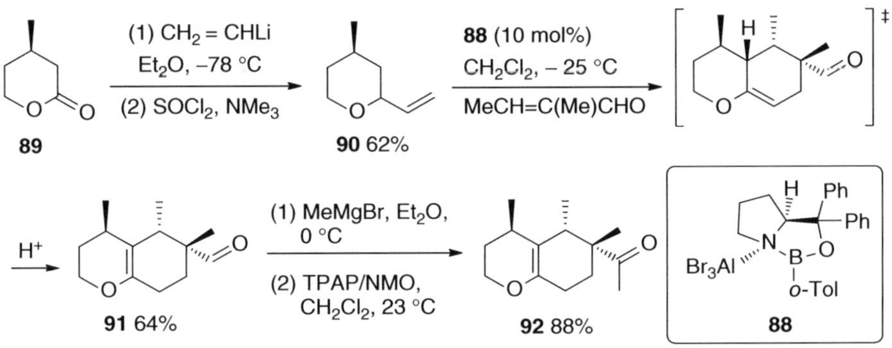

Scheme 9.8 Formation of a new ketone catalyzed by the B-containing catalyst **88** [22b].

Metallic reagents, such as Zn reagents, could play a good role in the ring formation. In absence of chiral catalysts, similar addition of derivative of allylzinc bromide to α-substituted ketone such as **94**, which could be obtained from **93** with 72% yield, could produce a new ring system with high diastereoselectivity (Scheme 9.9). The diastereoselectivtiy could reach up to 99 : 1 with a yield of 90% [23].

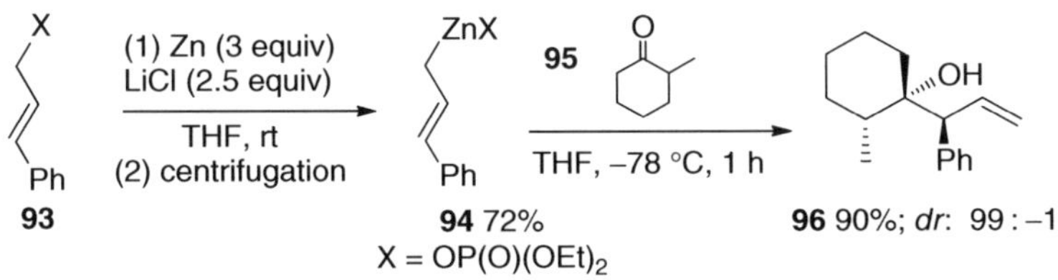

Scheme 9.9 Diastereoselective addition of a metallic Zn reagent to cyclic ketone [23].

Synthesis of a three-membered ring is a big challenge in organic synthesis. One method is to use the Kulinkovich reaction: that is to use EtMgCl to react with an ester functional group, such as in the procedure to synthesize (*S*)-cleonin (**99**, Scheme 9.10) [24]. However, this method has no diastereoselectivity.

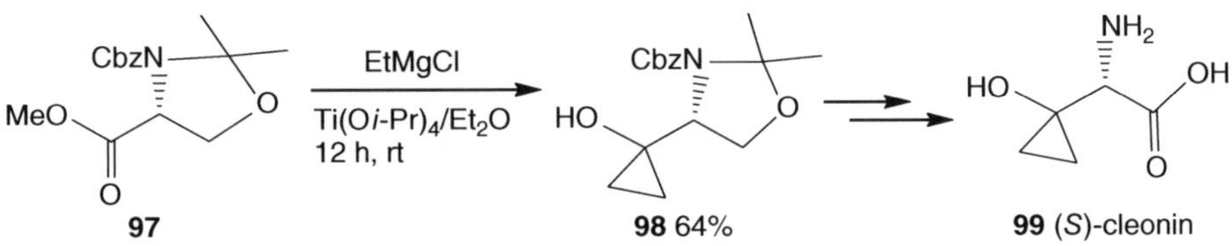

Scheme 9.10 Kulinkovich reaction used in (*S*)-cleonin synthesis [24].

To form products with a three-membered ring and different diastereoselectivity, use of zinc carbenoids is an effective protocol [25]. Its formation is similar to the synthesis of the C carbenoid from CH_2Cl_2 under a base condition. By using chiral catalysts, such as (−)-MIB (**100**), this method can be used for the construction of three-membered rings (Scheme 9.11). Groups R^1 to R^4 could be

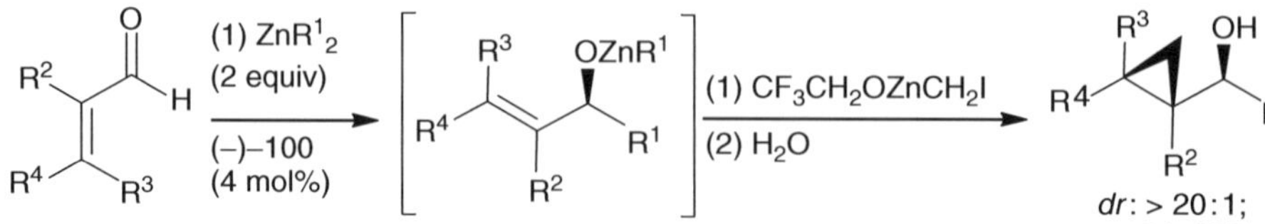

101

dr: > 20 : 1;
ee: 91–99%
102

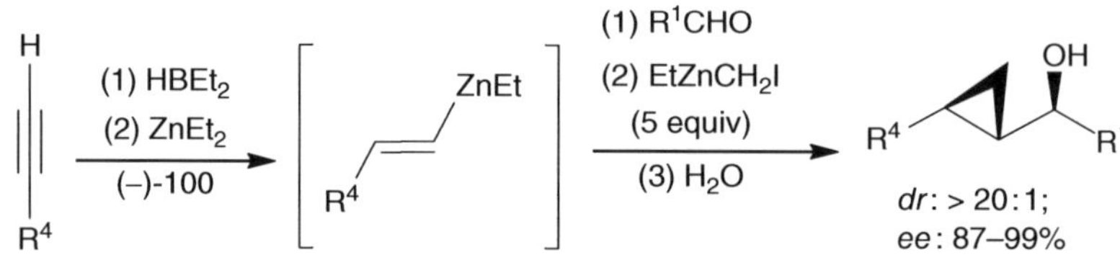

103

dr: > 20 : 1;
ee: 87–99%
104

Scheme 9.11 Zn carbenoid and its diastereoselective addition to aldehyde to afford three-membered ring compounds [25].

either alkyl or aromatic, or other substituents. High diastereoselectivity up to 91% (*dr* > 20 : 1) could be achieved.

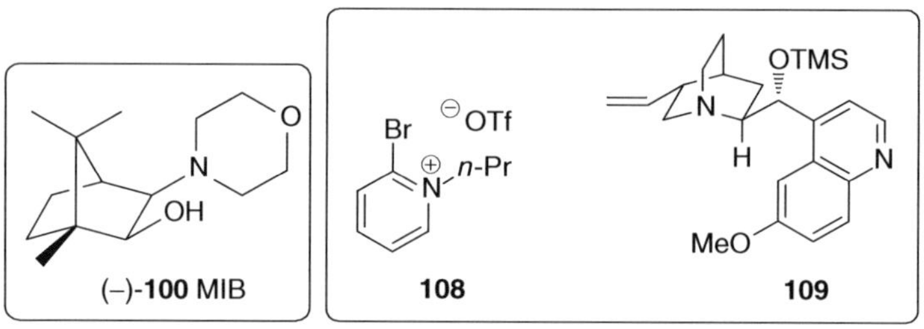

(–)-**100** MIB

108

109

The formation of the zinc carbenoid and the possible reaction procedure of its addition to olefin or alkyne is proposed and illustrated in Scheme 9.12. Substances with different substituents were used in the Simmons–Smith reactions.

To construct two rings in a molecule is a valid way in diastereoselective reactions. A double diastereoselective aldol reaction could afford this target using the catalyst **108** and the nucleophile reagent **109** [26]. One lactonization example is listed in Eq. (9.13). Its dr value is 1 : >19 for **111** and **112.** However, the yield is not high (23%).

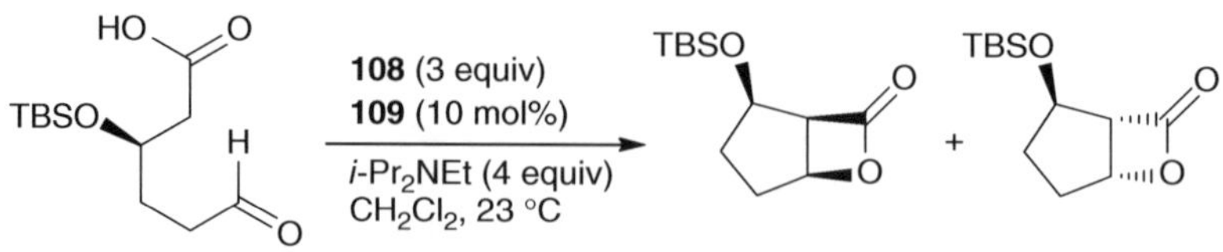

110 **111** **112** (9.13)

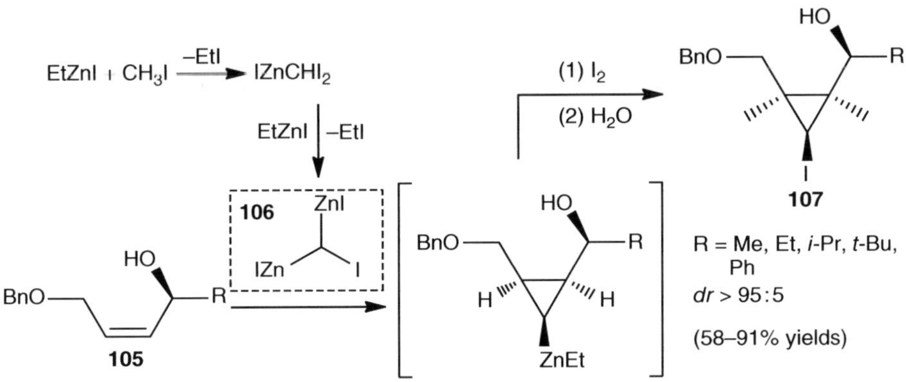

Scheme 9.12 Formation of dizinc carbenoids and their diastereoselective transfer to benzyl (*Z*)-allyl ethers [26].

The spiroketal moiety is an important motif in natural products. Its construction contains cyclic enol ether in some examples. In the presence of a chiral catalyst, the formation of the spiroketal moiety could be achieved in high diastereoselectivities and high yields. An example is shown in Eq. (9.14). Under the catalysis of **116** [27], high diastereoselectivity was achieved. For example, when (*S*)-**116** was used, the *dr* value was 9 : 1 with a yield of 86%; the *dr* value (**114** : **115**) became 13 : 87 when (*R*)-**116** was used, and 88% yield was recorded. If the methyl on the side chain in **113** did not exist, the dr value increased to 95 : 5 using (*S*)-**116**; however, the ratio decreased to 23 : 77 when (*R*)-**116** was used.

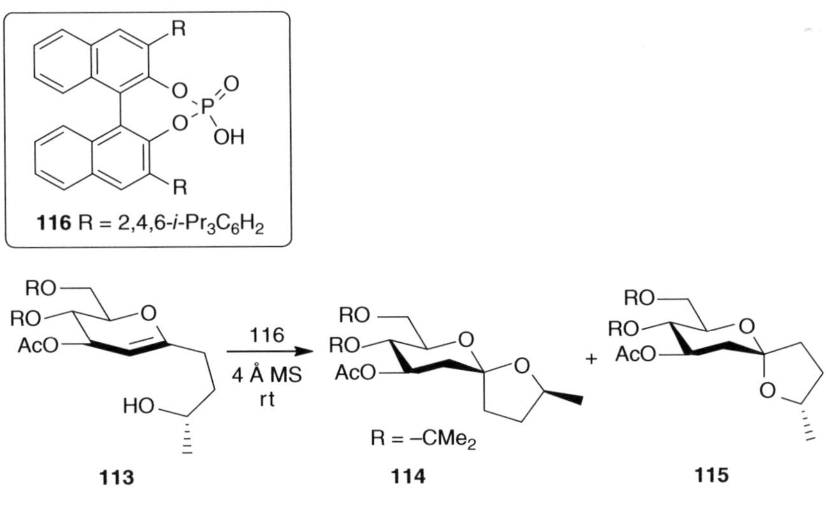

(9.14)

In addition to zinc carbenoid, Et_2Zn is also used in the olefin cyclization accompanying Pd(0) catalysis. This method was used to synthesize the alkaloid (−)-kainic acid (**120**) [28]. The formation of intermediate **119** was obtained with almost 100% yield (Scheme 9.13).

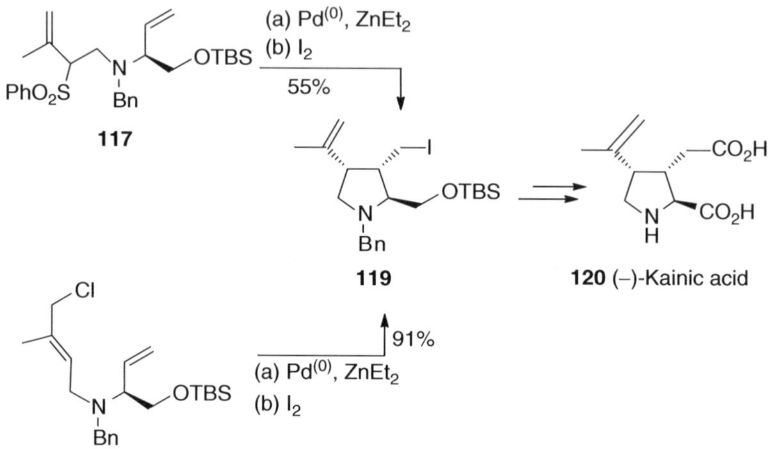

Scheme 9.13 Synthesis of the key intermediate **119** for (−)-kainic acid using diastereoselective addition [28].

Olefins can be converted into diorganozinc reagents [29]. Its diastereoselective conversion to the corresponding compounds was achieved with high diastereoselectivity [30]. CuCN–2LiCl was necessary in the procedure. The electrophilic reagent reacts with this organo-Cu intermediate to afford the final products. Scheme 9.14 provides more details in the conversions.

Construction of ring-containing compounds is one of the important targets in organic synthesis. At the same time, it can also be applied to produce linear chiral compounds and has been widely investigated. In rhodium-catalyzed cross-aldol reaction, it was found that the diastereoselective addition of aldehyde with allyloxyboranes resulted in a high selectivity (Eq. (9.15)) [31]. The major product formed in the addition was the syn compound. The results are listed in Table 9.11.

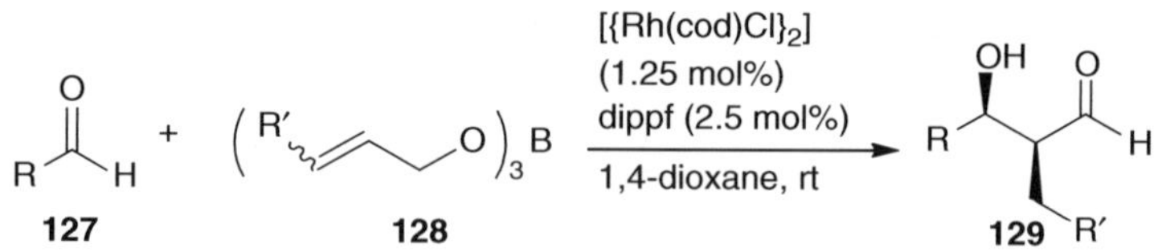

(9.15)

Inorganic $ZnBr_2$ can be used in organic reactions. Scheme 9.15 shows the transformations. The alkyne moiety becomes a 1,2-diene structure after the conversion, which is highly active in various reactions [32].

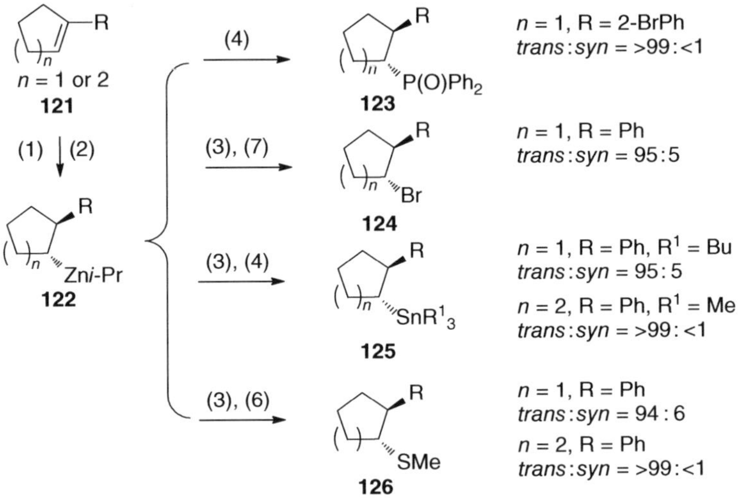

Scheme 9.14 The conversion plot for cyclic olefins under different conditions: (1) Et_2BH (3 equiv), 50 °C, 16 h; (2) $Zn(i\text{-}Pr)_2$ (3 equiv), 25 °C, 5 h; (3) CuCN–2LiCl (1 equiv), −78 °C, 0.5 h; (4) $ClPPh_2$ (4 equiv), 25 °C, 4 days, and then 30% H_2O_2; (5) $R^1{}_3SnCl$ (3 equiv), −40 °C, 16 h; (6) $MeSSO_2Me$ (3 equiv), −40 °C, 16 h; (7) $BrCl_2CCCl_2Br$ (3 equiv), −40 °C, 16 h [30].

The diastereoselective additions of metallated propargylic amines to aldehydes can be performed at a low temperature with high yields. High *dr* values were recorded in this addition (Eq. (9.16)) [32a]. Table 9.12 summarizes the partial results.

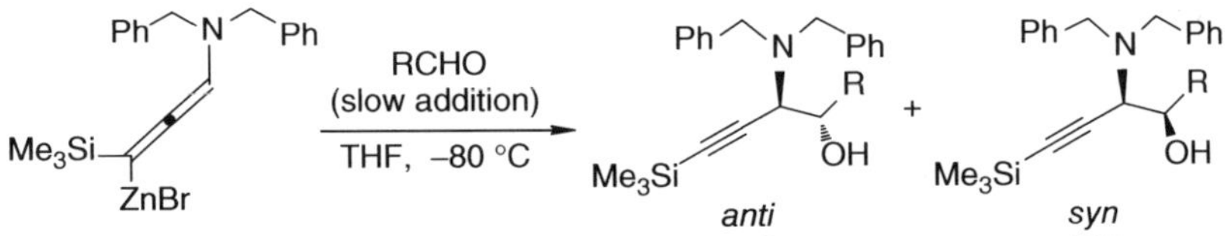

(9.16)

The salt of RZnBr is applied for its uses as an organic reagent, and can be used in syntheses. One preparation example is illustrated in Scheme 9.16. This is a typical method. Salt **138** could be used in addition to various aldehydes.

Its addition to aldehydes (**139**) formed two products, *anti* and *syn* geometries; the diastereoselectivities recorded in the addition (Eq. (9.17)) are summarized in Table 9.13. Catalysts **142** and **143** were used in the procedure [32b].

Table 9.11 Rh-catalyzed isomerization/cross-aldol reaction sequence with triallyloxyboranes [31].

Entry	R	R′	Time (h)	syn/anti[a)]	Yield (%)[b)]
1	2-BrPh	H	23	94:6	99
2	3-BrPh	H	36	93:7	72
3	4-BrPh	H	36	93:7	83
4	3-ClPh	H	36	91:9	95
5	4-FPh	H	36	93:7	87
6	4-NO_2Ph	H	36	94:6	90
7	2,6-diClPh	H	36	>95:5	85
8	2,4-diMeOPh	H	36	90:10	78
9	Ph	H	36	90:10	81
10	2-Naph	H	36	90:10	75
11	2-Furyl	H	36	94:6	60
12	Ph	Me[c)]	24	90:10	93
13	Ph	(*E*)-, Et	48	88:12	57
14	Ph	(*Z*)-, Et	12	87:13	84
15	Ph	(*Z*)-, Et	12	86:14	89
16	*n*-Pent	H	27	85:15	73
17	$PhCH_2CH_2$-	H	36	84:16	90
18	*c*-Hex	H	32	74:26	62
19	Et	Me	24	75:25	71[d)]

a) Determined by ^{1}H NMR analysis of the crude reaction mixture.
b) Yield of isolated product was determined after conversion into either the dimethylacetal with catalyst PPTS/MeOH or the 1,3-diol with $NaBH_4$, and purification by silica gel column chromatography.
c) It is a mixture of *E*- and *Z*-isomers in reaction.
d) Yield of the isolated *β*-hydroxy aldehyde form after careful purification by silica gel column chromatography. PPTS = pyridinium *para*-toluenesulfonate.

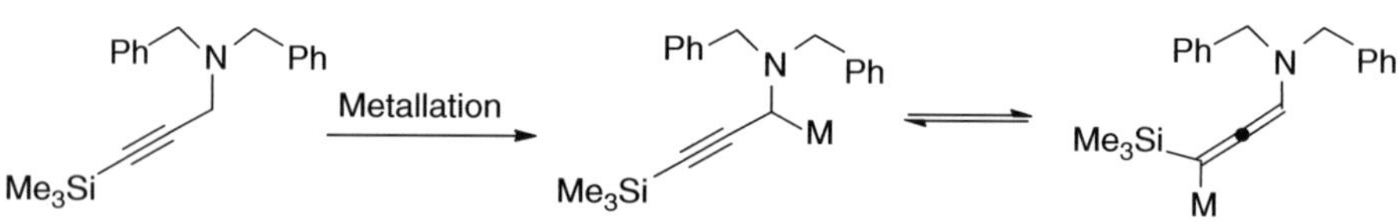

130 131 132

Scheme 9.15 Metallation of alkyne with metallic salts [32a].

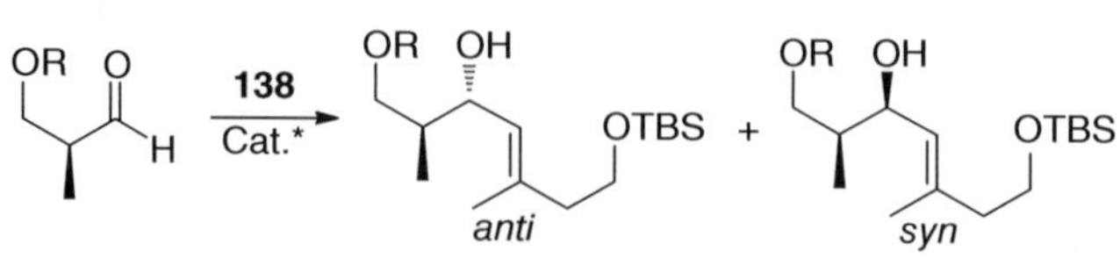

139 140 141 (9.17)

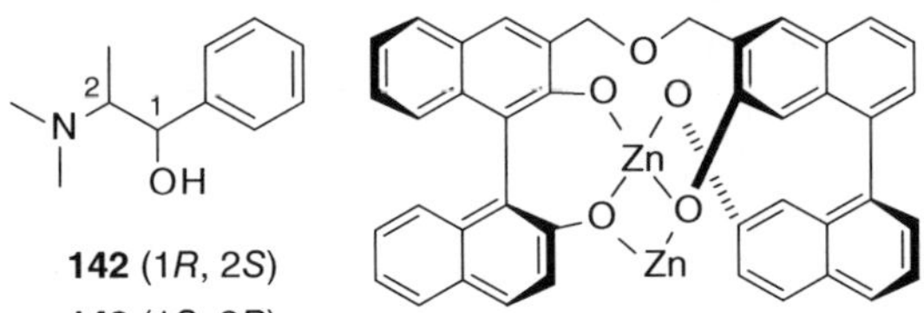

142 (1*R*, 2*S*)
143 (1*S*, 2*R*)
144

Catalytic enantioselective and diastereoselective aldol reactions using the dinuclear Zn complex **144** were investigated [33]. This Zn–Zn-linked binol complex

Table 9.12 Additions of Zn derivatives to aldehydes [32a].

Entry	R	dr[a)]	Yield (%)[b)]	Entry	R	dr[a)]	Yield (%)[b)]
1	*t*-Bu	>95/5	81	5	*i*-Pr	93/7	88
2	*c*-Hex	>95/5	81	6	Ph	68/32	79
3	*n*-Hex	85/15	81	7	Propenyl	88/12	83
4	*i*-Pr[c)]	90/10	86	8	Heptynyl	76/24[d)]	85

a) Determined by ^{1}H NMR of the crude product.
b) Isolated product.
c) Fast addition of the aldehyde.
d) Inseparable mixture.

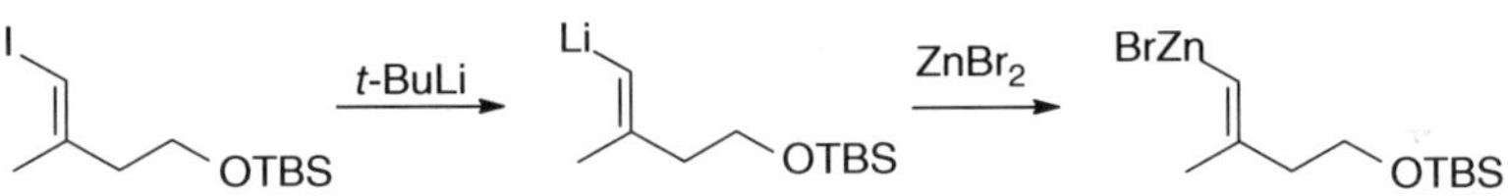

136 **137** **138**

Scheme 9.16 Preparation of Zn salt from *n*-BuLi and corresponding idiom derivative [32a].

Table 9.13 Addition of organic Zn reagent to aldehydes with (*S*)-AC [32b].

Entry	R	Catalyst	Yield (%)	anti : syn[a)]
1	TBS	None	74	45 : 55
2	TBS	**142**	70	>90 : 10
3	TBS	**143**	70	<10 : 90
4	Bn	None	77	75 : 25
5	Bn	**142**	68	95 : 5
6	Bn	**143**	70	75 : 25

a) ^{1}H NMR integration.

afforded a practical synthesis of *α,β*-dihydroxyl ketones from different aldehydes even when only 1 mol% of the catalyst was used. The *syn/anti* ratio could reach 97 : 3 and the ee value 99% (Eq. (9.18)).

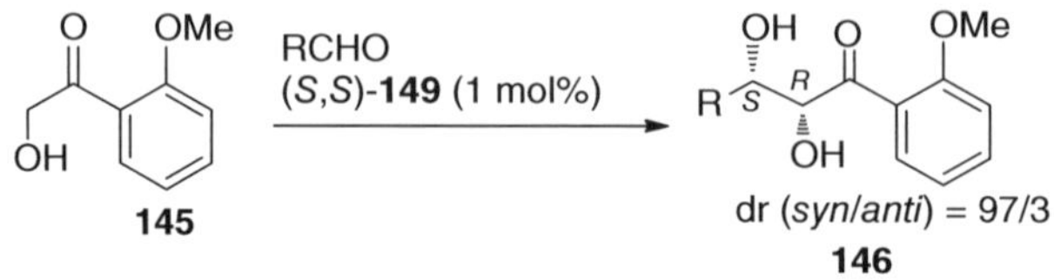

(9.18)

The use of chiral Ru complex, such as derivatives of diamine (*S*,*S*)-**147** (Ts = $SO_2C_6H_4$-*p*-CH_3) [34], afforded hydrobenoins **149** from benzil **148** in over 99% ee with diastereoselectivities from 90.4 : 9.6 to 98.6 : 1.4. The reaction temperature could vary from 30 to 60 °C.

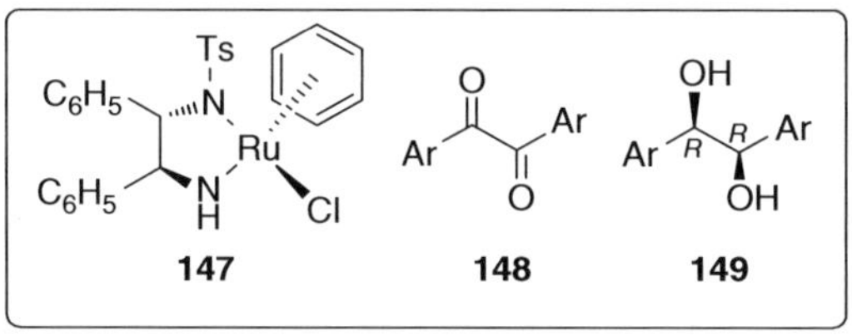

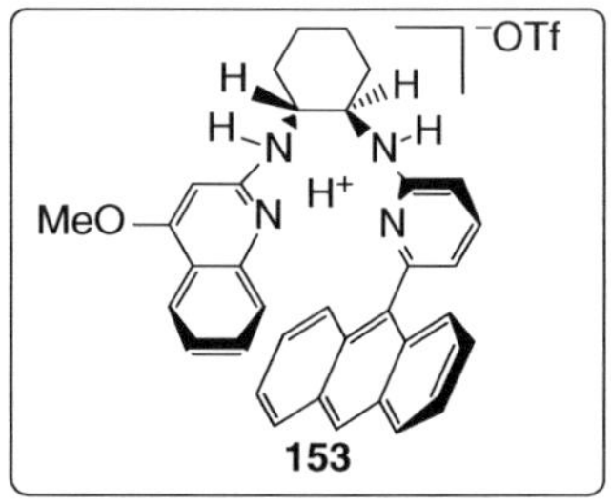

The addition of allylic nucleophiles to imines can be used for the synthesis of chiral homoallylic amines, which are important precursors to numerous organic compounds. One example is the synthesis of *cis*-*α,β*-diamino acids using the imine (**150**) and *α*-substituted nitro ester (**151**) (Eq. (9.19)) [35]. With the use of structurally complex aromatic chiral catalysts, such as **153**, which was found to have good conversion ability and high diastereoselectivity including enantioselectivity, high de values (up to 95%) could be achieved for the products.

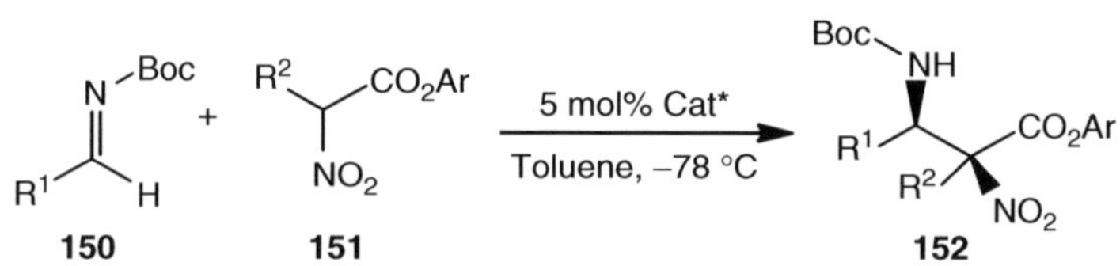

(9.19)

When Ar was a small group, such as Et, the *dr* and *ee* values were not high, but once it was a large group, the selectivity in the additions increased to about 98%. The experimental results are listed in Table 9.14 when Ar was 2,6-di(*i*-Pr)$_2$Ph.

In some cases, a diastereoselective reaction may accompany enantioselective reactions (Scheme 9.13). It was found that a regio- and diastereoselective allylic alkylation with an acyclic *α*-heteroatom-substituted ketone (**154**) furnished the secondary allylic alkylation adducts (**155** : **156**) in excellent yield, favoring **155**

Table 9.14 Different selectivity in addition of α-substituted nitro ester to imines catalyzed by **153** [35].

Entry	R^1	R^2	dr	ee (%)	Yield (%)
1	4-Cl–Ph	Et	>20:1	98	83
2	4-MeS–Ph	Et	13:1	—	81
3	4-PhS–Ph	Et	10:1	98	59
4	4-Me–Ph	Et	>20:1	96	61
5	4-MeO–Ph	Et	12:1	96	73
6	2-Furyl	Et	5:1	95	86
7	4-Cl–Ph	Me	12:1	94	82
8	4-Cl–Ph	*n*-Pr	15:1	99	82
9	4-Cl–Ph	*n*-Bu	16:1	97	88

(Eq. (9.20)) [36]. When R^2 was changed with different substituents with R^1 as -CH_2Ph, the *dr* values recorded were from 9 : 1 to 53 : 1. Most of them were near 20 : 1.

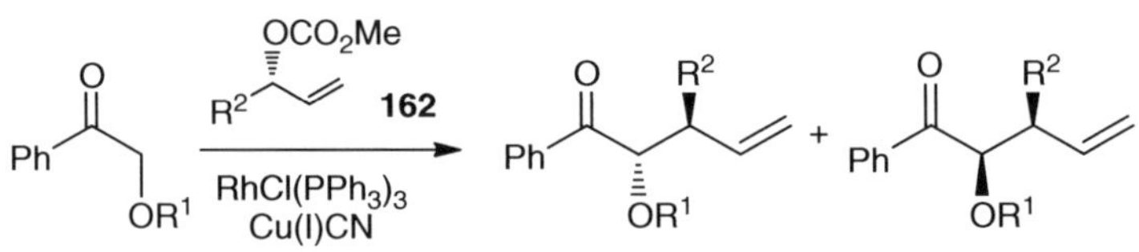

154 **155** **156** (9.20)

A change in the AC of a chiral catalyst can make product's AC change in most cases. It appears to be a general rule in the reactions. However, this may not be the case when different substrates or reaction conditions are used (not catalyst factor). Addition of the imine analog **157** could be performed by Zn-mediated allylation (Scheme 9.17). This reaction had two directions under different conditions [37]. One is the formation of **158** using THF and $In(OTf)_3$; the other is the formation of **159** by the use of HMPA (hexamethylphosphoramide) and water at room temperature.

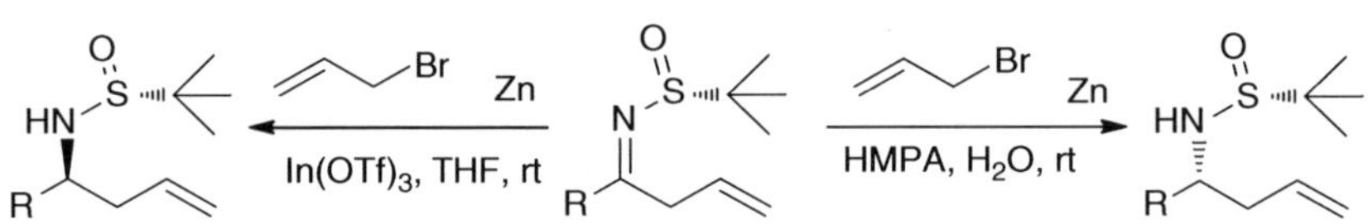

158 **157** **159**

Scheme 9.17 Diastereoselective allylation to (*R*)-*N*-*tert*-butanesulfinyl aldimines with two different reaction systems [37].

Table 9.15 Diastereoselectivity in allylation of imines under two reaction systems [37].

Entry	R	THF system		HMPA system	
		Yield (%)	*dr*	Yield (%)	*dr*
1	Ph	93	98 : 2	97	1 : 99
2	4-FPh	98	98 : 2	97	1 : 99
3	4-ClPh	98	98 : 2	96	3 : 97
4	4-MeOPh	91	95 : 5	81	2 : 98
5	2-MePh	95	98 : 2	89	3 : 97
6	3-BrPh	98	98 : 2	99	2 : 98
7	2-Naph	81	95 : 5	86	3 : 97
8	*c*-Pr	98	86 : 14	97	4 : 96
9	*c*-Hex	99	97 : 3	94	3 : 97
10	Et	93	88 : 12	92	5 : 95
11	*i*-Pr	96	94 : 6	94	2 : 98
12	Phenethyl	92	90 : 10	93	4 : 96
13	Phenylethenyl	83	91 : 9	87	6 : 94

This allylation exhibited very high diastereoselectivities using various substrates. Table 9.15 summarizes the results. This reaction led to very high *dr* values, such as 98 : 2 ((1*S*)-**158**: (1*R*)-**159**) in THF solution, or 1 : 99 in aqueous HMPA solution (Table 9.15, entry 1). The yields for almost cases were high (up to 98%). All reactions could be performed at room temperature.

9.4.2
Other Diastereoselective Reactions

The introduction above was on diastereoselective addition, which is the major reaction type. The other reactions include diastereoselective reductions, rearrangement, and so on.

Birch reduction followed by alkylation is a reaction type used in diastereoselective reactions. This reaction can afford relatively high yields if isoprene is added. For example, when **160** was used as a material, addition of isoprene could increase the yield from 36 to 61% when RX was allyl bromide [38]. Other alkyl halides could take part in the reaction with medium yields. However, EtI or 2-iodopropane could not react with **160**. The reaction time is about 40 min to 1.5 h. Scheme 9.18 summarizes some experimental results.

When diastereoselective oxidative carbon–carbon bond formation via silyl bis-enol ethers was performed, high diastereoselectivity up to 20 : 1 was recorded (Eq. (9.21)).

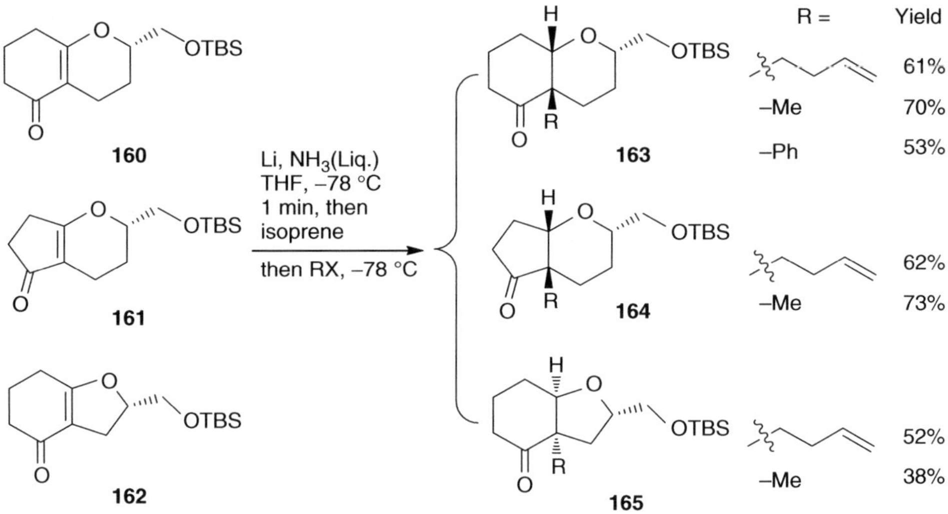

Scheme 9.18 Birch reduction-alkylation of bicyclic β-alkoxy-α,β-unsaturated carbonyl compounds [38].

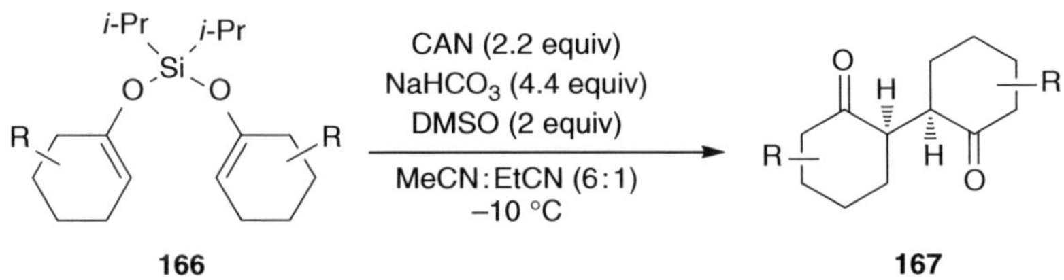

(9.21)

This diastereoselective oxidation-coupling reaction looks like a dimerization reaction of i-Pr_2Si-enol ethers [39]. The two substituents (R) could be different, and the highest *dr* values can reach up to 20 : 1. Partial experimental results are illustrated in Table 9.16. This method provides a good tool to synthesize 1,4-diketo derivatives, which are important intermediates.

In some cases of Claisen rearrangement, researchers have artfully and successfully used the differences in size and orientation of the substituents in a stereoselective reaction (Eq. (9.22)) [40].

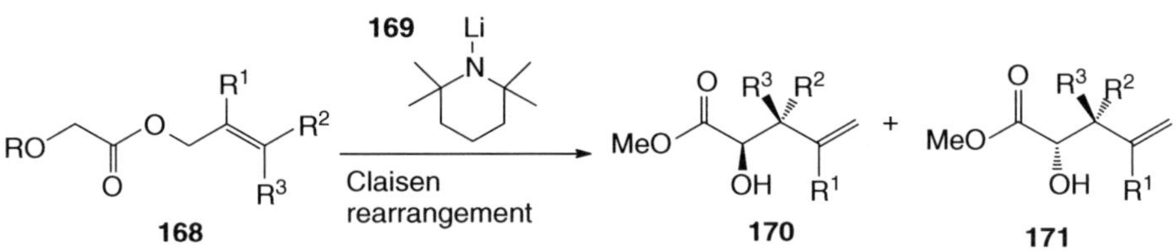

(9.22)

Table 9.16 Diastereoselectivity in oxidation-coupling reaction via silyl bis-enol ethers [39].

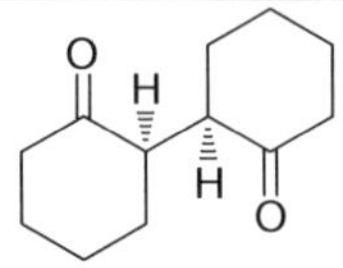 67% yield; *dr*, 14:1	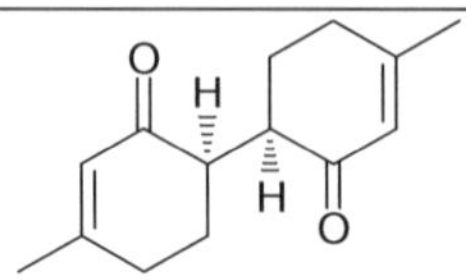 82%; *dr*, 2:1	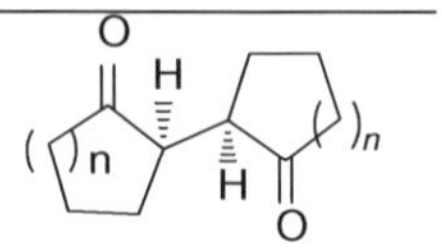 *n* = 1, 45%, *dr*, 7:1 *n* = 2, 50%; *dr*, 14:1
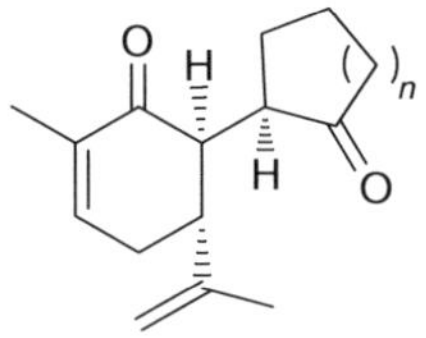	*n* = 1, 63%; *dr*, 16:1 *n* = 2, 58%; *dr*, 15;1	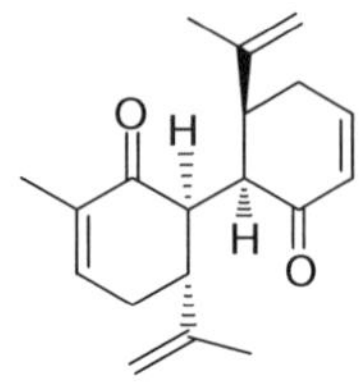 61%; *dr*, 5:1

In this reaction, different cis or trans isomers as starting materials afforded various products with different stereogenic centers and different ratios. Table 9.17 summarizes the results.

Notice

Indeed, hundreds of studies involving diastereoselective reactions have been reported, and many experimental results have shown very promising application ability. This is an important research area, and also an active field in current organic synthesis. Readers who have interests in this area can easily find many references from various journals and books. Only typical examples and representative methods are cited in this book, as also only the corresponding references directly involved in this study. Others are not cited in this book to save space.

9.5 Calculation Using Theoretical Protocol

Before 2000, it was quite a stereotyped viewpoint of most experimental chemists that the theoretical method is a tool of the theoretical chemist instead of a useful protocol for the experimental chemist. Most experimental chemists did not pay much attention to the use of theoretical methods in their study. The reasons may include the following: (i) Supercomputers were not as popular before 2000 as they are today. Except for the experts in United States and a few countries in Europe, thousands of experimental chemists in other areas had no chance to use theoretical protocols to handle their academic questions in their daily study. On the other hand, many theoretical studies published in various journals were generally using their own models instead of the real experimental molecules, which

Table 9.17 Effect of molecule with (*E*)- or (*Z*)-geometries on the ratio **170** : **171** [40].

Entry	Substituent 168	Method	Yield (%)	*dr*, 170:171
1	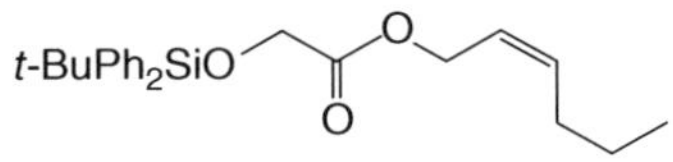 (*E*:*Z* = 2:98)	A	86	95:5
		B	80	3:97
2	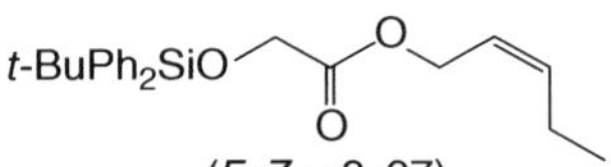 (*E*:*Z* = 3:97)	A	85	95:5
		B	80	4:96
3	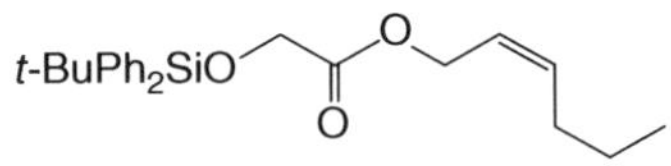 (*E*:*Z* = 2:98)	A	93	96:4
		B	85	3:97
4	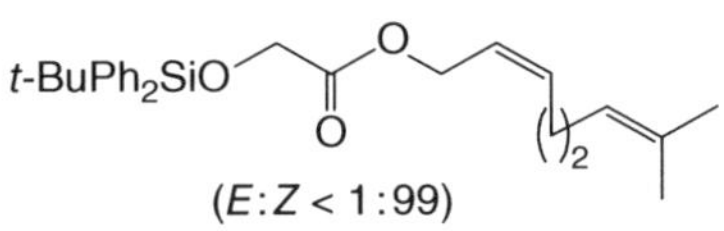 (*E*:*Z* < 1:99)	A	94	94:6
		B	81	4:96
5	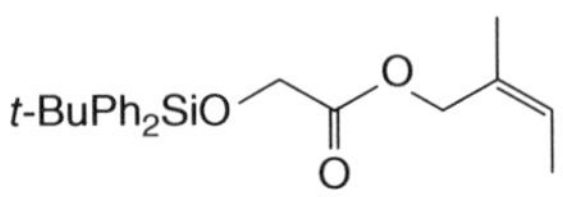 (*E*:*Z* < 1:99)	A	86	94:6
		B	80	1:99

Method A: Starting material **168** was put into the mixture of lithium 2,2,6,6-tetramethylpiperidin-1-ide (LTMP) and over chlorotrimethylsilane (TMSCl) at −100 °C, and then slowly warmed to room temperature for 4 h. Method B: Starting material **168** was put into the solution of THF−hexamethylphosphoramide (HMPA) (4:1) at −100 °C; then TMSCl was added to the solution for reaction, the mixture was then warmed up to room temperature.

might have been due to the computation limitation at that time. This affected the experimentalist's interests. (ii) There were not many researchers who are good at both computations and experiments. For example, when a person wants to do a TS study, he must keep two things in mind. The first one involves the stereochemistry of a specific reaction. All possible reaction TS geometries must be considered and constructed in mind at the first sight. This process is boring and ineffective sometimes. The second one is to correctly analyze the output results during the computations. Not every TS geometry that is used as input file in the calculation can afford a reasonable result. Thus, carefully analyzing the intermediate structure can provide many useful pieces of information to adjust the TS coordinate be used in further computations until a reliable result is produced. Another example is the AC assignment for complex chiral compounds. The current theoretical methods such as OR and ECD can provide many benefits in the study. Unfortunately, some experimental researchers ignore their usage range for many reasons, and finally the computational results easily lead to wrong conclusions.

However, with the development of supercomputer technology and the maturity of the corresponding software, a big opportunity has appeared for experimental chemists. Although theoretical methods may be questionable in their application for various targets, their accurate results in most cases can provide many new experimental insights. As the examples we listed in Chapter 6, the AC reassignment of bioactive natural products (+)-schizandrin and brevianamide M are the two extreme examples. Another example was presented in the discussion of the AC of (−)-*N*-(1-phenylethyl) aniline in the addition of enantioselective hydrosilylation of C=N with $HSiCl_3$. The AC of (−)-*N*-(1-phenylethyl)aniline was assigned as (*S*) in an early study. However, around 1997, its AC is reported as (*R*) in many experimental reports. Indeed, this change is wrong, which was established by our TS barrier calculations first, and then its AC was discussed using det(D), ECD, and VCD methods, respectively.

Therefore, the use theoretical methods in experiments is an important challenge. The introduction of some examples may break the ice and result in wide application. In diastereoselective reactions, it is an interesting topic to construct the relationship between the observed *de* values and the structure. Of course, this is related to its TS barriers as we mentioned many times. However, it is not easy to search out all reliable TS structures and their barriers. Thus, the construction of some useful relationship may provide experimental chemists more help in chiral catalyst design or in other studies. In this small section, the focus is on understanding the de, ee, magnitudes, and mechanisms for conversion in different reactions.

In the reduction of diketones to the corresponding alcohols catalyzed by catalysts **178–180**, the relationship between the observed de and ee was investigated (Scheme 9.19) [41]. This observation is based on the experimental data.

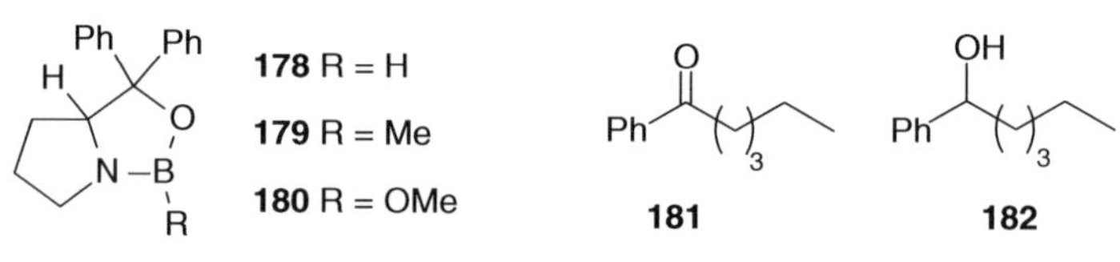

If ee_1 is defined as the enantioselectivity of the first step (formation of intermediate ketol, **173** and **174**) and de_2 as the diastereoselectivities of the second step (assumed identical), then de_{homo}, the preference for the formation of the homochiral versus meso diastereomer is simply the product of ee_1 and de_2 as illustrated below:

$$ee_1 de_2 = de_{homo} \tag{9.23}$$

When $ee_1 = de_2$, it becomes

$$ee_1{}^2 = de_{homo} \tag{9.24}$$

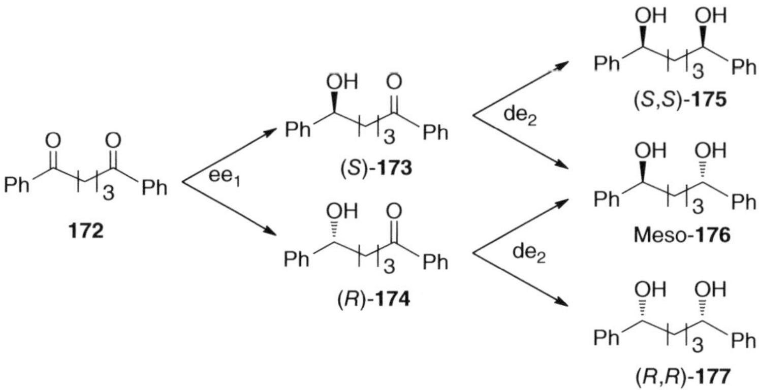

Scheme 9.19 Reduction routes and their different isomers [41].

Reduction of diketone **172** by enantiopure catalysts **178–180** afforded the chiral diol mixture of **173–175** in very high ee (close to 99%). At the same time, when ketone **182** was selected as a model for the study of the enantioselectivity of the monoreduction of **172**, the recorded ee was near 90%. The diastereoselectivity de_{homo} is very good, and the relationship between de_{homo} and ee_{homo} is described below:

$$ee_{homo} = 2(de_{homo})^{1/2}/(1 + de_{homo}) \tag{9.25}$$

Based on the assumption $ee_1 = de_2$, racemic catalysts were also used [41]. The results were calculated, and they are listed in Table 9.18. It can be found that it is possible to evaluate the enantioselectivity of racemic catalysts without equal mole ratio in reactions in some specialized cases.

The best method is to calculate the TS barriers for prediction of *ee* values. One successful example is the explanation of ee values observed in adol condensation reaction using catalyst **183** (Eq. (9.26)) [42]. Although the quantity of the catalyst was high (up to 20% mol), the highest ee was observed as 98% when R was the *c*-hexyl group. Most ee values recorded were over 95%. Prediction was made using HF/6-31G(d) method, and SPE correction was performed at the B3LYP/6-31G(d,p) level.

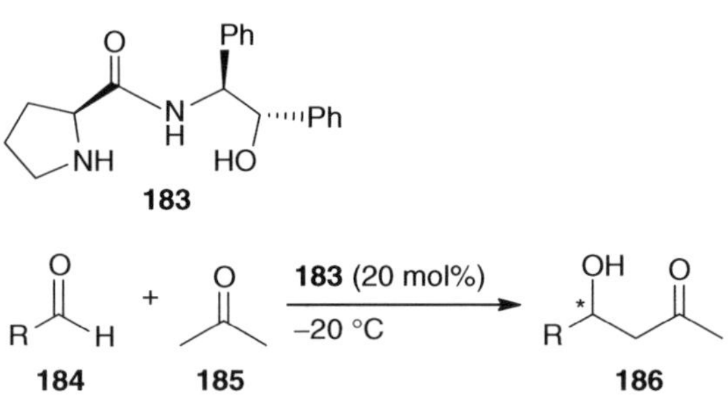

(9.26)

Table 9.18 Values of de and ee in reductions of 1-phenylhexanone, **172**, in THF [41].

Entry	Catalyst*	de_{homo} [a] (%)	ee_1 [b] (%) (ee% of **182**)[c]
1[c]	(*R*)-**178**	87	93(82)
2[d]	(*R*)-**178**	86	93(79)
3[c]	Racemic **178**	56	75(82)
4[d]	Racemic **178**	86	93(79)
5[c]	(*S*)-**179**	81	90(87)
6[d]	(*S*)-**179**	72	85(83)
7[d]	Racemic **179**	83	91(83)
8[c]	(*R*)-**180**	89	94(–)[e]
9[c]	Racemic **180**	82	91(–)[e]

a) Measured by chiral supercritical fluid chromatography (SFC) (Chiralcel OD-H). (*R*)-OAB catalysts give (*S*,*S*)-diol.
b) Calculated from the equation $ee_1{}^2 = de_{homo}$ (when $ee_1 = de_2$ in Formula (9.2)).
c) 0 °C.
d) 66 °C.
e) Not measured.

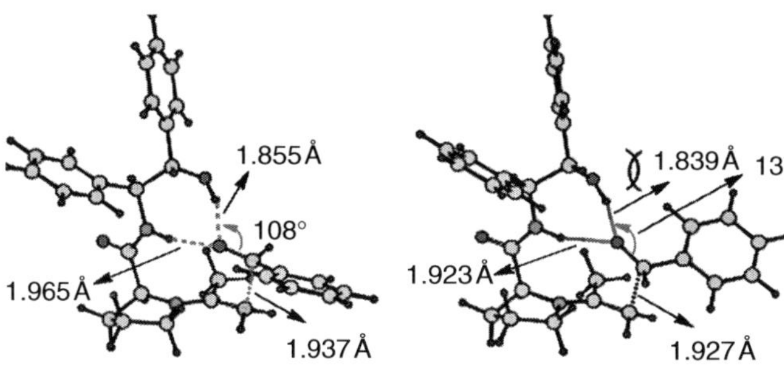

TS-1
0.0 (0.0) kcal mol^{-1}

TS-2
2.4 (3.0) kcal mol^{-1}

Figure 9.7 Two TS structures obtained using HF/6-31G(d) theory [42].

In the TS geometry, acetone forms a Shiff base first with the pyrrolidine moiety of **183**, and the $-CH_2-$ in the intermediate would attack carbonyl carbon of benzaldehyde to form the corresponding addition product **183.** The experimental ee value was 83%, and the predicted one was 96%. Both matched well. The predicted results are illustrated below the TS structure (Figure 9.7).

Obviously, product via **TS-1** should be the major one. The minor product via **TS-2**, based on the relative energy of 2.4 or 3.0 kcal mol^{-1}, occupied about 1.7%. Therefore, its predicted ee value should be about 96%.

References

1. Eliel, E.L. (1962) *Stereochemistry of Carbon Compounds*, Chapter 8, Mcgraw-Hill Book Company, Inc, New York.
2. Kagan, H.B. (1979) *Organic Stereochemistry*, p. 59.
3. Casarini, D., Rosini, C., Grilli, S., Lunazzi, L., and Mazzanti, A. (2003) *J. Org. Chem.*, **68**, 1815–1820.
4. Gasparrini, F., Lunazzi, L., Mazzanti, A., Pierini, M., Pietrusiewicz, K.M., and Villani, C. (2000) *J. Am. Chem. Soc.*, **122**, 4776–4780.
5. Yang, G., Han, X., Zhang, W., Liu, X., Yang, P., Zhou, Y., and Bao, X. (2005) *J. Phys. Chem. B*, **109**, 18690–18698.
6. (a) Jaramillo, C., de Diego, J.E., and Rivera-Sagredo, A. (2006) *Tetrahedron*, **62**, 12415–12419; (b) Gromova, M., Beguin, C.G., Goumont, R., Faucher, N., Tordeux, M., and Terrier, F. (2000) *Magn. Reson. Chem.*, **38**, 655–661.
7. Luo, X.D., Wu, S.H., Wu, D.G., Ma, Y.B., and Qi, S.H. (2002) *Tetrahedron*, **58**, 6691–6695.
8. (a) Barton, D.H. (1953) *J. Chem. Soc.*, **23**, 1027; (b) Cram, D.J. *et al.* (1952) *J. Am. Chem. Soc.*, **74**, 5828; (c) Bartlett, P.A. (1980) *Tetrahedron*, **36**, 15.
9. Zhang, Y., Wang, Y.Q., and Dai, W.M. (2006) *J. Org. Chem.*, **71**, 2445–2455.
10. Fukuzawa, S., Seki, K., Tastuzawa, M., and Mutoh, K. (1997) *J. Am. Chem. Soc.*, **119**, 1482–1483.
11. Zhu, H.J., Jiang, J.X., Saobo, S., and Pittman, C.U. (2005) *J. Org. Chem.*, **70**, 261–267.
12. Cho, C.W. and Krische, M. (2006) *Org. Lett.*, **8**, 3873–3876.
13. Li, L. and Herzon, S.B. (2012) *J. Am. Chem. Soc.*, **134**, 17376–17379.
14. (a) Wang, Y.Y., Wang, J.X., Su, J.C., Huang, F., Jiao, L., Liang, Y., Yang, D.Z., Zhang, S.W., Wender, P.A., and Yu, Z.X. (2007) *J. Am. Chem. Soc.*, **129**, 10060–10061; (b) Jiao, L., Yuan, C.X., and Yu, Z.X. (2008) *J. Am. Chem. Soc.*, **130**, 4421–4430.
15. (a) Ma, S.-M. and Zhang, J.-L. (2003) *J. Am. Chem. Soc.*, **125**, 12386–12387; (b) Chinnagolla, R.K. and Jeganmohan, M. (2012) *Chem. Commun.*, **48**, 2030–2032.
16. Lv, J., Zhang, L., Zhou, Y., Nie, Z., Luo, S., and Cheng, J.P. (2011) *Angew. Chem.*, **123**, 6740–6744.
17. Koscher, P., Lumbroso, A., and Breit, B. (2011) *J. Am. Chem. Soc.*, **133**, 20746.
18. Li, C. and Breit, B. (2014) *J. Am. Chem. Soc.*, **136**, 862–865.
19. Liu, C.R., Yang, F.L., Jin, Y.Z., Ma, X.T., Cheng, D.J., Li, N., and Tian, S.K. (2010) *Org. Lett.*, **12**, 3832–3835.
20. Lin, W.M. and Zercher, C.K. (2007) *J. Org. Chem.*, **72**, 4390–4395.
21. Tanaka, S., Tamba, S., Tanaka, D., Sugie, A., and Mori, A. (2011) *J. Am. Chem. Soc.*, **133**, 16734–16737.
22. (a) Wiedemann, S.H., Noda, H., Harada, S., Matsunaga, S., and Shibasaki, M. (2008) *Org. Lett.*, **10**, 1661–1664; (b) Hicken, E.J. and Corey, E.J. (2008) *Org. Lett.*, **10**, 1135–1138.
23. Dunet, G., Mayer, P., and Knochel, P. (2008) *Org. Lett.*, **10**, 117–120.
24. Esposito, A., Piras, P.P., Ramazzotti, D., and Taddei, M. (2001) *Org. Lett.*, **3**, 3273–3275.
25. Kim, H.Y., Salvi, L., Carroll, P.K.J., and Walsh, P.J. (2009) *J. Am. Chem. Soc.*, **131**, 954–962.
26. Morris, K.A., Arendt, K.M., Oh, S.H., and Romo, D. (2010) *Org. Lett.*, **12**, 3764–3767.
27. Sun, Z., Winschel, G.A., Borovika, A., and Nagorny, P. (2012) *J. Am. Chem. Soc.*, **134**, 8074–8077.
28. Chalker, J.M., Yang, A., Deng, K., and Cohen, T. (2007) *Org. Lett.*, **9**, 3825–3828.
29. (a) Knochel, P. and Stinger, R.D. (1993) *Chem. Rev.*, **93**, 2117–2188; (b) Hupe, E., Calaza, M.I., and Knochel, P. (2001) *Tetrahedron Lett.*, **42**, 8829–8831.
30. Hupe, E. and Knochel, P. (2001) *Org. Lett.*, **3**, 127–130.
31. Lin, L., Yamamoto, K., Matsunaga, S., and Kanai, M. (2012) *Angew. Chem.*, **124**, 10421–10425.
32. (a) Bernaud, F., Vrancken, E., and Mangeney, P. (2003) *Org. Lett.*, **5**, 2567–2569; (b) Marshall, J.A. and Eidam, P. (2004) *Org. Lett.*, **6**, 445–448.

33. Kumagai, N., Mataunaga, S., Yoshikawa, N., Ohshima, T., and Shibasaki, M. (2001) *Org. Lett.*, **3**, 1539–1542.
34. Murata, K., Okano, K., Miyagi, M., Wane, H., Noyori, R., and Ikariya, T. (1999) *Org. Lett.*, **1**, 1119–1121.
35. Singh, A. and Johston, J.N. (2008) *J. Am. Chem. Soc.*, **130**, 5866–5867.
36. Evans, P.A. and Lawler, M.J. (2004) *J. Am. Chem. Soc.*, **126**, 8642–8643.
37. Sun, X.W., Xu, M.H., and Lin, G.Q. (2006) *Org. Lett.*, **8**, 4979–4982.
38. Hiroya, K., Ichihashi, Y., Furutono, A., Inamoto, K., Sakamoto, T., and Doi, T. (2009) *J. Org. Chem.*, **74**, 6623–6630.
39. Avetta, C.T. Jr., Konkol, L.C., Taylor, C.N., Dugan, K.C., Stern, C.L., and Thomson, R.J. (2008) *Org. Lett.*, **10**, 5621–5624.
40. Hattori, K. and Yamamoto, H. (1993) *J. Org. Chem.*, **58**, 5301–5303.
41. Lagasse, F., Tsukamoto, M., Welch, C.J., and Kagan, H.B. (2003) *J. Am. Chem. Soc.*, **125**, 7490–7491.
42. Tang, Z., Jiang, F., Yu, L.T., Cui, X., Gong, L.Z., Mi, A.Q., Jiang, Y.Z., and Wu, Y.D. (2003) *J. Am. Chem. Soc.*, **125**, 5262–5263.

10
Total Organic Synthesis

10.1
Introduction

The history of organic synthesis shows the power of organic synthesis in early structure elucidation of natural products and the ability to make natural products. Milestones in early structural determination include the assignment of D-glucose (1891), morphine (1925), aspidospermine (1959), and patchouli alcohol (1962) [1]. Some important landmarks achieved in total synthesis are those of urea (1828), acetic acid (1845), α-terpineol (1904), quinine (1944), strychnie (1954), morphine (1956), penicillin V (1957), vitamin B12 (1978), and taxol (1994) [1]. The great achievements in total organic synthesis have made big contributions to the current pharmaceutical industry in the world, in addition to the organic chemistry part.

Total synthesis is an art. It requires the researchers to be familiar with most popular reactions and their usage ranges, especially some specific reactions that lead to abnormal results. The well-known reactions include hundreds of named reactions, such as Hofmann degradation (1851), Claisen condensation (1887), Diels–Alder reaction (1893), Pictet–Spenglar reaction (1911), Birch reduction (1944), and Sharpless epoxidation (1980). Many reaction types such as aldol reaction (1838) and olefin metathesis cross-coupling reaction (1967) [1] are frequently used in total synthesis. Because of the various reactions, organic synthesis has shown great ability in the preparation of a large number of organic compounds, including many natural products.

Discussion of total organic synthesis in this chapter should focus on solution synthesis. Solid-phase syntheses, such as polypeptide synthesis, combinatorial synthesis, and bioactive catalytic synthesis, are not discussed here. Interested readers may consult the corresponding references [1a, b].

10.2
Retrosynthesis Strategies

The core question in total synthesis is the retrosynthetic study. The "art" of synthesis lies in this step. It is already half success when a complex compound can be

Organic Stereochemistry: Experimental and Computational Methods, First Edition. Hua-Jie Zhu.

disconnected into several small fractions that can be obtained easily in practice. Different separation methods (disconnection, retrosynthesis) bring various synthetic routes.

The general strategy of retrosynthesis is to simply disconnect (fractionalize) a molecule into several sections (molecules) first, and then keep dividing them into different parts of much smaller sections, until the final sections, namely, synthons, are the ones that cannot be disconnected again and also be readily available.

During the procedure, the first job is to master the conversion reactions of the functional groups, such as C=C, which can be seen as the Wittig reagent reaction product of >C=O. In this case, >C=O is the equivalent functional group of C=C. Or it could be a product of two C=C bonds via a metathesis reaction. It could also be the result of loss of a water of one hydroxyl group of >CH–C(OH)H$^-$. Some frequently used transformations are illustrated in Scheme 10.1 [1].

Other useful conversions include ring formation methods, such as the Diels–Alder reaction, Pictet–Spengler cyclization, [4+1] annulation (high temperature), [2+3] (low temperature) annulation for heterocyclic compounds, and [2+3] (low temperature) for carbocyclic compounds. Some named reactions are important for special structure synthesis.

In a retrosynthesis procedure, the manner of disconnecting a target molecule (TM) can be based on the molecule's skeletal characteristics or its major functional groups. Today, with the development of computer skills, the retrosynthesis of a compound can be carried out by means of the appropriate software. For example, *Syngen*, which focuses on skeleton analysis rather than functional analysis, can cut a molecule into several fractions reasonably [2]. A *Chiron* then follows the retrosynthetic approach [2c]. No matter what the computer does for the retrosynthesis in the procedure, to optimize the computed synthetic route, more human intelligence is necessary.

There are other strategies such as pattern recognition [3]. It is different from the traditional disconnection methods based on the TM skeleton or major functional groups. It focuses on the similarity between a partial TM structure and a known structure. For example, in the retrosynthesis of the partial taxol structure of baccatin III [4], the key point is to look for a suitable pattern molecule as the starting material (SM). From its structure, the associated SM A [5] can afford the corresponding TM baccatin III (Scheme 10.2) via several steps.

In retrosynthesis analysis of eleutherobin [6], monoterpene **B** was selected as the pattern molecule in its total synthesis. By the reaction of **B** with 2,2-dichloroethenone, it afforded the intermediate which can undergo oxidation ring-opening and addition reactions to produce the TM eleutherobin (Scheme 10.3).

In the study of natural products, it is found that there are many structurally similar compounds isolated from the same species of plants or fungi or any other sources, such as animals; or several analogs could be found in different species. To produce the structural diversity in synthesis is interesting and valuable. This is the way to generate molecular shape diversity [7]. This procedure is visually illustrated in Figure 10.1.

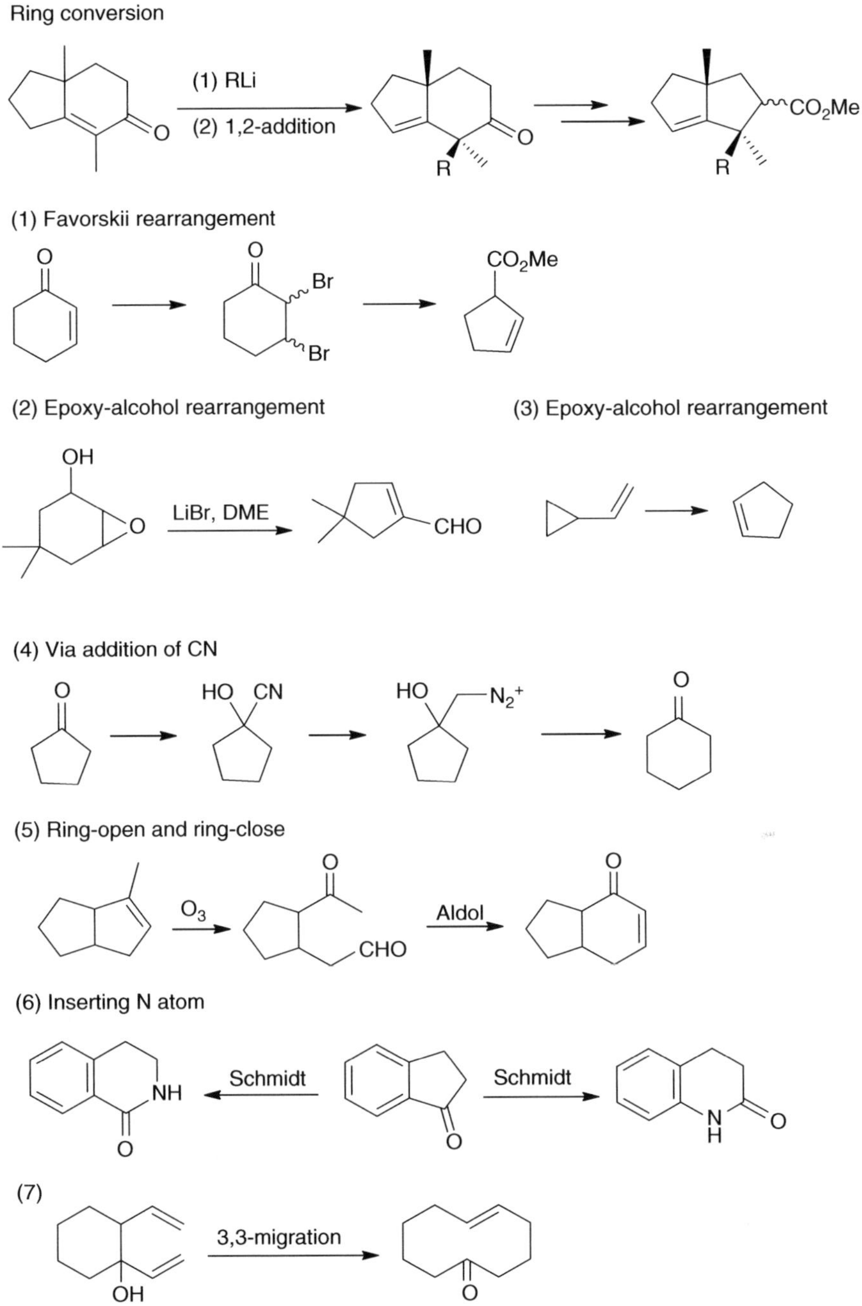

Scheme 10.1 Some conversions frequently used in total synthesis [1].

(8)

(9)

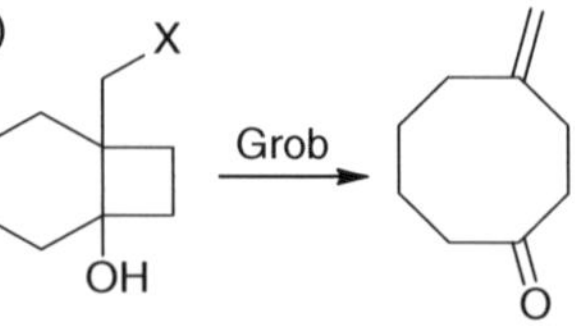

(10) Eschenmoser fragmentation

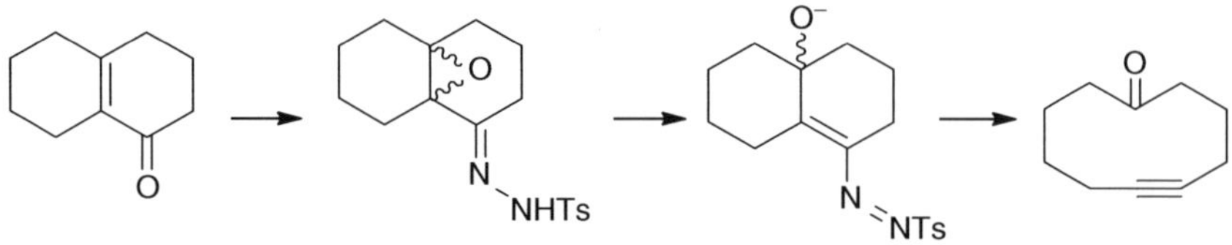

Chemoselective reaction

(11) Chemoselective reduction

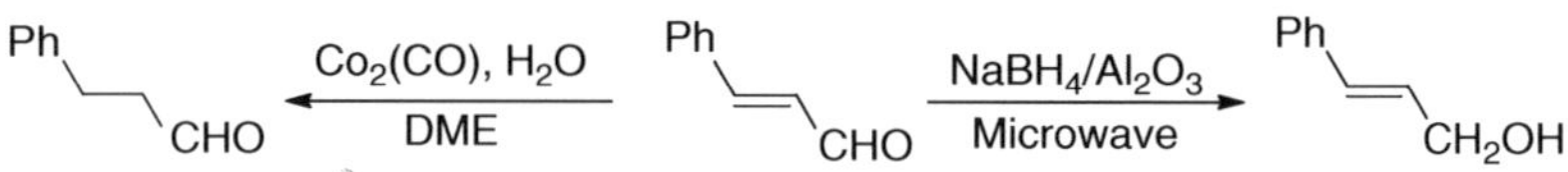

(12)

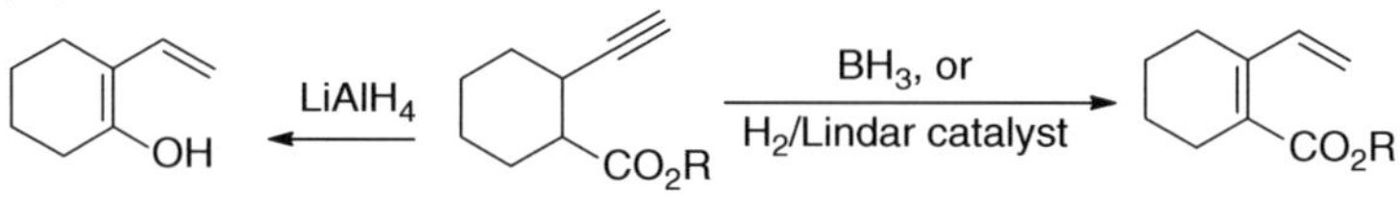

(13)

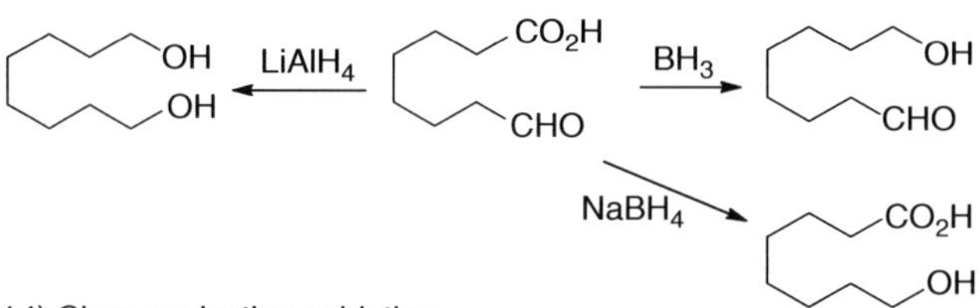

(14) Chemoselective oxidation

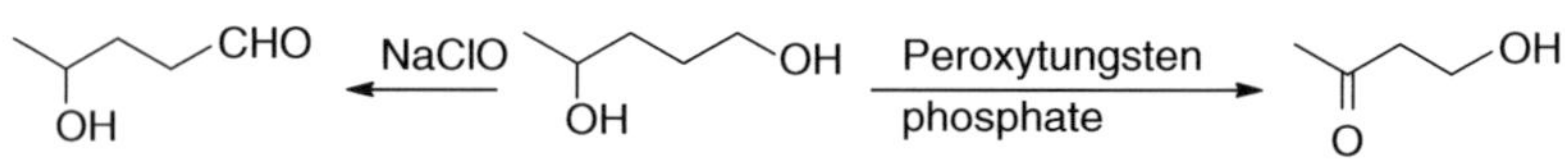

(15) Chemoselective hydrolysis

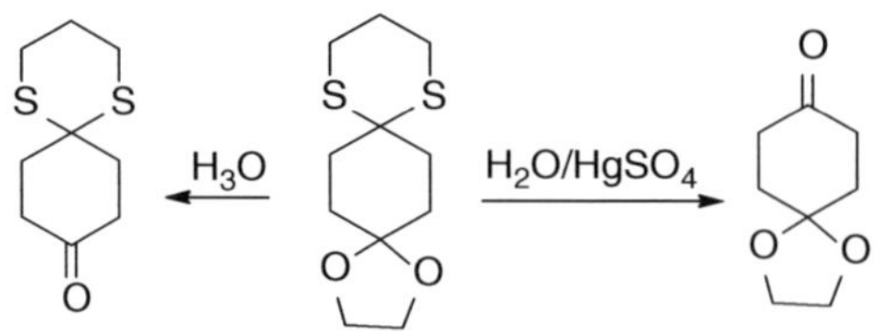

Scheme 10.1 (*Continued*)

(16) To use ketone to connect a substituent at its a-position
If stereochemistry is definite, the products should be definite based on Cram rule

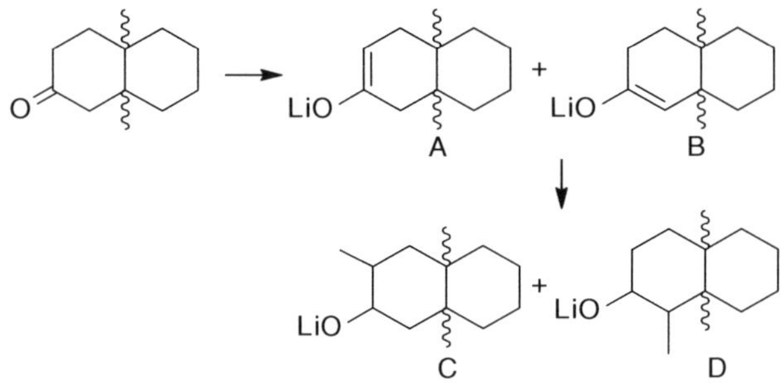

Scheme 10.1 (*Continued*)

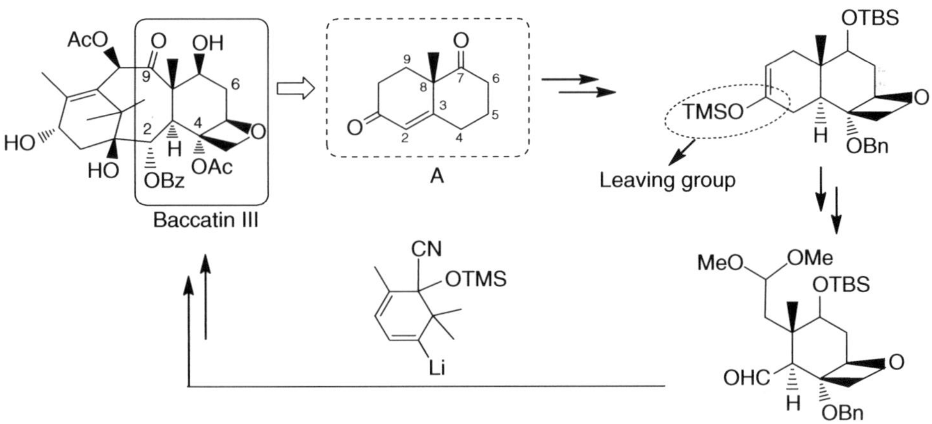

Scheme 10.2 Pattern recognition strategies used in synthesis of baccatin III [4, 5].

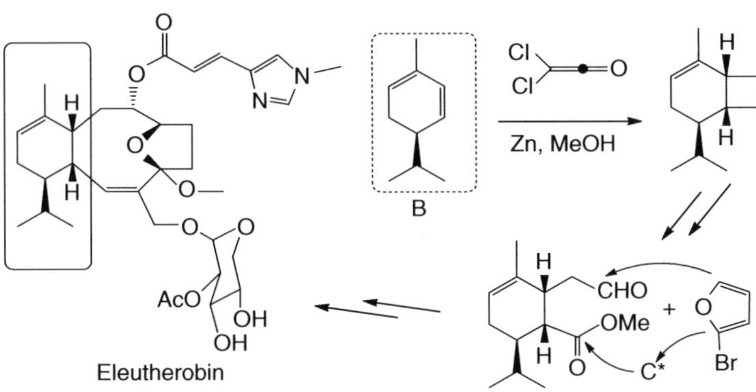

Scheme 10.3 Strategies for eleuthero biosynthesis using pattern recognition [6].

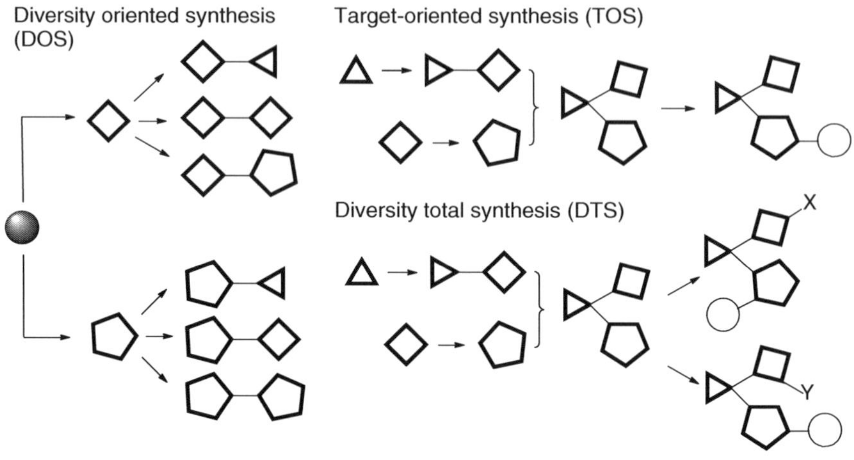

Figure 10.1 Comparison of three synthetic strategies: DOS, TOS, and DTS (the ball, triangle, quadrilateral, pentagon, circle. X and Y represent different molecules or molecular fragments) [7].

The procedure of diversity-oriented synthesis (DOS) is different; it can be used in forward synthetic analysis. Target-oriented synthesis (TOS) is a convergent procedure and may be suitable for retrosynthetic analysis. This is a frequently used model in practice. It can start from two or two more compounds as SMs, and afford one key intermediate that can be used for further synthesis of TM. Its structure can be confirmed by spectroscopic methods or X-rays. The diversity total synthesis (DTS) procedure is convergent at the first, and then it becomes divergent. This procedure is similar to the TOS before one key intermediate is obtained: it is convergent. However, after the key intermediate is obtained, it could be used for the synthesis of two or more compounds. This is the divergent procedure. This could be used for the synthesis of series of natural products with structural similarity.

Although the theoretical analysis for total synthesis of compounds looks simple, it is very difficult when facing a specific TM. Success or failure in a synthesis directly depends on the reasonability of the synthetic route, which is related to the having enough reactions and conversions in mind. One example is illustrated below (Scheme 10.4) to show the structural diversity of compounds in synthesis.

Details of total synthesis using pattern recognition or generation of molecular shape diversity using DOS or DTS will not be given in this chapter. It is one of the special examples in TOS where only one SM is used in the synthetic route. Interested readers may refer to the corresponding materials [3–7].

In synthetic procedure, oxidations, reductions, additions, and group protections may be involved for various transformations. High efficiency in each conversion is the key issue when attempting synthesis of a TM via 20 or more reaction steps.

All methods that were introduced in earlier chapters may be involved in total synthesis, such as enantioselective control using chiral catalyst or chemoselective

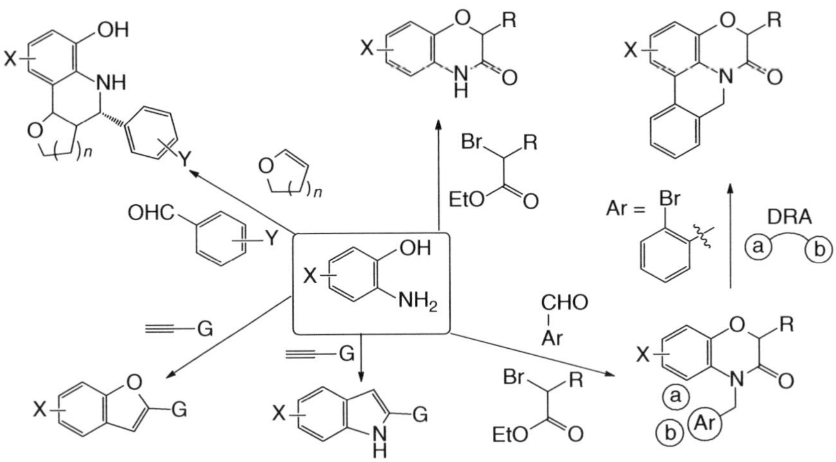

Scheme 10.4 Synthesis of privileged scaffolds from readily available 2-aminophenols and anilines [7].

reaction in some synthetic steps. The determination of chiral centers may be by means of various methods introduced in the earlier chapters.

As introduced in Chapter 1, there are different kinds of natural products with various bioactivities, like terpenoids, alkaloids, flavonoids, and others; their total organic synthesis would involve most types of natural products in practice as mentioned above. Until now, many hundreds of natural products have been synthesized. It is a big challenge to select them as examples from the reports. Many excellent works cannot be used here. The examples listed here are structurally both the simple and complex, or based on the reaction diversity in the synthetic procedure. It is not introduced as terpenoids or alkaloids or in a similar manner.

10.3
Examples in Synthesis

The structures listed in this chapter have been well determined by spectroscopic or X-ray studies. The corresponding nuclear magnetic resonance (NMR) data are the same as those reported in most cases except for the structure reassignment in these examples. This information will not be mentioned at the end of each example synthesis.

10.3.1
(+)-Hirsutene

The first example listed here, (+)-hirsutene (**1**) synthesis, uses a single SM for the synthesis. It is not a complex natural product and obtained from the metabolism of fungi [8], and it is relatively easy to carry out its total synthesis. This structure is used as a precursor of several antitumor compounds and brought its importance in synthesis. It has a five-membered ring in its neighbored. Its retrosynthesis is illustrated in Scheme 10.5 [8]. The first fragment molecule is **2**, which can afford **1** by a Wittig reaction; molecule **2** can be synthesized from **3** via a condensation reaction to give the α,β-unsaturated ketone structure in **2**. Further decomposition of **3** leads to the intermediates **4** and **5**, as illustrated in Scheme 10.5.

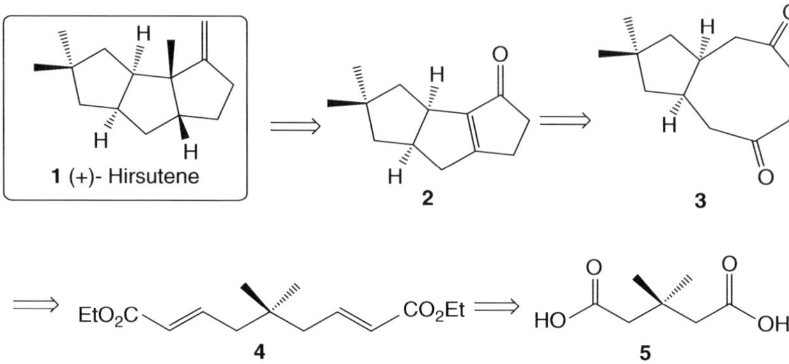

Scheme 10.5 Retrosynthesis strategy for **1** [8].

Clearly, compound **5** can be used as an SM. By reduction with BH_3, it forms a diol structure and then can be oxidized into dialdehyde using Swern oxidation. The aldehyde can undergo condensation reaction with $Ph_3P{=}CHCOMe$ to afford molecule **4**. This intermediate can undergo intramolecular condensation reaction to produce **6** with almost the same mole ratio of cis/trans (about 1.1 : 1). Hydrolysis under the base KOH gives the corresponding acid **7**. This diacid structure can be converted into an intermediate that can react with TMS-CHN_2 to produce the key intermediate **8**. Under catalysis of a Ru-containing ligand, an intramolecular cyclization takes place to form **9**. The formation of **9** generally indicates the successful synthesis of **1**, since its hydrogenation can lead to **10**, which can undergo the intramolecular condensation reaction to the intermediate α,β-unsaturated ketone **11**. Under strong base interaction, it can react with MeI to afford compound **12**, which can lead to the final product **1** via a Wittig reaction in a yield of 87% (Scheme 10.6).

Different viewpoints may lead to different synthetic routes. For the total synthesis of (+)-hirsutene (**1**), one may go through other ways to afford the key intermediate **13** [9], which can go through a cyclization to target **1**. One-step cyclization forms three circles (Eq. (10.1)).

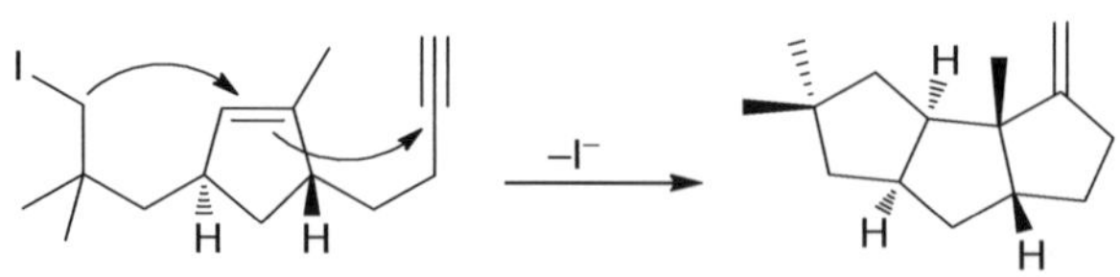

(10.1)

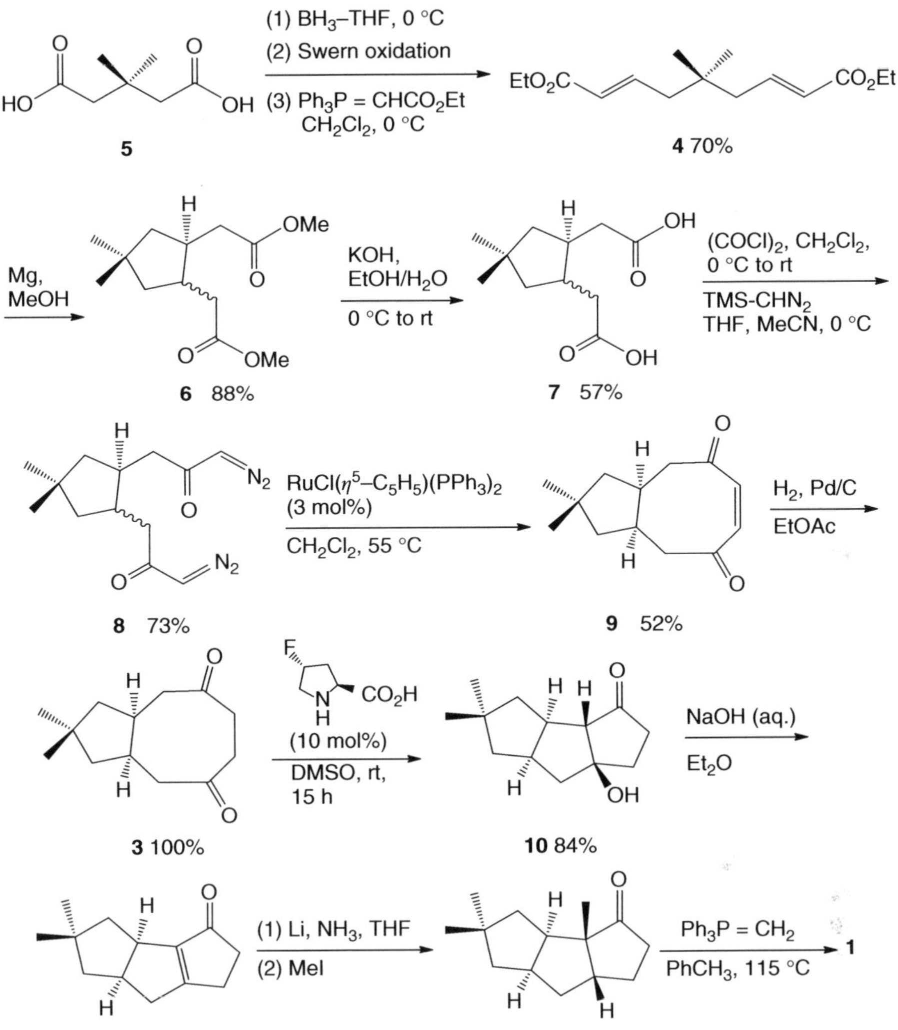

Scheme 10.6 Total synthesis of (+)-hirsutene (1) [8].

10.3.2
(2R,3S)-Rubiginone A_2 and Its Analog

Rubiginone A_2 (**14**), also named as Fujianmycins B or SNA-8073-A [10], exhibits anti-HIV or platelet aggregation inhibition bioactivity [11]. This has attracted the attention of organic chemists. Its total synthesis has been reported [12]. The routes listed here involve the preparation of two SMs. Its retrosynthesis route is illustrated in Scheme 10.7. The synthetic route looks quite simple in the procedure. The key step is to obtain the SM **19** and the Diels–Alder reaction

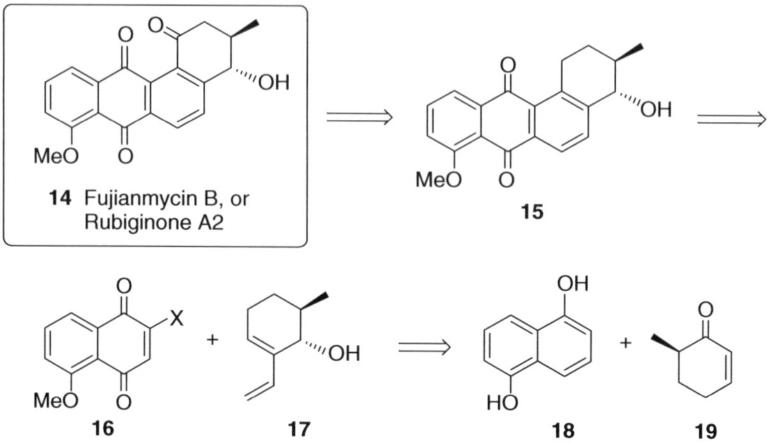

Scheme 10.7 Retrosynthesis analysis for the product preparation of **14** [13].

using **16** and **17** [13]. After the formation of the intermediate **15**, an oxidation reaction under light irradiation can convert it into TM **14**. This procedure needs two different SMs, **18** and **19**.

It is possible to use a natural product that has a similar stereogenic center for the preparation of **19**. The best candidate is the commercially available (2*R*,5*R*)-(+)-dihydrocarvone **20** as the enantiopure SM. It can be converted into **19** in a one-pot reaction in steps with a yield of 42% (Scheme 10.8). The yield can be improved up to 70%. It can react with an I_2−Py mixture at room temperature to afford the intermediate **21** in about 70% yield. Its reaction with vinyl magnesium bromide catalyzed by $ZnBr_2$ and $Pd(PPh_3)_4$ forms the intermediate **22**, which can be reduced to **17** by sodium borohydride.

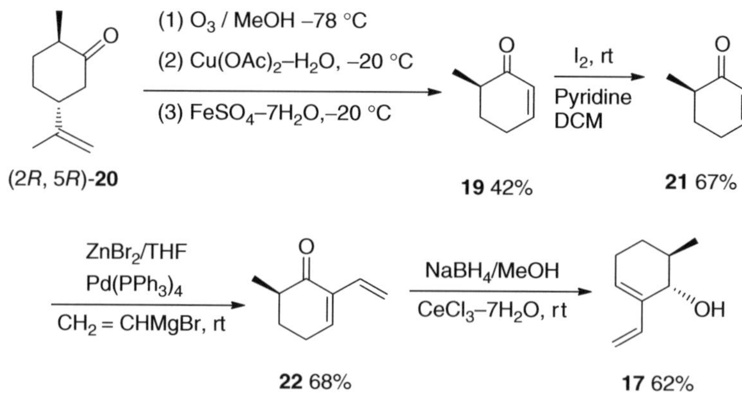

Scheme 10.8 Synthesis of intermediate **17** from the natural material (2*R*,5*R*)-(+)-dihydrocarvone [13].

Synthesis of the intermediate **16** is from SM **18** (Scheme 10.9). By esterification of **18**, the acetate **23** was obtained in 81% yield. Further reaction with

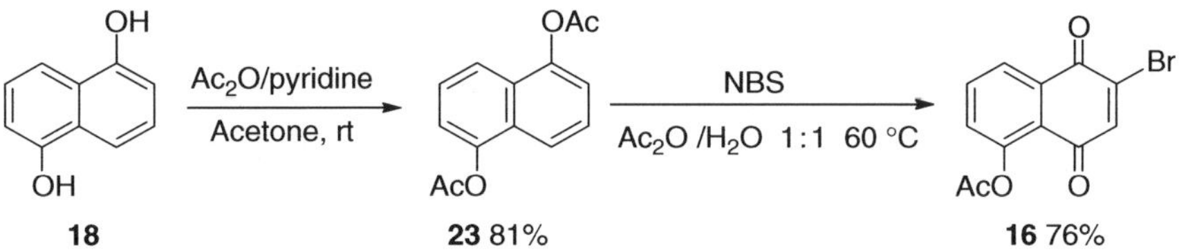

Scheme 10.9 Preparation of the intermediate **16** using naphthalene-1,5-diol [13].

N-bromosuccinimide (*N*-bromobutanimide) (NBS) produced the intermediate **16** using a mixture of acetic anhydride and water as the solvent under 60 °C [14].

Diels–Alder reaction can form the large aromatic ring system (Scheme 10.10). The first step is to protect the hydroxyl group in **17** using the methoxymethyl (MOM) group to afford **24**, and then to convert the –OAcof **16** to –OMe group to give **25.** Then compounds **24** and **25** are put together in toluene and heated to 90 °C for at least 1 h. Enough K_2CO_3 in methanol is added into the solution at room temperature after the reaction is finished (checked by TLC). It is then warmed to reflux. This procedure produces the intermediate **26**. By light oxidation, **26** can be converted to the last intermediate **27**, which can be hydrolyzed into the Fujianmycins B (**14**).

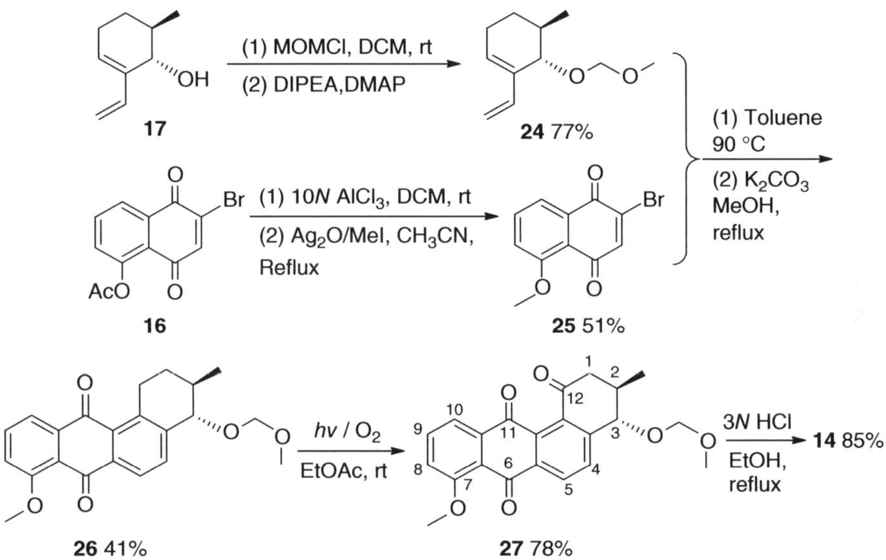

Scheme 10.10 Synthesis of rubiginone A_2 (Fujianmycins B) [13].

The hydrolysis of **14** can afford Fujianmycins A, which has no methyl group on –OH at C7. The acidic condition is 10 N HCl instead of 3 N HCl.

This example needs two kinds of SMs. Several parallel routes can afford two key intermediates and then they are combined together for the final product synthesis. This method is also frequently used statistics.

10.3.3 (+)-Brefeldin A

(+)-Brefeldin A (**28**) was reported early [15]. It has more than 30 total organic synthesis routes. Indeed, it is still attractive to use a novel method for its synthesis. For example, by use of an aldol condensation reaction, a different synthetic route becomes available [16]. In this molecule, there is one 5-memebred ring and one 13-membered ring structure. The synthesis difficulty lies in the construction of the big ring. It has two –OH groups; the –OH on C7 could be result of the reduction of a ketone, and the disconnection of C8a–C9 may lead to a B-reagent such as **30.** In the first step, cutting it into **29** and **30** are reasonably easy (Scheme 10.11). Keeping disconnection of **29** and **30** may involve the fraction molecules **31** and **32**. The key intermediate is fraction molecule **31**. It could be disconnected into smaller fractions such as **33** and **34**. At the same time, fraction **30** can also be separated into **35** and **36**.

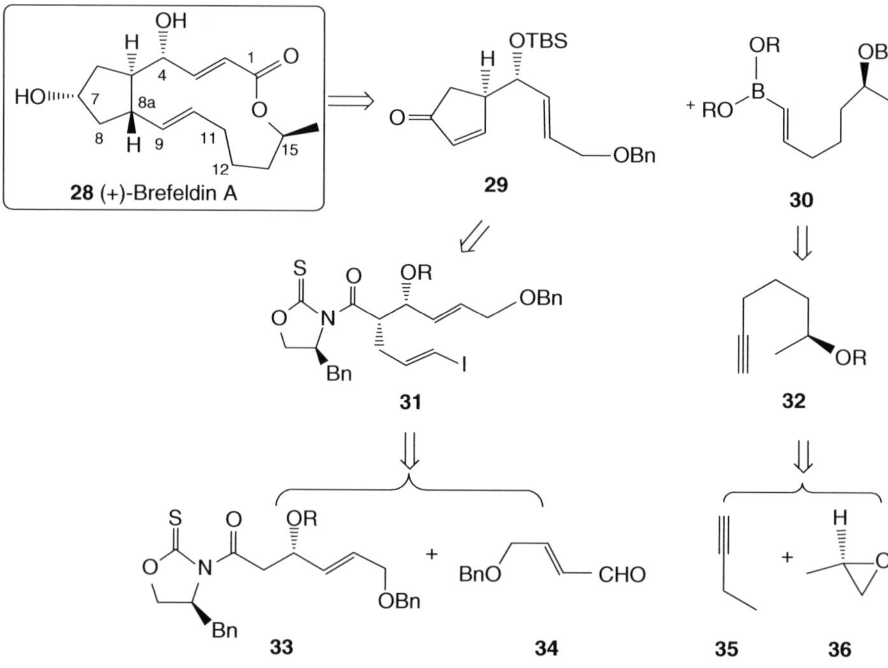

Scheme 10.11 Retrosynthesis of (+)-brefeldin A (**28**) [16].

This procedure contains two major intermediate syntheses and two SMs together for final TM synthesis. This procedure has a higher efficiency. First, an alkynol **37** was used as the SM to synthesize **29** (Scheme 10.12). By reduction with DIBAL-H in tetrahydrofuran (THF) solution and then with I_2, SM **37** afforded the cis intermediate **38**, which could be oxidized into the acid **39**. The intermediate **39** could react with (*S*)-4-benzyloxazolidine-2-thione and afford **40**, which

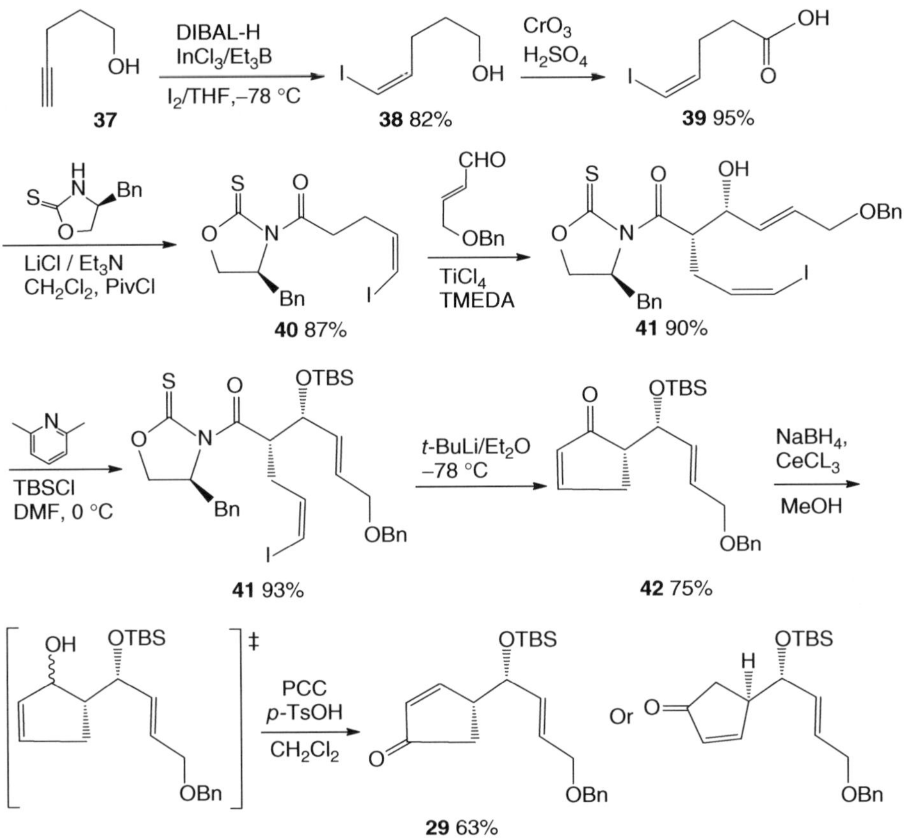

Scheme 10.12 Synthesis of the intermediate **29** [16].

could undergo aldol reaction with (*E*)-4-(benzyloxy)but-2-enal to produce the key intermediate **31**. After protection of the –OH of **31**, the exchange-cyclization of **41** formed **42**. By the isolation of the C=C double bond, **42** gave the intermediate **29**.

It was found that the exchange-cyclization from **41** to **42** is a key step. An effective way to carry out this conversion is to use *t*-BuLi instead of *n*-BuLi. The second key was to transfer **42** to **29**. If the oxidant is 2-Iodobenzoic acid (IBX/DMSO) dimethyl sulfoxide, this reaction could not be performed. It was possible only by using pyridinium chlorochromate (PCC)/*p*-TsOH, PCC/SiO_2, or PCC/4 Å MS (Scheme 10.12).

Then SMs **34** and **35** were used for the synthesis of **30**. This procedure was relatively simple. Under the base (*n*-LiBu) interaction, the terminal proton of the triple bonds could be removed and the anion would attack the carbon of propylene epoxide and lead to opening of the three-membered ring. Scheme 10.13 illustrates the whole procedure.

It is time to construct the TM (**28**) using intermediates **29** and **30.** The connection of **29** and **30** formed a ketone-containing compound (**46**) (Scheme 10.14),

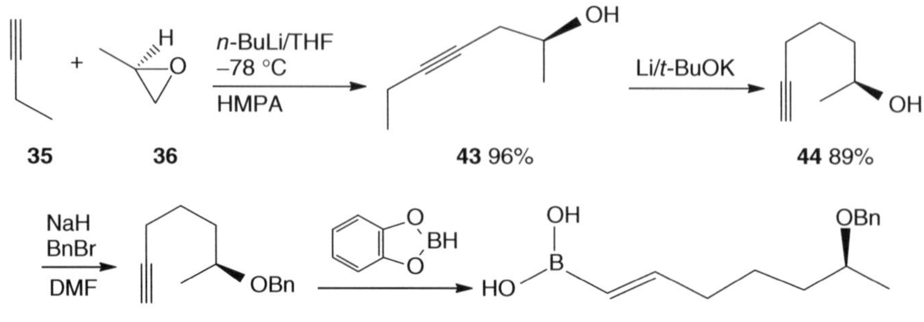

Scheme 10.13 Synthesis of the intermediate **30** [14].

which can be further reduced by DIBAL-H to afford two isomers with α-OH or β-OH orientations. The ratios of α-OH to β-OH were almost the same as those reported earlier. However, in this study, the ratio could reach up to 5 : 1. Therefore, the intermediate with α-OH structure (**47**) was used in further reactions.

The *α*-hydroxyl in **48** could be protected by *t*-butyldimethylsilyl (TBS) group and afford **49**, this intermediate could be hydrolyzed by metal Li in naphathene to remove protection group –Bn but keep TBS protection group to produce **50** (Scheme 10.14), it can take chemoselectively oxidation to **51** which has one $-CO_2H$ group and one –OH group. Under the condensation reagents 2-Methyl-6-nitrobenzoic anhydride, mixed with DMAP and used as a dehydrant (MNBA) and 4-Dimethylaminopyridine (DMAP), it can form the lactone **52**. After hydrolysis under weak acidic solution condition (2 N HCl in THF), it would produce the expected TM **28**.

10.3.4 Malyngamide U and Its AC Reassignment

Malyngamides were reported to possess wide bioactivities such as antitumor or anti-HIV activity [17]. For example, after malyngamide U (**53**) was reported, its structure attracted researchers' interests in its total synthesis [18]. It was found that its AC was not corrected in the original reports after its total organic synthesis; the stereochemistry was corrected [19]. The structure of this compound is not complex compared to that of Taxol. However, its total synthesis involved some important reaction types that are useful for other syntheses as well.

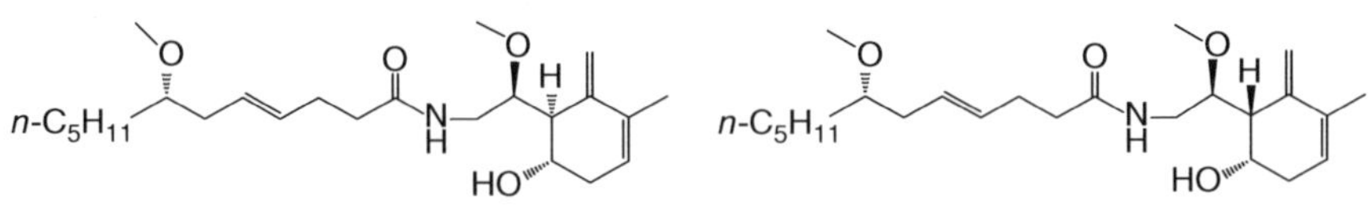

Reported **53**   Corrected **53**

Relatively, its retrosynthesis is not complex based on its structure. It could be disconnected as two fragmental molecules **54** and **55**. If (+)-**56** or (−)-**56** was

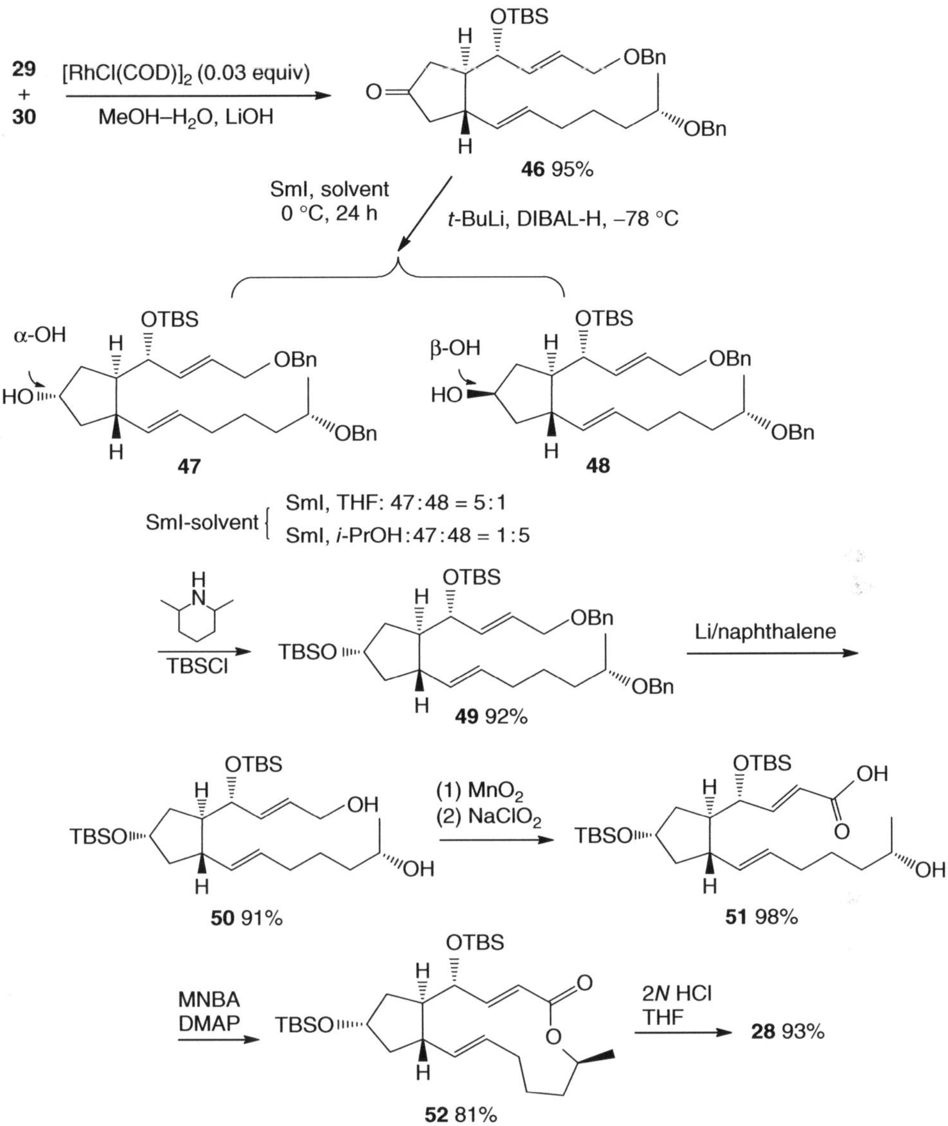

Scheme 10.14 Synthesis of TM (+)-brefeldin A **(28)** [17].

used as SM, it could provide two different products, and one of them must be the natural product. Scheme 10.15 summarizes the retrosynthesis routes.

Because of the earlier reported AC at C1 was (*S*), clearly, (*S*)-(+)-**56** was used as the SM (Scheme 10.16). By addition of CHO to the ring under lithium diisopropylamide (LDA) promotion, a $-CH_2OH$ was successfully introduced on the ring with (*S*) AC (**57**). It could be easily converted to **59** by reduction using sodium borohydride and $CeCl_3$. Epoxidation of **59** led to the formation of **60**, which could be

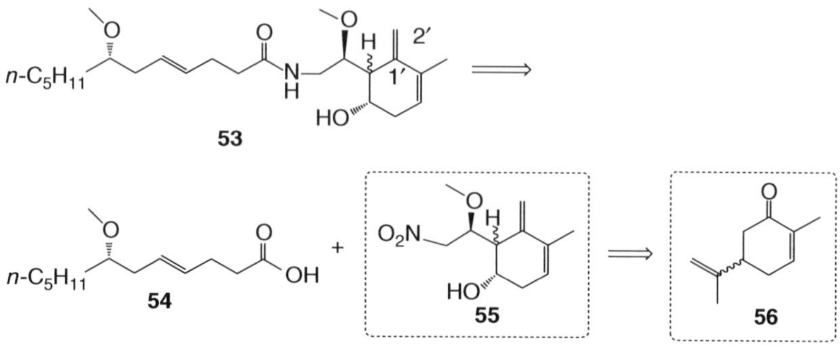

Scheme 10.15 Retrosynthetic analysis of malyngamide U (**53**) [19].

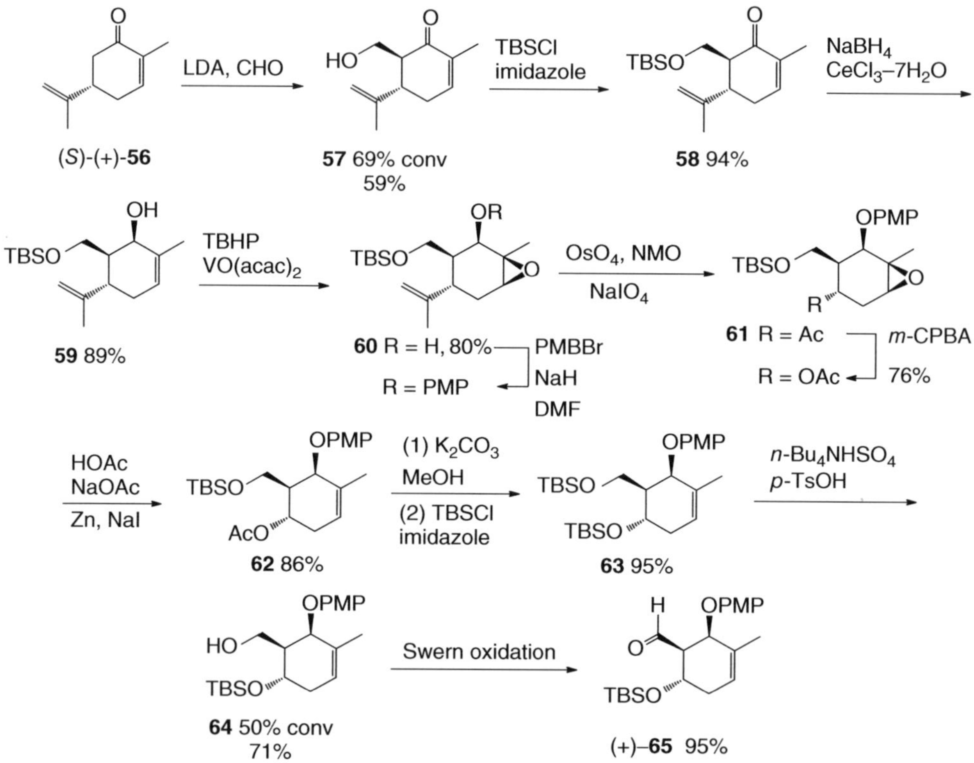

Scheme 10.16 Synthesis of the key intermediate (+)-**65** [19].

further oxidized by OsO_4 and $NaIO_4$ to give **61**. This intermediate could be converted into **62** by using Zn powder via an elimination reaction. After deprotection of the corresponding TBS from –CHOTBS in **63**, its derivative **64** could undergo a Swern oxidation to afford the key intermediate **65**.

By the use of two different chiral catalysts (+)- and (−)-**69**, the addition product of **65** by $MeNO_2$ could afford two isomers **66** and **68**, respectively (Scheme 10.17).

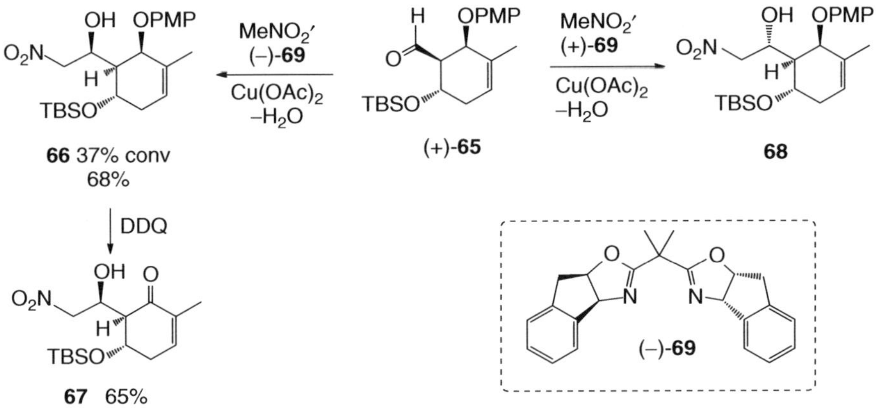

Scheme 10.17 Two stereoisomers in addition of $MeNO_2$ to (+)-**65** using chiral catalyst (+)- and (−)-**69**, respectively [19].

Therefore, the two isomers **66** and **68** could be used to react with fraction molecule **54**, and then the two compounds would have different ACs. The product derived from **66** should be the reported structure. The connection of **54** and **66** is illustrated in Scheme 10.18.

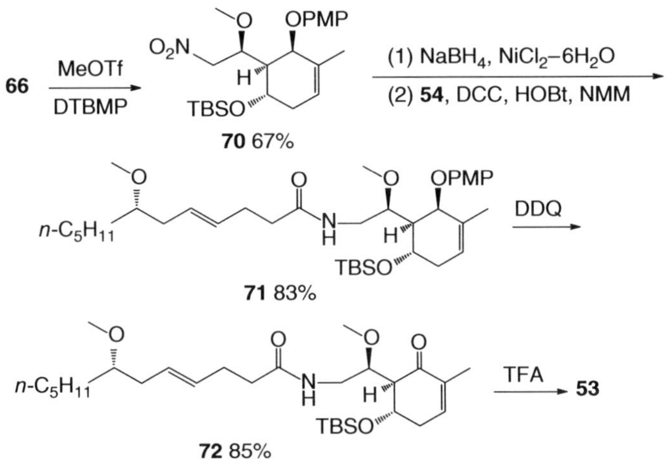

Scheme 10.18 Synthesis of the reported structure **53** [19].

Unfortunately, the evidences from spectroscopies such as NMR for the synthesized **53** did not agree with the original results. This suggests that the original structure might have given wrong results, especially in its AC assignment. Therefore, (*R*)-(−)-**56** was used as SM again for the synthesis of its another stereoisomer (corrected **53** structure). The obtained new stereoisomer had the almost same spectroscopic data. Now it could be concluded that the original reported structure of **53** was not correct. The corrected natural product should have the new AC.

10.3.5
Taxol Derivatives

Taxol (**73**), which was reported in the 1970s [20], is a famous star molecule. Its total synthesis includes its skeleton and side-chain syntheses. There are many reports about its synthesis [21]. Pattern recognition was used for its similar skeleton synthesis of baccatin III (**74**) introduced above [4, 5]. Synthesis of its side chain mainly involves the formation of the two stereogenic centers using various chiral catalysts.

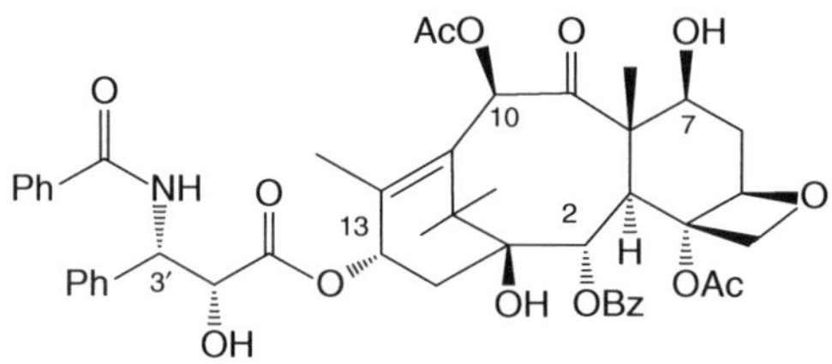

73

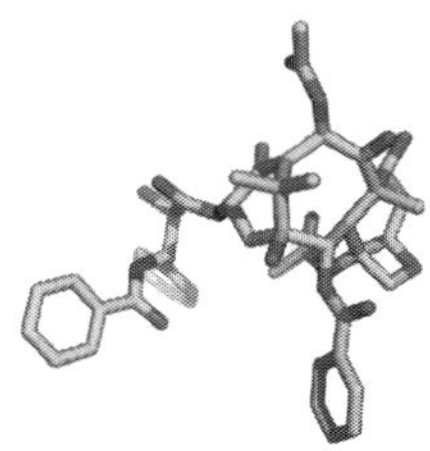

73a

As mentioned before, natural products are valuable lead compounds for medicines. Based on the natural product structures, some more active or important compounds can be discovered and become promising medicines. Taxol analogs are good examples used to illustrate this point. During the structure–activity relationship (SAR) study, an interesting discovery was that taxol had a T-conformation (**73a**) when it exhibited its strong activity [22].

Based on further investigation of the derivatives of taxol T-geometry, the analogs **75** and **76** were synthesized [22a]. It was found that compound **75** had much stronger cytotoxicity against A2780 tumor cells; for example, it was approximately 20 times more potent than taxol and about 300-fold more potent than its structurally similar analog **77**. The structures with strong activity illustrated here exhibited very similar T-conformation of taxol.

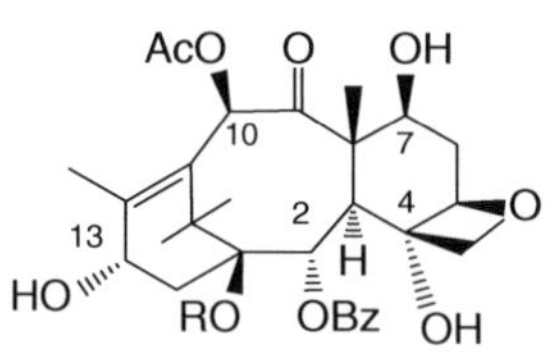

74 Baccatin III

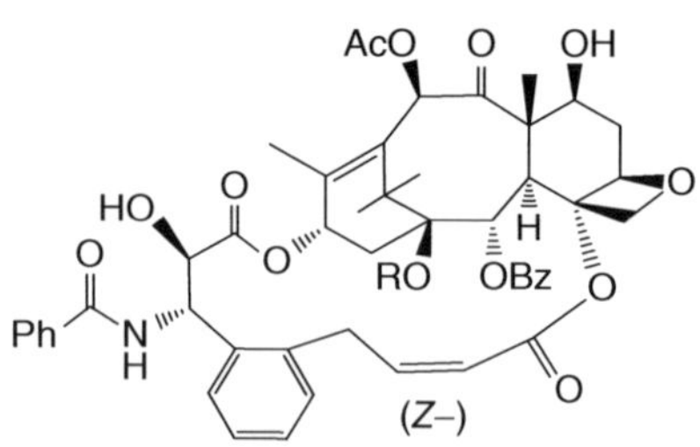

75

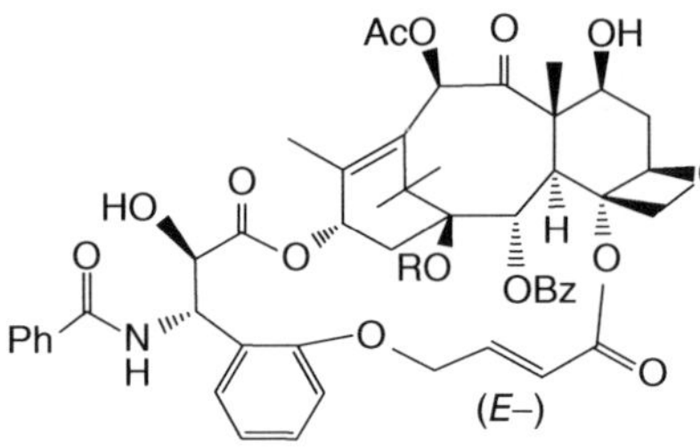

76

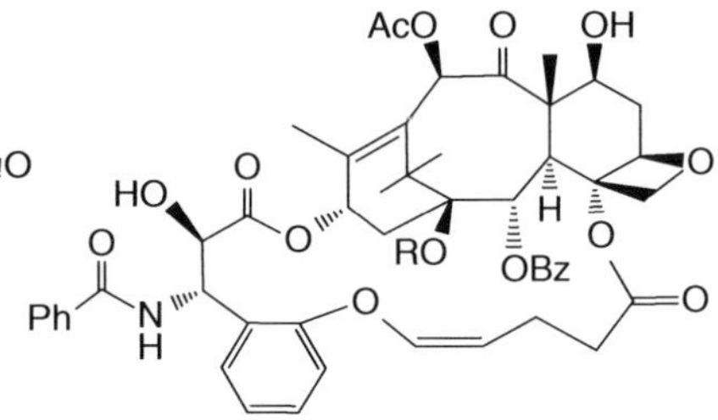

77

Total organic syntheses of taxol have been reported widely, and the references can be easily found [22c, d]. The examples listed here involve the synthesis of the derivatives **75** and **76**. As a special case, the retrosynthesis of the derivatives is not illustrated here. The synthesis contains two major steps. The first one is synthesis of the intermediates **84–87**. Scheme 10.19 illustrates the simplified procedure [22a].

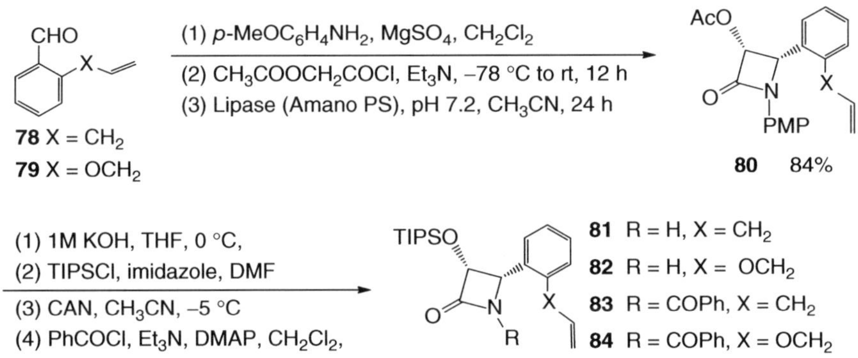

Scheme 10.19 Strategy for the synthesis of the key intermediates **81–84** [22a].

Therefore, based on the analogue of taxol (**85**), the intermediate **89** was designed and synthesized. In the procedure, once acryloyl chloride reacted with –OH on C4, cleavage of $-SiMe_3$ from **89** by using HF–Py in THF was necessary for acetylation of –OH on C10. Then, connection of $-SiMe_3$ on C7 afforded the key intermediate **89**. Scheme 10.20 illustrates the procedure.

For example, the intermediate **83** could be used in the reaction to connect the both sections together. It formed the diene structure, which could undergo metathesis reaction to give **75**. It is illustrated in Scheme 10.21.

Similarly, the same reaction conditions were used for the synthesis of **76**. The reduction of both **75** and **76** could afford products with double bonds in high (up to 99%) yield. The reduction could be performed using H_2 as reductant and catalyzed by Pd/C(10%) under the 35 psi pressure for about 2.5 h. Both reduced products showed strong cytotoxicity against PC3 and A2780 tumor cell lines, respectively.

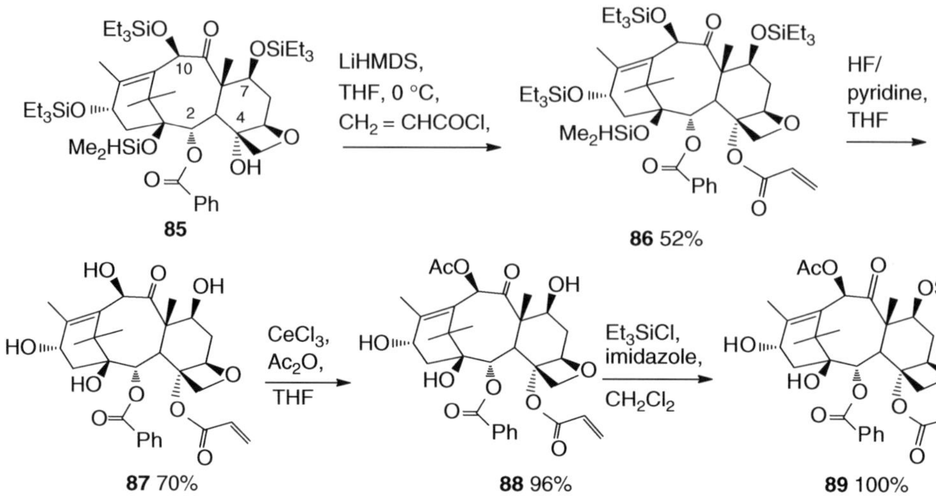

Scheme 10.20 Synthetic route to the intermediate **89** [22a].

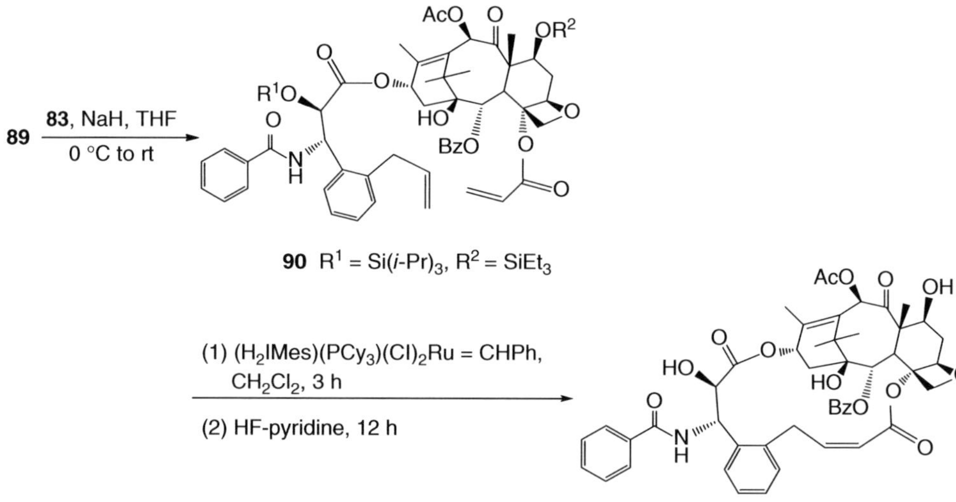

Scheme 10.21 Metathesis reaction to afford product **75** [22a].

10.3.6
Amphidinolide T2 and Its Derivatives

Many antibiotics have been developed, and one of them is macrolides. The difficulty in their total synthesis is how to efficiently increase the yield of the ring-closing reaction [23]. Its concise total synthesis represents a new study direction in synthetic methodology. Amphidinolides, such as amphidinolide T2(**91**), form a large family macrolides obtained from symbiotic marine dinoflagellates

Amphidinium sp. Its congeners have five structurally related 19-membered macrolides [24]. Cytotoxicity was reported for amphidinolide T analogs, with IC_{50} values of 7–18 μg ml^{-1} against murine lymphoma L1210 [25]. Compared to T2, structures of T1, T3–T5 have no stereogenic center at C21, and the side chain just has three carbons. Other differences are that they have different stereogenic centers at C12 and C13, while C=O at C13 becomes a hydroxyl group. The total syntheses of T1 and T3–T5 have been reported [26]. The retrosynthesis for **91** is illustrated in Scheme 10.22 [27]. First, it could be cut down between O1 and C18; then the second section could be between C12 and C13. This disconnection forms two fractions **92** and **93**.

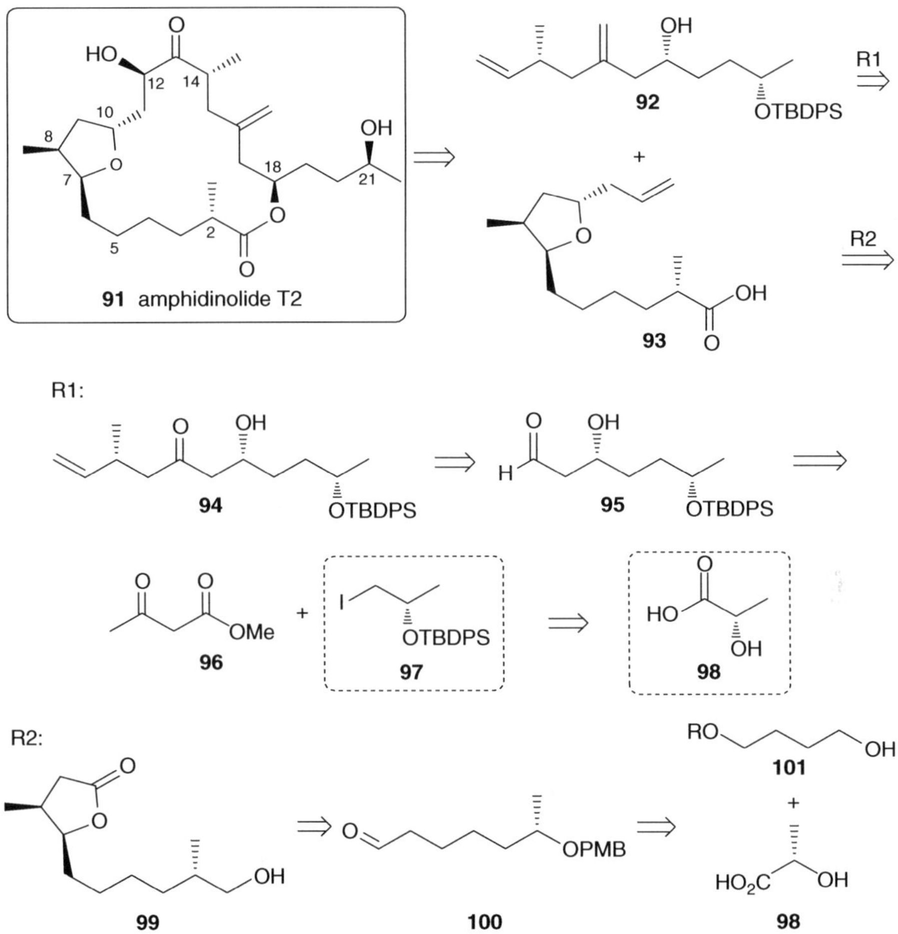

Scheme 10.22 Retrosynthesis of amphidinolide T2 [27].

The group >C=CH_2 in intermediate **92** could derive from >C=O (**94**), and the fragment molecule could be separated into **96** and **97**. Fragment **97** could derive from (*S*)-2-hydroxypropanoic acid (**98**). Another intermediate **93** could be from **99**, which can be further disconnected into the small diol derivative **101** and fragment **98**.

The synthesis of fragment **92** started from the methyl (*S*)-lactate **102**, which has natural chirality, obtained via esterification of **98** (Scheme 10.23). After protection of its –OH by the *t*-Butyldiphenylsilyl (TBDPS) group, it was reduced by BH_3 in refluxing THF. The newly formed –OH could be replaced by element I using a mixture of I_2, PPh_3, and imidazole in the ratio 4:2:4 in refluxing THF to form the intermediate **103**. Under catalysis of the chiral catalyst $RuBr_2$[(*R*)-BINAP] 2,2′-Bis(diphenylphosphino)-1,1′-binaphthalene, this key chiral fraction molecule could undergo a coupling reaction with methyl acetoacetate (**96**) to form a new (3*R*,6*S*)-methyl 3,6-dihydroxyheptanoate, which could be reduced by DIBAL-H to produce the aldehyde **104**. Before its further reaction, it is necessary to convert **107** to **108** via two major steps: the first is to use *p*-toluenesulfonyl chloride

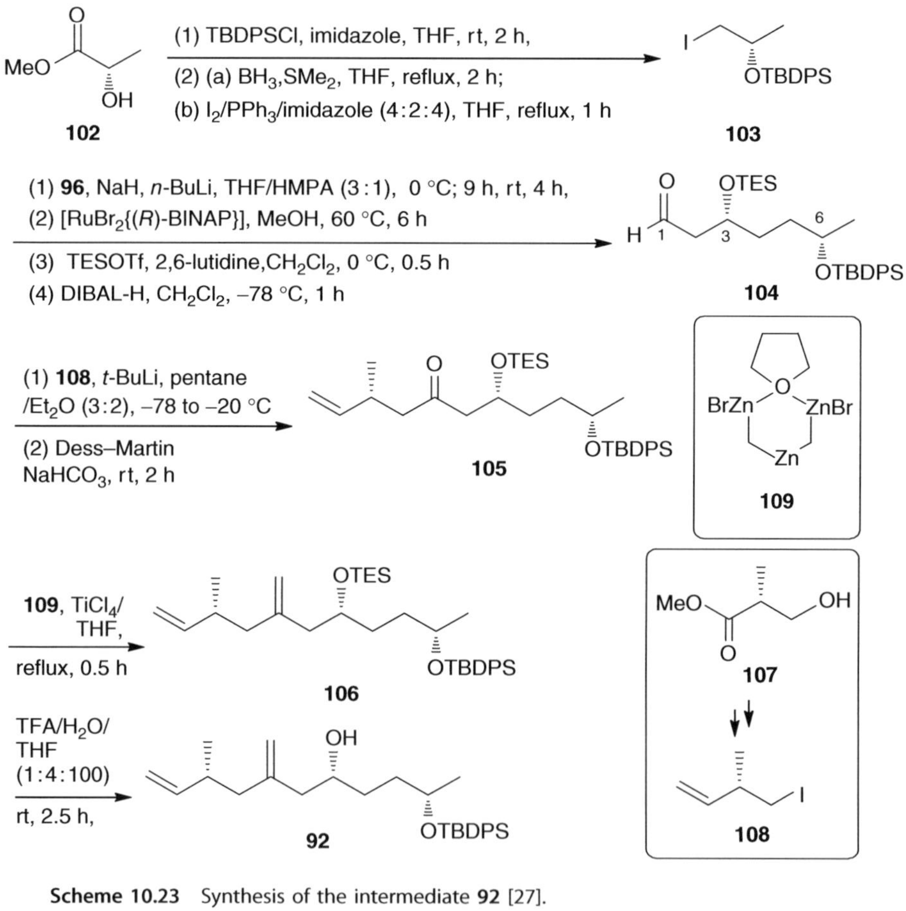

Scheme 10.23 Synthesis of the intermediate **92** [27].

reacting with –OH and then reduce its >COMe to aldehyde by DIBAL-H. LiI is then used to replace *p*-toluenesulfonyl in ether to form intermediate **108**.

The reaction of **104** and **108** could afford the key intermediate **105** in two steps. It could react with **109**, which could provide a CH_2- group like Wittig reagent to convert >CHO to the corresponding $>C{=}CH_2$, to afford **106.** It could be hydrolyzed to **92** in the mixture THF/H_2O under the catalysis of 1% TFA.

Synthesis of **93** requires the intermediate **117**. Therefore, key intermediate **117** was prepared first. In this procedure, SM **107** was used (Scheme 10.24). There are two routes for its synthesis. One is from butane-1,4-diol (**112**) (in dotted line frame), and the other is from **107** via **111**.

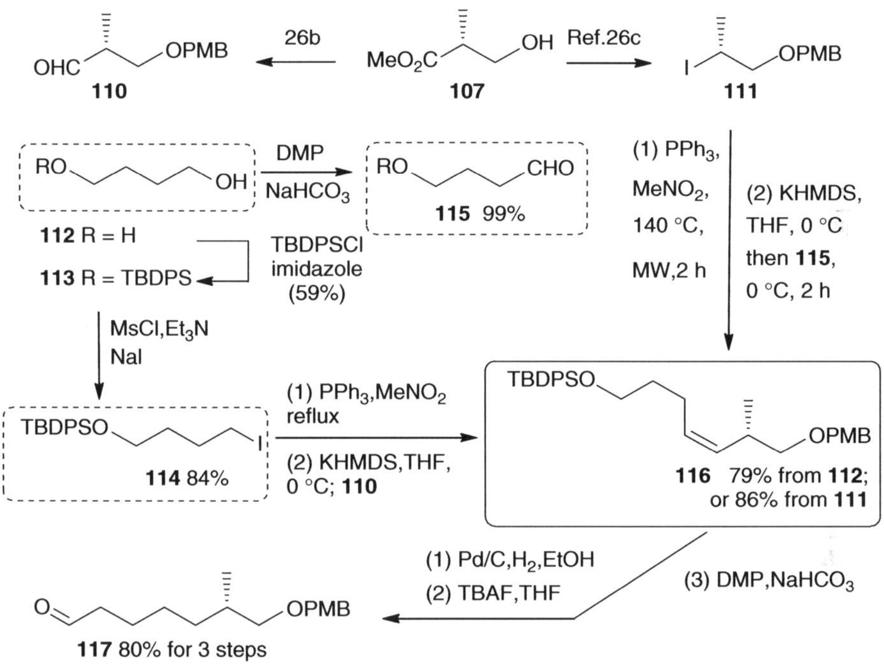

Scheme 10.24 Synthesis of the key intermediate **117** [27].

A SmI_2-mediated enantioselective reductive coupling reaction of **117** with (1*S*,2*R*)-*N*-methylephedrine-derived crotonate **118** afforded the desired *cis*-3,4-disubstituted γ-butyrolactone **119** with 97.1 : 2.9 dr. Further reduction provided the hemiacetal structure **120**. It reacted with allyltrimethylsilane to afford the corresponding intermediate **124**, which could undergo oxidation reaction with pyridinium dichromate (PDC) to form the product **93** (Scheme 10.25).

The intermediates **92** and **93** would react with each other under the catalysis of the condensation reagent DMAP to afford the key intermediate **122**, which could be used as an SM for metathesis reactions catalyzed by catalyst **123**. This catalyst could afford (*E*)-**124** and (*Z*)-**124** in about 2.7 : 1 ratio (Scheme 10.26).

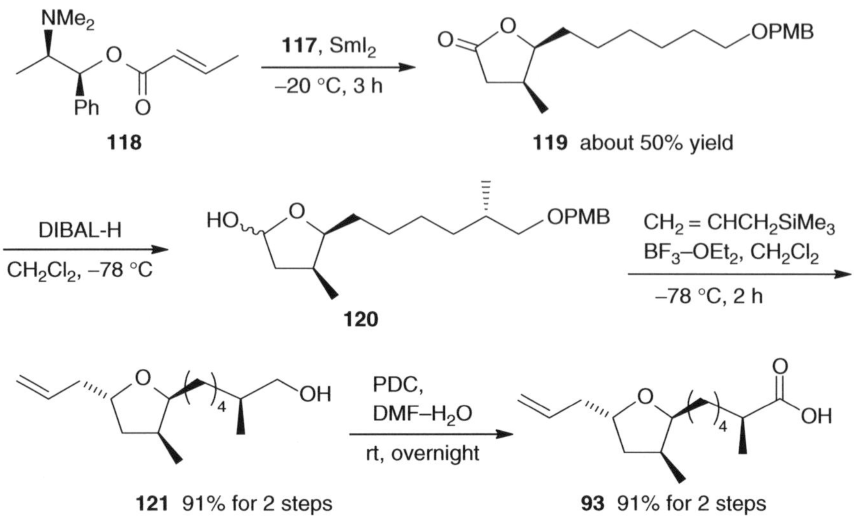

Scheme 10.25 Synthesis of the intermediate **93** [27].

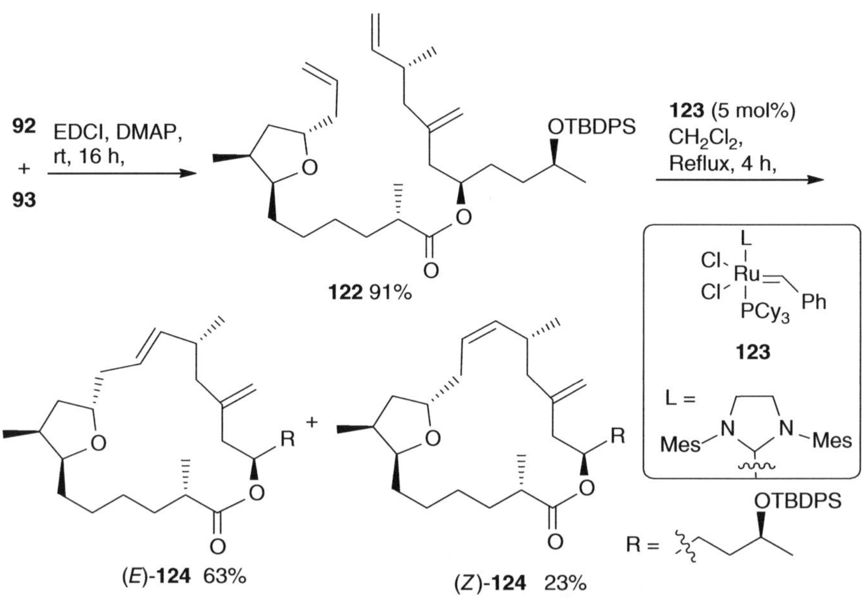

Scheme 10.26 Synthesis of the cyclic intermediate **124** with (*E*) and (*Z*)-structures (here Mes = 2,4,6-trimethylphenyl) [27].

The next challenge is the oxidation of C=C bond in (*E*)- and (*Z*)-**124** to form two hydroxyl groups in correct orientations. Experimental investigation of (*E*)- and (*Z*)-**124** in further experiments were performed (Scheme 10.27). It was found that only (*E*)-**124** could form the corresponding product with the same stereochemistry as **91**. Scheme 10.27 illustrates the procedure. In the procedure,

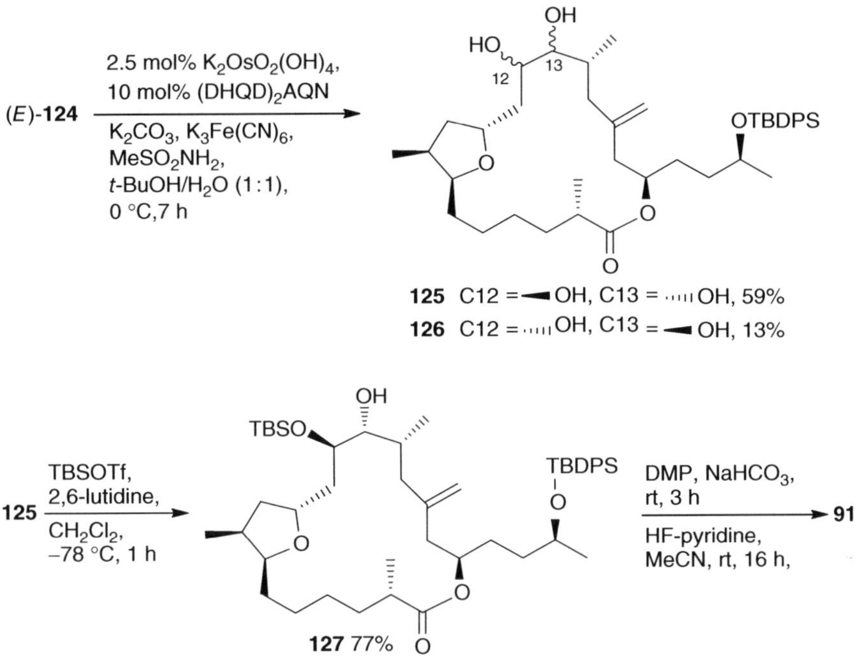

Scheme 10.27 Reaction routes to TM **91** [27].

2.5 mol% $K_2OsO_2(OH)_4$ was used for the oxidation reactions of >C=C<, and 59% yield of **125** was obtained. By protection of its –OH at C12, oxidation of –OH on C13 afforded the final TM **91**. Total yield from the whole procedure was about 8%. This is an excellent yield for a 16-step total organic synthetic route.

10.3.7
(+)-Vindoline

Indole alkaloids constitute a big family, and many of them have strong bioactivity. Because of their structural variances in nature, they are found in many folk medicines of various plants, and their structures have attracted the interests of organic synthetic chemists. Total synthesis of different indole alkaloids has been reported since 1950s; many of them have been obtained, for example, (+)-vindoline (**128**) [28], which belongs to the highly functionalized *Aspidosperma* alkaloid. It is found that this structure has a definite relationship with the indole-containing precursor. Its retrosynthetic analysis will not be described here. Different synthetic routes are introduced in this section.

(+)-Vindoline **128** was isolated in *Cantharanthus roseus* (L.) G. Don in trace quantities [29]. Its first total synthesis in 1975 used the tryptamine derivative **129** as a key intermediate [30]. It could undergo [4 + 2] cyclization reaction under BF_3

solution at 90 °C to form an intermediate with loss of the acetyl group on N atom and afforded **130**, which by Michal aldol condensation reaction with acrylaldehyde gave the skeleton molecule **131**; this intermediate could then be converted to **128** easily (Scheme 10.28).

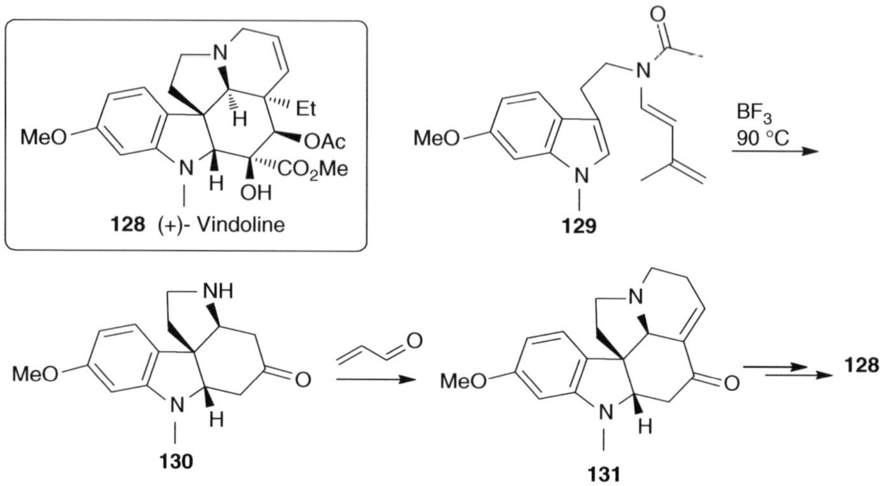

Scheme 10.28 Outline of the synthesis of vindoline [30].

A key intermediate **132** was also used in vindoline total synthesis [31]. Quaternary ammonium salt (**132**) could be cleaved at the C3 position into the corresponding compounds by cyanide (CN^-) attacking C3, and then CN could connect to C3 at the same time to give **133**, which could be used to react with MeI under the catalysis of mercuric acetate. This intermediate could form **134** and serve as another key intermediate to afford **128** (Scheme 10.29).

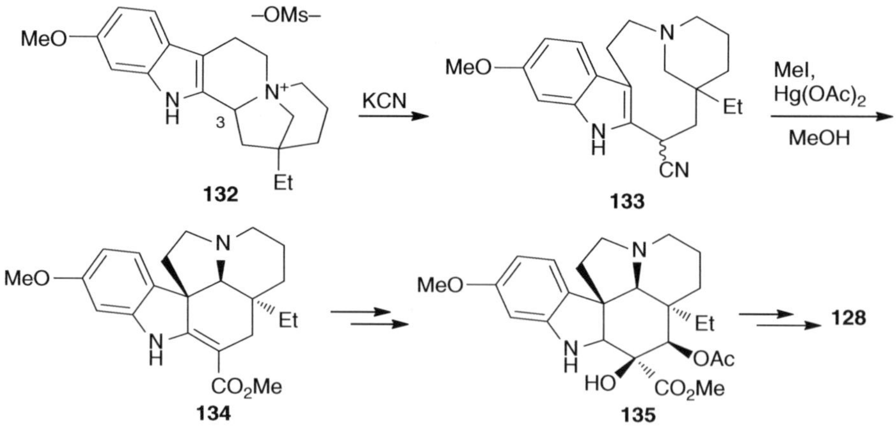

Scheme 10.29 Outline of the synthesis of vindoline**128** using quaternary salt [31].

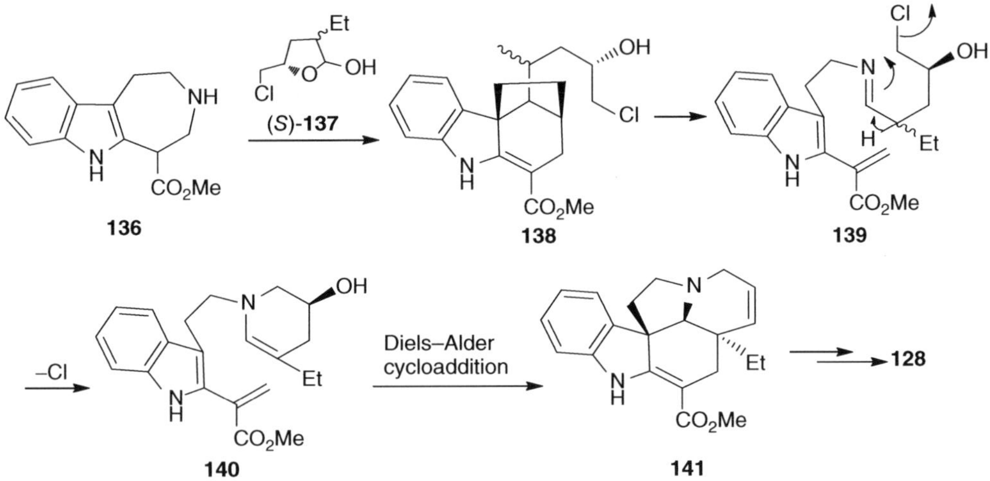

Scheme 10.30 Outline of the synthesis of skeleton of **128** [32].

Historically, indoloazepine (**136**) was used as a key intermediate for the synthesis of **128** (Scheme 10.30). This procedure required the use of a chiral lactol **137**, which was derived from (*S*)-epichlorohydrin. The bridged compound **138** could convert to **139,** which would undergo tautomerization to **140** via loss of a Cl anion [32]. By [4 + 2] cycloaddition to form tabersonin **141**, the TM **128** could be obtained.

The intermediate **142** derived from tryptamine was used for the synthesis of **128** too (Scheme 10.31). The strategy is to use cycloaddition cascade reactions [28].

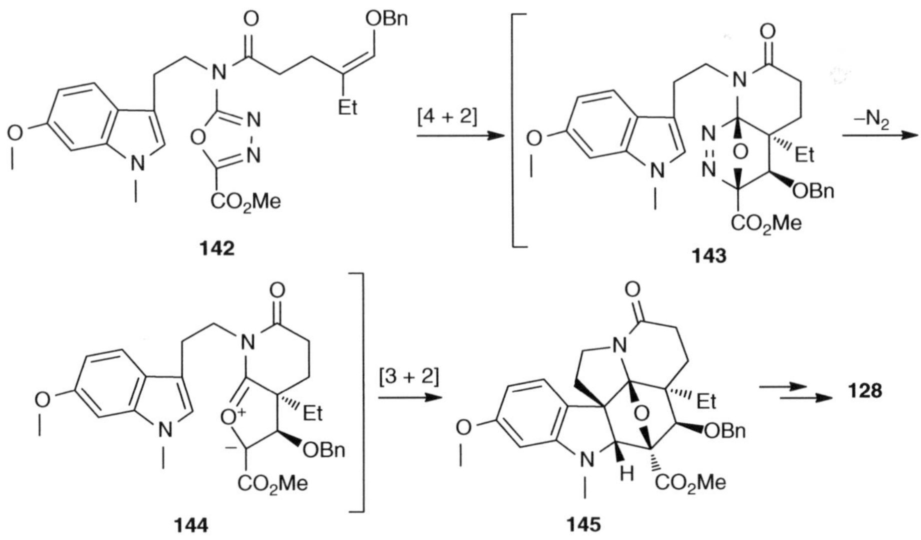

Scheme 10.31 Synthetic strategy of TM **128** [28].

1,3,4-Oxadiazoles were used in the procedure via sequential [4+2] and [3+2] cyclizations. (*Z*)-isomer of enol ether **142** could form the bridged ether **143** by heating in *o*-dichlorobenzene. This could generate the oxonium ylide **144**. This intermediate afforded bridged ether **145**, which could further undergo a series of reactions to give **128**.

An improved method to form **128** was reported using a similar method. The amide **146** (Scheme 10.32) in the new method was different from **142** listed in Scheme 10.31 [33].

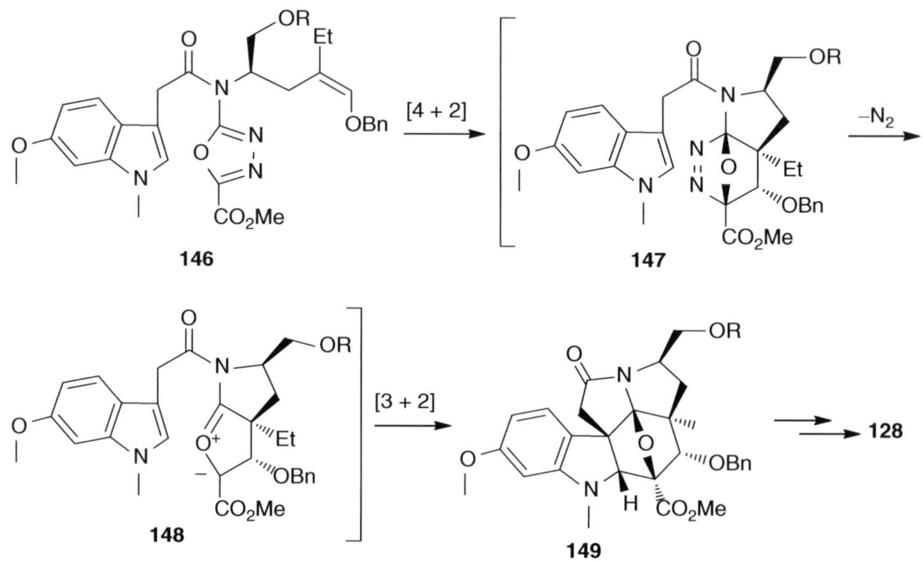

Scheme 10.32 Different synthetic strategies of TM **128** [33].

10.4 Calculation in Total Synthesis

Transition state (TS) calculations for total organic synthesis are rarely reported. This is expected. Since computation of TS in total synthesis is absolutely complex and time consuming, it may be out of the capacity of the computers in early time, although this is an aim of theoretical and experimental chemists. The second reason may be that no suitable TM is found yet. On the other hand, if there is any error in the calculation procedure, the whole procedure would fail. Other factors may involve the reliability of the quantum methods selected in TS calculations. However, to design a suitable theoretical total synthetic route is a big challenge.

In this procedure, every reaction should be carefully studied to examine its TS barriers if the reaction has two or more routes. The energy barrier sizes should tell us which route is favorable. Based on the route, the next possible reaction

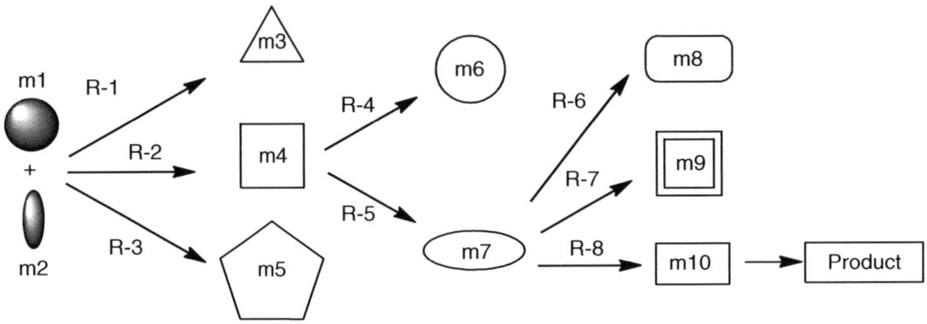

Figure 10.2 Diagrammatic sketch for selective reactions via different routes (the signs in this figure represent different molecules in reactions). In this sketch, the TS barrier via R-2 is lowest to the product **m4** (the square) from (**m1** + **m2**); the barrier via R-5 is lower than **m7** (oval); finally via R-8 is lowest to **m10** (oblong).

should be investigated by calculating its TS activation energy. This process should be repeated until the whole reactions are definite. If this procedure is divergent, the TS geometry investigation will be difficult because the number of TS structures may be more than the estimations. If it has reasonable TS geometry numbers, the TS barrier calculation is possible.

In many cases, TS barrier calculations involve chemoselective reduction or addition. Obviously, it is important to be familiar with various reductions and/or addition TS constructions. The whole procedure is illustrated in Figure 10.2.

The procedure is divergent theoretically at every step. However, only one or two procedures give low TS barriers. These barriers can be calculated for each step. After every TS barrier is predicted, the final product should be definite theoretically.

Apart from theoretical total synthesis, it could be used to determine reaction directions for some key series of reactions. In this way, the calculations are relatively easy. This is possible to be used in the study of biogenetic routes.

The strong cytotoxic product **150** was introduced through a biogenetic route, as illustrated in Scheme 10.33 [34]. The α-C would attack the carbonyl carbon under acidic condition, and then the intermediate would convert to the corresponding natural product. This enzyme-catalyzed biogenetic procedure is possible. However, discussion of possible organic synthetic routes is important from the viewpoint of theoretical chemistry and organic synthetic chemistry.

In viewpoint of theoretical chemists and organic synthetic chemists, the attack of α-C of 2-oxosuccinic acid of carbonyl carbon generally happens under a base catalysis instead of acidic conditions. The base would draw a proton from α-C and this would lead to α-C having one negative charge. This negative carbon can easily attach the positive carbonyl group. Secondly, chemists may want to know why α-C chemoselectively attacks C12 instead of C6 (both are carbonyl carbon), as illustrated in Scheme 10.33.

The first question was investigated by density functional theory (DFT) using the Gaussian package. Chemoselective addition of α-C to C6 or C12 in Scheme 10.33

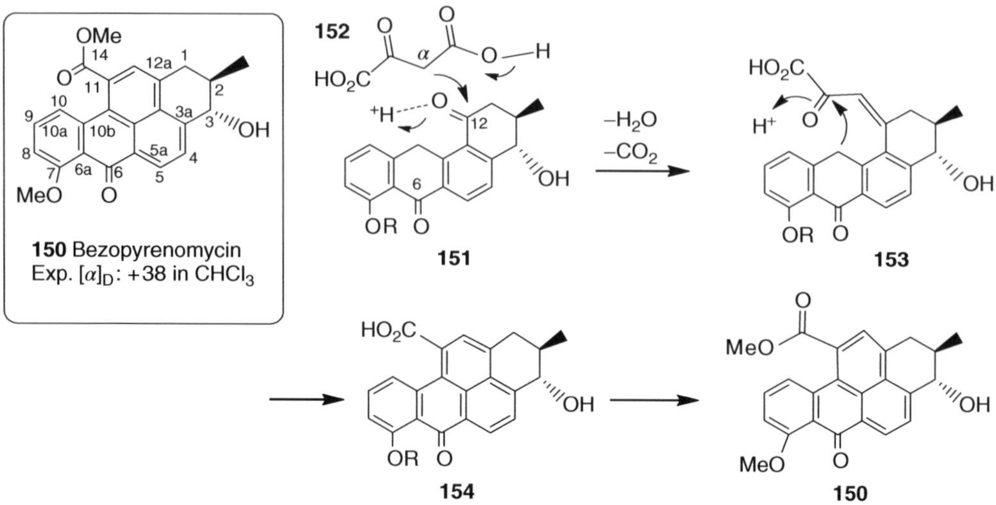

Scheme 10.33 Proposed biogenetic route for the formation of **150** from **151** in the original report [34].

is the key procedure. Only that the relative barrier in the addition of α-C to C12 is lower than that to C6, which could guarantee the next reaction leading to the isolated natural product. Because of a very flexible O–C bond in 2-oxosuccinic acid, the TS numbers would be more than computed value (H.J. Zhu, *et al.*, unpublished results). To reduce computational time, **155**, which is a reasonable model simplified from 2-oxosuccinic acid, was used in TS computations (H.J. Zhu, *et al.*, unpublished results). Under base conditions, a proton on α-C was removed. The negative α-C of **155** could attack the C12 or C6 of **151** via the front or the back, respectively (Table 10.1). The effect of the position of substituents such as –CHO and –MeO was investigated in TS barrier calculations. Total TS geometries were investigated at the B3LYP/6-31G(d) level. The TS geometries with lower barrier in each case were further computed at the B3LYP/6-311+G(d,p) level.

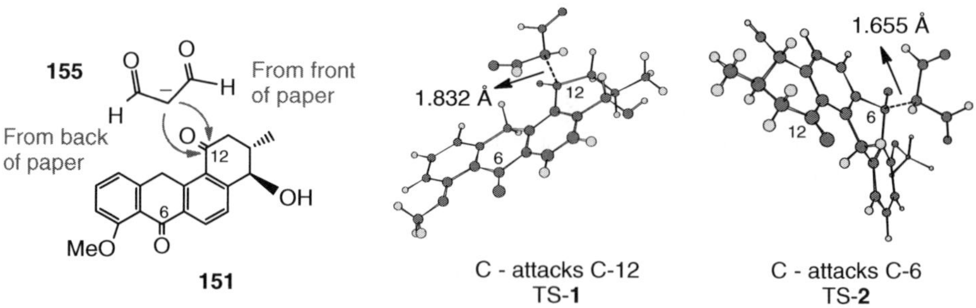

In this investigation, an enzyme effect was considered. However, it does not mean this enzyme would lead to the formation of a high-energy product. In this section, the normal effect is just considered from an enzyme. The enzyme may affect the barrier sequence. Since the exact enzyme is unknown, use of a solvent

Table 10.1 Predicted TS barriers for **TS-1** and **TS-2** (H.J. Zhu, *et al.*, unpublished results).

Entry	Item	TS-1	TS-2
1	ΔG_1[a)]	0.98	0.00
2	ΔE[b)]	2.09	0.00
3	ΔG_2[c)]	1.64	0.00
4	ΔE_0[d)]	1.99	0.00

a) The free energy difference obtained at the B3LYP/6-31G(d) level.
b) Mimic enzyme catalyst energy difference.
c) The free energy difference obtained at the B3LYP/6-311+G(d,p) level.
d) Zero-point energy correction difference at the B3LYP/6-311+G(d,p) level.

that has a dielectric constant (ε) close to that in the interior of enzyme ($\varepsilon = 4.0$) to represent an enzyme should be considered in TS barrier corrections. In this study, a neutral solvent, chloroform ($\varepsilon = 4.9$), was selected for the simulations of the effect of the enzyme on the barriers. The TS structures and energy barriers for the additions that were obtained at different levels are illustrated above. In addition, the zero-point energy (ZPE) correction data (ΔE_0) at the B3LYP/6-311+G(d,p) level were also used in the comparison of the relative energy differences.

Unfortunately, the computed barrier in procedure of α-C attacking C6 (**TS-2**) is lower by 0.98 kcal mol^{-1} in free energy than those in attacking C12 (**TS-1**) at the B3LYP/6-31G(d) level. Under the mimic enzyme catalysis, the energy difference between the two TSs became 2.09 kcal mol^{-1} at the B3LYP/aug-cc-pVDZ level using PCM model in chloroform. The lower energy barrier is still in **TS-2**. When calculations were performed at the B3LYP/6-311+G(d,p) level, the barrier difference between **TS-1** and **TS-2** increased from 0.98 to 1.64 kcal mol^{-1} in free energy. The barrier difference became 1.99 kcal mol^{-1} after ZPE correction. Thus, route from **151** to **150** is unlikely in organic synthesis via similar chemical procedure of α-C of **152** attacking C12 in **151**. This may suggest that other SMs could be used in the total organic synthesis, from the viewpoint of organic chemistry.

References

1. (a) Hudlicky, T. and Reed, J.W. (2007) *The Way of Synthesis*, Wiley-VCH Verlag GmbHAuthors introduced many summaries in this valuable book. In this book, the original citations are omitted for saving space; (b) Carlson, R. (1992) *Design and optimization in Organic synthesis*, Elsevier, Amsterdam.
2. (a) Hendrickson, J.B. (1990) *Anal. Chim. Acta*, **235**, 103–113; (b) Hendrickson, J.B. (1990) *Angew. Chem., Int. Ed. Engl.*, **29**, 1286–1295; (c) Helsen, H.E. and Jorgensen, W.L. (1994) *J. Org. Chem.*, **59**, 3841–3856.
3. Wilson, R. and Danishefsky, S. (2007) *J. Org. Chem.*, **72**, 4293–4305.
4. Danishefsky, S.J., Masters, J.J., Link, J.T., Young, W.B., Snyder, L.B., Magee, T.V., Jung, D.K., Isaacs, R.C.A., Bornmann, W.G., Alaimo, C.A., Coburn, C.A., and DiGrandi, M.J. (1996) *J. Am. Chem. Soc.*, **118**, 2843.
5. Wieland, P. and Miescher, K. (1950) *Helv. Chim. Acta*, **33**, 2215–2228.

6. (a) Chen, X.T., Zhou, B., Bhattacharya, S.K., Gutteridge, C.E., Pettus, T.R.R., and Danishefsky, S.J. (1998) *Angew. Chem., Int. Ed. Engl.*, **37**, 789; (b) Chen, X.T., Battacharya, S.K., Zhou, B., Gutteridge, C.E., Pettus, T.R.R., and Danishefsky, S.J. (1999) *J. Am. Chem. Soc.*, **121**, 6563.
7. (a) Dai, W.M. (2013) *Verista*, **10**, 11–20; (b) Burke, M.D. and Schreiber, S.L. (2004) *Angew. Chem. Int. Ed.*, **43**, 46; (c) Nicolaou, K.C., Hale, C.R.H., Nilewski, C., and Ioannidou, H.A. (2012) *Chem. Soc. Rev.*, **41**, 5185; (d) Szpilman, A.M. and Carreira, E.M. (2010) *Angew. Chem. Int. Ed.*, **49**, 9592.
8. Chandler, C.L. and List, B. (2008) *J. Am. Chem. Soc.*, **130**, 6737–6739.
9. Curran, D.P. and Rakiewicz, D.M. (1985) *J. Am. Chem. Soc.*, **107**, 1448–1449.
10. (a) Rickards, R.W. and Wu, J.P. (1985) *J. Antibiot.*, **38**, 513–515; (b) Kimura, K., Kanou, F., Koshino, H., Uramoto, M., and Yoshihama, M. (1997) *J. Antibiot.*, **50**, 291–296.
11. Kimura, K., Kano, F., Kurosawa, K., and Yoshihama, M. (1994) new isotetracenone-based substance and its production Jap. Patent, No. 06234693.
12. (a) Carreno, M.C., Somoza, A., Ribagorda, M., and Urbano, A. (2007) *Chem. Eur. J.*, **13**, 879–890; (b) Kesenheimer, C., Kalogerakis, A., Meiβer, A., and Groth, U. (2010) *Chem. Eur. J.*, **16**, 8805–8821.
13. Zhou, B.D., Ren, J., Liu, X.C., and Zhu, H.J. (2013) *Tetrahedron*, **69**, 1189–1194.
14. Van, T.N. and Kimpe, N.D. (2003) *Tetrahedron*, **59**, 5941–5946.
15. Singleton, V.L., Bohonos, N., and Ullstrup, A.J. (1958) *Nature (London)*, **181**, 1072–1073.
16. Wu, Y.K. and Gao, J. (2008) *Org. Lett.*, **10**, 1533–1536.
17. (a) Wan, F. and Erickson, K.L. (1999) *J. Nat. Prod.*, **62**, 1696; (b) Appleton, D.R., Sewell, M.A., Berridge, M.V., and Copp, B.R. (2002) *J. Nat. Prod.*, **65**, 630.
18. McPhail, K.L. and Gerwick, W.H. (2003) *J. Nat. Prod.*, **66**, 132.
19. Li, Y., Feng, J.P., Wang, W.H., Chen, J., and Cao, X.P. (2007) *J. Org. Chem.*, **72**, 2344–2350.
20. Wani, M.C., Taylor, H.L., Wall, M.E., Coggon, P., and McPhail, A.T. (1971) *J. Am. Chem. Soc.*, **93**, 2325–2327.
21. (a) Holton, R.A., Somoza, C., Kim, H.B., Liang, F., Biediger, R.J., Boatman, D., Shindo, M., Smith, C.C., Kim, S., Nadizadeh, H., Suzuki, Y., Tao, C., Vu, P., Tang, S., Zhang, P., Murthi, K.K., Gentile, L.N., and Liu, J.H. (1994) *J. Am. Chem. Soc.*, **116**, 1597–1598; (b) Nicolaou, K.C., Yang, Z., Liu, J.J., Ueno, H., Nantermet, P.G., Guy, R.K., Chairborne, C.F., Renaud, J., Couladouros, E.A., Paulvannan, K., and Sorensen, E.J. (1994) *Nature*, **367**, 630–634; (c) Wender, P.A., Badham, N.F., Conway, S.P., Floreancig, P.E., Glass, T.E., Gänicher, C., Houze, J.B., Jänichen, J., Lee, D., Marquess, D.G., McGrane, P.L., Meng, W., Tucciaro, T.P., Mühlebach, M., Natchus, M.G., Paulsen, H., Rawlins, D.B., Satkofsky, J., Shuker, A.J., Sutton, J.C., Taylor, R.E., and Tomooka, K. (1997) *J. Am. Chem. Soc.*, **119**, 2755–2756.
22. (a) Ganesh, T., Guza, R.C., Bane, S., Ravindra, R., Shanker, N., Lakdawala, A.S., Snyder, J.P., and Kingston, D.G.I. (2004) *Proc. Natl. Acad. Sci. U.S.A.*, **101**, 10006–10011; (b) Kingston, D.G.I. (2008) *J. Org. Chem.*, **73**, 3975–3984; (c) Holton, R.A., Somoza, C., Kim, H.-B., Liang, F., Biediger, R.J., Boatman, P.D., Shindo, M., Smith, C.C., Kim, S., Nadizadeh, H., Suzuki, Y., Tao, C., Vu, P., Tang, S., Zhang, P., Lisa, K.K.M., Gentile, N., and Liu, J.H. (1994) *J. Am. Chem. Soc.*, **116**, 1597–1598; (d) Masters, J.J., Link, J.T., Snyder, L.B., Young, W.B., and Danishefsky, S.J. (1995) *Angew. Chem. Int. Ed. Engl.*, **34**, 1723–1726.
23. Ghosh, A.K. and Liu, C. (2003) *J. Am. Chem. Soc.*, **125**, 2374–2375.
24. Review:(a) Kobayashi, J. and Kubota, T. (2007) *J. Nat. Prod.*, **70**, 451–460; (b) Isolation:Kubota, T., Endo, T., Tsuda, M., Shiro, M., and Kobayashi, J. (2001) *Tetrahedron*, **57**, 6175–6179.
25. Kobayashi, J., Kubota, T., Endo, T., and Tsuda, M. (2001) *J. Org. Chem.*, **66**, 134–142.
26. (a) Furstner, A., Aissa, C., Riveiros, R., and Ragot, J. (2002) *Angew. Chem. Int. Ed.*, **41**, 4763–4766; (b) Deng, L.S.,

Huang, X.P., and Zhao, G. (2006) *J. Org. Chem.*, **71**, 4625–4635; (c) Yadav, J.S. and Reddy, C.S. (2009) *Org. Lett.*, **11**, 1705–1708.

27. (a) Li, H., Wu, J., Luo, J., and Dai, W.M. (2010) *Chem. Eur. J.*, **16**, 11530–11534; (b) Smith, A.B. III, Qiu, Y., Jones, D.R., and Kobayashi, K. (1995) *J. Am. Chem. Soc.*, **117**, 12011–12012; (c) Heckrodt, T.J. and Mulzer, J. (2002) *Synthesis*, **13**, 1857–1866.
28. Ishikawa, H., Elliott, G.I., Velcicky, J., Choi, Y., and Boger, D.L. (2006) *J. Am. Chem. Soc.*, **128**, 10596–10612.
29. (a) Noble, R.L., Beer, C.T., and Cutts, J.H. (1958) *Ann. N.Y. Acad. Sci.*, **76**, 882; (b) Noble, R.L. (1964) *Lloydia*, **27**, 280.
30. Ando, M., Büchi, G., and Ohumam, T. (1975) *J. Am. Chem. Soc.*, **97**, 6680–6681.
31. Kutney, J.P., Bunzli-Trep, U., Chan, K.K., de Souza, J.P., Fujise, Y., Honda, T., Katsube, J., Klein, F.K., Leutwiler, A., Morehead, S., Rohr, M., and Worth, B.R. (1978) *J. Am. Chem. Soc.*, **100**, 4220–4224.
32. Kuehne, M.F., Podhorez, D.E., Mulamba, T., and Bornmann, W.G. (1987) *J. Org. Chem.*, **52**, 347–353.
33. Kato, D., Sasaki, Y., and Boger, D.L. (2010) *J. Am. Chem. Soc.*, **132**, 3685–3687.
34. Huang, X., He, J., Niu, X., Menzel, K.D., Dahse, H.M., Grabley, S., Fiedler, H.P., Sattler, I., and Hertweck, C. (2008) *Angew. Chem. Int. Ed.*, **47**, 3995–3998.

Index

a
absolute configuration (AC) 3, 31, 87, 107, 129, 247
acetonitrile 139, 213, 221, 229
actinoranone 12
acutumine 15
L-amino acids 3, 7, 8
β-amino alcohol catalyst 166, 189
amino alcohols 189–190
amphidinolide T2 298–303
androstan-3-one 16
9-anthranylmethoxyacetic acid (9-ATMA) 49
aquatolide 157
asymmetric Lewis acids 196, 197
asymmetric-axle-supported chiral catalysts 189, 194, 195
atropisomer 19, 116, 121, 243
auto-self catalysis 198–200

b
baccatin III 280, 283, 296
benzamide 235, 236
benzopyrenomycin 60, 144
betaine 97
[1,1′-binaphthalene]-2,2′-diol 19
bioassay-guided study 101
Birch reduction-alkylation 270, 271
Boltzmann statistics 36, 37, 39, 63, 95, 110, 112, 118, 145
brefeldin A 290–293
brevianamide M 43, 148, 274
Brewster model 59
(*R*)-butan 2-ol 33

c
carbon tetrahedron
– alkaloids 14, 15
– flavonoids 13, 14
– glycoside 17, 18
– steroids 16
– terpenoids 9, 10, 12, 13
carbon-13 NMR
– aromatic compounds 131
– conformational searches
– – Boltzmann statistics 36, 37
– – *cis*-structures 39
– – errors and coefficients 39
– – HF/6-31G(d)-optimized conformations 35
– – Me_4Si 36
– – MMFF94S force field 37
– – molecular mechanic force field 35
– – optical rotations 3, 36, 40, 59, 94, 107, 129, 165, 241
– – prediction 40
– – quantum theory 35
– conformations, stereoselective reaction 243
– NMR and atomic structure 31, 32
– stereogenic centers 4, 9, 10, 13, 19, 25, 40, 65, 68, 87, 95, 98, 99, 130, 132–133, 147, 206, 272, 299
(+)-catechin, 13, 14
chemoselective additions
– anhydrides 206
– ketone and aldehyde 205
– Me_2Zn/Et_3B 208
– spirocyclic 2-oxindoles 210
– *t*-BuOK 207
– TS barrier magnitude 209
– α,β -unsaturated nitrile 207
chemoselective oxidation
– BAIB-TEMPO 228
– DMP reagent 227
– $PhI(OAc)_2$ 226
– sodium azide 228

Organic Stereochemistry: Experimental and Computational Methods, First Edition. Hua-Jie Zhu.

chemoselective oxidation (*contd.*)
– sulfoether 229
chemoselective reduction
– AgNPs-CeO_2 223
– aldehyde 10, 20, 52, 163–167, 171–177, 183, 187, 190–193, 197, 203, 206–207, 209–210, 211, 212, 222–223, 225, 227, 229, 234–235, 245–246, 258, 263–266, 274, 284, 298, 301
– aromatic nitro compounds 223
– BH_3-SMe_2/$LiAlH_4$ 225
– Clemmensen reduction 213
– esters 213, 214
– α-F ester 215
– hydrides 211
– hydrogenation 211
– $InCl_3$/$NaBH_4$ 221
– intrinsic reaction coordinate 215
– iodotrichlorosilane (ITCS) reagents 213
– ketone 50, 52, 96–97, 113, 131, 171, 179, 189, 193, 203, 205–209, 211, 219–220, 222, 224–225, 249, 251–255, 258–259, 266, 272, 281, 284, 288–289
– metals/metallic salts 211
– *N*-methylacetamide 219
– $NaBH_4$ 217
– nitrines 217, 221
– sulfoxides 224
– tautomerization 219
– α,β-unsaturated aldehydes 211
– α,β -unsaturated ketones 222, 226
– $Zn(BH_4)_2$ 211
chirality
– D-amino acids 8
– L-amino acids 3, 7
– amplification 198, 199
– bioactivity 4
– carbon tetrahedron
– – alkaloids 14, 15
– – flavonoids 13, 14
– – glycoside 17, 18
– – steroids 16
– – terpenoids 9, 10, 12, 13
– chiral compounds 18, 19
– chiral compound separation 53
– chiral material study 6
– circularly polarized light (CPL) 3
– ^{13}C NMR spectra
– – NMR and atomic structure 31, 32
– – prediction and conformational search 34–37, 39–41
– CPL irradiation 8
– enantiomers 3
– ^{1}H NMR 34
– (-)-menthyl methacrylate 6
– Mosher method 44–51
– optical characteristics
– – ECD and definition 25, 26
– – OR measurement 24, 25
– – Raman optical activity 27
– – vibrational circular dichroism 26, 27
– organic acids 53
– organic alcohols 53
– organic bases 53
– recorded chiroptical spectroscopy 6
– Sharpless epoxidation 4
– stereogenic center
– – [1,1′-binaphthalene]-2,2′-diol 19
– – chiral ferrocenes 22
– – chiral structures 22
– – helix biopolymer 23
– – helix hexahelicenes 23
– – hydroxyl groups 21
– – N, P/S atom 19
– – sulfoxide's chirality 19
– D-sugars 3
– tertodotoxin 4
– transition state (TS) barrier 3
– transition state energy and chirality-selectivity 51, 52
– D-tryptophan 8
– L-tryptophan 8
– L-tryptophan methyl ester 8
– x-ray diffraction 41–44
Chiron 280
cholesterol 16
chromanes 152, 155
circularly polarized light (CPL) 3, 7, 64, 87, 107
circularly polarized Raman scattering 110
Claisen rearrangement, 271Clemmensen reduction 213
(*S*)-cleonin 261
codeine 34
concentricolide 79, 177
conessine 15
crotonide 17
cyclohexane 242
cycloisomerization 253, 255
cyclopropenyl ketones 253, 255

d

Davydov (energy) spitting 88
Dess–Martin periodinane (DMP) reagent 225, 227—229
diastereoselective reactions
– addition reactions
– – allylic alkylation 268, 269

– – chiral lactams 260
– – *α,β*-dihydroxyl ketones 268
– – homoallylic amines 268, 269
– – imines, allylation of 269, 270
– – Kulinkovich reaction 261
– – metallated propargylic amines 265
– – rhodium-catalyzed cross-aldol reaction 264, 266
– – Zn reagents 261
– Birch reduction-alkylation 270, 271
– Claisen rearrangement 271, 273
– oxidation-coupling reaction 271, 272
– theoretical methods *vs.* experiments
– – chiral compounds, AC assignment for 273, 274
– – de% values 274, 276
– – ee% values 274–276
Diels–Alder reactions 196, 197, 280, 287–289
1,5-di-F,2,6-di-NO_2-benzene 49
α,β-dihydroxyl ketones 268
4-dimethyl amino benzoic acid 89, 90
(*R*)-2′-((dimethylamino)methyl)-[1,1′-binaphthalen]-2-ol 19
N,N-dimethylformamide 219
1,2-diol dibenzoate 87–89
N,N′-dioxide chiral ligands 190
(-)-discorhabdin Z 99
DpenPhos 197

e
ECD, *see* electronic circular dichroism (ECD)
elatenyne 157
electronic circular dichroism (ECD) 23, 25, 36, 87, 107, 129, 134, 135, 241
– B3LYP/6-31G(d)-optimized conformers 97
– definition 87
– exciton chirality model, *see* exciton chirality circular dichroism (ECCD)
– quantum theory 94, 95
– theoretical curve *vs.* experimental curve
– – betaine 97
– – diincarvilone 97, 98
– – (-)-discorhabdin Z 99
– – incarvilone A 97, 98
– – mandassion A 99, 100
– – mandassion B 99, 100
– – psychotripine 99, 100
– – spirocurcasone 98, 100
– – streptocarbazoles A 101
– – streptosetin A 102
– UV correction method 102–105
enantioselective reaction
– addition reaction
– – allylic alkylations 174–176
– – asymmetric allylation 178, 179
– – bioactive compounds 177
– – ferrocene ligands 173
– – Henry reactions 177
– – intermolecular C-acylation 174, 175
– – Michael additions 173
– – phenylation 174
– chiral catalysts 191, 193, 195
– – amino alcohols 189, 190
– – asymmetric-axle-supported catalysts 194, 195
– – asymmetric Lewis acids 196, 197
– – N–O catalysts 190
– – organic-P containing ligands 197
– – Shiff base ligands 195, 196
– – solid-supported catalysts 193
– – spiral catalysts 193, 194
– diethylzinc addition reaction
– – acyclic enones 171
– – aldehydes 165–169
– – auto-self catalysis 199, 200
– – benzaldehyde 166
– – chiral axial catalysts 167–169
– – chirality amplification 198, 199
– – Cu-promoted conjugate additions 169, 170
– – *N*-diphenylphosphinoylimines 172
– – nitropropene acetals 170
– – odd-even carbon effect 199–201
– – phenylation 169
– oxidation reaction 184, 185
– reaction types 165
– reduction reaction
– – acetophenone 181
– – *N*-aryl ketoimine 183
– – aryl ketones 182
– – >C=N– group compounds 182
– – >C=C< group compounds 180
– stereogenic center formation 165
– TS barrier and selectivity prediction
– – B3LYP/6-311+G(d) level theory 188
– – B3LYP/6-31G(d) level theory 187
– – DFT theory 187, 188
– – HF/6-31G(d) theory 186, 187
– – quantum mechanics (QM)/molecular mechanism (MM) 185
ene/Prins cyclization reaction 190
enicillitone 16, 17
ephedrine 14
ethane 241, 242
5-ethyl 5-propyl-undecane 25, 73
exciton chirality
– circular dichroism (ECCD) 87

exciton chirality (*contd.*)
– – 4-dimethyl amino benzoic acid 90
– – 1,2-diol dibenzoate 87–89
– – 7-oxocholest-5-en-3 β -ol, chiral acetate of 91
– – (*S*)-PBA-containing Cu-complex 92, 93
– VCD 108, 109, *see* also vibrational circular dichroism (VCD)
exciton coupling 88, 90, 91, 108, 124

f
formestane 16, 17
Friedel–Crafts alkylation 256, 258

g
Gauge-independent-atomic-orbit (GIAO) 34
green chemistry 181–183
griseusins 159

h
Harada–Nakanishi function 94
harpagoside 10
Hatree–Fock (HF) 59
H-bonds 27, 76, 131, 136, 186
Henry reactions 177, 190
hexacyclinol 158
high performance liquid chromatography (HPLC) 55, 243, 244
(+)-hirsutene 285–287
17 β H-neriifolin 17, 18
^{1}H NMR 34, 45–47, 51, 144, 242, 267
Huang Min-lon reduction 213
(*S*)-3-hydroxydihydrofuran-2(3H)-one 108

i
indole alkaloids 232, 256, 303
indole derivatives 233
intrinsic reaction coordinate (IRC) 215
iodotrichlorosilane (ITCS) reagents 213
isoepitaondiol 156
isoschizogaline 134, 135
isovitexin 17

k
Kulinkovich reaction 261

m
(*S*)-(-) malic acid 53
malyngamides 292, 294, 295
mandassion A 99, 100
mandassion B 99, 100
matrix method 153, 154
matrix model 63–77, 132
L-menthol 9
menthyl hydrazine 54
menthyl isocyanate 53
mestanolone 16
2-(2′-methyoxy-1,1′-naphthyl)-3,5-di-chlorobenzoic acid (MNCB) 49, 50
N-methylacetamide 219
(*S*)-((4-methyl cyclohexylidene)methyl)benzene 22
1-methylhexane 243
methyloxirane 124, 125
Michael addition reactions 123, 173
molecular mechanic force field 35
monodentate hosphoramidite 197
morphine 14, 279
Mosher ester 44, 45, 48, 51, 129, 144, 251
Mosher method 44–48, 50, 51, 144
(–)-myrtenal 115, 116

n
naphthalene-1,5-diol 289
nitrines 217, 221
nobilisitine A 158
nonlinear effect, *see* chirality amplification
nuclear magnetic resonance (NMR), *see* carbon-13 NMR
nuclear Overhauser enhancement spectroscopy (NOESY) 34
nucleophilic addition 177, 208

o
odd–even carbon effect 198–201
oleane-3 β -ol 13
oligothiophenes 259
optical rotation (OR) 40
– AC assignment 131
– and ECD 144, 145
– quantum theory 131
optical rotatory dispersion (ORD) 25, 83
– AC assignment 131
– – mono-stereogenic center compounds 79
– – poly-stereogenic center compounds 82
– acetonitrile 139
– Brewster model 59
– comprehensive mass 68–70
– det(D) values 73–76
– electronegativity 72
– matrix basis 65–68
– matrix model 63–65, 80, 81
– *n*-propyl model 69
– quantum theory
– – bezopyrenomycin 60
– – (*R*)-2-chloropentane 62
– – OR calculation 60
– radius 70, 72

– substituents 70, 71
– symmetry factor 72, 73
– Van der Waal's radii 72
optical rotatory dispersion (ORD)
– AC assignment 83
– structural elucidation 77
organocatalytic Michael addition 123
oruwacin 157
1,3,4-oxadiazoles 229, 306
oxidation–coupling reaction 271, 272
7-oxocholest-5-en-3β -ol, chiral acetate of 91

p
paederoside 10
perdeuteriophenyl-phenyl-sulfoxide 116, 117
L-phenylalanine 148, 151
(-)-*N*-(1-phenylethyl) aniline 152, 154, 274
1-phenylhexanone 276
Pictet–Spengler cyclization 280
α-pinene 9
β-pinene 9
(*R*)-2-piperidine-1,1,2-triphenylethanol 185
plakotenin 158
plumericin 156, 157
preussidone 119
psychotripine 99, 100, 135
pyrimidine 220

q
quantum theory 28, 59–65, 80, 94–95, 97, 99, 109–113, 131, 136, 218

r
Raman optical activity (ROA) 23, 27, 36, 107 and 129
– CID spectra 111, 112
– methyloxirane 124, 125
– Raman scattering 27, 110–112
– theoretical *vs.* experimental spectra 125, 126
Raman scattering 27, 107, 110–112
reduction–coupling reaction 212, 251, 252
regioselective reaction
– 1,2- and 1,4-addition reactions 256, 257
– allenes 257, 258
– cyclization reaction 256
– [5 + 2 + 1] cycloaddition 254
– cycloisomerization 253–255
– Friedel–Crafts alkylation 256, 258, 259
– head-to-tail-type oligothiophenes 259
– reaction routes 252, 253
– reduction–coupling reaction 251
– reductive hydration 252
– zinc carbenoids 258
relative configuration (RC) 3, 31, 82, 98, 129
ring enlargement reaction 253, 258, 260
rosmaridiphenol 156
rotating frame Overhauser enhancement spectroscopy (ROESY) 34, 135
rotenone 14
rubiginone A2 287–289
rutin 14

s
saccopetrin A 46
schaffnerine 11, 12
(+)-schizandrin 146, 274
Schmidt reaction 236
8,14-*seco*-oleana-8(26),13-dien-3β-ol 13
Sharpless epoxidation 4, 185, 279
Sharpless oxidation 4, 163, 184, 185
Shiff base ligands 195, 196
Simmons–Smith reaction 258, 262
sodium borohydride 52, 186, 200, 215, 217, 288, 293
solid-supported catalysts 192, 193
spiral chiral catalysts 193, 194
spirocurcasone 98, 99
stereoselective reaction
– conformational study
– – B3LYP/6-31G (d) level energy optimizations 245–247
– – catalyst structure changes and ee% values 248
– – correct gesture collision 241
– – cyclohexane 242
– – energy difference and R% percentages 249, 250
– – ethane 241, 242
– – HF/6-31G(d,p) level energy computations 249
– – 1-methylhexane 243
– – MP2 energy 245, 246
– – NMR spectra 243, 244
– – remote chirality control reaction 247, 248
– – ring-opening and ring-closing changes 244, 245
– – softwares 249
– de% 241
– dr 241
stereostrein A 10
stereostrein D 10
streptocarbazole A 101
streptosetin A 102
sulfoxides 224, 225, 229, 230
Swern reaction 225
Syngen 280

t
D-(+)-tartaric acid 53
tartaric acid mono-anilide 53
taxol 11, 292, 296, 297
temperature gradient method 43, 44
terpenes 133
tertodotoxin 4
2,2,6,6-tetramethyl-piperidine 1-oxyl (TEMPO) 226, 227
(*S*)-thalidomide 3
total organic synthesis
– amphidinolide T2 and derivatives 298–303
– (+)-brefeldin A 290–293
– calculations
– – biogenetic routes study 307
– – density functional theory 307
– – reaction directions 307
– – TS geometries 308
– (+)-hirsutene 285, 286
– malyngamides and AC assignments 292, 294, 295
– retro-synthesis tactics
– – 2-aminophenols and anilines 285
– – baccatin III 280
– – DOS 284
– – DTS procedure 284
– – eleutherobin 280, 283
– – pattern recognition 280
– – ring formation methods 280
– – target molecule (TM) 280
– – Wittig reagent reaction 280
– rubiginone A2 and analog 287–289
– taxol derivatives 296–298
– (+)-vindoline 303–306
3,5-trichloro-1,3,5-triazinane-2,4,6-trione (TCTATO) 226
D-tryptophan 8
L-tryptophan 8, 178
L-tryptophan methyl ester 8

u
ursodeoxycholic acid 16
UV correction method 102–105, 153, 154

v
vannusal B 158
vapor diffusion method 44
VCD, *see* vibrational circular dichroism (VCD)
velocity rotatory strengths 94
vibrational circular dichroism (VCD) 23, 26, 36, 107, 129, 134, 241
– AC determinations 115
– calculated *vs*. experimental spectra 113, 114, 116–118, 120, 121, 124
– definition 107
– and ECD 146, 147, 149
– exciton chirality method 108, 109
– IR 109, 110, 113, 114, 118–120, 123, 124
– ROA, *see* Raman optical activity (ROA)
(+)-vindoline 303–306

w
Wittig reagent reaction 232, 233, 280, 286, 301

x
x-ray crystallography 147–148
x-ray diffraction 41–44, 151

z
zinc carbenoids 258, 261, 263

Printed and bound by CPI Group (UK) Ltd, Croydon, CR0 4YY